ISBN 978-1-5277-8548-9
PIBN 10895344

1 MONTH OF
FREE
READING

at
www.ForgottenBooks.com

By purchasing this book you are eligible for one month membership to ForgottenBooks.com, giving you unlimited access to our entire collection of over 1,000,000 titles via our web site and mobile apps.

To claim your free month visit: www.forgottenbooks.com/free895344

English
Français
Deutsche
Italiano
Español
Português

www.forgottenbooks.com

Mythology Photography **Fiction**
Fishing Christianity **Art** Cooking
Essays Buddhism Freemasonry
Medicine **Biology** Music **Ancient**
Egypt Evolution Carpentry Physics
Dance Geology **Mathematics** Fitness
Shakespeare **Folklore** Yoga Marketing
Confidence Immortality Biographies
Poetry **Psychology** Witchcraft
Electronics Chemistry History **Law**
Accounting **Philosophy** Anthropology
Alchemy Drama Quantum Mechanics
Atheism Sexual Health **Ancient History**
Entrepreneurship Languages Sport
Paleontology Needlework Islam
Metaphysics Investment Archaeology
Parenting Statistics Criminology
Motivational

_1503_ **No. 5037**

United States
Circuit Court of Appeals

For the Ninth Circuit.

_1497_

/

THE MISSION MARBLE WORKS, a Corporation,

Plaintiff in Error,

vs.

ROBINSON TILE & MARBLE COMPANY, a Corporation,

Defendant in Error.

Transcript of Record.

Upon Writ of Error to the United States District Court of the Western District of Washington, Northern Division.

No. 5037

United States
Circuit Court of Appeals
For the Ninth Circuit.

THE MISSION MARBLE WORKS, a Corporation,

        Plaintiff in Error,

vs.

ROBINSON TILE & MARBLE COMPANY, a Corporation,

        Defendant in Error.

Transcript of Record.

Upon Writ of Error to the United States District Court of the Western District of Washington, Northern Division.

INDEX TO THE PRINTED TRANSCRIPT OF RECORD.

[Clerk's Note: When deemed likely to be of an important nature, errors or doubtful matters appearing in the original certified record are printed literally in italic; and, likewise, cancelled matter appearing in the original certified record is printed and cancelled herein accordingly. When possible, an omission from the text is indicated by printing in italic the two words between which the omission seems to occur.]

Index.   Page

Index.   Page

NAMES AND ADDRESSES OF ATTORNEYS OF RECORD.

Messrs. GROSSCUP & MORROW, Attorneys for Plaintiff in Error,

2600 L. C. Smith Building, Seattle, Washington.

CHARLES A. WALLACE, Esquire, Attorney for Plaintiff in Error,

2600 L. C. Smith Building, Seattle, Washington.

J. O. DAVIES, Esquire, Attorney for Plaintiff in Error,

2600 L. C. Smith Building, Seattle, Washington.

H. A. P. MYERS, Esquire, Attorney for Defendant in Error,

1160 Empire Building, Seattle, Washington. [1*]

In the District Court of the United States for the Western District of Washington, Northern Division.

No. 10190—AT LAW.

THE MISSION MARBLE WORKS, a Corporation,

Plaintiff,

vs.

ROBINSON TILE AND MARBLE COMPANY, a Corporation,

Defendant.

*Page-number appearing at the foot of page of original certified Transcript of Record.

COMPLAINT.

The plaintiff complains and for cause of action against the defendant alleges:

I.

That the plaintiff at all times herein mentioned was and now is a corporation duly organized and existing under and by virtue of the laws of the State of California.

II.

That the defendant at all the times herein mentioned was and now is a corporation duly organized and existing under and by virtue of the laws of the State of Washington. That the defendant at the time of entering into the contract hereinafter alleged was doing business as such corporation under the style and name of: Robinson Tile and Supply Company.

III.

That this action is wholly between citizens of different states, the plaintiff being a citizen of the State of California, and the defendant a citizen of the State of Washington. That the amount involved in this action is the sum of Seven Thousand Two Hundred and Ninety and 32/100 Dollars ($7,-290.32), exclusive of interest and costs.

IV.

That on or about the 30th day of January, 1924, the plaintiff and the defendant entered into a contract in writing wherein [2] and whereby the plaintiff agreed to furnish and sell to the defendant, and the defendant agreed to purchase from the plaintiff, building marble, which was to be used by the defendant in the construction and finishing of that certain building in the city of Seattle, Washington, known as the Dexter Horton National Bank Building, of the agreed price and value of Forty-Six Thousand Dollars ($46,000.00). That the plaintiff has fully kept and performed all of the covenants and conditions of said contract on its part to be kept and performed.

V.

That under and by virtue of said contract, the plaintiff at the special instance and request of the defendant also furnished to the defendant in the construction and finishing of the said Dexter Horton National Bank Building, other building marble, in addition to that required to be furnished by the aforesaid contract for the said sum of $46,-000.00, of the fair and reasonable price and value of One Thousand Three Hundred and Seventy-one and 56/100 Dollars ($1,371.56). That the said sums became due on or about April 1st, 1925. That no part of said sums or either of them has been paid except the sum of Forty Thousand and Eighty-one and 24/100 Dollars ($40,081.24).

WHEREFORE plaintiff demands judgment against the defendant for the sum of Seven Thousand Two Hundred and Ninety and 32/100 Dollars ($7,290.32) with interest thereon from the first day of April, 1925, to date of judgment at six per cent per annum, and costs of suit.

> GROSSCUP & MORROW,
> CHAS. A. WALLACE,
> J. O. DAVIES,
>   Attorneys for Plaintiff.   [3]

[Endorsed]: Filed Nov. 25, 1925.   [4]

[Title of Court and Cause.]

ANSWER.

Comes now the defendant and for answer to the complaint says:

I.

Answering paragraphs I, II and III of the complaint, defendant admits the same.

II.

Answering paragraph IV of the complaint, defendant admits that on or about the 30th day of January, 1924, it entered into a contract with plaintiff whereby plaintiff was to furnish the marble for the Dexter Horton National Bank Building at the agreed price of $46,000.00, but defendant denies that said marble sold and delivered to this defendant for said bank was of the fair and reasonable value of $46,000.00; and denies all other allegations of said paragraph, and especially denies that plaintiff has kept and performed the covenants and conditions of said contract with defendant.

III.

Answering paragraph V thereof, defendant admits that it has paid the sum of $40,081.24 upon said contract, and admits that defendant furnished some other extra marble, but denies all other allegations of said paragraph. [5]

For a further separate and independent answer to said complaint, and by way of counterclaim, the defendant alleges:

I.

That during all the times herein mentioned the defendant was and is now a corporation, organized, existing and doing business under and by virtue of the laws of Washington, and that it has paid its last annual corporation license fee due the State of Washington.

II.

That prior to the making of the contract in writing between plaintiff and defendant, and to wit, on the 25th day of January, 1924, this defendant entered into a written contract with the Dexter Horton National Bank to furnish and install all the marble work in the Dexter Horton National Bank Building (except in toilet-rooms and elevator lobby), and plaintiff herein at the time of making this contract with the defendant was fully aware of the terms and conditions of said contract with said bank and had a copy thereof.

III.

That the contract between plaintiff and defendant provides that:

"The Mission Marble Works (hereinafter called the Company) proposes to furnish as required by the specifications and drawing prepared by JOHN GRAHAM, Architects (acting for the purposes of this agreement as the Agents of the Purchaser) except as hereinafter modified, for the DEXTER HORTON NATIONAL BANK building at Seattle, Wash., the following building marble, viz.:

Per Sheets Nos. 2, 2A, 5, 6, 8 & 10.

"According to the plans, specifications, terms, and conditions of same, and as covered in purchasers contract with said Bank in order to enable the purchaser to faithfully and promptly fulfill their contract with the above bank." [6]

IV.

That the plaintiff herein has failed to keep and perform its said contract with this defendant in particulars hereinafter set forth, resulting in damage and loss to this defendant as follows:

(a) That the plaintiff failed to furnish shop drawings for the marble of said bank building, for the architect's approval, before the marble was fabricated, and on account of the fact that plaintiff did not get out said marble properly according to the plans and specifications, this defendant was required by the architect to do certain necessary extra work and labor upon said marble to make the same conform to the plans and specifications, which extra work and labor was and is of the fair and reasonable value of ....$893.88.

(b) That the plaintiff failed to furnish the marble for said bank on time or within the time as required by said contract and on account of said delay, plaintiff shipped a portion of the marble by water in piece-meal instead of in carload lots by rail. This shipment by water necessitated boxing, at a cost of ....$484.78, and cartage at a cost of ....$ 60.00 which said charges were unnecessary and caused wholly by plaintiff's delay in getting out said marble to the damage of this defendant in the sum of ....$544.78.

(c) That on or about November 1, 1924, the architect was pressing this defendant for a more vigorous prosecution of the marble work and defendant had to lay off some of its employees for lack of marble, and after repeated exchanges of letters and wires, defendant was obliged to send A. P. Robinson, its president, to plaintiff's plant in San Francisco, in order to expedite the shipment of said marble, to the damage and expense of this defendant in the sum of ....$250.00. [7]

(d) That on or about November 13, 1924, the architect for said bank threatened to cancel the defendant's contract on account of lack of marble necessary to prosecute said work, and in this situation it became necessary to send some representative of the Bank to the plaintiff's plant to expedite shipment of marble and F. W. Grant, representative of the owner and architect, made a trip to San Francisco for that purpose, the reasonable cost and

expense of which was paid by this defendant in the
sum of ....$300.00.

(e)   That on or about November 1, 1924, defendant was in great need of more marble for said bank
building, and demanded of plaintiff that it immediately put on extra labor to get out the marble
which plaintiff refused to do unless defendant
would agree to advance in cash the cost for said extra labor, and under these circumstances the defendant was compelled to and did advance the same
to plaintiff under protect in the sum of ...$600.00.

(f)   That a portion of the marble sold by the
plaintiff for use in the counters of said bank building was defective in color, and a portion thereof for
use in the check desk tops was defective in waxing,
and contained numerous artificial holes, and none
of same was according to specifications, and on account thereof, none of the marble in this paragraph
mentioned was accepted or approved by the architect; and by reason thereof, and on account of
plaintiff's failure to furnish the marble under its
contract in time to enable this defendant to faithfully and promptly fulfill its contract with said
bank, the final completion of said bank building
was delayed until the 28th day of February, 1925,
and the architect refused to approve the final payment of $6,000.00, due this defendant on its con-
tract, until said defective marble is replaced, and
the bank's damages for delay paid, all of which
was well known to the plaintiff prior [8] to the
commencement of this action, all to plaintiff's damage in the sum of ....$6,000.00.

That none of the damages, counterclaims or demands in this paragraph set forth have been paid.

WHEREFORE, defendant prays judgment that plaintiff take nothing by this action, and that the same be dismissed, and that defendant have judgment for its costs and disbursements herein to be taxed.

H. A. P. MYERS,
Attorney for Defendant.

[Endorsed]: Service hereof by receipt of a copy admitted this 8th day of January, 1926.

GROSSCUP & MORROW,
Attorneys for —————.

[Endorsed]: Filed Feb. 1, 1926. [9]

———

[Title of Court and Cause.]

REPLY.

Comes now the plaintiff and for its reply to the answer of the defendant on file herein admits, denies, and alleges as follows, to wit:

I.

Admits the allegations contained in Paragraphs I, II, and III, of that portion of the answer designated as "a further separate and independent answer to said complaint, and by way of counterclaim."

II.

The plaintiff for its reply to Paragraph IV of said answer admits that the plaintiff shipped a portion of the marble by water instead of by rail.

That the marble shipped by water was boxed at a cost of about $484.78 and was carted at a cost of about $60.00.

Admits that A. P. Robinson visited the plaintiff's plant in San Francisco.

Admits that F. W. Grant visited the plaintiff's plant in San Francisco.

As to whether or not the defendant paid the costs and expenses of said trips the plaintiff has no knowledge or information   [10]   thereof sufficient to form a belief.

Admits that the servants of the plaintiff, at the request of the defendant, worked overtime in the manufacture of said marble, and that the defendant paid to the plaintiff the sum of $600.00 on account of said overtime work.

Admits that none of the sums mentioned in said paragraph IV have been paid, and denies each and every other allegation, and all other allegations contained in said Paragraph IV.

Plaintiff denies generally and specifically each and every allegation and all allegations contained in the answer of the defendant herein not hereinbefore specifically admitted.

WHEREFORE, plaintiff, having fully replied to the answer of the defendant on file herein, demands judgment in accordance with the prayer of its complaint.

GROSSCUP & MORROW,
CHAS. A. WALLACE,
J. O. DAVIES,
              Attorneys for Plaintiff.

[Endorsed]: Filed Jan. 29, 1926. [11—20]

[Title of Court and Cause.]

JUDGMENT.

This cause came on regularly for trial on June 17, 1926, before Honorable Edward E. Cushman, Judge in the above-entitled court, without a jury, a jury having been waived by the parties in writing; the plaintiff appearing at the trial by its president, George W. Eastman, and its attorneys, Grosscup & Morrow, Charles A. Wallace and J. O. Davies; the defendant appearing at the trial by its president, A. P. Robinson and its attorney, H. A. P. Myers. Thereupon testimony of witnesses and evidence of the parties was introduced. And the court having filed its findings of fact and conclusions of law herein in favor of the defendant, now, in accordance therewith, the Court being fully advised in the premises, it is

ORDERED, ADJUDGED and DECREED That the defendant, Robinson Tile & Marble Company, do have and recover of and from the plaintiff, The Mission Marble Works, Six Hundred Eighteen and 32/100 Dollars ($618.32), with interest thereon from this day until paid at six per cent (6%) per annum, besides defendant's costs and disbursements herein to be taxed.

Done in Open Court on this 1st day of November, 1926.

EDWARD E. CUSHMAN,

Judge.

[Endorsed]: Filed Nov. 1, 1926. [21]

[Title of Court and Cause.]
PETITION FOR WRIT OF ERRORS.

To the Honorable EDWARD E. CUSHMAN,
 Judge of the Above-entitled Court:

Comes now The Mission Marble Works, plaintiff above-named, and plaintiff in error, and by Messrs. Grosscup & Morrow, Chas. A. Wallace, and J. O. Davies, respectfully shows that on or about the first day of November, 1926, the jury having been waived in the above-entitled cause, the Court made and entered its findings of fact and conclusions of law in the said cause in favor of the defendant and against this plaintiff, your petitioner, directing that judgment be entered thereon in favor of the defendant and against your petitioner in the sum of $618.32, and a final judgment was entered on said findings of fact and conclusions of law in favor of the said defendant and against this plaintiff, your petitioner, in said sum on said day. Your petitioner, feeling itself aggrieved by said findings of fact and conclusions of law and judgment, entered herein as aforesaid, herewith petitions the Court for an order allowing it to prosecute a writ of error to the United States Circuit Court of Appeals for the Ninth Circuit under the laws of the United States in such cases made and provided, and presents and files herewith its assignment of errors.
[22]

WHEREFORE, by reason of the premises, your petitioner, plaintiff above named, prays that a writ of error be issued in its behalf from the United States Circuit Court of Appeals for the Ninth Circuit, sitting at San Francisco in the State of California, in said circuit for the correction of the errors complained of and herewith assigned in the accompanying assignment of errors be allowed, and that an order be made fixing the amount of security to be given by this plaintiff, plaintiff in error herein, conditioned as the law directs, and upon the giving of such bond as may be required that all further proceedings be suspended until the determination of said writ of error by the said Circuit Court of Appeals.

> GROSSCUP & MORROW,
> CHAS. A. WALLACE,
> J. O. DAVIES,
> > Attorneys for Petitioner.

[Endorsed]: Filed Dec. 9, 1926. [23]

[Title of Court and Cause.]

ASSIGNMENT OF ERRORS.

Comes now The Mission Marble Works, a corporation, plaintiff in the above-entitled cause, and in connection with its petition for a writ of error in this cause assigns the following errors which plaintiff herein avers occurred at the trial thereof:

I.

The Court erred in overruling the plaintiff's objection to the following question asked the witness, Francis W. Grant, on direct examination by the defendant, and allowing the said witness to testify concerning damages sustained by the bank on account of delay in furnishing marble:

"Q. Referring to the two items upon which the architect declined to make payment on this job until they are corrected, one being for defective marble, and that has never been corrected, and that is the item which you have estimated to the Court would cost $2,500.00 and the other is on account of delay in the performance of the contract; what was the item of damages under that?"

II.

The Court erred in overruling the plaintiff's objection to the following question asked the witness, John Graham, by the defendant on direct examination, and in allowing the said witness to testify concerning demages received by the bank on account of [24] delay in installing marble:

"Q. What was the other item, besides the item of defective marble, and on account of adjustment of damages sustained by the bank on account of that delay?"

III.

The Court erred in refusing to adopt the following finding of fact tendered by the plaintiff:

"That on or about the 30th day of January, 1924, the plaintiff and defendant entered into a contract in writing wherein the plaintiff agreed to furnish and sell to defendant, and defendant agreed to purchase from the plaintiff, building and finishing marble which was to be used by the defendant in the construction and finishing of the banking quarters of that certain building in the city of Seattle known as the Dexter Horton National Bank Building at the agreed price and value of $46,000.00. That the plaintiff has fully kept and performed all of the covenants and conditions of said contract on its part to be kept and performed, and has furnished all of the marble called for in said contract, and the said marble has been installed in the banking quarters of said building, and is now in the same and being used by the Dexter Horton National Bank.''

IV.

The Court erred in refusing to adopt the following finding of fact tendered by the plaintiff:

''The Court further finds that under and by virtue of said contract and the terms and provisions thereof plaintiff at the special instance and request of the defendant also furnished to the defendant for use by defendant in the construction and finishing of said banking quarters of the said Dexter Horton National Bank Building, other building marble, in addition to that [25] required to be furnished by the contract above mentioned for the contract price $46,000.00, of the fair and reasonable price and value of the sum of $826.78.''

V.

The Court erred in refusing to adopt the following finding of fact tendered by the plaintiff:

''The Court further finds that under the contract entered into between plaintiff and defendant the marble was to be shipped by plaintiff from the City of San Francisco to the City of Seattle, Washington, in carload lots f. o. b. San Francisco, California. That at the special instance and request of the plaintiff forty tons of the marble shipped by the plaintiff to the defendant under said contract was shipped by water in pieces as manufactured by the plaintiff. That under this arrangement it was necessary for the plaintiff to crate said marble. That the reasonable cost of crating said marble was and is the sum of $484.78.''

VI.

The Court erred in refusing to adopt the following finding of fact tendered by the plaintiff :

"The Court further finds that the change in shipping, as above mentioned, necessitated the carting of said marble from plaintiff's plant to a dock in San Francisco. That the reasonable cost and charge for carting said marble is the sum of $60.00. That the plaintiff paid out the said sum of $484.78 above-mentioned, and the sum of $60.00 for the crating and cartage of said marble, and the same is a reasonable and just charge and should be paid by the defendant.'' [26]

VII.

The Court erred in refusing to adopt the following finding of fact tendered by the plaintiff:

"The Court further finds that the total sum which became due and owing to plaintiff for marble furnished the defendant for installation and completion of the banking quarters of the said

Dexter Horton National Bank Building was the sum of $47,371.56. That of the said sum defendant has paid the plaintiff the sum of $40,081.24. That there is now a balance due and owing the plaintiff from the defendant, of the sum of $7,290.-32. That said sum became due on the first day of April, 1925, and plaintiff is entitled to interest thereon at the rate of six per cent per annum from and after the said first day of April, 1925, to date hereof."

VIII.

The Court erred in refusing to adopt the following finding of fact tendered by the plaintiff:

"The Court further finds that all of the allegations of plaintiff's complaint were and are true."

IX.

The Court erred in refusing to adopt the following finding of fact tendered by the plaintiff:

"The Court further finds that the banking quarters of said Dexter Horton National Bank Building were not completed and ready for the installation of marble until the 22d day of September, 1924, and that the said banking quarters were not turned over to the defendant for installation of marble until said date. That under the contract existing between defendant and Dexter Horton National Bank for the installation of marble therein, the defendant was to have three months in which to install the marble after the [27] said banking quarters were turned over to the defendant for the installation of the same. That the Dexter Horton National Bank moved into said banking quarters

on the 20th day of December, 1924, and before the
time had expired for the installation of said marble.
That there was no delay on the part of the plaintiff in furnishing the marble. That neither the defendant nor the said bank suffered any damages by
reason of any claim of delay on the part of the
plaintiff in furnishing marble for the banking quarters of the said building.''

X.

The Court erred in refusing to adopt the following finding of fact tendered by the plaintiff: •

''The court further finds that all of the marble
furnished to defendant by the plaintiff for installation in said bank building was inspected by
the architect acting for the bank before the same
was installed as required by the contract between
the defendant and the bank and the plaintiff and
defendant and was accepted by said architect and
permitted to be installed.''

XI.

The Court erred in refusing to adopt the following finding of fact tendered by the plaintiff:

''The court further finds that all of the marble
furnished by the plaintiff to the defendant for installation in said banking quarters conformed to
the requirements of the plaintiff's contract with
the defendant in all respects.''

XII.

The Court erred in refusing to adopt the following finding of fact tendered by the plaintiff:

''The court further finds that all claims of the defendant [28] and the bank for defective marble were waived by the defendant and the bank when that architect for the bank permitted the same to be installed without any objection or protect or claim to the plaintiff that the same was defective.''

XIII.

The Court erred in refusing to adopt the following finding of fact tendered by the plaintiff:

"The court further finds that there was no objection made to the plaintiff by the defendant or the bank, or the architect acting for the bank that the marble was defective in color or in any other respect prior to the completion of the installation of the marble in the banking quarters of the said bank building.''

XIV.

The Court erred in refusing to adopt the following finding of fact tendered by the plaintiff:

"The court further finds that there was no evidence introduced in this case and no competent proof that the defendant or the bank sustained any damages by reason of any claim of defect in the color or finishing of the marble furnished to defendant by the plaintiff under the contract between the plaintiff and the defendant.''

XV.

The Court erred in refusing to adopt the following finding of fact tendered by the plaintiff:

"The court further finds that the defendant has wholly and entirely failed to prove by competent evidence, or otherwise, each and all of the affirma-

tive defenses set out in the defendant's answer
herein.''

XVI.

The Court erred in refusing to adopt the following finding of [29] fact tendered by the plaintiff:

''The court further finds that the plaintiff at all
times prior to the bringing of the present action
furnished the defendant all of the marble requested
by the defendant to take the place of any marble
claimed to be defective by the defendant.''

XVII.

The Court erred in refusing to adopt the following finding of fact tendered by the plaintiff:

''The court further finds that the plaintiff has
never failed or refused to furnish any marble to
take the place of any marble claimed to be defective
by the defendant, or by the bank, or by the architect acting for the bank.''

XVIII.

The Court erred in refusing to make findings of
fact in accordance with each of the above specified
findings of fact tendered by the plaintiff.

XIX.

The Court erred in refusing to adopt the following conclusions of law tendered by the plaintiff:

''That the plaintiff is entitled to a judgment in its
favor against the defendant in the sum of $7,290.32
with interest thereon at the rate of 6% per annum
from April 1st, 1925, to date hereof.''

XX.

The Court erred in adopting the following finding of fact tendered by the defendant:

"The court finds that in addition to furnishing the marble for said building, the plaintiff, at the instance and request of the defendant, furnished certain extra marble and material, to the extent and of the value of $657.92 and no more; and that all other items of extras claimed by the plaintiff have not been established by the evidence and are disallowed. [30]

"The original contract for said marble work amounted to the sum of $46,000.00 and, together with the extras herein allowed, amounts to the total sum of $46,657.92, upon which the defendant has been paid the sum of $40,081.24, leaving a balance unpaid upon said original contract and extras in the sum of $6,576.68."

XXI.

The Court erred in adopting the following finding of fact tendered by the defendant:

"The court further finds that the plaintiff did not furnish said marble according to said contract but that it greatly delayed the furnishing of said marble and the defendant in performing its said contract with the Dexter Horton National Bank, although defendant urged plaintiff for a speedier delivery of said marble, by numerous letters, telegrams and cablegrams, and personal visits to San Francisco; that a portion of said marble was defective and not according to contract, especially in color, workmanship and material; and the Court also finds that the defendant was required to do certain extra work and labor, hereinafter referred to, in connection with said contract, on account of a portion of said marble not being fabricated according to the plans and specifications."

XXII.

The Court erred in adopting the following finding of fact tendered by the defendant:

"The court further finds that the defendant is entitled to and is allowed the following counterclaims and offsets, as shown by the evidence:

(a)   Damages caused by delay on plaintiff's part in furnishing said marble according to contract......$4,500.00

(b)   For defects in said marble.......... 2,500.00

(c)   For extra work and labor, consisting of 32 hours of labor in cutting triangles to fit the plans, and 104 hours of labor in drilling holes in balustrade,—at $1.25 per hour....   170.00

(d)   For wooden hand balustrade on open stairway to the basement....   25.00

TOTAL.................$7,195.00

and all other items in defendant's counterclaims and setoffs have not been proven by the witnesses."
[31]

XXIII.

The Court erred in adopting the following finding of fact tendered by the defendant:

"The Court further finds that the defendant has

fully kept and performed all of the covenants and agreements of its contract with the plaintiff.''

XXIV.

The Court erred in adopting the following conclusions of law tendered by the defendant:

''That the counterclaims and offsets to which the defendant is entitled exceed the plaintiff's claim in the sum of Six Hundred Eighteen and 32/100 Dollars ($618.32), and that the defendant is therefore entitled to a judgment against the plaintiff in that sum.''

XXV.

The Court erred in adopting each and all of the above specified findings of fact and conclusions of law for the reason that the same are contrary to the evidence and without evidence to support the same, and are contrary to the law applicable to the case.

XXVI.

The Court erred in entering a judgment in favor of the defendant and against the plaintiff in this cause.

WHEREFORE, plaintiff prays that the judgment herein of the District Court in favor of the defendant and against the plaintiff may be reversed.

GROSSCUP & MORROW,
CHAS. A. WALLACE,
J. O. DAVIES,
Attorneys for Plaintiff.

[Endorsed]: Filed Dec. 9, 1926. [32]

[Title of Court and Cause.]

ORDER ALLOWING WRIT OF ERROR.

WHEREAS, on the 9th day of December, A. D. 1926, the petition of The Mission Marble Works, a corporation, plaintiff in the above-entitled action for a writ of error, came on to be heard, and it appearing to the Court from the petition filed therein, and from the assignment of errors filed therewith, that its application should be granted and that a transcript of the record, proceedings, papers and exhibits upon which the judgment of the Court was rendered, properly certified, should be sent to the United States Circuit Court of Appeals for the Ninth Circuit to be held at San Francisco, California, as prayed, in order that such proceedings may be had as may be just and proper in the premises.

NOW, THEREFORE, IT IS ORDERED, that a writ of error be allowed upon supersedeas bond being furnished by The Mission Marble Works, conditioned according to law, in the sum of One Thousand ($1000.00) Dollars, and a true copy of the records, assignment of errors, and all other proceedings in the case, together with Plaintiff's Exhibit No. 3 for identification, shall be duly certified and transmitted to the United States Circuit Court of Appeals [33] for the Ninth Circuit to be held at San Francisco, California, in order that said Court may inspect the same and take such action thereon as it deems proper according to law.

Granted this 9th day of December, A. D. 1926.

EDWARD E. CUSHMAN,

Judge.

[Endorsed]: Filed Dec. 9, 1926. [34]

[Title of Court and Cause.]

STIPULATION RE SUPERSEDEAS BOND AND PRINTING OF RECORD.

IT IS HEREBY STIPULATED AND AGREED by and between the parties hereto, through their attorneys, that one bond in the sum of One Thousand ($1,000.00) Dollars, will be sufficient for the purposes of cost bond and supersedeas bond on writ of error to the United States Circuit Court of Appeals for the Ninth Circuit, in the above-entitled cause. Title of the court and cause may be omitted on all papers on error proceedings, except the complaint.

DATED this 30th day of November, 1926.

GROSSCUP & MORROW,

CHAS. A. WALLACE,

J. O. DAVIES,

Attorneys for the Plaintiff,

H. A. P. MYERS,

Attorney for the Defendant.

[Endorsed]: Filed Nov. 30, 1926. [35]

[Title of Court and Cause.]

WRIT OF ERROR BOND.

KNOW ALL MEN BY THESE PRESENTS: That the Mission Marble Works, a corporation, as principal, and the Aetna Casualty and Surety Company, a corporation, as surety, are held and firmly bound unto Robinson Tile & Marble Company, a corporation, in the full and just sum of One Thousand ($1,000.00) Dollars, to be paid to the said Robinson Tile & Marble Company, its successors and assigns, to which payments well and truly to be made we bind ourselves and our successors and assigns, jointly and severally, by these presents.

DATED this 8th day of December, A. D. 1926.

WHEREAS, lately and at the regular term of the District Court of the United States for the Western District of Washington, Northern Division, sitting at Seattle in said District, in the suit pending in said court between The Mission Marble Works, a corporation, as plaintiff, and Robinson Tile & Marble Company, a corporation, as defendant, Cause No. 10,190, on the Law Docket of said court, final judgment was rendered against the said The Mission Marble Works, a corporation, for the sum of $618.22 and costs of suit, and the said The Mission Marble Works has obtained a writ of [36] error and filed a copy thereof in the Clerk's office of the said court, to reverse the judgment of said Court in the aforesaid suit, and a citation directed to the said Robinson Tile & Marble Company, a corporation, defendant in error, citing it to be and appear before the United States Circuit Court of Appeals for the Ninth Circuit, to be holden at San Francisco, California, according to law, within thirty (30) days from the date thereof.

Now, the condition of the above obligation is such that if the said The Mission Marble Works, a corporation, shall prosecute its writ of error to effect and answer all damages and costs, if it fail to make its plea good, then the above obligation is to be void; otherwise to remain in full force and effect.

THE MISSION MARBLE WORKS.

By GROSSCUP & MORROW,

Its Attorneys.

THE AETNA CASUALTY AND SURETY COMPANY.

By A. R. CLOSE,

Resident Vice-President.

[Seal]   Attest: CHAS. W. DIAL,

Resident Assistant Secretary.

12/8/26.

Bond is O. K.

H. A. P. MYERS,

Atty. for Defendant.

Approved as a cost bond on appeal this 9th day of December, 1926.

EDWARD E. CUSHMAN,

Judge.

[Endorsed]: Filed Dec. 9, 1926. [37]

ORDER EXTENDING TIME TO AND INCLUDING DECEMBER 1, 1926, TO SERVE AND FILE BILL OF EXCEPTIONS.

Upon stipulation of the parties and good cause therefor appearing, it is hereby

ORDERED that the plaintiff be, and it is hereby given until the 1st day of December, 1926, in which to prepare, serve and file its bill of exceptions herein.

Dated this 1st day of November, 1926.

EDWARD E. CUSHMAN,

Judge.

[Endorsed]: Filed Nov. 1, 1926.   [38]

[Title of Court and Cause.]

BILL OF EXCEPTIONS.

BE IT REMEMBERED that on the 16th day of June, 1926, at the hour of 2 o'clock P. M., the above-entitled cause came on for trial before the Honorable Edward E. Cushman, Judge, in the Federal Court and Building at Seattle, King County, Washington.

Plaintiff was present and represented by J. O. Davies, Esq. Defendant was present and represented by H. A. P. Myers, Esq. Both sides having announced themselves ready for trial, thereupon the following proceedings were had and the following evidence was introduced; and thereupon the following proceedings were had and done and no other proceedings were had or done and the following evidence was introduced and no other evidence was introduced.

By written stipulation of the parties to the cause, a jury was waived and the cause was tried before the court without a jury.

TESTIMONY OF GEORGE B. EASTMAN, FOR PLAINTIFF.

GEORGE B. EASTMAN, a witness produced on behalf of the plaintiff, being first duly sworn, testified as follows:

Direct Examination by Mr. DAVIS.

My name is George B. Eastman. I am president and manager of plaintiff, The Mission Marble Works, a corporation. Its place of business is 209 Mississippi Street, San Francisco. I [39] was such president in the year 1924. I am acquainted with the defendant Robinson Tile & Marble Co. I know Mr. Robinson, the president of that company. I had, as such president and manager, business dealings with the defendant in the year 1924.

Plaintiff's Exhibit No. 1 is a contract to furnish marble f. o. b. cars, San Francisco, to the defendant for the Dexter Horton National Bank Building of Seattle.

Mr. DAVIES.—We offer this contract in evidence.

Mr. MYERS.—No objection.

The COURT.—May be admitted.

Contract received in evidence and marked Plaintiff's Exhibit One.

Exhibit One is as follows:

PLAINTIFF'S EXHIBIT No. 1.

"THE MISSION MARBLE WORKS,

Importers, Quarry Producers and Dealers in
MARBLE

FOR

BUILDING CONSTRUCTION.

PROPOSAL FOR BUILDING MARBLE F. O. B.
CARS. (A)

San Francisco, Calif., Jan. 30, 1924.

To Robinson Tile & Supply Co.

(Hereinafter called the Purchaser)

219 Marion St., Seattle, Wash.

The Mission Marble Works (hereinafter called the Company) proposes to furnish as required by the specifications prepared by John Graham, Architect (acting for the purposes of this agreement as the Agents of the Purchaser), except as hereinafter modified, for the Dexter Horton National Bank Building at Seattle, Wash., the following building marble, viz.:

Per Sheets Nos. 2, 2A, 5, 6, 8 & 10. [40]

According to the plans, specifications, terms, and conditions of same, and as covered in purchasers contract with said Bank in order to enable the purchaser to faithfully and promptly fulfill their contract with the above bank.

Said drawings and specifications are made a part hereof and are identified as follows:

¼″ Scale, ¾″ Scale and 3″ Scale plans.

The grade of the marble to be furnished hereunder is Napoleon Gray, Belgian Black, and Golden Vein St. Genevieve.

Said marble will be delivered F. O. B. cars at our works in San Francisco, Calif.

The price for said marble delivered F. O. B. cars as aforesaid is ($46,000.00).

Forty six thousand and No/00 Dollars, payable in cash as follows:

On the tenth day of each month 90% per cent of the amount of marble delivered hereunder during the previous calendar month and the 10 per cent before retained within thirty-five days after the full completion of this agreement by the Company.

The marble is to be cut and finished by the Company ready for setting, except such fitting as is usually done at the building, in a good workmanlike manner, and unless otherwise herein stipulated, in strict conformity with said drawings and specifications and to the satisfaction of said Architects. The Company is not to cut lewis holes or anchor holes.

The Purchaser shall furnish or cause to be furnished such further details or explanations as may be necessary to delineate the plans and specifications and to enable the Company to perform said work as herein provided. Such models, if any, as may be required for the performance of said work shall be furnished by the Purchaser. The Purchaser shall make or cause to be made at the building and furnish to the Company all necessary detail measurements.

If any modifications are made in the work or material shown or described in said drawings or specifications, or if the further details or explanations furnished should call for work or material more elaborate or of greater value than is warranted by said drawings and specifications, the value of the work added or omitted shall be added or deducted from the contract price at a fair and reasonable valuation. In case of disagreement with respect thereto, or with respect to any matter arising under this agreement, the question at issue shall be referred to three disinterested arbitrators, one to be appointed by the Purchaser and one by the Company and the third by the two thus chosen, and the decision of any two of them shall be final and binding and each of the parties hereto shall pay one-half of the expense of such reference.   [41]

It is a well-known fact that fancy or colored marbles are more or less unsound, and it is hereby agreed and understood that should any of the fancy or colored marbles to be furnished under this proposal require waxing, filling, cementing or backing in order to present a smooth surface, that we will have the right to do such waxing, filling, cementing or repairing as may be necessary, and this marble, when so treated, is to be accepted.

The Company, shall not in any event be held responsible or liable for any loss, damage, detention or delay caused by the Owner, the Purchaser, or any other contractor or subcontractor upon the building, or by delays in transportation, fire, strikes, lock-outs, civil or military authority, or by insurrection or riot, or by any other cause beyond its reasonable control. If the financial responsibility of the Purchaser is or becomes unsatisfactory, the Company reserves the right at any time during the life of this agreement to require the purchaser to give satisfactory security. Any delay on the part of the Purchaser to furnish such security or in making any payment hereunder as and when it becomes due and payable shall operate to that extent as an extension for the time of completion, and such delay if continued for more than ten days shall at the option of the Company be held to be prevention by the Purchaser of the performance of this agreement by the Company. Deferred payments shall be subject to interest.

The Purchaser shall during the progress of the work maintain or cause to be maintained full insurance on all marble or other work herein called for, incorporated in the building or in or about the premises until paid for, with loss, if any, payable to the parties hereto as their interests may appear.

AGREEMENT.

All previous communications between the parties hereto, either verbal or written, contrary to the provisions of this proposal are hereby withdrawn and annulled, and this proposal duly accepted and approved constitutes the agreement between the

(Testimony of George B. Eastman.)

parties hereto, and no modification of this agreement shall be binding upon the parties hereto, or either of them, unless such modification shall be in writing duly accepted by the Purchaser and approved by an executive officer of the Company or the Manager of its office.

The foregoing proposal is subject to the approval of an executive officer of the Company or the Manager of the office, and shall not be binding upon the Company, until so approved nor unless accepted by the Purchaser within sixty days of the date hereof.

THE MISSION MARBLE WORKS.

By A. S. HUNTER.

To The Mission Marble Works,

Your proposal as above is hereby accepted this 30 day of Jan'y, A. D. 1924. [42]

ROBINSON TILE & SUPPLY CO.

By A. P. ROBINSON,

Pres.

Approved at San Francisco, Cal., January 30th, 1924.

THE MISSION MARBLE WORKS.

By GEO. M. EASTMAN."

Under this contract we furnished the marble work for the Dexter Horton National Bank building and shipped part of the same by rail and part by boat. We fabricated and shipped all the marble required under this contract and all of the marble was received at Seattle and has been installed in the bank and is in the bank at the present time. I have seen it there. At the request of the defendant we

(Testimony of George B. Eastman.)

shipped other marble besides that covered by the contract. We furnished marble for information desk in addition to that called for by the contract, of the value of $440.00. We furnished marble for change between columns 61 and 66, requested by letter of October 31, 1924, of the value of $7.85. We furnished some St. Genevieve marble for the same location for the sum of $154.85. We furnished mould and cove labor in the sum of $47.25. We furnished Belgian Black Coved Base per requisition #351 in the sum of $81.94. We furnished replacement pieces broken in transit, 55 in number, for $15.60. We furnished St. Genevieve marble at Pier #75 and Pier #72, at a value of $63.20. We furnished a piece of St. Genevieve apron, including labor, at a value of $16.00. The boxing of shipments in less than carload lots, from November 3, 1924, to February 19, 1925, amounted to the sum of $484.78. The cartage on shipments in less than carload lots, amounted to $60.00. We also furnished some full-sized slabs to take the place of others, which are not charged for in this invoice, and we make no charge for them now. [43]

The contract calls for shipment of this marble in carload lots, f. o. b. cars San Francisco. When shipped by boat in pieces it is necessary to crate the same and cart them to the terminals.

The defendant requested all marble shipped by boat to be shipped in this manner. Request was made by defendant by telegram. I have a confirmation of the telegram, sent by letter.

Mr. DAVIES.—We offer this as Plaintiff's Exhibit 2, in evidence.

The COURT.—It may be admitted and marked Plaintiff's Exhibit No. 2.

Plaintiff's Exhibit No. 2 is as follows:

PLAINTIFF'S EXHIBIT No. 2.

"Seattle, November 26, 1924.

Mission Marble Works,
    209 Mississippi Street,
        San Francisco, California.

Make all shipments by Pacific Steamship boats Notice shipment nineteenth shipped Southern Pacific May take two weeks to arrive Stop Please rush revised schedule showing dates various parts will be shipped so we can arrange our program here Stop When will main floor ramps be shipped

ROBINSON TILE AND MARBLE COMPANY,
    600 8th Avenue North

Charge our account
                    Confirmation
    10190

Ptf. Ex. 2."

The total amount of those extras is the sum of $1,371.76. All of those extras that I spoke about were shipped and received here and installed in the bank of Seattle, with the exception of the check desk. I do not know whether that was installed in the bank or not, but it was shipped as an extra at defendant's request.

(Testimony of George B. Eastman.)

The amount of money which was paid the plaintiff by the defendant on the whole contract is the sum of $40,081.24. There is a balance due on the contract, according to our figures, in [44] the sum of $7,290.32, and that amount is still due.

Cross-examination by Mr. MYERS.

WITNESS.—The amount of the claim that we are suing for, for extras, is the sum of $1,371.76.

I am pretty familiar with the marble. There were no holes in this marble when it left our shop. I know from correspondence that the bank is making some claim for defective marble.

Q. And the fact that they had to patch it and some of it was broken.

A. Referring to the claim of the Dexter Horton Bank saying one or two check desks that some holes filled in it by ourselves. That is what I have no knowledge of. I would not know about the breakage except he sent letters down to replace a piece once in awhile; so I replaced it. I don't know any other breakage.

Q. Weren't the holes filled in your shop?

A. Yes, sir.

We have you charged with $544.78 for boxing and cartage of this marble. Our basis for charging that sum to you is the telegram, Plaintiff's Exhibit No. 2.

We understood from that telegram that shipments had to be made that way because they could not wait for the train.

(Testimony of George B. Eastman.)

Q. Didn't you also understand that that would be a saving to you and Mr. Robinson from damage by bank in delaying building; you thought that was Mr. Robinson's proposition.

A. I was not fearing any damages. I knew they were urging us all the time to get the marble up here. I received numerous and many telegrams urging us to hurry up with this marble. I went east but I did not leave the plant without anyone to take [45] care of it. It is our custom to charge for boxing and cartage when required. I think that bill for boxing and cartage is right. I think it is right to collect it. I have no memory or recollection of writing a letter to Mr. Robinson telling him I would forget it. I told him if he would pay us the money due we would forget those charges. I don't remember writing to that effect; but he did not pay us so, therefore, the charge for boxing and cartage still stands.

We understood that the reason the marble was shipped by water was to furnish some marble here so they could keep on working. We were advised by various letters of certain items that were found short. I make no charge for them. I did not intend to. I do not believe we have. It is only where marble was broken in transit or where defendant wanted some change made.

Q. In your account of extras, we find an item short, of $81.94, of which you were advised. We find that you still have those in your bill. How do

(Testimony of George B. Eastman.)
you account for that. Do you know anything about that?

A. I would have to refer to the shipping sheets, to our orders from the office.

Q. Do you want the sheet that you just had?

A. No, the original shop orders.

Mr. DAVIES.—Here they are.

WITNESS.—That is the office copy of the shop order and the invoice was made from that.

Q. Mr. Eastman, it is a fact that you allowed other jobs that you had taken to interfere with the furnishing of this marble in such a manner that the Robinson Tile & Marble Company were delayed and the bank was delayed in this job, isn't that true?

Mr. DAVIES.—I object to this as not being proper cross-examiantion. [46] I think they should be required to prove their counterclaim as a part of their case.

The COURT.—The objection is overruled; exception is allowed.

A. Not until they had notified us to cease making shipments. We diligently worked on the job until they told us not to ship.

Q. Who told you not to ship?

A. First, the man at the bank that I visited with Mr. Robinson at the time we got the second check. I never knew his name. He asked us and told us that the weight of the building would not stand. It was too green concrete. We had two cars of marble already shipped, and he did not want us to

(Testimony of George B. Eastman.)

store any more. I told him it would be impossible for us to fabricate our material, some eight or nine thousand pieces, and not have it shipped. He said the building won't stand the weight. It was quite a weight. I explained to him that if we stopped proceeding with the shipment we would have to stop the proceeding of fabrication. I took it up and talked with him in the presence of Mr. Robinson, and then he told me,—I went back to my boys, and I told my foreman that the job had been delayed, as we had been told to lay off of that job and to take on something else, because we could not go any further, except as the thing progressed. We got our marble in large blocks, fifteen to twenty tons; and we would saw it in gang saws and it cuts in one end of the material and goes through there, and when it gets into the machinery we had to keep going on this.

The man who told me not to ship, from the Dexter Horton National Bank, is the man who gave Mr. Robinson the check. I never knew his name. It was at the Dexter Horton National Bank. [47] It was on June 23, 1924, before they moved the bank.

Yes, sir, someone in the Dexter Horton Bank told me expressly to let up sending marble; that they would not take any more shipments. I think I am correct in the date; that it was June 23, 1924.

The marble that we had sent was stored in the rear of the bank. Our contract called for the storing of marble in the bank where it was to be set, but they stored it in the subbasement. Instead of

(Testimony of George B. Eastman.)
being sublower it was subhigher, where it runs up.
I saw the marble stored there.

Mr. Graham agreed when he visited us that there
was no haste; that the job had been delayed. That
was on July 16, 1924. On July 16th, Mr. Graham,
the architect, visited us. I told him then we could
hasten further delivery. He said there was no
haste, that the job had been delayed and we were
just finishing then the only marble that was coming
through, and it would come through from one
end and goes out the other. We were finishing the
material then and we were slacked up from that
time; that is, we did not do any more; perhaps a
little, but we did not pay particular attention to the
job.

We had our contract with Mr. Robinson. I don't
believe Mr. Robinson ever told me to let up on the
work. Mr. Robinson during the latter part of the
job wired us many, many times, urging us to send
the marble on here. That was after he told us
again to start. Instructions were to have the first
carload here by June 1st and I spent the extra
freight and was charged eight hundred and some
odd dollars to have the car come over extra; and
we had the second car shipped in June, the 4th, I
believe, if I remember the right date, because when
I was here on the 23d the car was here, but after
they told us to stop [48] shipment we didn't
ship any more at that time. They urged us in Sep_
tember to start shipment again; the building was to

(Testimony of George B. Eastman.)

be ready and he was to have everything finished in three months.

When I signed the contract with Mr. Robinson I saw a copy of the contract that Mr. Robinson made with the bank. I understood that I was to furnish this marble so as not to delay the prosecution of the work on Mr. Robinson's part. We did so until we got authority to cease. We understood that Mr. Robinson had three months' time to finish the job. I worked very diligently to furnish this marble so that he could finish it within the three months, until they notified us not to ship any more. On Sept. 17th we were notified by telegram to go ahead and start shipping again and after that we prosecuted the work just as fast as we could. We did everything in our power. We went to work and got another firm to help us get the material out. We hired extra men. Mr. Graham, Mr. Grant and Mr. Robinson visited us. Mr. Robinson asked us to employ extra men to work Sundays and extra times to get the marble out, and we complied. After we were notified to cease shipping we started to work on other jobs.

Mr. MYERS.—I will read this into the record and ask you if you recognize it:

"George B. Eastman, Mission Marble Works, San Francisco, California. After my visit with you had formed high opinion of yourself and company. I felt that we were going to have a good job, completed on time, but find now, ten months after contract signed that you have fallen down so com-

(Testimony of George B. Eastman.)

pletely and left us in the lurch, it is a very great disappointment. I had assured the bank directors that Mission Marble Works were a fine [49] firm and would complete their engagements on time. The bank have all arrangements made to open on December 6th and conditions here are very serious. Will you let me know personally exact condition? Signed, John Graham." That telegram is November 11, 1924.

A. I was away at the time. I went away October 10th. I have examined the correspondence in this case. I don't recall the telegram, but there was a similar telegram.

Q. Now, what was done by you in response to that telegram, or by your company?

A. As I stated, I think I went away and returned about that time, and every effort was made to continue to hurry up with the job. I left for the east October 10th and got back November 11th. I was gone about a month and a day. I remember while in the east of getting a very unpleasant telegram from Mr. Robinson.

Q. You were advised, were you not, of the bank's claim. You knew that the bank had refused to pay on this contract $6,000.00, and on account of the color of the marble and the holes in the marble and defects in the marble, and also on account of the delay in going in the bank, did you not?

A. Two items, I believe, one was the coloring of one panel, one certain panel was pointed out and one check desk had holes in, that was the only thing

(Testimony of George B. Eastman.)

I knew of that there was any defects ever presented. I did not know anything else. That was called to my attention by Mr. Grant on the visit here on June 25th, 1925.

I don't recall Mr. Robinson ever telling us to let up on sending marble. Our contract was with Mr. Robinson.

Q. Why did you pay any attention to anybody else.

A. Because Mr. Robinson was present with me at the time; we [50] were there with the gentlemen when he told us not to ship any more marble and I wrote Mr. Robinson afterwards, when Mr. Graham visited us, confirming what Mr. Graham said—and I wrote him a letter.

Q. What did Mr. Robinson tell you about that conversation, if anything?

A. Mr. Robinson did not say; I don't recall anything that he said.

I am positive that Mr. Robinson did not, at that time, tell me not to let up on sending marble. If he had I would have continued shipping.

Redirect Examination by Mr. DAVIES.

The conversation I had with the gentleman at the bank, who paid the money, and who told me not to ship any more marble, was on or about June 23, 1924. At that time I explained to them that if we stopped shipment we would have to stop manufacturing; but what was started through the process of manufacture had to continue through. I was afraid some other things would come up and cause

(Testimony of George B. Eastman.)

delay and I explained about the process of finishing marble the best I knew how. I explained this to the gentleman at the bank. He had a moustache and light whiskers. I don't know his name. Whoever it was, I explained to him at that time that I would have to keep my plant going and that we would have to take on other contracts and other work would get ahead and we would have to finish that before we could start something else again. That is the process.

Q. You explained to them at that time that if you stopped fabricating marble for the bank and if you were actually engaged in fabricating marble for somebody else at the time you got your orders from them, everything that you started you would have to [51] complete?

A. Just as I explained to them, thoroughly, the best I knew how. We can't buy our material finished. We have to manufacture and thaw it, and rub it, and polish it, and it can't stop when it once goes through, unless to throw it back into a block yard again.

Mr. Robinson was present when I had that conversation with the bank. He did not make any protest and so far as I know, it was agreeable with him. I was trying to please his client and he was too. We were both trying to please his client.

Recross-examination by Mr. MYERS.

Q. Who gave you orders to resume the prosecution of this contract and this marble?

(Testimony of George B. Eastman.)

A. Some telegram of September 16th, I think it was, saying that they would be ready in about a week to start to set the job, and asked if I could not get some setters for them. We had a large portion of the marble on the job already. We had three cars shipped. The first car was shipped about April 26, 1924. The second car was shipped June 4, 1924. The third car was shipped July 28, 1924, and the fourth car was shipped October 11, 1924, and the L. C. L. shipments started there. When Mr. Graham visited us on July 16th, a car was going on the way through the mill and we had to ship it to get rid of it on the 26th, and we did not ship again until October, because we did not have a carload.

Q. The trouble you encountered, and what made the trouble with the bank and with the architect, was your failure to deliver this material and marble at the latter part of the contract?

A. It referred to the stairway portion. The other part of the marble we were pretty near all filled, if not all. I think it was the stairway portion, the heavy grooving material. [52]

Q. You knew that the bank could not very well move in the bank and have the workmen working around the stairway and hammering and working around there while they had the bank in there, didn't you?

A. We do work around a bank, the same as we do any place. We make it convenient for the people in the bank. We do it while they are there as

(Testimony of George B. Eastman.)

well as out of there; but moving into a new place they don't want the noise. I understand they worked in the hallway or downstairs or something. I was not there.

Q. You were warned, both by Mr. Robinson and by Mr. Graham, that there would be big damages against you on account of this delay, weren't you? You knew it would cost the rent?

A. I did not know.

Q. You knew that a big bank, occupying large quarters would cost rent if they did not own it?

A. I did not know about the rent. I was not fearing the rent. I did the best I could: After they stopped us and started in again, I did the very best that we could for everybody. It is our policy to do so.

Mr. DAVIES.—Have you got that telegram that you sent, telling them to start work?

Mr. MYERS.—I don't have it. It may be among the files.

Plaintiff rests.

TESTIMONY OF FRANCIS W. GRANT, FOR DEFENDANT.

FRANCIS W. GRANT, a witness on behalf of the defendant, being first duly sworn, testified as follows:

Direct Examination by Mr. MYERS.

My name is F. W. Grant. I am employed by John Graham, Architect. I am an architect.

(Testimony of Francis W. Grant.)

John Graham was the architect for the Dexter Horton building. I have been with Mr. Graham four [53] years. I have followed the business of an architect twenty-five or thirty years, I guess. I am familiar with this contract and the work on the Dexter Horton building. The Robinson Tile & Marble Company had the contract for the marble in that building.

Q. What difficulties did you encounter on account of fulfilling this contract with reference to the marble? A. They were rather numerous.

The Robinson Tile & Marble Company began the work of setting the marble on September 22, 1924. There was insufficient marble to proceed with the setting of it continuously at the time they were ready to set—not enough marble in Seattle, delivered by the Mission Marble Works, to permit the continuous work of setting it in place. The setting was dependent upon the other conditions of the building and the commencement of setting. Shop drawings were never furnished complete. The result of not furnishing shop drawings was a great deal of confusion as to the conduct of the work and slowed up the work and led to some errors that I am not prepared to name specifically. This refers to the whole marble job. We wrote several times to Mr. Robinson, requesting the setting plans or shop drawings. We made our demands upon the contractor with whom we were dealing; that is, the Robinson Tile & Marble Company. We never wrote the Mission Marble Works

(Testimony of Francis W. Grant.)

on the subject of shop drawings or setting plans. I did, however, personally demand of them shop drawings on a visit of mine to San Francisco in the last week in July, I think. On July 27th I made an informal visit to San Francisco, in vacation time, and urged the completion of the shop drawings, and was shown shop drawings then in the course of preparation; also some places which were incident to the making of shop drawings, [54] the stairways. We should have had those shop drawings so as not to interfere with the prosecution of the work. At any time before the marble came here would have been time to have avoided trouble, provided the shop drawings needed any correction, and the purpose of the shop drawing is to enable the architect to correct them in case the contractor fails to properly interpret the drawings.

Q. When did you get shop drawings?

A. I never did get them.

The reason for some of the extra work that Mr. Robinson did was that he had to erect the forms on the balustrade on the stairway because of failure to show on the shop drawings how it should have been done. I went to San Francisco in reference to this bank job, at the request of Mr. Robinson. The object of my visit was to see if I could expedite the furnishing of marble. I returned from my visit on November 20th. I have no record of the day I left. I was absent about a week. I talked to Mr. Eastman about expediting the work. I visited the shop of the marble concern to

(Testimony of Francis W. Grant.)

whom he had sublet a portion of the work in order
to get it done faster. He had exceeded the capacity
of his own place and had sublet to this other concern the making of the balustrades, and I think the
stair rail and stair stringer. I am not sure what
particular part, except the balusters and the rail.
I know they made the rail. I visited that place
and saw the work in progress and endeavored to
expedite it. I asked them to rush it as much as
possible. Mr. Robinson paid my expenses on that
trip. I don't remember in what sum. He bought
the transportation and Pullman accommodations.
I don't remember the amount that he gave me.
The marble that was furnished by the Mission
Marble Works for the job was supposed to be in
color [55] corresponding with the sample that
was approved, and was so until a certain time, when
a marble of decidedly lighter and yellowish color
commenced to come. When this contract was made
there was a sample of marble that was broken in
two, one-half going to Mr. Robinson and we kept
the other. A considerable quantity of the marble
that was delivered was very much yellower than
the sample, so much so that it could not properly
be placed where the setting plan indicated it should
be placed and put close to the marble of the other
color. It was condemned. We told Mr. Robinson
that we would not accept it. In view of the fact
that the bank was very anxious to get in their
premises, a compromise was made as to some of it
and there was put in some with the understanding

(Testimony of Francis W. Grant.)

and promise of Mr. Robinson to take it out and put in other corresponding with the contract sample. Some of that has been done and some has not. This light color in the marble could have been avoided. This marble is delivered to Mr. Eastman by the quarry in large blocks and on the side of the block is a painted line, indicating the direction in which the saw should cut the marble into slabs. If the marble is not cut according to that instruction of the quarryman the result will be a different color altogether; in other words, we get a yellower color by reason of it being sawed the opposite way from the marks given by the quarryman. I presume this was done to save stock, or he would not do it. I can conceive no advantage to Mr. Eastman other than to make the blocks of a certain size better fit into the work and get his pieces out. They had a limited number of blocks to get this out of. It may have been, had he sawed the line under the instruction of the quarryman he would have run out of marble. I think there were five check desks. The tops of these check desks are very large. They weigh about 3,000 lbs. [56] apiece. They are about 5″ thick and three feet wide. I don't remember the length. They were all broken in the process of manufacture. They were broken across. Each one was broken, due to, as I was told, while I was in San Francisco, because of an accident while on the saws. We have not accepted them yet. They were put together again with wax, reinforced on the bottom with steel and had the

(Testimony of Francis W. Grant.)

waxing been sufficient to deceive the ordinary eye we probably would have accepted them. This is visible to anyone now. One of the check desks was patched. Certain holes left by drilling at the quarry were filled and the patch is quite evident and shows that the holes were there, several holes about an inch and a half in diameter. The edge of this check desk top shows some portions of the cylinder and of the wax and chips of marble. By portions of the cylinder I mean the hole cut down from the edge, a round hole, and filled with wax and chips of marble to make its surface complete. Just the same as boring a hole through the top of the desk and plugging it up. That is visible to anyone. It could be pointed out. I could not very well leave this desk out. The bank had to do business and I set it up and put plate glass on top of that and on top of the plate glass there was a piece of bronze work and cabinet. The Robinson Tile & Marble Company had a man working there several days to remedy these defects on this check desk, waxing and filling and trying to conceal the false work. I do not recall that Mr. Robinson made any effort to get us to accept this check desk. He tried to remedy the defect the best he could. We have never accepted this check desk. [57]

Q. Have you made a figuration of what it would cost to replace the defective work in the bank, according to the plans and specifications?

A. Yes, sir, Mr. Graham and I attempted to compute that, and in our judgment it would cost

(Testimony of Francis W. Grant.)

about $2,500.00 to replace and correct all of the defects.

The letter which you hand me, marked Defendant's Exhibit "A-1," for identification, is a letter written to the Robinson Tile & Marble Company explaining the amount that they would be entitled to; that a certificate would be given for the balance due them by the bank, minus the sum of $6,000.00.

Mr. MYERS.—I offer it in evidence.

The COURT.—It may be admitted in evidence as Defendant's Exhibit "A-1," which exhibit is as follows:

DEFENDANT'S EXHIBIT "A-1."

"JOHN GRAHAM
Architect
Dexter Horton Building, Seattle.

March 19, 1925.

Robinson Tile and Marble Company,
600—8th North,
Seattle, Washington.

Gentlemen:

Referring to your statement of February 28th showing a balance due you on your contract for marble work in the Dexter Horton National Bank of $18,706.00.

Confirming our conversation with you relative your charge for extras and allowance for credits, these are entirely eliminated by mutual consent, it being agreed that debits are offset by credits which leaves the balance under the contract as $18,431.00.

(Testimony of Francis W. Grant.)

Of this sum $6,000.00 will be retained pending the correction by you of all defective marble and the satisfactory adjustment by you of damages sustained by the bank on account of your delay in performance of your contract. [58]

A certificate is handed you herewith for the remainder of $12,431.00.

Yours very truly,

JOHN GRAHAM.

Per: G.

FWG. HM.

10190.

Deft. A–1 Adm.''

Q. Referring to the two items upon which the architect declined to make payment on this job until they are corrected, one being for defective marble, and that has never been corrected, and that is the item which you have estimated to the Court would cost $2,500.00 and the other is on account of delay in the performance of the contract; what was the item of damages under that?

Mr. DAVIES.—We think that it is wholly immaterial. 'The architect would not have any right to estimate damages to the bank by reason of any delay. He might have a right to point.out any defects in the marble and give his estimate as to how much it would cost to remedy it, but would not have any right to say that the bank has been damaged so much by failure to get in there or to use it, or anything of that kind.

(Testimony of Francis W. Grant.)

Mr. MYERS.—It is to show good faith of the architect in refusing and declining to pay this $6,000.00.

The COURT.—Objection overruled.

Mr. DAVIES.—Note exception.

A. We did not attempt to assess the damages to the bank nor [59] the cost of storing the defective marble. We had arrived at the sum of $7,300.00.

Q. How did you arrive at that sum?

A. By using $2,500.00 for the correction of the defects and $4,500.00 for the delay. I think I have a memorandum of what the figures were. Our original computation contained Twenty-five Hundred for defects, Forty-five Hundred for damages and $327.00 for some plastering and painting that had to be done after the marble workers had finished their work. That is from a memorandum and the slip that I had that on. It does not seem to be here. I think it is about right. At any rate we did not use that figure in making this deduction.

Q. You deducted less than that amount?

A. We deducted $6,000.00, which is Thirteen or Fourteen Hundred Dollars less than the other.

The bank moved into its present quarters where this marble was put in on December 20, 1924. The job was not entirely finished at that time. All of the stairs leading to the safety deposit vaults below the first floor level and probably one-third of the stair railing above that level, had not been completed. The treads were in so they could walk up

(Testimony of Francis W. Grant.)
and down.   The work was finished March 3, 1925.
The work was prosecuted night and day time both.
The last marble was received from the Mission
Marble Works February 24, 1925, and the bank
moved into its quarters on the 20th of December,
1924.

Defendant's Exhibit "A–2," for identification, is
a contract between the Dexter Horton National
Bank and the Robinson Tile & Marble Company for
this bank marble.

Mr. MYERS.—I offer it in evidence.

Admitted in evidence and is as follows: [60]

DEFENDANT'S EXHIBIT "A–2."

THIS AGREEMENT, entered into this 25 DAY
of JANUARY 1924, by and between the ROBIN-
SON TILE AND SUPPLY COMPANY, a corporation under the laws of the State of Washington,
hereinafter designated the CONTRACTOR and the
DEXTER HORTON NATIONAL BANK, a corporation under the laws of the State of Washington, hereinafter designated the OWNER.

WITNESSETH:

That the Contractor agrees to furnish and install,
in the Bank quarters of the Owner in the Dexter
Horton Building at the corner of Second Avenue
and Cherry Street in the City of Seattle, all marble
work (except toilet rooms and in elevator lobby) as
shown on drawings #2, 2–A, 5, 6, 8 and 10, prepared by John Graham, Architect, and to the entire

satisfaction of said John Graham, Architect. Said drawings and the proposal of the Contractor dated January 21, 1924, are hereby made a part of this contract.

The marble is to be of the several kinds and colors as follows: The floor field to be Napoleon Gray marble, the base, border and deal plates to be Belgian Black marble; all other marble to be Golden Veined St. Genevieve.

The quality and finish of all marble installed shall be equal to that represented by samples on file in the office of the Architect and identified by the signature of the parties hereto. Skilled mechanics shall be employed and the work throughout shall be first class in accordance with the best practice of the marble working and setting industry.

It is agreed that all structural steel supports for marble work will be furnished and set in place by the owner, but that the Contractor shall furnish all required masonry backing and all necessary copper anchors for securing marble to structural steel and masonry backing.

It is agreed that the Contractor shall take measurements at the building prior to the cutting of marble and shall be wholly responsible for the accuracy of all dimensions and shall make good at his expense all errors due to inaccuracy of said measurements or conflicts between the drawings and actual conditions affecting dimensions. The Owner will provide concrete slab under marble floors to within 2″ of finished floor level.

It is agreed that the Contractor shall submit setting plans showing all jointings and anchorage for marble and that the same shall be amended as may be directed by the architect and finally approved by the architect before marble is cut.

It is agreed that the owner shall have the right to add to or deduct from the quantity of marble work included in this contract at any time before the work is started, and that additions to or deeuctions from the contract sum hereinafter named, by reason of said charges shall be as appraised and determined by John Graham, Architect.

It is agreed that the Architect shall be the sole judge of the quality and quantity of the work performed under this contract [61] and that all operations hereunder shall be under his direction and control.

It is agreed that the work under this contract shall be commenced immediately after the date hereof and that the first material suitable for immediate installation shall be delivered at the premises not later than three months after setting plans have been approved by the architect and that the remainder of the marble shall be delivered at such subsequent date or dates as shall not interfere with continuous progress with the installation of the marble work, and it is further agreed that the entire contract shall be completed within three months after the first delivery of material as above provided. It is agreed, however, that the Contractor shall not be held liable for delay or failure to perform by acts of the Owner or of any other contractor or subcontractor upon the building, or by fires, storms, accidents, strikes, boycotts, lockouts, riots, lack of transportation facilities, delays in transportation, government regulations, acts of God or by any other causes beyond the Contractors reasonable control.

It is agreed that the Contractor shall have possession of the space where marble is to be installed for necessary storage of material after the time for commencement of the work as herein provided, and that the Contractor shall furnish all necessary hoisting facilities, power, gangways, scaffolding, heat, water, light and labor incident to the performance of this contract.

The Owner agrees to insure the marble work and protect the Contractor's interest against loss or damage by fire. The Contractor's interest shall for the purpose of fire insurance, consist of all labor and material incorporated in the building for which the contractor shall not have been paid and all material delivered at the premises and not incorporated in the building, but shall not include tools, apparatus and equipment.

In consideration of the faithful performance of the covenants herein contained, the Owner agrees to pay to the Contractor the sum of SIXTY THOUSAND ONE HUNDRED AND ONE DOLLARS ($60,101.00) in installments on or about the 5th and 20th of each month, equal to 90% of the value of material delivered to and labor performed at the premises, as ascertained and computed by the Architect and upon his certificate, and the remainder upon certificate of the Architect that the work is

(Testimony of Francis W. Grant.)
satisfactorily completed and upon a satisfactory showing that all bills and claims of whatsoever nature are fully paid.

The Contractor and the Owner for themselves, successors and assigns hereby agree to the full performance of these covenants and in witness thereof have hereto set their hands the day and year first above written.

ROBINSON TILE & SUPPLY CO.

By A. P. ROBINSON.

DEXTOR HORTON NATIONAL BANK OF SEATTLE.

By D. H. PARSONS PT.

Witnesses:

FRANCIS W. GRANT. [62]

The Robinson Tile & Marble Company gave a bond for the faithful performance of its contract. We notified the Bonding Company that there was evidence of default.

Q. Of what did this consist?

A. Delay, that is all.

I was right there on the job all the time except for the brief period that I was in San Francisco. The work was delayed for lack of marble.

Q. Did you make any objections to anything so far as,—when I say "you," I mean the architect,— so far as Mr. Robinson was concerned, or anything in his personal control, aside from the absence of this marble? Did you have anything in the way of an objection to Mr. Robinson except lack in furnishing materials?

(Testimony of Francis W. Grant.)

A. Well, I don't know. I think there was some complaint made the first week of operations for lack of men. That was at the very beginning of the work. I have a note on that here. "Started work on the 22d of September; on the 30th of September our crew had only got up to six men," and I have a note here to the effect that they had great trouble in getting men. There was no complaint made to Mr. Robinson of his work or the way of conducting the work other than that one early in the game of lack of men. There were complaints made that we did not know at first whether they were attributable to Mr. Robinson or not, but later found out that they were attributable to the Mission Marble Works, with defective co-ordination of the parts due to a failure of shop drawings to properly show how they should co-ordinate.

Q. I did not understand.

A. There was some lack of co-ordination of the parts of the marble work, which we criticised, not knowing at the time whether it was Mr. Robinson's fault or someone else's. We later learned [63] what the fault was, that it was due to designing of the shop drawings that the work had not been gotten out as it should have been gotten out.

I know Mr. Eastman and I know Mr. Robinson very well.

Q. To refresh your recollection I will ask you if you recall that Mr. Robinson told Mr. Eastman in your presence and in your office, not to make any delay of any shipments? A. Yes, sir.

(Testimony of Francis W. Grant.)

Q. Do you remember what reply Mr. Eastman made, if any?

A. No, sir, I do not. I did not take him very seriously at the time. I do not think either party could have meant to stop the marble. I do not think either party had any serious intention of stopping work at any place.

Q. Had you ever come to that conclusion at any time,—you say you did not have, at the time,—have you ever at any time come to any contrary conclusion?   A. That we wanted to accept it?

Q. Yes, sir.

A. I did not know what to make of it. They stopped it; they seemed to stop. The occasion of that conversation I am quite familiar with. The marble was stored in a basement of the Dexter-Horton Building. That was not confined to the bank premises, and the space was filled, and we did not consider that any affair of ours. That was Mr. Robinson's business, to take care of the marble, and I took it for granted that he would take care of it. When it came I found a place for it; and a place was found for all the marble that came, when it came. There was always a place for it.

All payments made for the work were approved by us. The [64] trouble that culminated in the delay in getting in the bank was in the latter part of the performance of the contract. That is when the trouble came up.

Q. When should that contract have been completed?

(Testimony of Francis W. Grant.)

Mr. DAVIES.—I object to that. The contract speaks for itself.

The COURT.—It is in the contract; but the witness may answer.

Exception noted.

A. I think it was three months after the first delivery of the marble.

Q. You are familiar with that more than I am; see if you can find it.

A. Installation. It provides that the marble should be delivered after the submission of shop drawings and entire contract completed within three months after delivery.

Q. How did you understand that?

A. I understood it to be six months after the time they submitted and had their shop drawings approved. The delay was not caused by not getting the shop drawings here. The delay was caused by not getting the marble here. We could go ahead without the shop drawings. We did have difficulty and we proceeded without them; that is, without a complete set. We got part of the shop drawings. In reference to the shop drawings of the stairs, I have no data on which I can say just what drawings we had. On September 24th a letter was written to the Robinson Tile & Marble Company, giving the numbers of the drawings that we had and the numbers missing of shop drawings. [65]

The COURT.—Q. This marble, would not this be prepared by the plant from drawings and then by Mr. Robinson?

(Testimony of Francis W. Grant.)

A. That would be a matter of contract between Mr. Robinson and the Mission, of which I have no knowledge, for Mr. Robinson to make the shop drawings.

Q. How would the plaintiff know what to get out, except by drawings or directions that came from Mr. Robinson?

A. The plaintiff is supplied by drawings made by the architect, complete drawings, showing the general way, the ultimate object sought and finally the finished marble, and, the contract provides that the manufacturer shall reduce those drawings to a drawing in his own language, giving the length of each individual piece of marble, its dimensions complete, making proper provisions for joints and place the joints wherever they should come. The architect never does carry his drawings to that extent. It is the custom, and seems to be necessary, that the manufacturer work the drawings of the architect into what is called shop or setting drawings.

The COURT.—Q. Those are the ones you referred to? A. They are the ones we are talking about.

Q. (The COURT.) The architect would only furnish a detail on request?

A. He starts out with what is thought to be complete, and upon request for anything missing,—he makes the original set of drawings, upon which the contract is based, and is supposed to be complete. If, through oversight, anything is omitted or if there be a doubt as to the meaning of any drawing

(Testimony of Francis W. Grant.)
the architect can furnish another drawing or other drawings, to the extent that is [66] necessary to make the subject clear.

Q. And you say this marble stairway—do they design the railing and length and dimensions to prepare it? It does not specify the length of each piece of stairway, does it? A. No, sir.

With reference to the marble stairway—the architects do not design the railing and length and dimensions to prepare it. The drawings would show that.

I recognize the telegram from John Graham to Mr. Robinson, while Mr. Robinson was down there at San Francisco.

Telegram received in evidence as Defendant's Exhibit "A–3," and is as follows:

DEFENDANT'S EXHIBIT "A–3."

"195FD   BG   134 Collect NL

Seattle Wn Nov 5 1924

A T Robinson
    Care Mission Marble Works
        San Francisco Cal

Bank directors very much disturbed at not having marble ready to move bank on Thanksgiving day Stop Have agreed to postpone to December sixth but no later Stop If your work is not finished then will hold you and your bondsmen for all damages caused by delay Stop Everyone else will be finished on time Stop It is outrageous that after all this time you have had since January twenty

(Testimony of Francis W. Grant.)

fifth that your work should not be ready Stop Bronze workers have had to stop their work today because of missing marble Stop Will you wire immediately definite statement as to when your work will be completed as printed Invitations are being made for bank opening on December sixth and will have to be cancelled if that date cannot be kept.

JOHN GRAHAM

10190

Deft. Ex A–3. Adm."

The attitude of the bank in reference to opening the bank without completion of the stairs was: They refused absolutely to consider that proposition when it was first suggested to them. The bank was just entering upon the safety deposit business as a new venture and the safety deposit is connected with the basement to which these marble stairs led, and they were particularly anxious the opening day should put the safety deposit department [67] in the most favorable light before the public.

On the day of the opening we had a wooden affair protecting the guard-rail and bunting.

Q. I show you a telegram that was sent to Mr. Robinson at San Francisco from your office on November 7, 1924.

A. I recall the telegram—that telegram was sent all right.

Mr. MYERS.—We offer this in evidence.

Mr. DAVIES.—We object to that as being

(Testimony of Francis W. Grant.)

wholly immaterial and having no bearing upon the case whatever. It is not addressed to us.

The COURT.—Do you expect to show that this was communicated to the plaintiff?

Mr. MYERS.—Certainly.

The COURT.—Under that promise it will be admitted.

Exception noted.

Which said telegram was received in evidence and marked Exhibit "A-4," and is as follows:

DEFENDANT'S EXHIBIT "A-4."

"Seattle, November 7th, 1924

Mr. M. P. Robinson

Palace Hotel

San Francisco Calif

No means of knowing damages at this time but will be very heavy   Stop   Impossible to open without stairs

Collect

JOHN GRAHAM

10190.

Deft. Ex. A-4 Adm." [68]

Cross-examination by Mr. DAVIES.

The bank building was in condition to start the work, as far as installing the marble in the bank proper was concerned, on the day the work was started and it was not in condition before that day.

Q. The banking-room itself was not in condition

(Testimony of Francis W. Grant.)
either, prior to the time you started work there for
storing marble in there, was it?

A. It was all being used right along.

Q. There was no place for storing marble in
there, was there?   A. When?

Q. Until the time you started work, September,
1922?

A. No marble was stored in there; no one wanted
to store marble in there.

Q. It was not in condition to store?

A. It was not; it was not built for that purpose.

Q. It was not in condition to be used for storing
marble, was it?   A. No.

The marble used in the bank is known as St.
Genevieve. It is a fancy grain marble, highly
polished. This kind of marble does not have cracks
running through it. The piece of marble you show
me—I cannot recognize it as half of the sample that
we furnished Mr. Robinson. I will say that it is
just as good. This sample has streaks running
through it that is necessary to polish and wax. All
marble of that kind has them.

Mr. DAVIES.—I will not mark it at this time.
I will introduce it later.

The COURT.—We will take an adjournment until 10:00 o'clock to-morrow morning.   [69]

The Court resumed pursuant to adjournment.
All parties present.

Mr. DAVIES.—Mark this sample of marble as
Plaintiff's Exhibit No. 3 for identification.

Q. Mr. Grant, taking it for granted that this is a

(Testimony of Francis W. Grant.)

sample of the marble furnished Mr. Robinson by you, have you the other half that was cut in two?

A. We have the other half that was cut in two where we can produce it. I did not bring it with me—it is not in our possession now.

Mr. DAVIES.—I request that you bring it where we can get it.

WITNESS.—I am willing to testify that it is the same as that.

Q. I wish you would produce the other half of this.

A. This sample of marble came originally from Tompkins & Kiel, representatives of the quarry producing that marble. This marble is known as the St. Genevieve Rose. It is highly decorative. It is what is known generally in the trade as a highly decorative marble. All highly decorative marbles have seams running through them. That is one of the things that adds to its beauty—all of which is well known; and those seams running through make the marble fragile.

With reference to the check desk tops. I mean what I said—that they were broken accidentally at the factory.

Q. Are you sure, about that?

A. Yes, sir. I was not present when they were broken, but I saw the marble before it had been joined together and I was thoroughly satisfied that they were broken while in the gang-saw. All the [70] breaking is in the same place, or nearly so, breaking crossways of the grain. The break was

(Testimony of Francis W. Grant.)

not where one of those seams run through. It did not break there in the direction of the main seams of the marble. It broke crossways of the main seams of the marble. In this particular marble the seams run in every direction; to some extent only there is a direction in which they predominate. The seams do, however, run in every direction, to some extent. They are, however, stronger in one direction than in another. In all of the marble there in the bank, that was supplied by the Robinson Tile & Marble Company, and that out in the hall and vestibule leading to the elevators, it shows frequently where it has been broken, where it has come apart, where the seams are. It shows this all over and this is true of the marble set in the hall as well as in the bank proper. It is not generally true, with reference to highly decorative marbles, that in the manufacture of them they have come apart where those seams run through. They differ. Highly decorative marbles, more or less, are liable to show seams and they will come apart, of course. If they are handled properly they won't come apart. They don't always come apart.

I had no experience in the actual fabrication of the marble. I stated yesterday that there were some pieces of marble that were off color in the bank.

Q. Those, I believe, are how many? Are there three or four of them that you say that you would not accept?   A. Oh, twenty or thirty.

Q. When you mention twenty or thirty, those are

(Testimony of Francis W. Grant.)

some you had some dispute about, but did let them go in?

A. Let them go in under protest, with the understanding that they would be taken out and others put in their places. [71]

Q. Twenty or thirty, would you say?

A. I never counted them, but one or the other, twenty or thirty rather than three or four, as you suggested.

On or about June 25, 1925, when Mr. Eastman was in Seattle, trying to get his money after the work on the bank was completed, I recall going down to the bank with him and Mr. Robinson. I recall that I pointed out some marble to him at that time, which I claimed was off color.

Q. Whereabouts was that? Can you tell the Court?

A. I pointed it out in so general a way it would be impossible to definitely locate it. The whole balustrade I pointed out, the whole side, and he could pick it out for himself and see it. It was obvious; anybody could see it. I did not specifically name, at that time, any specific piece referred to.

Q. Isn't it a fact that at that time you pointed out to him some four or five pieces of marble on the left-hand side as you go into the bank?

A. As specimens only of the whole, I did, but not as limited to quantity.

Q. Didn't you point out three or four pieces and say those were the pieces you objected to as being off color and no more? A. I did not.

(Testimony of Francis W. Grant.)

Q. Did you point out three or four pieces to him?

A. I did.

Q. Whereabouts were those?

A. I think they were near the entrance from the corridor. There are two entrances to the corridor, however, three entrances to the building, or there were dissimilar defects at either entrance.

Q. Did you point out any particular defects, if so, state to the Court what that was.   [72]

A. I don't know whether I pointed out any defective marble at the westerly entrance or easterly entrance. It depends on where I was standing. It depends on where I was standing at the time. Either one could have been used to point out, for both were defective.

Q. You don't know where it was that you pointed out?   A. No, sir, I do not.

Q. With reference to the check desk tops, the only objections I understood you to say yesterday, which you have at this time, to those check desk tops are they are defective in waxing—is that true, and if the waxing was properly done you would approve of them at this time?

A. I don't recall testifying to that effect.

Q. That was my understanding; that is true, isn't it, that the only objection you have to the check desk tops is to the waxing.

A. Not being a mechanic I don't know what could be done to those tops to make them right. If there is a way of using waxings and making them look like real marble instead of botchery, I don't

(Testimony of Francis W. Grant.)

know. They now look like botch work. I could not make them better. If anyone else could, I don't know.

Q. Have you any objections to them now, except the waxing; have you any objection to the waxing?

A. The objections to them I would have now are the appearance of being artificial and not whole pieces of marble. Whether you want to consider them to be waxed or the fact that they are broken, I don't know. They are not now whole pieces of marble. They are obviously made up of pieces of marble manufactured by action of the workmen, accidentally or otherwise, and not natural fractures that we would find in the quarry, the same as you have been talking about. [73] They are not the correction of natural seams, they are the correction of an entire fractured block crossways of the vein.

I do not know Mr. Boucher's relation to Mr. Robinson. At the time the check desks arrived he was there overseeing the work of installing the marble for Mr. Robinson. When those check desks arrived I examined them before they were installed. I made my objection to those check desks to Mr. Boucher. That is, I presume I did. I do not know whether it was Mr. Boucher or Mr. Robinson.

Q. Isn't it a fact that when those check desks arrived you went to Mr. Boucher and made the objection, the only objection you made to them at that time, the objection to them being installed, saying that the waxing was defective and one desk showed

(Testimony of Francis W. Grant.)
the mark of a drill hole made at the quarry, which
had not been properly filled and polished to satisfy
you, and you wanted it filled up better and polished
better?

A. I did not attempt to say that waxing would
correct the defect.

Q. Then you will say that is not true?

A. I will say that is not true, as you put it.

Plaintiff's Exhibit No. 3 for identification, is intended as a general sample—a general representative of what the marble should be.

Q. You knew as a matter of fact, that some of
the pieces would be darker and some of the pieces
would be lighter?

A. We knew that the deviations from that sample
should not be more than the architect said would be
permissible.   The architect was given entire authority.   To what extent it would deviate from the
contract, what is made a matter of contract, I don't
know.   [74]

Q. It was a representation of the average marble
to be furnished, conforming generally to that
sample?

A. A large portion of the marble installed was
accepted, using that sample as a basis, and of course
there are no identical pieces; that is impossible.
It is impossible to produce two pieces of marble that
are identical.

Q. It was understood that some of them, therefore, would be darker and some would be lighter?

A. Slightly, yes, sir, very slightly.

(Testimony of Francis W. Grant.)

Mr. Robinson started to lay this marble just as soon as the building was ready for him.

Q. You spoke of this marble, some of it being lighter than others, too light, you said yesterday, because it was not sawed according to the marks which the quarryman put on it. You don't know anything about whether this marble came there with marks on it, or not, do you?   A. Came where?

Q. Came from the quarry to Mr. Eastman, the plaintiff here?

A. I did not see the marble delivered at Mr. Eastman's yard or sawed.

Q. You don't know whether it had any marks on it or not?

A. Only as I was informed by the manufacturers of the marble, the Tompkins & Kiel Company, who produce the marble, and from blocks I saw at the other yards in San Francisco at the time I was investigating the matter, I found in the case of blocks at other yards that they were invariably marked. I was told at this quarry that they were always marked, showing which way to saw it.

Q. You are testifying to hearsay knowledge?

A. As to the marking itself, yes, sir. As to the necessity for sawing in a certain direction, no. That is obvious by examination [75] of a block of marble. The marble has splotches of yellow in it—some relatively thin—spread out, and it is sawed along the major axis of the splotch, which will make a very large yellow spot in the marble; while, if it is sawed across the splotch it will make a narrow

(Testimony of Francis W. Grant.)
strip of yellow, which is not objectionable in this
marble. In some cases the yellow covered three-
fourths of the entire piece of marble, showing that
they sawed longitudinally, the long way in the
splotch of the yellow in the quarry deposit.

Q. However, you don't know anything about
whether this was marked at the quarry or whether
it was not, for any particular sawing?

A. I would not say. I don't know anything about
it.

Q. Do you know—   A. I know the custom.

Q. Do you know whether they were or not?

A. I do not.

Q. Now, you spoke about—or the contract men-
tions—between the bank and Mr. Robinson, men-
tions setting plans; those setting plans are what?
What are setting plans?

A. A setting plan is a manufacturer's drawing
prepared to show in his language and in greater
detail, how the work should conform, after a study
of the architect's drawings, and which under our
contract were to be provided by Mr. Robinson,
showing the way that the marble should be set.

Q. Can you at this time, state definitely how many
pieces of marble went in there that were not inspected and accepted by you?

A. They were all inspected.

Q. And how many can you say right now that
you did not permit to go in, that you did not give
your permission to go in?

(Testimony of Francis W. Grant.)

A. It has not been testified yet that we forbid the introduction [76] of any.

Q. You did not forbid them to set up any of it at all?

A. We already testified that we permitted it to go in under protest.

Q. Are there any pieces of marble in there now that you contend that you rejected, and told them they could not put in?

A. No, sir. I say I have not testified that they could not put it in, and I told you yesterday that we had to have it in to open the bank.

Q. I understood you to say yesterday—is why I am asking this question—that you got into some controversy with Mr. Robinson or his superintendent, about the color of this marble, and some of them you accepted, a great many of them you accepted, and some you told them they could not put in. I was mistaken about that?

A. We accepted under protest, a considerable portion of the marble that is still there. There were some pieces of marble rejected which were not put in. You asked what pieces were put in after being rejected. There were some pieces of marble rejected that never were put in.

Q. Those that were rejected were not put in; that is true, isn't it, they were not put in at all?

A. Those that were rejected.

Q. By you, as representing the architect, you represented the architect, didn't you?

A. I represented the architect, yes, sir.

(Testimony of Francis W. Grant.)

Q. Those that were rejected by you were not put in at all?

A. Some were. I will qualify that. I have already attempted to make it clear. We rejected some, but with the understanding that they would be taken out later. Others were rejected because there was a possibility of getting other marble for it. Mr. [77] Robinson got a piece of marble up from San Francisco and made some substitution, but in other cases he was unable to make those substitutions.

Q. I wish you would try and answer the question and we will get along faster.

A. You ask your questions clearly.

Q. Mr. Grant, I will try to make myself clear. Were there any pieces of marble that they put in and you said, "No, don't put that in"?

A. No, sir.

Q. And you were there and inspected all of the marble?   A. Yes, sir.

On December 6th, the date Mr. Graham, the architect, set for the opening of the bank, the plumber had completed his work, except for the drinking fountain, which he could not complete until the marble was there. That marble was at San Francisco at that time. The decorators were still working, kalsomining, on December 6th. The electricians may have been working on the telephone connections, on December 6th, but on no work that would interfere with the occupation of the bank. If they were working, they were working because they had been prevented from finishing the work sooner on account of the lack of marble. That is true of all the crafts, if there were any working at all. They were working simply for the reason that the marble had been in their way all the time previous and they could not get through.

The Tompkins & Kiel Marble Company are people who represent this particular marble.

The letter you showed me, addressed from Tompkins & Kiel Marble Company to the Mission Marble Works—I have read the same and what is stated in this letter refers to the kind and [78] quantity of marble sent, is true.

This letter was in evidence as Plaintiff's Exhibit No. 4, and is as follows:

PLAINTIFF'S EXHIBIT No. 4.

"TOMPKINS-KIEL MARBLE CO.
San Francisco.

March Eleventh, 1926.

Mission Marble Works,
209 Mississippi St.,
San Francisco, Calif.

Gentlemen:

With further reference to telephone conversation with regard to St. Genevieve Golden Vein Marble, would say that the stock furnished you for the Dexter Horton Bank, Seattle, was from the same quarry as we have furnished hundreds of other jobs in the United States and Canada, and it is meeting with good success in all parts of the Country.

(Testimony of Francis W. Grant.)

It is considered a highly decorative material and like most colored marbles, is subject to a small percentage of filling and waxing. All colored marbles are subject to variation and all manufacturers take this into consideration in the working of same.

The order you placed with us for St. Genevieve Golden Vein was filled with the best material we had and promptly and eight of the blocks we furnished you were taken out of our consignment stock, which we had on the Pacific Coast and delivered to you on March 27, 1924.

Yours very truly,

TOMPKINS–KIEL MARBLE CO.,

By Robert Connell.   (Sgd.)

10190.

Ptf. Ex. 4.''

Q. I also did the inspection upon what is called the lobby, or entrance to the bank building proper, and I passed that.

Mr. DAVIES.—I am asking these questions because I expect your Honor will want to view the premises. We would like to have your Honor see the premises. [79]

The COURT.—If either side requests it I will do it.

Mr. MYERS.—Both sides are willing to have the Court do so.

Redirect Examination by Mr. MYERS.

I have been dealing with various kinds of marble for thirty years. I had charge of the construction

(Testimony of Francis W. Grant.)

of this building we are now in, for the Government. I am the same Mr. Grant who was formerly Superintendent of Buildings of this city. I know that gentleman there—(pointing to him). He was superintendent for Mr. Robinson in that bank job. His name is Mr. Jenks. He was in charge for Mr. Robinson in the construction of the marble work for this bank.

When the bank opened, it was on the 20th day of December, 1924, the railing on the stairs leading from the bank down to the basement, was not completed, and the side walls of the stairs and there was some work at the base of the stairs; a portion of the ballustrade on the stairs on the first floor was also incomplete. The only portion that was not completed pertained to the stairs leading from the bank proper down to the basement. That includes the drinking fountain and the steps of the stairs. The marble on those was in place so people could walk on them.

Redirect Examination by Mr. MYERS.

The stairs that I referred to are in the main lobby of the bank, in the center of the main lobby, leading down into the safety deposit department of the bank. When the bank was opened the safety deposit vaults were ready for use. Mr. Robinson provided temporary wooden railing and covered it with bunting. [80]

TESTIMONY OF JOHN C. JUENGST, FOR DEFENDANT.

JOHN C. JUENGST, produced as a witness on behalf of the defendant, having been first duly sworn, testified as follows:

Direct Examination by Mr. MYERS.

ᵕ live in Seattle and am acquainted with the Robinson Tile & Marble Company.

I superintended the marble construction for Mr. Robinson in the Dexter Horton National Bank. I acted as superintendent and foreman combined. I found it necessary to make quite a lot of changes and corrections in the work of the marble.

The various slips of paper which you hand me, marked Defendant's Exhibit "A–5," for identification, are in my handwriting. There are fifteen in number.

Q. State generally what that is and when you made it and how you came to make it—what was it for?

A. Well, it is customary for us to have a shop detail when we go to work, a detail from the concern that we operate out of or from, and we install the marble according to our detail. Also we have a set of architect's details, that is, if we need them we get them. We go up to the architect's office and get them, and we install them according to the detail that is got out from the shop. Then the inspector of the architect, if he happens to see something that

(Testimony of John C. Juengst.)

varies from his drawings or his details, he informs you that it is not right, and it has to be changed. Myself, as foreman, or superintendent, in certain instances I have no authority to change them without calling attention of the people that we work for. We notify the contractor and the contractor submits to your idea or rejects it; that is, he tells you to change it, and I change them then and I keep time on them.

The purpose of this is what we call in the marble business, in the setting end of it, "to extra." This is a statement of the [81] work I made at that time and submitted to Mr. Robinson. That is all,— not according to the form of the original plan.

Q. (The COURT.) That is not according to the original details?

A. That is not according to the original details; I would not say all. There were cases that were eliminated, the cutting according to their own detail; but most of it is on account of not being in compliance with the architect's drawings.

The marble in all cases did not come in accordance with the shop drawings. That necessitated changes in part of it.

In explanation of manner of keeping time, if a man would make a bill for 98¢ we would charge him a dollar for our time. If our time is an hour and ten minutes, we make it an hour. If our time would show twelve minutes or ten minutes we make it fifteen minutes. If it is twenty minutes, it is a half hour. We don't go right by the hour or

(Testimony of John C. Juengst.)

minute. We never do that. If it were an hour and four minutes, we would call it an hour; and ten minutes we would call it an hour. I never make that fine charges.

Defendant's Exhibit 5–A for identification, was received in evidence and is as follows:

DEFENDANT'S EXHIBIT No. 5–A.

(Page 1)   No. 959
"ROBINSON TILE & SUPPLY CO.
Seattle

Charge to Mission Marble Co.

Cutting mitres on 4 pieces upper mould on
     Cages 9–10–11–12   Time ....Hrs. 2

Sawing upper front mould & lower front
     mould of Pier 40 for vent. space shop
     & cordge time.....................Hrs. 5

Piece #22 Pier #39–5" short.   Should be
     2–1, sent 1–8 used from right side ele-
     vator front.

Recutting Ret. of Mould—Pier #75......Hrs. 2½

To make (?) return on same had to bring
     out

Recutting same on Pier #66..........Hrs. 2½

Mitreing piece on right side of elevator..Hrs. 1

[82]

"(Page 2) No. 961

ROBINSON TILE & SUPPLY CO.

Seattle

Date Sold 11/20 1924

Received 16 Boxes of Marble on the Nov.
11th

Nov. 18th received 7 boxes marble

Nov. 19th received 13 boxes marble

Nov. 20th received 6 boxes marble

Pier 65 #66 Missing.

Pier 66 #73 Not on job

Elevation of elevator opening #15 Apron
2–8¾. Not on job.

(Marble floor 4" from face of Marble to

(Slab are rough concrete.

A(All grill apron 2–5⁄8" thick is short

(Extra labor—Time.................Hrs.   13

Extra on putting temporary stringer for
stairs—time ....

Cutting tile that is not according to dimension given & out of—Time........."

"(Page 3) No. 962

ROBINSON TILE & SUPPLY CO.

Seattle

Date Sold 11/20 1924

(Cutting counter tops for iron at

(savings ....   (24)

A(Statements & other places—Time

(about .... 2 Days

Cutting corner top moulds at all grill
openings from Pier to counter

Elev. as—to allow for iron. Have
you noticed your iron drawing—
time ....2 Days (24)
(Making pieces and installing same
(for all apron
B(Returning to grill on all cages from
(piers. Same too short. Error
(on Mission draftsmen not list-
(ing same properly.—Time.....1 Day
          8 Helpers & 8 Setters.''

"(Page 4)   No. 960
          ROBINSON TILE & SUPPLY CO.
                    Seattle
                         Date sold 11/11 1924
2 Base Mould # Bet. Col. 47 & 52
To short as you figured 3" on front mould
     & sent only 2½". Had to make 2 new
     ones out of extra piece—Time.......2½ Hrs.
Recut end piece of mould between 74 &
     75, complete to member to make re-
     cess
Apron #53 broken in 3 pieces
Rodel has written.
     Received 24 crates & 6 crates Nov. 8th
     Received 24 crates of marble Nov. 11th.''
[83]

"(Page 5) No. 963

ROBINSON TILE & MARBLE CO.
Seattle

Date sold 11/22 1924

(Piece #65 Apron. Had to cut Ret. on
A(back of mould and polish—Time...... 2 Hrs.
Received crates of marble
(Cutting one row of tile clean through
(each panel on job and a whole lot
B(of others on account of tile not
(rubbed to size and out of.........74 Hrs.

(Duplicated T 961) (See on other
sheet same) 9 days.................... 2 Hrs.

Mould apron #38 should return itself on one
place to bronze. Was returned on two places, to
fix Scaffold taken out of south side bank on 17/22
for marble work."

"(Page 6) No. 964

ROBINSON TILE & SUPPLY CO.
Seattle

Date sold 11/26 1924

Cutting holes in 12 pieces of check desk
      bases and sub-base for conduit—24
      hours—Time ....25 Hrs.

Received Nov. 21st—8 crates marble
Received Nov. 25th—4 crates
Stair Base #103 4—9x5 Broken in transit
Basement Base #7 Broken in transit
Basement stile #42 Broken in transit
1 Piece Cap 7x9x2—Corner off
#G Ashler 2—11x1—3½—Broken in
      transit

Short 4 pieces base 3–2–3/8x5″x7/8 for under seats. Never sent them. Ordered from shop. All check desk cap not according to drawing. Mr. Bergsteth wants explanation. Take up with Mission. Why the error on margin.''

''(Page 7)   No. 965

ROBINSON TILE & SUPPLY CO.
Seattle
Date sold 11/29 1924

Bal. Black bases for seats sent 1—5
Plan shows 1–5–3/8 have to recut scrolls
    and other work to suit—Time ...... 4 Hrs.
Nov. 26th   Received 2 lots 16 crates—9
    crates
Basement Pilaster panel #48 Broke
Stair freeze broke 6 pcs. #30–33–32–37–
    18–28
Broke in transit Broke one in 3 pieces.
Received 10 crates marble 11/29/24. Short 4
pcs. tile 2–4–3/8x2–2–9/16 Made in shop. Taking
polish off Deal Plates & Rehorning & rubbing to
right size as Mission has not allowed 1/32 for set-
ting in place. They made same net. Shop
time—'' [84]

''(Page 8)   No. 966

ROBINSON TILE & SUPPLY CO.
Seattle
Date sold 12/5 1924

2 Pieces Ashler cut ¼″ short on rake.
    Made 2 new pieces .#J. Basement 4
    pcs. Bal. Black Base under seat never

sent got 4 new pieces from shop.
Charge to Mission.

A(Two pieces #115–#117 sub-base to
fit pilaster posts never cut to receive—Time .... 2 Hrs.

Received 10 crates marble 12/2

Received 7 crates marble 12/2

Received 1 check desk top 12/3

Received 14 crates marble 12/5

Made 5 tile for panel A. Checked tile,
recutting #149, balustrade corner no
good on 4 places & polishing, recutting all moulds to suit. A terrible
job from Mission.................. 64 Hrs.''

''(Page 9)   No. 967

ROBINSON TILE & SUPPLY CO.
Seattle

Date sold 12/18 1924

Recutting check desk tops to plant pieces
so to get over rejection by Architect
& resticking & polishing—Time

Recutting all upper & lower moulds
that come in contact with balustrade
posts & ramps not according to
Architect's drawing. A serious error on expense time.

(Refurnishing 16 pieces of marble to
get over

A(mistake of Mission on account of
changing dimension of returns on
basement floor. Pilasters from 4″
to 2″. Setting time............ 29 Hrs.

Drilling holes in all balustrades rail T
Sub-base. Shows on Architect's
Plan, but not in marble. A serious
error.

Average 30 min. ea. Helper..102 Hrs.''

"(Page 10)   No. 968
ROBINSON TILE & SUPPLY CO.
Seattle
Date sold 12/18 1924

12/1 Cutting new hole for conduit to be
changed to panel in facia belt, putting on rosette by drinking fountain,
safe deposit sign—Time............   3 Hrs.
12/1 Drilling 48 holes for 6 doors to banking room spaces Time to Architect
—Bank ....   8 Hrs.
12/4 Repolishing counter top between
Col. 72 & 73
Charge to Telephone Co.—See Grant....   8 Hrs.
12/18 Drilling holes & putting on handles
for check desk doors—Time .... 5 Hrs.
12/18 Cutting all border down to fit thresh-
old at door opening to building corridor, on account of door's being too
low. Time—charge to Bank, per
Grant ....   8 Hrs.''

[85]

"(Page 11.) No. 969.

ROBINSON TILE & SUPPLY CO.

Seattle Date Sold

Dexter Horton Bank.

Replacing floor panel. Panel H., in place where information counter was. Check from plan feet.

See Gen. plan from Mission, for actual feet.

NOTE: Juengst left this. Says he overlooked it before.

12/29/24."

"(Page 12.) No. 970.

ROBINSON TILE & SUPPLY CO.

Seattle Date Sold 1/12/25

Cutting shell to receive conduit supply pipe and fitting in to drinking bowl two different profiles. This should have been fitted in by Mission shop before sending. Time..8 hrs., 2 Min.

16
———

Charge to Puget Sound Con. Co.

Cutting holes for floor hinges at 6 gate openings. Time for Helper.................... 2 Hrs.

Setter's Time ....20 "

Drilling holes for straps to strengthen door hinge (Not done yet). Time........

(Dave) T2597.

"(Page 13.)   No. 971.

ROBINSON TILE & SUPPLY CO.
Seattle   1/12.

#46	4–0 x 4¾x1¼	Broke	MKS.
43	"	"	"
49	3–0 "		
37	4–7–⅝ "		
46	4.0 '		
40	3–11–⅜ "		
44	3–7–8/16 "		
29		"	"
8		"	"
97		"	"
52		"	"
193	Strutt Pier Stile "		"
7	Stairs 4–7–⅝ "		" '

[86]

"(Page 14.)   No. 973.

ROBINSON TILE & SUPPLY CO.
Seattle.   Date Sold.

2 Pieces Stair Soffett #11–1'–08"x3x3¾.

Never sent—had two new made.

Charge to Mission Marble Co.

2 Pieces #200 Stair Column.   Sent 9½ in length.
Should be 11".   Mission's mistake.

(Recutting Landing Ramp.   Mistake on Wash
Concise Goose Neck.

A(never dropped same to 6x1–1.   Rake ⅜" too
high on lower turn.

(Have to cut all work intersecting to same and
black ramp.

(Time . . . . .2 days (16)

(Testimony of John C. Juengst.)

Recutting 48 Returns of Moulding on rail and sub-
base. Not according to plans and polishing
same. Will cut and polish the rest when I get
stock. Time for all .... "

"(Page 15.)   No. 974.

ROBINSON TILE & SUPPLY CO.

Seattle   Date Sold

Recutting lower ramp as rail would not fit plumbing
¼ 11. Time .... 2 Hrs."

[87]

We were waiting there for marble quite often;
lots of times. A couple of times we had to let men
go and another time we kept men that would leave
town on the job when there wasn't work for them
to do. I was on the job from the start until it was
finished. On December 20, 1924, when the bank
moved in, all the marble was finished that we had.
Every piece we had was installed. We did not have
marble for the stairway. After the bank moved in
we kept installing marble for the stairway as the
stock came in. It came in three or four boxes at
a time. We had two or three men on the stairway.
We did that work during banking hours until the
last end of the job, then we worked overtime to
satisfy the bank. As soon as the last crates came
we finished it up Saturday and Sunday.

Referring to Plaintiff's Exhibit 5–A, it says:

"The regular check desk top, to blend pieces in
to get over rejection by architect and re-stick and
polish."

(Testimony of John C. Juengst.)

I haven't got the amount of time we put in on that check desk. When the check desk came they objected to the conspicuous veinage; that is, the breakage was not uniform with the general form of the marble. They made objection to the check desk because it was broken and sagged and not in conformity with the rest of the marble. The veinage of the marble was conspicuous and it was off at right angles from the general vein. They made complaint to me particularly, also to Mr. Robinson, and the whole office, I guess. We stuck it with shellac and waxed it with shellac to make it look like a vein, to resemble the natural veinage of the marble. This work was done to obviate the objections. We tried to remedy that defect of veinage as much as possible. We never guaranteed to remedy it, because the veinage was opposite from the general routine of the marble. We could not guarantee it, [88] because it was almost impossible to take a piece of marble and stick it opposite from the veinage. It is too conspicuous. Lengthwise with the vein we can do it, but opposite we cannot. You would have to imitate the body, and that is almost impossible.

Mr. Robinson told us to do the work on the check desk. I never talked to the architect myself about it. All my business was mostly through Mr. Robinson.

I have been in the marble business between nineteen and twenty years. I never worked on this kind of marble before; that is, St. Genevieve—but

(Testimony of John C. Juengst.)

I did work in nearly every bank in Seattle, and constructed lots of banks throughout the country. The veins or seams of this particular kind of marble run exactly as shown on this sample; they run in every direction. The majority of that kind of marble does not have veins running in any particular direction. I won't say that about the check desks.

TESTIMONY OF FRANCIS W. GRANT, FOR DEFENDANT (RECALLED—CROSS-EXAMINATION).

FRANCIS W. GRANT, being recalled for further cross-examination, testified as follows:

Mr. DAVIES.—We never received shop drawings or setting plans, so we could not approve them. They were never submitted—only partially.

Redirect Examination by Mr. MYERS.

Q. Do you know about how the marble was broken—any information on that score?

A. I was told by the superintendent that the blocks, while on the gang saw being sawed, were insufficient to hold in place and they tipped over and broke the whole batch. I was told this by the Mission Marble Company when I was there in November. Mr. Boucher was the gentleman who told me. [89]

TESTIMONY OF JOHN GRAHAM, FOR DE-FENDANT.

JOHN GRAHAM, produced as a witness on be-half of defendant, being first duly sworn, testified as follows:

Direct Examination by Mr. MYERS.

My name is John Graham. I am an architect and have been an architect about thirty or forty years. I have done a great deal of work in and about this city on some very large buildings. I signed the telegram marked Defendant's Exhibit No. 4–A.

Q. Mr. Graham, I wish you would explain to the Court, also, the meaning of the word "shop draw-ings" as applied to a job like this, or any other job, and the relationship to the architect.

A. Your Honor, in the development of a set of drawings for a building, first of all, in the archi-tect's office we make some small scale drawings, for instance, if we are designing this room, there would be small drawings of a design generally in our mind, and after that we proceed to make larger scale drawings for the purpose of getting a pre-liminary set, and then we make full-sized drawings to give the contractor, so that the work would be properly made. The full-sized drawings show a drawing of ornamentation or detail of all mould-ings, details and contours of it. In order to do that we make a drawing for the purpose of illus-tration, like this one, which is a part of the marble

(Testimony of John Graham.)

drawing for this work under discussion. That shows the forms of moulding, with the size of the pieces, the sections. This shows the construction, how it is fastened to that masonry behind, and the general form and shape, and that is supplemented by other drawings showing all these in full size. The contractor has that and from that he has to tell the manufacturer as to his own methods. From this he gets his information which he gives to his own workmen in turn. This coming from the Mission Marble Works, this is what we call the shop or setting drawing, and we take a portion of this and detail it [90] and work it out. This shows all the joints of his work and shows the size. These are all finally lettered and numbered and his workmen and his company will get out these pieces of marble from this drawing, and in turn they send them up here to the men on the job and the men on the job set them according to the drawings.

The COURT.—Q. In the joints there is some judgment exercised in placing these joints, for appearance and strength both?

A. That is exactly the point. For instance, he might have a block of marble in the shape which he wants to make work in, and we aim to work with him as much as possible. For instance, you get a piece of marble as long as that, and if he comes in and shows a joint here and there we check it over, and if it is right we let it go. For instance, there are some pencils, and this drawing that is sent to us, we have crossed out these joints at this point;

(Testimony of John Graham.)

they are not permissible, and other joints they are approved, and that particular joint is red penciled by us and sent back to him. This is the usual custom, and the manufacturer, in turn, corrects it as sent back and sends it back to us, corrected, which we approve or possibly he might come back and say, "Haven't got a piece of marble that will do that. Let's do it something like that," and we might finally approve it the other way. In this case these crosses were made and sent back to the Mission Marble Company, and in this particular case they were never acknowledged and never returned and nothing more was done, and it was such an ignoring of the instruction as to cause delay, and frequently, —for instance, you can see, your Honor, they have that piece joined together with one side of that piece of marble, that prominent part. That can't be matched. If he puts it in the center it might have been allowed, but to put a piece like [91] that on one side, a little grain there, and another different grain here, we can't permit it. We say, "That can't be done." Those are the things that shop drawings are made for. Shop drawings are made by the manufacturer, steel manufacturers, terra cotta manufacturers, masonry manufacturers, and all. It is *usage* throughout the building trade. It is our rechecking of the instructions, the manufacturer gives to his own workingmen. That is what it amounts to.

In this particular case, "shop drawings" and "setting plans" are identical and synonymous.

(Testimony of John Graham.)

That job has not been accepted by the architect. The ground for refusal is defective workmanship and material not being in accordance with the contract and delayed delivery of the work. There is a great deal there of defective marble.

In reference to the check desks, I will illustrate the point: Assume by a piece of wood, which is common to us all. Assume that one is making a desk like this; this is made of quarter-sawed oak and that oak is cut to bring out the grain in a certain way; had the wood been cut in the way that we call slashed grain, the wood would be cut different. The quarter-sawed oak being a superior article, the way the oak is sawed, for a good job, and it is not slashed grain, it is more or less inferior, it does not look as well; but if you see some in quarter-sawed and some slashed grain, you would get a mess. That is what we got in this marble. We got some slashed grain and some quarter-sawed both together, and it is nothing like the marble that we stipulated for in the contract. In fact, we haven't got what we purchased.

Q. What about the check desks?

A. The check desks were very defective. They were full of [92] defects, in which they had cut pieces out. Going back again to the woodwork, we will assume that there had been knots and sawed pieces in this desk, and the carpenters come along and had taken pieces out and put a patch in. That is the condition that we had in the marble work, defects had been cut out and other pieces put in.

(Testimony of John Graham.)

The grain did not match. It was very poorly installed, and very inferior, and having paid money for a good job, to have to put up with a job like this, is unfair.

Q. You are speaking of the desk tops?

A. I am speaking of the desk tops and I am speaking of the whole desks, the stands underneath them, as well as the tops.

Q. Mr. Graham, why did you permit that defective material to go in?

A. We did not permit the defective material to go in. I never have permitted it to go in. The bank originally talked of opening up about Labor Day. The contract with Mr. Robinson was made with that in view. Owing to delays the building was not ready at the time. It was impossible to go in it on Labor Day, and finally the building got ready at such time they might have moved in, and after all the moving in, some time early in November, then they were confronted with the fact that the marble was not all ready, a lot of the marble had not even been started. I had been down myself, personally, to San Francisco. I made three visits there personally. I saw Mr. Eastman, who showed me over his plant and assured me how he was getting along, and the stuff would be ready, and further on my third visit I went down and found it was not ready and no preparations made, and I was astonished.

Q. On the third trip you were down and saw no preparation, and you were astonished; I will ask

(Testimony of John Graham.)

you about that lack of preparation. You did not go down there at Mr. Robinson's expense? [93]

A. No, sir. I went down on my second visit, and visited the Mission Marble Company's plant, and at that time they just had the contract a short time, two or three weeks, and they were getting ready and preparing the drawings. I saw Mr. Eastman, who assured me he was going to get the work over fine, and push it through and make it a good job, and so forth. That seemed to be all right. I was next down there on July 14th and then Mr. Eastman showed me a number of blocks of marble which he had, and he showed me some work which was in his mill in the process of construction and showed me a lot of tiling that he had ready.

Q. (The COURT.) What?

A. That was tiling, marble tiling for the floor; and he said he was pushing it on, and everything was going to be ready on time. At that time, everything looked all right. The next time I was there was in October, towards the end of October. I was down again. Then they calmly told me that the work was not ready and they could not possibly ship it for six weeks.

Q. Who told you that?

A. His foreman down there. At this time I think Mr. Eastman was not there. I don't think I saw Mr. Eastman at that time. I am not swearing to that, whether I saw Mr. Eastman at that time, or not, but I do not think I did, but as to the work, I asked what they had been doing. They were prac-

(Testimony of John Graham.)

tically no further advanced from the time previously. "Oh, well," they "had got some other work there" and they had "been busy with some work there." As a matter of fact, ours had been entirely neglected and they had done practically nothing on our work from the time of my previous visit. I was very much put out, because I had been assured by Mr. Robinson that everything was going to be ready at that time. At that time they had already, I think they had two [94] months more than their contract called for. They would be given that, and there was no reason why their work should be delayed. They assumed it was all ready. It was a great shock to us to find they had neglected it, and let it go. From then on it was a constant source of annoyance until the work was finished, I think, March 3d or 4th, when the work was finally completed, about five months after the date that the contract called for.

Q. Did you see Mr. Eastman on your last trip?

A. That is the trip—he had been away in the east, I think, this time.

Q. Did you talk to someone in charge?

A. I did.

Q. Did you impress upon them the necessity of getting the marble in?

A. I certainly did, and that was a funny thing,—from then on we could neither get answers or telegrams in answer to our requests. I never run into anything like that. We run into a standstill. They said, "We will get it as soon as we can get it, get

(Testimony of John Graham.)

it up.'' In response to telegrams, letters and entreaties of all sorts, we could not even get the common courtesy of a reply, and finally the banker, Mr. Parsons, himself, took it up. He could not realize, when I told him that the firm was a responsible firm and would not acknowledge letters and telegrams, and he took it upon himself, Mr. Parsons, telegraphed back and he in turn did not get an acknowledgment or reply and we were up against a ground-hog case. The bank had not been moved. There were great issues at stake, and we were up against it. Mr. Eastman was in San Francisco, and we were here and we had no marble and he had, or whatever it was, and it was a difficult situation. We could not even get the common courtesies of business. [95]

Q. Did you, together with Mr. Grant, make a figuration and estimate of what it would cost to replace this marble with substitute work?

A. Yes, sir.

Q. And material?

A. Yes, sir. We estimated it would cost about $2,500.00.

I sent the letter you showed me, marked Exhibit ''A-6.'' The real cause of this trouble was delay in getting the marble.

Defendant's Exhibit ''A-6'' received in evidence, as follows:

(Testimony of John Graham.)

DEFENDANT'S EXHIBIT "A–6."

"November 22d, 1924.

U. S. Fidelity & Guarantee Co.,
    208 Marion Bldg.,
        Seattle, Wash.
    Attn. Mr. John C. McCollister, Attny-in-fact.
Dear Sir:

Referring to your bond given as surety for the Robinson Tile & Marble Co., dated Jan. 25th, 1924, and particularly to that portion requiring written notice to you in the event of default by the contractor, you are advised that the contractor is not showing due diligence in the performance of his contract.

Yours very truly,
JOHN GRAHAM.
By —————.

FWG/GB."

The letter you showed me, marked Defendant's Exhibit "A–1," was written by Mr. Grant. The $6,000.00 mentioned in the letter that would be retained pending the correction of defective marble, refers to defects which I have already spoken of, and this sum of $2,500.00 is the cost of correcting the same. I refused at the time this letter was written, and have at all times since, refused to make any further payment on that contract, by virtue of those two items.

Q. What was the other item, besides the item of defective [96] marble, and on account of adjust-

(Testimony of John Graham.)

ment of damages sustained by the bank on account of that delay?

Mr. DAVIES.—That is objected to for the reason he is not competent to estimate the damages to the bank.

The COURT.—I do not see how the Court can go into the bank's damage, or why it should go into it.

Mr. MYERS.—No more than to justify his refusal and in making payment.

The COURT.—I overrule the objection.

Mr. DAVIES.—Note an exception.

The WITNESS.—At that time the bank had expended three hundred and twenty-seven dollars in making good defective plaster; that is, plaster damaged—caused by the marble workers. The damages I speak of were damages caused the plaster of the bank by the workmen of the tile and marble company, caused by the necessity of erecting the marble after the plaster had been done, which should have been done beforehand. The other item, the check desks, and marble that was rejected and mussed up, being replaced and the damages for the delay. Now, in order to minimize that as much as possible, while the work was five months behind them, we concluded, that without any shadow of a doubt there could not be any quibble with the bank, that the bank had been delayed one month by this marble contract, to get away from any element of controversy at all, we figured that nobody could

(Testimony of John Graham.)

dispute the fact that we were delayed one month; therefore we added a month's rent.   [97]

That was the basis of my writing this letter. I never gave Mr. Eastman any authority or said anything to him that could in any way be construed as authority, or twisted into authority to stop, slacken or let up in sending marble here. I did not interfere with the contract between Mr. Robinson and Mr. Eastman. I looked upon Mr. Eastman as a subcontractor under Mr. Robinson and all our business details were carried out with Mr. Robinson. I know my assistant, Mr. Grant, went to San Francisco. He was sent to see if he could expedite the work, not being able to get any communication from the Mission Marble Works, no answer to telegrams or letters and no information, and Mr. Robinson suggested that Mr. Grant might go down and satisfy himself as to what was being done. Mr. Robinson was much perturbed about it, because naturally he was the man to whom we looked. He went down at Mr. Robinson's expense.

Cross-examination by Mr. DAVIES.

Mr. Grant was the one who had active charge and supervision of inspecting the marble and passing upon it. Mr. Grant was superintending all that work for me and had all the duties.

Q. Isn't it a fact that when you were down visiting the Mission Marble Works in San Francisco, on July 14th, as you say you were there, that you told Mr. Eastman at that time that the building

(Testimony of John Graham.)

was not in condition to receive marble and not to ship?

A. No, that is not true. You are partially true and partially untrue. I told Mr. Eastman at that time that the building was not in condition to receive marble and neither was Mr. Eastman in condition to ship it.

The COURT.—Let me get first of all what he told him.

A. That the building was not in condition to receive the [98] marble. This was July 14th. Of course there is no pretense that it was. The building was in the course of construction.

Q. I will ask you if you did not tell Mr. Eastman when you were on your visit there that the building was not in condition to receive or store marble, and did you not tell him not to ship any more marble?

A. If you must have an answer definitely to that I say no, nothing of that sort; but your question is half true and half not true.

The COURT.—You may make any explanation you desire now. If there is any explanation regarding the answer, all right, go ahead.

A. In the discussion with Mr. Eastman at that time there was a lot of tiling there, and Mr. Eastman said, "I am going to have this ready about the end of the month and we are going to be ready to ship them." This is July 14th, long before there was any expectation of having the stuff ready, or long before it was required by the contract to be ready. I told him, when I left there, that the

(Testimony of John Graham.)

building was not ready to receive that tiling and when I got back I would see if they could make some arrangements and if he was going to be ready at that time to wire me and let me know, but not to ship it without wiring me, and I says, "That is very kind. I am awful glad to see you so far ahead; don't delay"; the very words I said. He said, "We can slack up." I said, "Don't delay, don't delay, push it. You are doing fine work, and keep it going." And he made the remark, he said, "I would rather pay for warehousing your tiling than to have the slightest delay."

Q. You did tell Mr. Eastman not to ship any more marble?

A. No, sir, I did not tell Mr. Eastman not to ship any more [99] marble. Mr. Eastman had not shipped any at all at that time. I told Mr. Eastman he must not delay anything, but to send me a wire before actually shipping it.

The COURT.—Mr. Graham, can you tell me when the first marble was shipped?

WITNESS.—Some of the tiles for the floor arrived on May 10th and the first marble for the work proper arrived some time between the latter end of August—I can't give the exact date.

Q. When this conversation was had in July no marble had been shipped?

A. No, sir. Mr. Eastman told me that he got that tiling, he did not manufacture that himself, that he got that already made and he had it already to ship, but the stuff that was manufactured in his

(Testimony of John Graham.)
own plant, I did not think any of it was shipped, at any time before the last week in August.

Q. It was the understanding, the expression put in that question, not to ship any more marble, in counsel's question, that is what I want to get at.

A. The building was in such condition, of course, we could not take care of any.

TESTIMONY OF ALBERT P. ROBINSON, FOR DEFENDANT.

ALBERT P. ROBINSON, a witness produced on behalf of the defendant, having been first duly sworn, testified as follows:

Direct Examination by Mr. MYERS.

My name is Albert P. Robinson. I am president of the Robinson Tile & Marble Company, that had this contract with the Dexter Horton National Bank.

Referring to Defendant's Exhibit "A–5," the extra work mentioned therein was performed under the direction of Mr. Juengst. [100]

I know what occasioned that extra work, that is mentioned in there—extra labor and hours mentioned in that exhibit. In some cases it was caused because the marble was not gotten out according to the setting plans. In other cases it was because it was gotten out according to the architect's plans, without setting plans. In other cases, because there had been an error made in dimensions, by the Mission Marble Works, the manufacturer.

(Testimony of Albert P. Robinson.)

Q. Was all that extra work that is mentioned in this exhibit, the extra hours' time, caused by the things which you have mentioned?

A. Some of it was caused by having to re-wax the marble. We endeavored to, and did, re-wax the marble and I believe, caused that portion of it to be accepted instead of rejected. If it were rejected that means that the Mission Marble Company would have to replace it with new marble and we, in turn, would have to furnish the marble to replace it and it would make an added expense to the Mission Marble Works and to ourselves, and in spending this little bit of money to re-wax it or remedy it would cost less than ten per cent, or less than twenty per cent of what it would cost to replace it with new marble. We did not want the Mission Marble Works to have to furnish any more marble, than would be absolutely necessary. In other words, we treated them the way we would appreciate their treating us if the occasion were reversed.

Defendant's Exhibit "A-7" is a summary of the extra work and hours, as shown by Defendant's Exhibit "A-5."

Defendant's Exhibit "A-7" was introduced in evidence and is as follows: [101]

DEFENDANT'S EXHIBIT "A–7."

ITEMIZED STATEMENT OF THE ITEM OF
$893.88, CONTAINED IN SUBDIVISION
"A" OF PARAGRAPH THREE OF DE-
FENDANT'S COUNTERCLAIM.

64	hrs.	Labor, marble setter, cutting Triangles to fit per plans,
5	"	Labor, cutting Return Mould on Piers 39–40 from 12″ to 11″
8	"	Labor, cutting stiles at Gates to 3″
2	"	Labor, recutting Mitres on Mould at Cages 9–10–11–12
1	"	Labor, cutting Mitre on Mould at right side of Elevator
2½	"	Labor, making new Base Mould for between Col. 47–52, account ½″ short
24	"	Labor, cutting Corner Top Moulds for Grill openings,
8	"	Labor, making and installing pieces of Marble on all aprons returning to Grills on Cages, at all Cages; some were too short; not listed properly
40	"	Labor, recutting all upper and lower Return Mould at Pier Intersection, all hand Balustrade Moulding on Stairs— not according plan of Architect
24	"	Labor, cutting Counter Tops for Iron at Windows
2	"	Labor, cutting Return on back of Mould piece #65, Apron, and polishing,

74   "   Labor, cutting Tile, account not rubbed
         in size, and out of square

11   "   Labor recutting Seat Bases, sent 1–5,
         plan shows 1–5⅜

2   "   Labor, cutting 2 pieces Sub-base to re-
         ceive Pilaster Posts,

20   "   Labor, refurnishing marble, account mis-
         take of changing dimensions of Re-
         turns on basement Floor Pilasters

16   "   Labor, cutting Shell to receive Pipe and
         fitting into Drinking Bowl two profiles

16   "   Labor, recutting Landing Ramp

13   "   Labor, making correction on grill Apron
         which was short

2   "   Recutting lower Ramp, account rail ¼″

2½   "   Labor, recutting return Mould, Pier 75.

2½   "   "   "   "   "   " 66.

339½ "   .... @ $1.75 $594.13

75   "   Labor, Polishing Check Desk
         Tops and Balustrades, where
         cut @   1.25..............   93.75

104 "   Labor, drilling holes in marble
         Balustrades, 204 holes @ 1.25   130.00

         [102]

Brought Forward ....

518½ $817.88

8 Hrs. Labor, making and installing
         pieces for all apron returning
         to Grill on all Cages, from
         Piers, account same sent too
         short—Setter—$1.75.........   14.00

8   "   Labor, same as above—Helper
         —125 ....   10.00

(Testimony of Albert P. Robinson.)

12 "	Labor on machine, cutting moulds, Pier #10, for Vent spaces, 2.00	24.00
	Cartage, two ways	3.00
	Wooden hand balustrade on open stairs to basement (because marble wasn't here).........	25.00

$893.88

10190.

Deft's. Ex. "A–7" Adm.

[103]

We had considerable difficulty in getting this marble up here in order to transact this work for the bank on time, as will be shown by, at least one hundred telegrams and letters, asking them to please rush the marble. I would say further about this difficulty, if I may, that I am the *only who* had authority to tell the Mission Marble Works to delay shipment. Delayed shipment commenced in October. The flooring was shipped from Missouri, the treads, I believe, were shipped from San Francisco, and all the bank counter and store work was shipped from San Francisco. I sent Mr. Grant down there, the occasion being—the Dexter Horton National Bank and Mr. Graham, Mr. Graham is the man whom I was working with, and he could not understand or could not conceive of why there was such a delay and why they could not get any more satisfaction, and they were getting to the

(Testimony of Albert P. Robinson.)

point of turning it over to the bonding company. I knew if they turned it over to the bonding company that the bonding company would spare no expense in endeavoring to complete the contract on time. They would even go to the extent of going anywhere in the United States or Canada for that matter, if possible to have it manufactured, by running twenty-four hours a day, and they would do that, ship it by express, is another thing that they would do. I realized those points and I wanted to keep the expense down, for the benefit of myself and the Mission Marble Works, because the Mission Marble Works were bounden to me as I was bounden to the bank.

I sent Mr. Grant down there to let him see the situation and try to reconcile the bank, and Mr. Graham, as far as possible, and keep this out of the hands of the bonding company. I told Mr. Grant to do anything that he could to expedite the future progress of the work. I authorized Mr. Eastman to work overtime, [104] that is, two hours a day, and any working of overtime we had to pay for. When we speak of overtime they had to pay time and a half, and the extra half time I was supposed to pay. Mr. Eastman at that time demanded that we send $300.00 a week, in advance. Mr. Grant asked him, he said he would have to have his money in advance,—asked him what it would be and so forth, and he said, "Three hundred dollars a week." We sent down $300.00 a week for two consecutive weeks. It was my un-

(Testimony of Albert P. Robinson.)

derstanding from Mr. Grant that we were to pay, as I stated, the overtime only.

Q. You were willing. You wanted to get this work here at all hazards, and the object of paying any overtime at all was to expedite the shipments which were at that time several months late?

A. I could not say. He estimated $300.00 a week; $300.00 a week for the overtime. I never had any accounting with Mr. Eastman. As far as I understand he had not spent over half that amount. My request for an accounting of that has been ignored. I paid them that sum in two checks of $300.00 each.

Defendant's·Exhibit "A–8" for identification, is the two checks—one is dated November 22d and the other December 2d, each for $300.00. That is for that $600.00 extra work overtime in San Francisco.

Mr. MYERS.—I offer them in evidence.

Mr. DAVIES.—They are objected to. He went down there, he sent a man down there with authority to do anything that he could, and the man says, "You work overtime and we will pay you for your cost for the overtime," and he did that and he paid him, and that is the end of the situation. [105]

The COURT.—I will let it go in and hear you at the final argument, about it. [105–A]

We also wrote a letter and stated that this was paid under protest, nevertheless, that is damages to us. I wrote a letter with the first check.

Witness excused.

TESTIMONY OF JOHN GRAHAM, FOR DE-
FENDANT (RECALLED).

Mr. GRAHAM, recalled for further examination, testified as follows:

Mr. MYERS.—Q. I understand that you wish to make some explanation concerning the date of your visit in San Francisco.

A. Yes, sir, I spoke of the date of my conversation with Mr. Eastman, that was referred to this morning, and I stated this morning that it was on my visit in July, but that was a conversation that occurred in my visit of May. I wanted to make that correction.

Q. You were there in San Francisco to see him about this job on three occasions?

A. I was there May 2d and on July 14th and October 8th; but the conversation that I spoke of this morning I had with Mr. Eastman in which he told me he had some marble ready to ship, and I asked him to telegraph me before shipping any; that conversation took place on the second of May.

Witness excused.

TESTIMONY OF A. P. ROBINSON, FOR DE-
FENDANT (RECALLED).

A. P. ROBINSON, recalled for further examination by Mr. MYERS, testified as follows:

I have seen Defendant's Exhibit "A-9," for identification, before. That is the Mission Marble Works' final statement and invoice for all charges

(Testimony of A. P. Robinson.)

against the Dexter-Horton National Bank job. I found some things incorrect in that statement. Those that I checked and made no further notations after them are correct, no question about the items. The first two items are [106] correct, the first item is as marked, was short of, that means we checked short, never received. That is the item of $7.85. The next item, $81.94 was also short. The item of $16.09 is marked short. We disputed this and marked that short for the reason that we had no shortages in shipment to the transportation company; that is to say, by the bill of lading showing a certain number of cases, that many were received and consequently we could not claim any shortage against the transportation company; so that is usually done as a matter of fact, that it has been an oversight from the original shipment, which is from the customer. It very often happens where there so many pieces. We have one item of $63.20, condemned marble. The statement says to furnish St. Genevieve marble ordered by A. P. Robinson. I told him to ship us the rough slabs and we would finish them up here to take the place of the marble which had been condemned, which they did, and they charged us for this marble. Later they notified us to return the marble which had been condemned and we did that. That item which is charged, we returned and those slabs that were sent up, we finished them at our expense and made no charge for them. There was a run like in front of the counter or wainscoting, like from the

(Testimony of A. P. Robinson.)

corner to that pier, where the marble panels were condemned, and in order to help them out and expedite the matter, I says, "Ship the rough slabs and we will finish it up here and put them in." They charged us for the rough slabs and when we returned the condemned marble they neglected to credit us. Therefore, there is an item of $63.20 which was marked "Condemned and returned." That is charged in the bill against me. The next item is boxing and cartage—two items aggregating $544.78. This boxing and cartage I claim should not have been charged to us, due to the fact that they did not have the marble [107] out to ship it in the usual and customary manner by rail, to load in a box-car at the time planned. I do not deny that I told them to ship the marble by boat, and we had a shipment coming on practically every boat. Had we waited until all the marble was completed, it would have left too much of the building in an incompleted state and I figured that the bank would sue us for damages' and would get them. I mean on account of the delay because we had contracted to have the job done within a reasonable time and if we had not brought those piecemeal shipments in and put them in as they came, we would have had the whole thing hung up until the end of February and then it would have been some considerable time taken to install the marble. It would have been much cheaper for the Mission people to pay this boxing and cartage themselves than to cause so much delay. The two items making up this amount

(Testimony of A. P. Robinson.)

are $484.00 for boxing and $60.00 for cartage. Had this marble been gotten out on time there would have been no occasion for boxing and cartage. Therefore, I claimed that the delay is a damage to me, to that extent, and which I have charged for.

I have a letter from Mr. Eastman, suggesting that he would forget those items if I would pay him the amount due.

Defendant's Exhibit "A–10," for identification, consists of five checks, all being drawn by the Robinson Tile & Marble Company. One was for $50.00, one for $100.00, one more for $50.00, one for $160.00 and one for $40.00. These checks were drawn for expense money on account of my trip and also Mr. Grant's trip to San Francisco. I made the trip to San Francisco in the first part of November. Mr. Grant left about the 12th.

Mr. MYERS.—I offer these checks in evidence. [108]

Mr. DAVIES.—I object to them as being immaterial.

The COURT. — Objection overruled; exception allowed. They may be admitted in evidence and marked Defendant's Exhibit "A–10." [108A]

My expenses, which I computed some time ago, were $250.00. These checks that I cashed and got cash on were checks used in expenses of myself and Mr. Graham. These checks were part of my expense; part of the $250.00. These checks amounted to $400.00. · I don't know how much Mr. Grant's expenses were.

Defendant's Exhibit "A–11," consisting of telegraph wires, was introduced in evidence.   The said exhibit is as follows:

DEFENDANT'S EXHIBIT "A–11."

"Seattle, August 26, 1924.

Mission Marble Works,
    209 Mississippi St.
       San Francisco, Calif.
Rush all setting plans for Dexter Horton Bank especially stairs

      ROBINSON TILE & MARBLE CO.
         600—8th Avenue North

10190

_____

Defendant's Ex. 'A–11.'   Adm."

"Oct. 15, 1925

Mission Marble Works
    209 Mississippi St.
       San Francisco, Calif.
Did you ship car Saturday   Did it contain remainder Bank Marble and Travertine wire today

      ROBINSON TILE & MARBLE CO.
         600—8th Ave. North

   Charge our account"

"Seattle, Oct. 21, 1924

Mission Marble Works
    209 Mississippi St.
       San Francisco, Calif.
You have not sent necessary material for finish-

ing counter and screen work as we go Rush it quick

ROBINSON TILE & MARBLE CO.

600—8th Ave. North

Charge our account" [109]

. "Seattle, Wash. Oct. 31, 1924.

Mission Marble Works

209 Mississippi Street

San Francisco, California

Rush shipments by boat for marble work wherever bronze work comes in contact must have this on job before next Saturday so bronze and cabinet work at cages can be installed Stop If you do not rush shipments for above will be forced lay off crew and then will certainly be charged with full damages as set forth in contract Stop Do not understand why you shipped open counter work first instead of grill counter work Stop You will be ahead if you run crews nights and Sundays if this program will not finish marble sufficiently soon you better secure assistance other shops Stop We will do everything possible avoid liquidated damages and you must do your part or suffer consequences accordingly Stop By sending ledge for space between column fifty eight and sixty nine remainder stock between forty four and fifty eight for cages eleven and twelve also sixteen Savings Department countor work between columns fifty two and sixty one in one or more boat shipments immediately situation would be partially relieved Stop Have had eight mechanics working but forced to lay off five and the

three now on job working great disadvantage account your not sending stock to complete units as we go Stop Why indefinite shipping date stair work Do you not have stock on hand Stop Bob Connell stated he was not in position to state whether you had sufficient stock Stop Wire by noon Saturday giving full information
ROBINSON TILE & MARBLE CO."

"San Francisco, California
November 6, 1924

Geo. M. Eastman,
Commodore Hotel,
New York City.

Your firm making miserable failure bank contract I found stair plans not yet complete on fifth this month Advise you authorize Hunter sublet sufficient portion work to effect our execution contract within period grace granted final limit December sixth Graham filed notice fifth owner will hold us and our bondsmen responsible all damages caused by failure conclude contract by December sixth Stop We hereby notify you and Mission Marble Works that according terms contract between us we will hold you liable for all damages assessed and incurred by reason your failure fulfill contract You can ill afford loss reputation Graham greatly disappointed
ROBINSON TILE & MARBLE CO.
By A. P. ROBINSON." [110]

"San Francisco, Nov. 7, 1924.

To Geo. M. Eastman, Commodore Hotel, New York
    City

If you would authorize Miller or Hunter sublet
parts of stairs railing to Vermont spindles to
American and possibly some to Cook your shop
should be able finish rest on time to avoid Bank
assessing damages I find it impossible get satisfactory dates completion your employees no initi-
ative head in charge they refuse work employees
overtime or endeavor rush completion as above suggested Stop If you do not wire your office or me
authority before noon today my efforts expedite
completion your work cease and delay will far more
expensive than handling as above suggested

A. P. ROBINSON,
Palace Hotel

Room 6131"

"Seattle, Nov. 11, 1924

Mission Marble Works
    209 Mississippi St.
    San Francisco, Calif.

Die between piers sixty five and sixty six also at
right side of pier seventy five rejected account excessive light veining send new marble We have
architects permission to use condemned die in stair
Ashler Stop Your absolute stupidity ignoring
telegrams simply adding more fuel to flame We
are accountable to Mister Graham He must have
prompt reports Stop Your compliance with our request of the tenth in getting finished stock here also

sending rough slabs for us to finish for Ashler and die where required would show sincere desire to co-operate Stop Is Eastman home If not where is he and when will he be home This delay is bound to cost considerable money as it is can't you see that it will cost considerable more if we don't show some speed your unbusinesslike methods and inactivity in handling the favrication of this marble costing us exorbitant expense installation If you don't heed our request and continue to ignore our telegrams which should be answered immediately in all cases we will charge you with our said expense in addition to damages assessed against us by the owners.

ROBINSON TILE & MARBLE CO.
600—8th Avenue North
Charge our account'' [111]

"Seattle, November 12, 1924.
R. A. Boucher,
c/o Mission Marble Works,
San Francisco, California.

Will Mission ship Ashler according program Why are they not working nights Stop Understand Eastman coming home Thursday Pay particular attention his attitude and wire immediately his intentions and program Stop Mailing check tonight Graham thinks best you stay on job at present will await action Eastman Stop Are they rushing stair facie and soffeit

ROBINSON TILE & MARBLE CO.
600—8th Ave. N.
Charge our account
Copy to Mr. Graham''

"November 13, 1924

Geo. M. Eastman
    c/o Mission Marble Works,
      209 Mississippi St.
        San Francisco, Calif.

Inform yourself fully status Bank marble work and phone Mr. Graham immediately Stop Duplicate mouldings twenty five and thirty one north side savings department seventeen must be dark to match other stock Stop Rush die between forty seven and fifty two Stop Are you working night crew Stop When will you ship Ashler

    ROBINSON TILE & MARBLE COMPANY.
Chg. acct.
Robinson Tile & Marble Co."

"Seattle, November 15, 1924

Mr. F. W. Grant
    Palace Hotel,
      San Francisco, California

Received wire Eastman stating they had repeated requests delay marble as building not ready Robinson notified Eastman when in your office about May twenty fourth to not delay fabrication of marble Stop Boucher anxious come Seattle Please go over situation with him and advise me how long you think he should stay there

    ROBINSON TILE & MARBLE CO.
Charge our account" [112]

"November 28, 1924.

Mission Marble Works,
    209 Mississippi Street,
        San Francisco, California.

We understand you are only working four cutters on ramps and that they will not be completed before February first We have promised main floor completed December fifteenth Ramps January fifteenth program must be followed out Put yourself on record by wire immediately when you will ship ramps.

ROBINSON TILE & MARBLE CO.
600—8th Avenue North.

Charge our account"

"Seattle, December 1, 1924

Mission Marble Works,
    209 Mississippi Street,
        San Francisco, California.

We have today uncrated and erected first check desk top and Architect has rejected it. Our foreman believes he can refill them with marble so that Architect will accept in order to hurry completion If all other work except four lower ramps can be completed by December thirteenth for opening on morning of fifteenth we advise you send expert mechanic to do this filling or wire us order to do it at your expense and guarantee shipping dates of various parts now short Stop Refer die panels above seats What is reveal with relation posts full size plans conflict wire

ROBINSON TILE & MARBLE CO."

(Testimony of A. P. Robinson.)

"Seattle, Washington, March 8, 1925
Mission Marble Works,
209 Mississippi Street,
San Francisco, California.

Referring your wire seventh Bank had previously refused any further payment until rest marble on job We inferred from this that they would give us payment soon as marble landed however in order please them and get in their better graces we worked big crew of men Saturday afternoon and all day Sunday double time expense amounting almost two hundred dollars then issued final bill as February twenty eighth Now they are pondering over amount they think they can hold out account delay We are handling matter as diplomatically as we can endeavoring keep *i*riginal alleged damages low as possible.

ROBINSON TILE & MARBLE COM-
PANY." [113]

Q. Handing you a bunch of letters, marked Defendant's Exhibit "A–12" for identification.

A. I didn't write the second one. The secretary of our Company wrote the second one.

Q. I mean from your office; it came from your office?   A. Yes, sir.

Mr. MYERS.—I offer these in evidence.

Mr. DAVIES.—We make the objection that they are wholly incompetent, immaterial and irrelevant. They do not tend to prove any single item of damage or counterclaim which is set up.

(Testimony of A. P. Robinson.)

The COURT.—You are not objecting because they are copies?

Mr. DAVIES.—No, your Honor.

The COURT. — Objection overruled; exception allowed.

Letters in Defendant's Exhibit "A–12" admitted in evidence and are as follows:

DEFENDANT'S EXHIBIT "A–12."

"March 27, 1924.

Mission Marble Works,
  209 Mississippi St.,
    San Francisco, Cal.

Gentlemen:

Your night letter of March 25 received and we wish to thank you. It was very well worded and, we think, very satisfactory in results. Mr. Graham's office seemed to be satisfied. You know how it is; they get to wondering about these things and want to be reassured. So far as we are concerned, you know we are satisfied.

Referring to your letter of March 22, fifth paragraph, you ask for a schedule of time in order to plan the finishing of the marble. We were informed yesterday that the plasterers will start within the next thirty days, and that they will complete their work within thirty days. This means that they will turn the bank building over [114] to us the first of June. We note that you will have the floor tile here about that time. We should like to have the remainder of the marble which is coming from you in the order of its erection, the cove, base and work above the same. We think we should set this before the floor.

We have six good workmen lined up and should like to give real service on this job. Making prompt deliveries and finishing it up in a short time will be a great big feather in your cap and also in ours, especially since the opposition have made so many bold and biased statements. If you can get all the marble here before the middle of June, so much the better. You will have it out of your way and will get your money that much quicker.

Yours respectfully,
ROBINSON TILE & SUPPLY CO.
By ————.

APR:S
10190

Dft. Ex. A–12"

"May 5, 1924.

Mission Marble Works,
209 Mississippi Street,
San Francisco, California.
Dexter Horton Bank.

Gentlemen:

In case you do not happen to have the deal plates cut the architect would like to have them made three-fourths instead of one and one-fourth inch thick.

The millwork man was complaining that using one and one-fourth inch marble would crowd his shelving too much. Nothing serious, however.

Mr. Bergseth would like a working plan of the information counter. Mr. Boucher will understand what he wants. Please send this at your earliest convenience.

We had a wire last week from Bob Connell saying the tile was ready for shipment. We wired him to let it come ahead and it should be here between the 15th and the 20th.

Would like for you to give us a line up on how the materials are coming thru so we can make a report to the architect before he gets nervous and asks us for it. We have been thinking it would be a good idea if you could let Boucher come up here when we started the job so he can interpret the plans to all parties interested. We will gladly pay his expenses and his salary and in that way keep the good reputation of both you and ourselves at, or above, par. Understanding his business as he does, Mr. Boucher can line the marble setters up in a short order so that all they will have to do is work, instead of doing as is sometimes the case—arguing about details which were worked out by the absent man. [115]

When Boucher was up here we were figuring on the Metropolitan Medical Building. Our price was $50,000; Udenses price was $51,000; Drake Marble and Tiling Company was $48,000. It was awarded to the latter.

Yours respectfully,

ROBINSON TILE & MARBLE CO.

By ————.

APR:ES"

"October 30, 1924.

Mission Marble Works,
   209 Mississippi Street,
      San Francisco, California.
         Re: Dexter Horton Bank.

Gentlemen:

Where the 5-inch step was omitted in the Senior Office and other spaces, the die on the piers was naturally affected. In some of your details the change has been made, showing the die 2 feet, 10 inches in place of 2 feet, 5 inches; whereas in others no notation was made. We are wondering whether this has been taken into consideration in getting out the stock.

We took this matter up with the architect, and in case this had been overlooked, suggested that an additional 5 inches of Belgian Black be put on top of the base to take up this discrepancy, but he would not stand for this. He insists on the St. Genevieve, but he will allow an extra piece of St. Genevieve, to save getting out all new pieces. If it is necessary to get these pieces out, get them out 5 1/16 inches, and on the returns to the counter where it shows rough, approximately 1 foot, 7 inches. These pieces should be not less than 1 foot, 9½ inches. Be sure that you match up your work mitres, and remember that we are forced to work very close on these piers. We would appreciate a letter from you with regard to this point, explaining just how it has been taken care of.

We stated in our wire yesterday, we will be out of work for lack of material in about a week or ten days, so we trust that arrangements have been made to ship your car to reach us by that time.

Yours respectfully,

ROBINSON TILE & MARBLE CO.

By ————.

EHR:G'' [116]

"San Francisco Cal. Nov. 6, 1924.
Mission Marble Works,
　209 Mississippi Street,
　　San Francisco, California.

Gentlemen:

Further referring to your failure to make delivery of marbles for the Dexter Horton National Bank, Seattle, on time, below is a copy of a telegram from John Graham, the architect:

'A. T. Robinson,
　Care Mission Marble Works,
　　San Francisco, Cal.

Bank directors very much distrubed at not having marble ready to move bank on Thanksgiving Day stop Have agreed to postpone to December sixth but no later. stop If your work is not finished then will hold you and your bondsmen responsible for all damages caused by delay stop Everyone else will be finished on time stop It is outrageous that after all the time you have had since January twenty fifth that your work should not be ready stop Bronze workers have had to stop their work today because of missing marble stop Will you wire immediately a definite statement as to when your work will be completed as printed invitations are being made for bank opening on December sixth and will have to be cancelled if that date cannot be kept.

JOHN GRAHAM.'

We have today wired George M. Eastman, Commodore Hotel, New York City, as follows:

'Your firm making miserable failure bank contract, I found stair plans not yet complete on fifth this month advise you authorize Hunter sublet sufficient portion work to effect our execution contract within period granted find limit December sixth Graham filed notice fifth owner will hold us and our bondsmen responsible stop We hereby notify you and Mission works that according to terms contract, between us, we will hold you liable for all damages assessed and incurred by reason your failure fulfill contract you can ill afford loss reputation Graham greatly disappointed.'

This is also to serve as further notice to you and impress upon you the necessity of completion and shipment of this marble in due time for us to have it installed complete by December 6th.

You will notice that in Mr. Graham's telegram he has asked the writer to give him a definite date as to when said work will be completed. Kindly give me this information immediately so that a wire can be sent to Mr. Graham to-night. [117]

Mission Marble Works, #2   November 6, 1924

Kindly notice that the period of grace allowed is December 6th and that after that date we and our bondsmen will be responsible for damages caused by delay.

According to the terms of our contract whatever damages may be assessed against us by the Bank will in turn be charged against you.

Yours very truly,

ROBINSON TILE & MARBLE CO.

By ——————.

(COPY)"

"COPY

ROBINSON TILE & MARBLE COMPANY

San Francisco, California, 11/8–24.

Mission Marble Works, City—Mr. Miller:

Kindly let Mr. Boucher have access to correspondence file concerning matters taken upon the bank job, so he can attend to any unfinished business and give necessary further info. Also let him see all future correspondence re this job. Also let him have access to list showing shipping dates, various unfinished members and parts.

Mr. Boucher is instructed to keep us informed as to whether you have as many men on the work as you should, and if you are also working as much over and double time as you should in this emergency. He is also instructed to insist upon your rushing work along, using extra help as above mentioned to expedite completion and shipments, all according to or better than shcedule time.

Finishing and making shipments per said schedule, or better, and using extra help and working overtime will materially assist in lessening trouble with the owners account this long delay, which is due to carelessness and neglect on the part of your

firm. At the present time you do not have sufficient men on our work. We know it is possible
for you to have more men on our work, as other
shops in the city are not very busy, except Am.
M. & M. Co.

Yours respectfully,
ROBINSON TILE & MARBLE COM-
PANY.

By A. P. ROBINSON, Pres. (Sigd.)''
[118]

"November 10, 1924.
Mission Marble Works,
San Francisco, Cal*k*fornia.

Re: Dexter-Horton Bank

Gentlemen:

We had a conference with the Bank people and
the architect this morning, and we are under the
impression that if we follow out this program, and
on December 6th the only uncompleted part is the
ramps and stair railings, we will avoid a lot of
trouble, litigation and expense, and, last but not
least, our reputation.

In any event, we must live up to this program
which has been promised them, as they can open
their bank according to schedule December 6th.

You will have to rush the fountain, so it can be
filled. By giving us templets looking down on
ramps, we will make necessary allowance for same.
With Boucher's help you should be able to give
this to us very quickly.

Don't spend any time saying it can't be done. Make up your mind it has to be done, and do it.

Yours respectfully,

ROBINSON TILE & MARBLE CO.

By ————.

APR:G

INC."

"November 12, 1924.

Mission Marble Works.

San Francisco, California.

Re: Dexter-Horton Bank

Gentlemen:

Die between Piers sixty five and sixty six also at right side of Pier seventy five rejected account excessive light veining send new marble we have architect's permission to use condemned die in stair ashler Stop Your absolute stupidity ignoring telegrams simply adding more fuel to flame. We are accountable to Mister Graham. He must have pormpt reports Stop Your compliance with our request of the tenth in getting finished stock here also sending rou*ch* slabs for us to finish for *Sh*ler and die where required would show your sincere desire to cooperate Stop Is Eastman home if not where is he and when will he be home This delay is bound to cost considerable money as it is cant you see that it will cost considerable more if we dont show some speed Your unbusinesslike methods and inactivity in handling the fabrication of this marble costing us exorbitant expense installation If you dont heed our request and continue to ignore our telegrams which should be answered immediately in all cases we will charge you with our said expense in addition to damages assessed against us by the Owners. [119]

We sincerely trust that you will have gotten into high gear before this reaches you. It seems you should be able to comprehend how exasperating your actions are.

You should also realize that the more trouble you cause a person, the less said person will endeavor to help you keep out of trouble. The writer is in conference with Mr. Graham two and three times a day, and naturally he gets wild when we have to tell him that we are not able to get prompt answers to our telegrams. He wants to be friendly and help us keep out of trouble, instead of getting further in all of the time, but the way you do, aside from all other reasons mentioned is showing scandalous neglect to Mr. Graham, who has always thought so well of you and really plugged for you.

Yours respectfully,

ROBINSON TILE & MARBLE CO.

By ————.

APR: G

Copy to Mr. Graham"

"Nov 15, 1924

Mission Marble Works,

San Francisco, Calif.

Att. Mr. Ostemeyer and Mr. Wyrick

Gentlemen:

The following apron pieces are not yet received:

No. 65 for Pier $\overline{65}$ NOTE should be 1–10 5/8

No. 65 for Pier 65

No. 73, between Piers 65 and 76, should be 7″ long, you made error in dimensions this per pier.

Apron No. 30 for Pier 39 should be 4″ instead of 3″

Apron No. 30 for Pier 40 should be 4″ instead of 3″

Please rush new pieces.

Please send plan showing section through balustrade in relation to fountain so can see dimensions of fountain return also give description of hole cut through for plumbing for fountain.

When are you shipping corrected stair treads?

Top riser, or first one below main floor-architects plan shows 7½″ face of same to spring line. You show 6⅝″. Please check up on this and advise.

Please rush plans for check desks. Should have had these long ago account electrical outlets.

Thanking you in advance for prompt attention to above, we remain,

Yours respectfully,
ROBINSON TILE & MARBLE CO.

By ——————.

EHR: AR" [120]

"April 1, 1925.

Mission Marble Works,
209 Mississippi Street,
San Francisco, California.

Mr. Eastman:

Dear Sir:

Attached we hand you copy of letter from the

Architect's office, which plainly explains the position we are in. We have said nothing before, owing to the fact that we were waiting to hear from you as you mentioned when here, and also as you suggested to let the matter ride along awhile.

Referring to the check, I had put in a stop payment the morning I mentioned it to you for the reason as explained to you, and which you understood, that the bank was not going to pay us anything at all until the whole thing was settled up, notwithstanding the fact that the previous day they had led me to believe that they were going to give us a payment. However, on the date shown on the architect's letter we were able to get them to make payment on account. We ordered the nonpayment withdrawn at the bank. The teller, being unable to find it, said he would take care of it, so we presumed he would. However, when the check came through the cashier called us up, and we told him that it had been withdrawn. He then had the check recalled and paid, and we now have the check in our office, the only trouble being that owing to the bank's error your bank was notified by wire that the payment was stopped, but they did not receive further notice that it was recalled and paid. We hope that this has not inconvenienced you.

Referring to the last paragraph of your letter where you speak about the way I rode your boys when down there, had I not done so when, on November 5th I found you did not have even the stair plans completed, much less the marble, the situation up here would have been much more serious than it is. I was simply trying to help you and ourselves out of a bad hole. Inasmuch as we were in a bad position, I was trying to get out of it as gracefully as possible, the trouble being due to your failure to get the marble out on time.

You speak about the building being held up here. We think it one of the luckiest things that ever happened to you that it was held up. Had it gone through on schedule, the bank certainly would have had a larger claim for damages, as I am firmly of the opinion that you could not have had the stair work completed and on the job in July or August, or for that matter, a lot of the other work.

Referring to my disposition with regard to your boys, you must know that it cost me considerable time and money to take this trip down there to get you to fulfill your promise. As to taking it upon myself to order the Vermont to finish regardless of cost, I do not understand what you mean. I understand that you wired your office to use any legitimate means to rush the work out, and on the strength of this some of the work was turned over to the Vermont to finish, but I of course had no authority to do this for you. [121]

George, if you would put yourself in my position of having placed an order with someone to finish up a job like this and seeing that they were falling down scandalously and that they did not even take it seriously, would you not get up in the air too?

I consider a man a friend who can be growled at for his shortcomings, and still be a friend. You

(Testimony of A. P. Robinson.)

may be assured there was no personal antipathy toward you or any of your boys.

Yours respectfully,

ROBINSON TILE & MARBLE CO.

By ——————,

President.

APR:G

Inc.''

Mr. MYERS.—The first one being March 27th, 1924, and the last one April 1st, 1925.

WITNESS.—That is the American Marble & Mosaic Company.

Q. Mr. Robinson, what do you know yourself about the marble that was put in there and the work that was put in there that the architect refused to pay for, that has been referred to in this trial?

A. Well, just about as Mr. Grant and Mr. Graham explained it this morning. There were four or five check desk tops that had been broken crossways. A top is supposed to be in one piece, and they are about four feet by ten feet, and five inches thick and they were broken in two; nevertheless some were broken in two places where they were waxed together. It seems to be impossible to conceal that waxing so it would not be plainly shown. There are two or three runs in the counter front. I don't refer to the pieces. I refer to the runs which have considerable yellow in them, far more than the samples show.

(Testimony of A. P. Robinson.)

Q. Do you know how that extra yellow is caused?
[122]

A. That may be caused by sawing the block the wrong way of the bed. The bed is the way the rock lays in the ground, different strata. We speak of the strata in rock or in marble, and the different strata will be different colors. If you go straight through that strata you have combinations, if you cut it the other way you are likely to have large panels of one color of strata instead of mixed, and again, there may be parts of the marble where it would be practically all yellow, and again, in one end of the block or one side of the block or the entire block might be yellow instead of gray, like the samples on the table.

Q. Mr. Robinson, what did you get from the Mission Marble Works by way of response to those telegrams that I have read to the Court here today?

A. We didn't get much response. The context of the letters will show my anxiety there.

Q. Is there anything else you want to explain?

A. A further point I think about referring to this invoice down here at the bottom—

The COURT.—What exhibit number is that?

A. That is Exhibit "A-9." I will put a circle around that so you can find it. On May 25th we gave them a check for $5,955, and on June 23d for $7,895 and on August 22d, $12,600 and on October 27th paid them $8,500. The reason I mention those is to show that we paid promptly for the ship-

(Testimony of A. P. Robinson.)

ments that were made, and, had the contract been fulfilled they would have had the money perhaps before it was set, or about ninety per cent of it, and it would have saved us a lot of trouble and expense and everybody else involved.

Cross-examination by Mr. DAVIES.

I have handled marble as a salesman for fifteen years and [123] in looking after its installation. I never had any experience in the manufacture of marble until I opened up my own plant. I think it was on May 23, 1924. I never did any marble work prior to this job, other than I explained, as to selling marble and looking after the installation.

Referring to item found in Defendant's Exhibit "A-5"—"Labor, marble setter and triangles to fit proper plans." That is explained in this way: The floor is laid out in pattern and there is a diamond-shape tile and to fit around that you have to have four triangular pieces, the unit being practically square, or an oblong square, you might say. That does not always require work for the marble setter. It is not the *the* usual rule to spend 64 hours time cutting a few diamonds. We had lots of them.

Q. Hundreds of them there that you had to fit?

A. Well, I would not say hundreds.

Q. That is work usually done by the marble setter? A. No, sir; that is done by the mills.

Q. The factory can't furnish those so that you don't have to do any cutting on them?

(Testimony of A. P. Robinson.)

A. We don't make any kick if they send them out so we have to do some edging on it, on some of them, but if you have to edge all of them, heavy pieces of marble as sent out and 11½ inches is sent for and it is sent out 12 inches, it is a half-inch that has to come off, and that is a mistake. If we have an eighth of an inch off on one side, we usually take that off ourselves.

If we have to take any off of these ourselves we have to charge for it. They were wrong and we had to charge for them. They were too big. I don't know how much too big they were, but I can find out. There is nothing here in this exhibit from which I can tell. The second item—"Labor, cutting return molding on piers $39.40, from 12 inches to 11 inches"? It is not usual to [124] have to cut an inch off. I think that they came defective. I think the factory should have made them according to the plan.

Q. Did they?   A. No.

Q. Where have you anything to show that they did not do it according to the plan?

A. I do not know whether it is on here or not. What is the number of the item?

Q. It is five hours labor cutting return molding on piers, thirty-nine and fourth, from 12 inches to 11 inches.

A. Sheets disconnected this morning, as a summary of this and shows the different items it was taken from.

(Testimony of A. P. Robinson.)

Q. You can't tell us anything about it from these sheets?

A. Yes, sir, if I take the time, and everybody here can read through all of them.

Q. If there is any question about it, let us settle it now.

A. Let's see those three preliminary sheets in pencil notes that were disconnected this morning.

In reference to the item of labor, drilling holes in marble balustrade, 204 holes. I claim that is not a part of my work, drilling those holes, because it is customary to drill small dowel holes and anchorage holes on the job, but it is not customary, and that is not the rule, to send them out that way, because they had to have big holes drilled in there, about five-eighths of an inch in diameter and an inch and a half to two inches deep in the very hard marble, which to drill that by hand it would take a man over an hour to drill one of those holes by hand, in some of the marble. I got the idea that this was a proper charge against the Mission Marble Works before we did them.

Q. When did you first inform Mr. Eastman or the Mission Marble [125] Works, the plaintiff, that you had any such idea or counter-charges against him, as these?

A. I don't remember. I did not consider it material to inform him. When I ship some materials and do not send them right and they have had to reject them, I expect to pay for them, and that is customary in the trade.

(Testimony of A. P. Robinson.)

Mr. Eastman did not ask me to go to San Francisco, nor did he ask me to send Mr. Grant down there, or ánybody else.

Q. Do you know, as a matter of fact, that Mr. Eastman rushed the job? Did Mr. Eastman sublet considerable of the work to the Vermont Marble Company?

A. I know that I went down there and raised a lot of trouble.

Q. Answer the question.

A. I had to qualify my answer. I would not say that Mr. Eastman did or did not sublet a good deal of this marble to the Vermont Marble Company, but his company did, at my request, and after I requested it, but he took it away from them before they finished it.

Q. Are you sure now?   A. I don't know.

The last date that I made any payment to the Mission Marble Works on this contract was March 16, 1825. I gave them a check for $4,000. That is the last payment we made on it.

The signature on Plaintiff's Exhibit No. 5, marked for Identification, was my signature.

Letter was received in evidence as Plaintiff's Exhibit No. 5 and is as follows:

(Testimony of A. P. Robinson.)

PLAINTIFF'S EXHIBIT No. 5.

"ROBINSON TILE & MARBLE COMPANY.

May 25, 1925.

Mission Marble Works,

San Francisco, California. [126]

Re: Dexter-Horton Bank Contract.

Gentlemen:

Referring to our letter of May 14th, the second paragraph states that according to our records we owe you $5976.46.

The book-keeper had the $600 item for overtime charged against your account. This you probably did not understand. No considering this $600 item, our books would show a balance of $6576.64. We have not given you credit for the items on your invoice which the writer checked over with Mr. Eastman while he was here.

Yours respectfully,

ROBINSON TILE & MARBLE CO.

By A. P. ROBINSON (Sgd.)

10190

Ptf. Ex. 5–Adm.''

Q. Mr. Robinson, you have purchased some blocks of marble from the same quarry that this marble came from, have you not? A. Yes, sir.

Q. Quite frequently it has happened, has it not, that there are seams through it, which, when they come in the blocks, so when you come to manufacturing them, fall apart and come apart?

(Testimony of A. P. Robinson.)

A. That sometimes happens with any fancy marble.

Q. You are not answering my question, Mr. Robinson. Speaking about this particular marble, I will ask you if you did not buy some, and if you did not find it with seams in it so that they fall apart?

A. We bought some of it; I don't recall that any of it fell apart.

Q. Haven't you got a block now that was of that character?

A. Yes, sir, we have a part of a block.

Q. And that very frequently happens in all kinds of fancy marble, doesn't it, that is expected by everybody that deals in it?

A. Yes, sir, it is expected that it will come apart in seams where the different strata come together; that is quite true.

Redirect Examination by Mr. MYERS.   [127]

Q. This letter to which you referred, shows a correct statement of the account for the marble as billed to you and the last payment which you made?

A. When the invoices come in and the bookkeeper enters them on the books, we don't detach them. We don't have to receive the invoice and make any record on them. The fact that we did that on our books, we don't also say we admit liability, so if we did, it is limited or contingent upon the conditions which may arise.

Q. Mr. Robinson, did you ever have a lawsuit before?   A. No, sir.

TESTIMONY OF HAMILTON L. MERRITT, FOR DEFENDANT.

HAMILTON L. MERRITT was produced as a witness on behalf of the defendant, and having been first duly sworn, testified as follows:

My name is Hamilton L. Merritt. I am connected with the Dexter-Horton National Bank. I am vice-president. I have been connected with the bank since 1907. I might have seen the gentleman sitting here, Mr. Eastman, but I do not recall Mr. Eastman. I know that the Mission Marble Works was placing the marble in the bank, but I do not recall Mr. Eastman.

Q. Mr. Eastman, during this trial, testified that while there he was here in this city some time in 1924, that he was present with an officer of your bank at a time when the check of $5,900 was handed by this officer to Mr. Robinson in his presence, and that this particular officer said, in substance to Mr. Eastman that he should be in no hurry, or let up on sending this marble, that they would not need it now, or something like that; you did not say that or in substance like it? A. No, sir.

TESTIMONY OF GEORGE B. EASTMAN, FOR PLAINTIFF (RECALLED—CROSS-EXAMINATION).

GEORGE B. EASTMAN, recalled for further examination, testified as follows: [128]

Cross-examination by Mr. MYERS.

I do not know the witness who was just on the

(Testimony of George B. Eastman.)
stand, no more than he probably knows me. I
could not identify him, no, sir. I could not say
that he is the man that said to me in substance that
I could delay, let up on sending marble. I don't
identify the man as the gentleman that talked to
me at the bank. The man that spoke to me was
the man who was negotiating with Mr. Robinson
about getting his money. I don't know whether
it was the cashier, but it was some man.

TESTIMONY OF A. P. ROBINSON, FOR DE-FENDANT (RECALLED).

A. P. ROBINSON, recalled for further examination, testified as follows:

Direct Examination by Mr. MYERS.

Q. I want to refer to the time when some officer
of the bank early in 1924 passed a check over for
I think $5,900, in your presence, and in the pres-
ence of Mr. Eastman; what officer of that bank was
that?

A. That was Mr. Merritt. The reason I say that
is because he is the only one of the officers there
that I ever talked to about getting payment after
I had received a certificate from the architect's office. I do recall however, that there was something said about having too much weight on one
floor, where we had this piled down in the bank or
on the bank's premises.

Q. Where was it?

A. It was in the Dexter-Horton Bank Building

(Testimony of A. P. Robinson.)

basement, upper subbasement; but that was not the bank proper. However, up in Mr. Graham's office I recall mentioning that point to Mr. Grant and I says should we hold up shipments or not and he says, "No," and I at that time told Mr. Eastman, "Don't hold them up; you might hold them up too long." [129]

Q. You said that to Mr. Eastman?

A. Yes, sir; I was up in Mr. Graham's office. I says, "Don't hold them up, because you might hold them up too long." I was afraid he might do the same as I would or anybody else, taking on another bunch of work.

Q. What do you know about his taking on another bunch of work?

A. Well, when I was down there in November they had considerable other work in the shop.

' Cross-examination by Mr. DAVIES.

I remember the bank representative saying that he did not want us to pile any more marble on that floor because it would make it settle. It is not a fact that we could not store marble in the bank proper, prior to September 22, 1924.

Q. When did you store marble in the bank proper? A. I don't remember.

We may have called up Mr. Graham's office, Mr. Eastman and I, and I might have mentioned to Mr. Grant that there was something said about it, and I says, "You think we better hold them up, do you?" and he says, "No." I told Mr. Eastman,

(Testimony of A. P. Robinson.)

"Don't hold them up; you might hold them up too long." I told Mr. Eastman at this meeting in Mr. Graham's office, not to hold up the shipments. Notwithstanding the bank had told Mr. Eastman not to ship because the bank could not store it. Mr. Eastman knew that the bank had no authority to talk to me or tell me anything about it. He was not dealing with the bank direct, or taking orders from them. He was dealing with me. I was the fellow that he was doing business with. I stood by while Mr. Eastman and the bank representative were talking. Whatever they said I did not object to it at that time, while the bank representative was there; I did not consider it necessary to object or to pay any attention to it. [130] He did not say not to ship the marble. He said there was too much weight on the floor in one room and that somebody might complain because the floor had settled.

TESTIMONY OF F. W. GRANT, FOR DEFENDANT (RECALLED).

F. W. GRANT, recalled for further examination, testified as follows:

I don't think you want to identify that sample of marble, because I can't identify it.

Q. I asked you to bring up here a sample from which—

A. It does not matter what you asked me; I did not bring it.

Q. You did not bring the sample?

(Testimony of F. W. Grant.)

A. I did not bring the sample that I could identify as being the other portion of the sample that the plaintiff had exhibited.

Q. I want the sample that was kept in your office; under the contract you were to keep a sample in your office and another was to go to the person who was to supply the marble.

A. I have not brought the sample. I failed to find it.

The defendant rests.

TESTIMONY OF ROLAND E. BOUCHER, FOR PLAINTIFF.

ROLAND E. BOUCHER, produced as a witness on behalf of the plaintiff, being first duly sworn, testified as follows:

Direct Examination by Mr. DAVIES.

My name is Roland E. Boucher. My business is estimating. I have been in the marble business all my life; always in the west. I am familiar with the St. Genevieve marble. It is a clay and lime formation, the marble formation. It has no veins, as a rule. It comes in blotches. It is found in Missouri. It is a mud pressed together by nature, and the yellow you see in there is clay and the gray is the lime. The veins are silica, which holds it together. That is a formation of three kinds of stuff. There are other foreign matters along the seams, but it is purely lime or mud; in other words, clay. This kind of marble does not [131] have any veins running through it in any definite direction. It comes—on one side of the block there may be yellow—it comes either brown or yellow and you can turn the block upside down and it is the same. You can cut the block many ways and you get almost the same effect. It is a mass formation and not a vein formation. Whatever seams it has in it will run in every direction.

I was employed by the Mission Marble Works during July and August, 1924. I was there when the check desk tops were fabricated. They came apart during fabrication. They did not fall over or break, or anything like that. The block was put in the gang saw, in a big mass, and we have blades and we dress them up on the side with two big extensions and this particular block was sawed in five pieces at a time, four inches thick, ten feet long and about five feet three wide. It was a special block at that time for this job, and in sawing it down we discovered, two-thirds of the way, almost two-thirds in the block, we saw it was opening up at the seam. It did not fall over or break but it came apart at the seam, which it does in all cases, which opens it, where the vein is in what we would call a formation with the yellow. That is what took place in this piece.

I did not tell Mr. Grant at any time when he was down there while we were sawing those blocks for the check desks, that they fell over on the table and broke or anything like that. That is not true; it did not happen.

(Testimony of Roland E. Boucher.)

Q. How many check desk tops came apart, if you know, out of that block?

A. I believe the whole five. No, there were about three really that fell apart, opened up. The other two still had a crack, opening on the other two as well, but they were n ot clear open; they were not loose. [132]

Q. What did you do?

A. We customarily, in cases like that, being so much weight,—they weigh*t* about pretty nearly a ton apiece, and we have to put a channel iron through the bottom and gouge it out and put sulphur and a band about an inch and a half wide and a quarter of an inch thick and put sulphur through it and burn it into the marble to keep it rigid in one mass. It has happened in all cases of the best of the marble that size. The marble would not hold its own weight, that size of a piece of marble.

We did a good workmanlike job on it, cementing it together to make it strong. After we got the blocks fastened together, we took a chisel and where the clay or yellow is, where it is loose, we took it out. We treated it and waxed it and filled up the joints in every case. We call that waxing. There are two ways of doing it. One way we put it together with German cement and take the block and leave it twenty-four hours and wax it on top of that. Another way is to put a composition of rosin and shellac and burn it in; but in this case we put rosin and shellac at that point. It is a very common

(Testimony of Roland E. Boucher.)

thing for highly decorative marbles to have seams in them and fall apart or come to pieces while they are being fabricated. It is understood by the contract. It is a common thing for highly decorative marbles to have seams in them, which allows them to come apart during fabrication. The quarries will not ship them and they never guarantee a fancy marble for soundness. When we give samples to the architect, it always says—"This is not guaranteed for soundness."

I came to Seattle and worked on the job of installing this marble. I arrived here about the first part of December, about three weeks before the bank moved into its quarters. I was here when the fabricated check desks arrived. Mr. Grant and Mr. Morgan [133] inspected them. I came up here as superintendent of the marble end of the job. As superintendent for Mr. Robinson I undertook to and did remedy those defects that were pointed out at that time. Mr. Robinson had three or four men rubbing and polishing and filling and putting little slivers, little blocks where the holes would show up. He put in two or three weeks there off and on, to fix up those check desks, I think the three of them. No objections were afterwards made to me about them, by Mr. Grant, as to whether or not they were actually accepted by the architects. No objections were ever made to me personally after I did this work. At all times while I was working there as superintendent of the marble work for Mr. Robinson, finishing up this bank, I complied with

(Testimony of Roland E. Boucher.)

any and all suggestions by the architect. We tried in every way to satisfy the architect.

Q. Well, did you?   A. Yes, sir.

Q. Did you actually remedy any defects which the architect showed you?

A. I will tell you, Mr. Davies, Mr. Robinson instructed all hands running to rub the floor and clean it over and over again quite a number of times to satisfy them so that they could open that bank.

Q. Did you at any time put in any pieces of marble after they had been rejected? If they did reject any, and tell you not to put it in?

A. No, sir, no, by no means.

Q. State generally, in your opinion, as to the character of the marble and the work on this job—, as to whether or not it is a good job or incomplete, or a poor job.

A. I should call it a good, fair job, on account of marble used—that quality of marble—the job is general. [134]

Q. How is it as to workmanship?   A. Good.

At this time I am working for the Robinson Tile & Marble Company. I have been working for them ever since I came to Seattle.

Cross-examination by Mr. MYERS.

I said I came here from San Francisco as superintendent of the marble for the Robinson Tile & Marble Company. Mr. Juengst was foreman of the job. He had charge of it as foreman. I did not work under him. At the factory it come from

(Testimony of Roland E. Boucher.)

time to time and things had to be fixed, fabricated over again and jointed over again, and I had to go and do it to suit the conditions. I was running the shop here.

I know a man by the name of Hunter in San Francisco. Mr. Hunter's position with the Mission Marble Works is that of estimator, I guess. My position with the Mission Marble Works during November, 1924, was draftsman and estimator.

Defendant's Exhibit "A–13" for identification, is a telegram I sent to Mr. Robinson.

Telegram admitted in evidence.

Defendant's Exhibit "A–13" is as follows:

DEFENDANT'S EXHIBIT "A–13."

"Night letter 'phoned from Western Union

11/14/24

They wont ship ashler until finished Railing and ramps not ready. One check desk done also some cap to wicket opening Not working night shift only two men two hours overtime each night something must be done to compel them to let out part work Eastman back today Will report his attitude Two facia or soffeit done Drawings mailed today Afraid job never done until January first if then Writing tomorrow

R. A. BOUCHER.

10190.

Deft. Ex. "A–13." [135]

Q. The statement in that telegram is true, is it?

(Testimony of Roland E. Boucher.)

A. These initials—I was representing Mr. Robinson, to get the work out, and I sent the telegram.

I don't know the exact date I quit the Mission Marble Works, but it was three weeks and a half before Mr. Robinson came down there. About three weeks after I quit the Mission Marble Works I went to work for Mr. Robinson. Mr. Robinson employed me in San Francisco, to help rush the work of the Dexter-Horton National Bank. Mr. Robinson explained to me that he was anxious to get the marble up here. When marble comes from the manufacturer there are some little things that have to be done. There always some closures or joints; you have to generally do that, in the corners, by hand, to fit it nicely.

I am connected with the Robinson Tile & Marble Company now. That is the way we get the work out now. That is understood; one end rough. I can't get the point—what are you driving at? I want to understand. One piece of marble is in the rough.

The COURT.—He is asking if you are doing the same way now at the Robinson Tile & Marble Company that you did with the plaintiff?

A. Oh, yes, sir. It is not customary to send marble on the job and require the ones that are to put up the marble to do cutting that was properly for the shop. I am in a peculiar position, because I am working for this firm now.

Mr. MYERS.—The truth is all we want, Mr. Hunter. I knew Mr. Hunter in San Francisco. We

(Testimony of Roland E. Boucher.)

ranked the same, as far as wages were concerned, but I had more authority. I did not [136] *I did not* hear Mr. Hunter talking to Mr. Grant down there, or any conversation with Mr. Grant.

Q. Are you well acquainted with Mr. Grant?

A. I am well acquainted with Mr. Grant.

Q. Mr. Grant testified here, that you, or, I think it was you and some other man in possession of the Mission Marble Works in San Francisco, when he was down there, had advised him this marble at the top of those check desks had fallen down and broken.

A. They opened, that is right.

(The COURT.) Actually fell down?

A. No, they never did that; they opened up. It was not in the fault of sawing.

Q. When you say the clay formation opens up,— you did not tell him that, did you?

A. No, sir, I didn't tell him that.

Q. I am telling you that Mr. Grant so testified,— that he thought you and some other man told him that?

A. No, sir, I did not say to Mr. Grant that it fell down and broke.

Q. You heard Mr. Robinson testify that notwithstanding any conversation that you had with the bank that when he was down there with you at the architect's office that he told you not to discontinue; you heard Mr. Robinson's evidence in that regard, did you not?   A. I did.

Q. Did you hear him say that at that time?

(Testimony of Roland E. Boucher.)

A. No, sir.

The COURT.—I think his exact words were, ''not to let up.''

Q. You did not hear that?

A. I don't recall that at all, honestly I do not.

Q. You were here in the courtroom to-day and heard me read those letters and telegrams to the Judge?   A. Yes, sir.

Q. And you got them all, didn't you?

A. No, sir.

Q. What is your answer to it?   A. To what?

Q. To all those letters and telegrams?

A. Why should I make an answer to you. What reasons have I got? I got the telegrams.

Q. What answer did you make to them?

A. That we were doing the very best we could. 1 understand you now.

Q. You haven't offered here to the Court any response that you made to those repeated and numerous demands we made upon you, you haven't offered any to the Judge.

A. I am working with my attorney's advice. My attorney represents me in this matter.

(Witness excused.) [137]

TESTIMONY OF GEORGE B. EASTMAN, FOR PLAINTIFF (RECALLED IN REBUTTAL).

GEORGE B. EASTMAN, recalled, testified in rebuttal as follows:

After getting the order to stop shipping we did

(Testimony of George B. Eastman.)

not stop fabrication altogether. We had to continue on what was in process, but we did not continue to do any more when he told us not to, that the building would not be ready, and to stop shipment. When we were told to stop shipping, we finished what was in the process of manufacture and then we quit manufacturing this marble. After the order to stop shipping we got an order to go ahead, by a telegram from Mr. Robinson, stating that in about a week's time he would be ready to start setting.

The telegram you show me, marked Exhibit 6, is the telegram.

The telegram marked Plaintiff's Exhibit No. 6, received in evidence and is as follows:

PLAINTIFF'S EXHIBIT No. 6.

"1924 Sep 17 PM 11 51

EAB708 64 NL
        Seattle Wash 17
Mission Marble Works
        209 Mississippi St
                San Francisco Calif

Have remainder marble for bank ready to ship on short notice is request of Architect Graham Stop We have not yet started but room is actually ready now Stop Grant demands complete working plans Stop Please send whatever additional plans required soon as possible procure three or four marble setters make best arrangements you

(Testimony of George B. Eastman.)
can and we will be grateful No news mausoleum yet

ROBINSON TILE & MARBLE CO.

10190

Ptf. Ex 6 Adm."

After receiving this telegram I instructed my foreman to proceed and to continue fabricating the material. At the time we got this telegram our plant was in operation, working on other jobs. We had to work continually. We set aside as much as possible of the work we were doing and tried to get the work out that was going through shop so that we could start on this job. The foreman is the man who did that, I only instructed him. When the telegram was received we were engaged on other jobs. You can't shut your plant down—you must keep your men going or you will lose your employees. When we received this telegram I instructed my foreman to proceed with this job and he did that. We got extra help and then we got outsiders to do some work—the Vermont helped us. We got more employees and got matters done as they were planned. We paid the Vermont about eight hundred and some odd dollars for the work they did. We made no charge against the defendant for that money. We furnished the material and the Vermont planed and did some of the polishing. I sent extra [138] slabs to Mr. Robinson for this job, besides those called for in the contract, to take

the place of the pieces rejected. I made no charge
for those and they are not in my final statement to
him.

I was in Seattle at the time the last payment was
made—the sum of $4,000. At that time I went to
the bank to look over the work with Mr. Grant and
Mr. Robinson. We went to the bank to look over
the marble work installed. When we got down
there we looked around—the three of us together.
At that time Mr. Grant referred to the wall as you
go in—the north wall. He said it was too light,
and he spoke about one of the check desks that
would have to be remedied. He said that it would
have to be re-waxed. At that time he pointed out
five or six panels in this wall that he mentioned and
he said they were too light. We were standing in
front of the hand rail of the staircase, the three
of us, and he pointed to this wall desk, looking to
the north. At that time the only defects he pointed
out were those five or six panels and one check
desk, that I have any knowledge of. At that time
I offered to send other slabs to take their place and
he spoke about that work would have to be done,
if it was changed, on nights and Sundays. I explained to him that it was not my part to do the
work, that it was simply my part to furnish the
marble, and I would send them other slabs to take
the place of those if they would give me the size to
go by. Mr. Robinson was right there at the time.
Nobody has sent me any dimensions or anything

(Testimony of George B. Eastman.)

to go by. Nothing on that subject at all, that I know of. We were ready and willing to do that at any time. So far as supplying material or anything that was my portion to do, I have been at all times ready and willing to do anything in my power to furnish any marble or fix up any defects which have ever been pointed [139] out to me. No one has asked me to furnish anything at any time for that bank that I have not furnished. I have no knowledge of anybody asking me to supply anything since the time I was there to get that check. At all times while we were manufacturing the marble, all the way through, we furnished other pieces of marble to take the place of any that was claimed to be defective. We had items of replacement many times. When I was up here about March 16th, the time the last payment of $4,000 was made, I ran over our account with Mr. Robinson. At that time Mr. Robinson did not present to me any such claim as that contained in Defendant's Exhibit "A–7," the summary of the counterclaim of defendant. He made no such claims and I never saw anything like that until I came into your office on this trip. At that time he did not make any claim against me for any work which he claimed should have been done in our factory, that he had to do up here. He mentioned that he had a little hand labor on some pieces that had to be returned and it was pointed out to me in the bank. He made a memorandum of what it was to take

(Testimony of George B. Eastman.)

back to our office. What I understood that he referred to, and what he showed me, I would not imagine it would be a very great amount. I made no complaint of it because I thought perhaps it was just. He referred to my final invoice and told me the amounts of the invoice that he thought were not chargeable. The only discussion we had was with reference to some of the charges I had on my invoice that he thought were not proper. I have a memorandum of the items that he claimed were not proper. They are as follows:

The item of $7.85 he claimed was short—not shipped originally. He claimed the item of $81.99 was unjust and not in the contract. He claimed the item of $16.09 was not an extra; that was covered by the contract. He also objected to the item of $540.00, boxing [140] and cartage. He came to no agreement as to the items we had a dispute over. I told him it would be easy to adjust those things if he was ready to pay us our bill. He did not pay us our bill and we did not do anything more. He gave us only $4,000.00 on account. The boxing was necessary as to those shipments which were made by water. You cannot ship marble that way unless it is boxed. I think every effort was made to expedite the shipment of this marble after we got that order to go ahead. We did everything we possibly could—hired the extra men and did everything we could in our capacity. We had some other work on hand and of course we had to go through with that, as I explained before. When.

(Testimony of George B. Eastman.)
you got part of the work on the different machines
it has to continue through.

Cross-examination by Mr. MYERS.

The first carload was shipped from Kansas City
on April 26, 1924. The second carload was shipped
June 4, 1924. The third carload was shipped July
28, 1924. The fourth carload was shipped October
11, 1924; and from that date on the L. C. L. shipments, so we had shipped three carloads of marble
before September 17th. I could not say whether or
not there was 80% of the marble made up before
I got the telegram of September 17th. We shipped
many pieces in L. C. L. The estimated data shows
that there were 141 tons.

Without any question, when I made this contract,
I understood that it was my duty to fill that contract, to furnish that marble to Mr. Robinson so
as to enable him to carry out his contract with the
bank, and I certainly felt my obligation to the Dexter-Horton bank just as much as though I had myself signed the contract. Every effort was made
and I had a personal interest in trying to do so.

Q. That was just what you stated in your contract, that you were [141] to furnish that according to the plans and specifications, terms and
conditions of the same, in order to enable the purchaser, Mr. Robinson, to faithfully and promptly
fulfill the contract?

A. That is the writing that he put in. I cer-

(Testimony of George B. Eastman.)

tainly intended to do so. Until they stopped us I did so.

What I have heretofore detailed is what I mean by the term "they stopped us"; and the agent told us not to ship any more marble—this architect and the other agent. I so notified my foreman.

I heard Mr. Robinson testify, that at that time, at the architect's office, he told me not to ship—to discontinue. I did not hear him say that at that time.

The COURT.—I think his exact words were "not to let up."

I did not hear him say that. I was in the courtroom to-day and I heard you read those letters and telegrams to the Judge. I got the telegrams, and the answer I made to them was "that we were doing the very best we could."

Q. You haven't offered here to the Court any response that you made to those repeated and numerous demands we made upon you, you haven't offered any to the Judge.

A. I am working with my attorney's advice. My attorney represents me in this matter.

I have told you many times that I stopped the fabrication of this marble because I can't continue the fabrication without shipments.

I left San Francisco October 10th, I believe, and I went to New York on business, partially so. I went one way in order to get a boat ride, instead of by train. I went by the Canal—that was all, to New York. I returned November 11th. I do not

(Testimony of George B. Eastman.)

know [142] whether the setting plans were completed on November 5, 1924 or *no* not. The marble was not fabricated on that date. Our shipping sheets show later shipments. I got the wire that you showed me, at the Commodore Hotel in New York City, from Mr. Robinson. I did not make any response directly to Mr. Robinson. I sent advice to my boys—you understand. I advised my boys. They telegraphed me also. I advised them to secure additional help from the outside if necessary. I do not know why the shop drawings were not ready on or before November 5, 1924.

(Witness excused.)

Testimony closed.

Thereafter, and on or about June 28, 1926, the parties entered into a stipulation which after title of court and cause, is as follows:

"It is hereby stipulated between the parties, as follows:

I.

That paragraph five (5) of plaintiff's complaint may be amended by adding the words, 'work, labor and material' after the word 'marble' in the fifth line of said paragraph.

II.

That the final statement from plaintiff to defendant dated February 25, 1925, may be introduced in evidence as Plaintiff's Exhibit 7.

III.

That the letter written to plaintiff by Dexter-Horton National Bank, dated January 17, 1925, and attached to this stipulation, may be introduced in evidence as Defendant's Exhibit 'A–15.'

IV.

It is stipulated that the work and labor referred to in subdivision 'A' of paragraph three of defendant's further and independent answer as paid for in cash at a cost to the defendant of $784.24, and defendant charged plaintiff the sum of $983.88, being actual cost plus profit of $109.64.

J. O. DAVIES,   (Sgd.)

Attorney for Plaintiff.

H. A. P. MYERS, (Sgd.)

Attorney for Defendant." [143]

Defendant's Exhibit No. 7, mentioned in said stipulation, is as follows:

DEFENDANT'S EXHIBIT No. 7.

"San Francisco, Cal., Feb. 25, 1925.

ROBINSON TILE & TILE CO,

Our Order No. 976

Seattle, Wash.

Job No. 644

Bought of

THE MISSION MARBLE WORKS

DEXTER HORTON NATIONAL BANK

To furnishing marble in the above job per
  contract dated Jan. 30, 1924............. $46000.00

To furnishing marble for Information Desk
  per letter dated Mch 17, 1924 ....   440.00

To finishing marble—change bet ween column 61–66 per your letter Oct. 31, 1924....

BELGIAN BLACK.

1–0 2 pcs. base 1–3½x0–7½x0–7/8" 2–8 @ 2.45.. $ 6.53

2 shoes 0–6x0–2x0–2½" 1–0 @ 6.55... $6.55

13.08

Less 40% .   5.23

7.85

ST. GENEVIEVE.

1–0–0 2 Die 0–6x1–9½x0–⅞,	1–10	@ $ 3.85	7.05
1–1–0 2 M. Cap 1–3¼x0–4¾x0–3	0–2	@ 27.25	4.54
2–1–R 1 Top 7–9x0–2x2–0	17–0		
2–1–L 1 Top 7–9x0–2x2–0	17–0		
	34–0	@ 7.25	246.50

$258.09

Less 40% .   103.24

154.85

Mould and Cove Labor 47.25

To furnishing Belgian Black Coved Base
per your requisition #351

3 pcs. shoe 4–5x0–2x0–2½ 6–8
4 pcs. shoe 3–7x0–2x0–2½ 7–2
                      ———
           13–10 @ 6.55.. 90.60
    Less 40% 36.24
                          ———
                          54.36
Labor coving 27–1 @ $1.00 ..... . 27.58   81.94
                                  ———

To replacing pcs. #28 & 31 on pier #38 &
pc. #13 on pier #55. Broken in transit.

ST. GENEVIEVE.

1–0–0 1 Die 2–6x1–3½x0–⅞	3–4	
1–1–0 1 Rail 2–6x0–3x0–⅞	1–6	
1–1–0 1 Rail 3–4x0–3x0–⅞	1–11	
	6–9 @ 3.85	26.00
	Less 40%	10.40
		15.60

To furnishing St. Genevieve marble at pier #75 ordered by A. P. Robinson in person to our foreman. [144]

1-1-R #28 1 Die 1-6x2-101/16x7/8 4-7
1-1-L 29 1 " 0-1x " x " 1-8
1-1-R 30 1 " 0-9x " x " 2-5
1-1-L 31 1 " 2-0x " x " 6-1
                              14-9 @ 3.85..   62.19

Pier #72
1-1-R #20 1 Die 2-0x2-101/16x7/8 6-1
1-1-L 23 1 " 0-11x " x " 2-11
                              9-0 @ 3.85.   34.65
Quirk Mitres 17-0 @ .. ..@ .50.   8.50

                                             105.34
Less 40% .   42.14

                                              63.20

ST. GENEVIEVE.

1-0-0 #53 Apron 4-4 ¾x0-5x11½ 2-7 @ 6.10...	15.76	
Less 40% ..	6.30	
	9.46	
Labor 4-5 @ 1.50	6.63	16.09
Boxing on less C/L shipments from Nov. 3, 1924, to Feb. 19, 1925.		
Sup. ft. 1781-6 @ .12 ... .. 213.78		
Cub. ft. 271-0 @ $1.00 .. . 271.00		484.78
Cartage 40 Tons @ .	$1.50	60.00
		$47371.56

CREDITS.

Apr. 29, 1924, Allowance on check desk..	940.00	
May 16, 1924, Difference of freight ..	190.24	$ 1130.24
		$46241.32
May 26, 1924, Check	$ 5955.62	
Jun. 23, " "	7895.38	
Aug. 22, " "	12600.00	
Oct. 27, " "	8500.00	$34951.00
Balance due ..		$11290.32"

Defendant's Exhibit "A–15," mentioned in said stipulation, is as follows:

DEFENDANT'S EXHIBIT "A–15."

"Jan. 17, 1925.

Mission Marble Company,
    San Francisco, Calif.

We are extremely desirous of knowing how you can excuse yourselves for the treatment you have shown the Dexter Horton National Bank in attempting to fill the order of marble for our banking room. The original order was placed with you over a year ago and we were kept out of the occupancy of our new banking room for a long period due entirely to our inability to get marble to finish the interior of the room. We took possession of our new room on December Twentieth and have had to get along the best we could with the center of our lobby all torn up waiting for the balance of the marble to arrive. It is nearly a month since we moved in and still you have not shipped the balance of order required to complete job. Will you not in fairness and decency see that the order is completed and shipped immediately. [145]

    "Please answer.

    DEXTER HORTON NATIONAL BANK.

    Paid

    Day Letter

Defendant's Exhibit 'A–15.' "

Thereafter the cause was submitted to the Court on briefs and oral argument; and thereafter and on

or about the 4th day of October, 1926, the Court
made and entered its memorandum decision in the
above-entitled cause, and which said memorandum
decision after title of court and cause, is as follows:

"Plaintiff, a cub-contractor, sues the defendant,
the contractor, for a balance alleged to be due for
marble and labor furnished and done under a contract for the marble to finish the interior of the
Dexter Horton banking rooms of the Dexter Horton Building in Seattle, and for certain extras.
The defendant admits that the marble was furnished under the contract, but denies the indebted-
ness, setting up certain items of damage claimed
for alleged failures in performance on plaintiff's
part.

The contract between plaintiff and defendant
provides:

"The Mission Marble Works (hereinafter called
the Company) proposes to furnish as required by
the specifications and drawings prepared by John
Graham, Architect (acting for the purpose of this
agreement as the Agents of the Purchaser) except
as hereinafter modified, for the Dexter Horton National Bank building at Seattle, Wash., the following marble, viz.:

Per sheets Nos. 2, 2A, 5, 6, 8 & 10.   According to
the plans, specifications, terms, and conditions of
same, and as covered in purchasers contract with
said Bank in order to enable the purchaser to faithfully and promptly fulfill their contract with the
above bank,   *   *   *   [146]

In defendant's agreement with the bank it is provided:

(1) " * * * That the contractor agrees to furnish and install in the Bank quarters of the Owner in the Dexter-Horton Building at the corner of Second Avenue and Cherry Street in the City of Seattle, all marble work (except toilet rooms and in the elevator lobby) as shown on drawings #2, 2–a, 5, 6, 8 and 10, prepared by John Graham, Architect, and to the entire satisfaction of said John Graham, Architect. Said drawings and the proposal of the Contractor dated January 21, 1924, are hereby made a part of this contract.

(2) The marble is to be of the several kinds and colors as follows: The floor field to be Napoleon Gray marble, the base, border and deal plates to be Belgian Black marble; all other marble to be Golden Viened St. Genevieve.

(3) The quality and finish of all marble installed shall be equal to that represented by samples on file in the office of the Architect and identified by the signature of the parties hereto. Skilled mechanics shall be employed and the work throughout shall be first class in accordance with the best practice of the marble working and setting industry.

(4)   *   *   *   *   *   *   *   *   *

(5) It is agreed that the contractor shall take measurements at the building prior to the cutting of marble and shall be wholly responsible for the accuracy of all dimensions and shall make good at his expense all errors due to inaccuracy of said

measurements of conflict between the drawings and actual conditions affecting dimensions. The Owner will provide concrete slab under marble *fll*ors to within 2″ of finished floor level.

(6) It is agreed that the contractor shall submit setting plans showing all jointings and anchorage for marble and that the same shall be amended as may be directed by the Architect and finally approved by the Architect before marble is cut.

(7) It is agreed that the Owner shall have the right to add to, or deduct from the quantity of marble work included in this contract at any time before the work is started, and that additions to or deductions from the contract sum hereinafter named, by reason of said changes shall be appraised and determined by John Graham, Architect.

(8) It is agreed that the architect shall be the sole judge of the quality and quantity of the work performed under this contract and that all operations hereunder shall be under his direction and control.

(9) It is agreed that the work under this contract shall be commenced immediately after the date hereof and that the first material suitable for immediate installation shall be delivered at the premises not later than three months after setting plans have been approved by the architect and that the remainder of the marble shall be [147] delivered at such subsequent date or dates as shall not interfere with continuous progress with the installation of the marble work, and it is further

agreed that the entire contract shall be completed within three months after the first delivery of material as above provided; It is agreed, however, that the Contractor shall not be held liable for delay or failure to perform by acts of the Owner or of any other contractor or subcontractor upon the building, or by fire, storms, accidents, strikes, boycotts, lock-outs, riots, lack of transportation facilities, delays in transportation, government regulations, acts of God or by any other causes beyond the Contractors reasonable control.

(10)   It is agreed that the Contractor shall have possession of the space where marble is to be installed for necessary storage of material after the time for commencement of the work as herein provided and that the Contractor shall furnish all necessary hoisting facilities, power, gangways, scaffolding, heat, water, light and labor incident to the performance of this contract.

(11)   *   *   *   *   *   *   *   *   *

(12)   In consideration of the faithful performance of the covenants herein contained, the Owner agrees to pay to the Contractor the sum of SIXTY THOUSAND ONE HUNDRED & ONE DOLLARS ($60,101.00) in installments on or about the 5th and 20 of each *of* month, equal to 90% of the value of material delivered to and labor performed at the premises, as ascertained and computed by the architect and upon his certificate, and the remainder upon certificate of the architect that the work is satisfactorily completed and upon

a satisfactory showing that all bills and claims of whatsoever nature are fully paid.

(13)   *   *   *   *   *   *   *   *   *

The bank has withheld Six Thousand Dollars from the defendant, under this contract, on account of alleged defects in marble, and delay. The defendant contends that in both respects the bank is in the right and that the plaintiff is to blame for both of these failures. This contention forms the chief ground of dispute.

Under written stipulation, this case has been tried to the Court without a jury. By Sec. 9 of the contract last above quoted it was agreed the entire contract should be completed within three months after the first delivery of material. This provision was, by reference, made a part of plaintiff's contract with the defendant. The first car of marble was shipped by plaintiff April 26th, [148] 1924. The bank moved into its rooms December 20th the same year, before the marble had been all installed, a part of which plaintiff did not furnish until after the latter date. The architect, Mr. Graham and his superintendent, Mr. Grant, testified that this delay had damaged the bank to the extent of $4,500. Aside from the rent that the bank was paying while kept out if its rooms, probably the exact amount of this damage it would be impossible to establish, a considerable item of which would be the injury to its business and prestige caused by carrying on the former in unfinished quarters in the construction of which workmen were still engaged.

The main contention has been that the plaintiff did not cause the delay, but that the bank with the acquiescence of defendant, after plaintiff began the delivery of the marble, requested plaintiff to ship no more marble until requested. The evidence does not make this matter entirely clear. There evidently was a conversation between Mr. Eastman, president and manager of plaintiff, and Mr. Robinson, president of defendant, and some officer of the bank, in which the dilemma in which the bank and defendant found themselves because of the lack of a place in the building to store the marble in accordance with par. 10 of the above contract, was discussed. This situation, in part, was no doubt caused by the condition of the concrete floor in the basement, which the "Owner" was to provide, in compliance with Par. 5 of such agreement. The plaintiff has not sustained the burden of showing any modification of the contract as to the time allowed for completion, as now claimed by it. Plaintiff's contract was to deliver marble f. o. b. cars at its work in San Francisco, so it was not concerned with the storage of marble in Seattle, and it must have been obvious at all times that the expense of storage elsewhere than in the building, would be a small fraction of that caused by any substantial delay in the performance of the contract. Under the [149] circumstances, and in the light of the conduct of the parties, it is not reasonable to conclude that such an agreement was ever made.

Plaintiff also contends that it was delayed because of defendant's failure to furnish setting plans and shop drawings. While as between the defendant and the bank it may have been defendant's duty under Par. 6 and 9 to furnish them, this contract did not contemplate a subcontract between the plaintiff and defendant. The obligation growing out of these provisions was owing from plaintiff to defendant, and was in no way lessened by the provision of its contract with the defendant providing that the latter should furnish.

> " * * * such further details or explanations as may be necessary to delineate the plans and specifications * * * "

and that the defendant should

> " * * * make or cause to be made at the building and furnish to the Company (plaintiff) all necessary detail measurements."

While the plaintiff may not have been responsible for the entire delay, the Court finds that it caused damage on account of default in delivery in the amount fixed by the architect, to wit, $4,500.00.

The architect and his superintendent testified that the damage from defects in the marble would amount, in their opinion, to $2,500.00. A number of check desk tops, each formed of single slabs three inches thick and three feet wide, of Golden Grain St. Genevieve marble, came apart and could not be restored satisfactorily. In plaintiff's contract there was the following provision:

> "It is a well known fact that fancy or col_ored marble are more or less unsound, and it

is hereby agreed and understood that should any of the fancy or colored marbles to be furnished under this proposal require waxing, filling, cementing or backing in order to present a smooth surface, that we will have the right to do such waxing, filling, cementing or repairing as may be necessary, and this marble, when so treated is to be accepted." [150]

It is plaintiff's contention that the marble separated along the seams, and having done all that skill could accomplish in the way of restoration, that because of the foregoing provision it is not liable. Defendant contends that the marble did not separate along the seams, but that the slabs fell and were broken across in plaintiff's plant. The testimony is not clear, but the weight of it is in defendant's favor upon this point.

The defendant also complains of holes in the marble,—somewhat as knot-holes would occur in lumber; that part fell out of the slabs of marble and it could not be satisfactorily restored. The evidence on this head is not sufficient to take any imperfections of this nature in installed marble out of the above quoted provisions, and the Court finds no liability on plaintiff's part on account thereof.

Defendant complains of a lack of uniformity in the coloring of the Golden Grain St. Genevieve marble. There was opinion testimony upon the defendant's part that this was probably caused in the manner in which the marble was sawed along the grain, instead of across; that when sawed along the grain yellow splotches would show up that would not when sawed across. There was testimony upon plaintiff's part that such marble has no veins, as a rule. That it comes in splotches; that the marble was originally clay and lime; that the clay shows yellow and the lime shows gray, and the veins are of silica; that it has no definite veins; that a block of this marble can be cut in many ways and have the same appearance; that this character of marble is of mass formation and not a vein formation; that vein marble is formed by water dropping; that the character of marble here in question is thrown up in mass. [151]

The effect of the pieces of marble splotched with yellow when brought in proximity with the grayer pieces of marble produces a clashing effect,—not as pleasing to the eye as it would be if more uniform in color. It is not necessary to determine the cause that produced this effect.

The contract between defendant and the bank, Par. 3, provided that the "quality and finish" of the marble should be equal to those represented by samples on file in the office of the architect. The Court concludes, in view of the use to which this marble was to be put,—the interior finish of such banking-rooms, that "quality and finish" in this character of marble includes coloring. The sample filed with the architect was not produced in evidence, but a sample of the marble was offered by plaintiff. This sample shows no indication of any such want of uniformity in coloring as appears in the installed marble. The Court finds the defect in the marble, in this respect, damaged the bank, and defendant, in the amount to which the architects testified, to wit: $2,500.00.

The plaintiff has contended that the marble having been installed that it has been accepted, and the defendant cannot complain of these defects. Acceptance, in the sense claimed, has not been established.

Plaintiff claims extra to the amount of $1,371.76. Among these is a charge of $544.78 of which $484.78 is for the expense of boxing marble, and for cartage $60,00. This expense was incurred by plaintiff in shipping, at defendant's request, marble by boat instead of by carload lots. The shipments by water avoided further delay. These items are disallowed. Of the other extras claimed by plaintiff, the following are disallowed as not established by a preponderance of the evidence: $81.94; $7.85; $18.09; $63.20. The remaining extras claimed by plaintiff are allowed.

While the prayer of the answer is for the dismissal of [152] plaintiff's action, and for costs and disbursements to be taxed, yet there are other of its counterclaims to be considered.

Defendant's claim for $544.78 is disallowed. This item is for the boxing and cartage already disallowed plaintiff, and was evidently inadvertently asserted.

The items of $250.00 and $300.00 claimed for the expense of sending Mr. Robinson and Mr. Grant from Seattle to plaintiff's plant in San Francisco in an effort to expedite the work of getting out the marble, are disallowed. Plaintiff's delay doubtless occasioned these visits, but they cannot be considered, in law, as a reasonable result of plaintiff's default.

The item of $600.00 claimed by defendant, paid plaintiff's workmen for working overtime, is disallowed. This was done through an agreement by defendant with plaintiff, and plaintiff in consenting evidently did so upon the assumption that it was a concession by defendant. The Court is of the opinion that it was justified in such assumption.

Defendant also counterclaims for expense of labor required of it at the building, by the architect, in order to make the marble conform to the plans and specifications. The amount claimed on this account is $893.88. All of this amount in excess of $784.24 is profit, claimed by the defendant on the expenditure of the latter amount. The question being one of damages, it is entitled to no part of such profit. The evidence upon the items entering into this claim is meager and unsatisfactory. The Court finds that the defendant is entitled to recover one-half of the amount paid for sixty-four hours labor required in cutting the triangles to fit the plans; and finds that the holes drilled in the marble balustrade were neither the "lewis" holes nor "anchor" holes of the contract, and that plaintiff should have drilled them, and that the defendant is entitled to recover in full for the one hundred four hours labor required for this work. [153]

Defendant is also allowed $25.00 for the wooden hand balustrade on the open stair to the basement, put in when the bank opened because the marble balustrade was not in place. All other items making up this counterclaim are not supported by sufficient evidence to entitle the defendant to recover.

Findings and judgment in accordance with the foregoing will be presented by either party, upon notice to the opposing party. [154]

Thereafter and on the first day of November, 1926, the above-entitled cause came regularly on before the above-entitled court for the settlement and adoption of findings of fact and conclusions of law. This plaintiff tendered to the court its findings of fact and conclusions of law, which were each and all rejected by the court, which said findings of fact and conclusions of law, tendered by the plaintiff and the order of the Court rejecting the same and allowing the plaintiff an exception thereto, after title of court and cause, are as follows, to wit:

"I.

The Court finds that the plaintiff at all times mentioned in its complaint was and now is a corporation duly organized and existing under and by virtue of the laws of the state of California.

II.

That the defendant at all times mentioned in plaintiff's complaint was and now is a corporation duly organized and existing under and by virtue of the laws of the State of Washington. The court further finds that the defendant at the time of entering into the contract alleged in plaintiff's complaint, was doing business as such corporation under the firm name and style of Robinson Tile & Supply Company.

III.

That the above-entitled action is wholly between citizens of different states. That the plaintiff is a citizen of the state of California and the defendant is a citizen of the state of Washington. That the amount involved in the above-entitled action is the sum of Seven Thousand, Two Hundred Ninety and 32/100 ($7,290.32) Dollars, exclusive of costs and interest. [155]

IV.

That on or about the 30th day of January, 1924, the plaintiff and defendant entered into a contract in writing wherein the plaintiff agreed to furnish and sell to defendant, and defendant agreed to purchase from the plaintiff, building and finishing marble which was to be used by the defendant in the construction and finishing of the banking quarters of that certain building in the city of Seattle known as the Dexter Horton National Bank Building at the agreed price and value of $46,000.00. That the plaintiff has fully kept and performed all of the covenants and conditions of said contract on its part to be kept and performed, and has furnished all of the marble called for in said contract, and the said marble has been installed in the banking quarters of said building, and is now in the same and being used by the Dexter Horton National Bank.

V.

The Court further finds that under and by virtue of said contract and the terms and provisions thereof plaintiff at the special instance and request of the defendant also furnished to the defendant for use by defendant in the construction and finishing of said banking quarters of the said Dexter Horton National Bank Building, other building marble, in addition to that required to be furnished by the contract above mentioned for the contract price of $46,000.00, of the fair and reasonable price and value of the sum of $826.78.

VI.

The Court further finds that under the contract entered into between plaintiff and defendant the marble was to be [156] shipped by plaintiff from the city of San Francisco to the city of Seattle, Washington, in carload lots f. o. b., San Francisco, California. That at the special instance and request of the plaintiff forty tons of the marble shipped by the plaintiff to the defendant under said contract was shipped by water in pieces as manufactured by the plaintiff. That under this arrangement it was necessary for the plaintiff to crate said marble. That the reasonable cost of crating said marble was and is the sum of $484.78.

VII.

The Court further finds that the change in shipping, as above mentioned, necessitated the carting of said marble from plaintiff's plant to a dock in San Francisco. That the reasonable cost and charge for carting said marble is the sum of $60.00. That the plaintiff paid out the said sum of $484.78 above mentioned, and the sum of $60.00 for the crating and cartage of said marble, and the same is a reasonable and just charge and should be paid by the defendant.

VIII.

The Court further finds that the total sum which became due and owing to plaintiff for marble furnished the defendant for installation and completion of the banking quarters of the said Dexter Horton National Bank Building was the sum of $47,371.56. That of said sum defendant has paid the plaintiff the sum of $40,081.24. That there is now a balance due and owing the plaintiff from the defendant, of the sum of $7,290.32. That said sum became due on the first day of April, 1925, and plaintiff is entitled to interest thereon at the rate of six per cent per annum from and after the said first day of April, 1925, to date hereof. [157]

IX.

The Court further finds that all of the allegations of plaintiff's complaint were and are true.

X.

The Court further finds that the banking quarters of said Dexter Horton National Bank Building were not completed and ready for the installation of marble until the 22d day of September, 1924, and that the said banking quarters were not turned over to the defendant for installation or marble until said date. That under the contract existing between defendant and Dexter Horton National Bank for the installation of marble therein, the defendant was to have three months in which to install the marble after the said banking quarters were turned over to the defendant for the installation of the same. That the Dexter Horton National Bank moved into said banking quarters on the 20th day of December, 1924, and before the time had expired for the installation of said marble. That there was no delay on the part of the plaintiff in furnishing the marble. That neither the defendant nor the said bank suffered any damages by reason of any claim of delay on the part of the plaintiff in furnishing marble for the banking quarters of the said building.

XI.

The Court further finds that all of the marble furnished to defendant by the plaintiff for installation in said bank building was inspected by the architect acting for the bank before the same was installed as required by the contract between the defendant and the bank and the plaintiff and defendant and was accepted by said architect and permitted to be installed. [158]

XII.

The Court further finds that all of the marble furnished by the plaintiff to the defendant for installation in said banking quarters conformed to the requirements of the plaintiff's contract with the defendant in all respects.

XIII.

The Court further finds that all claims of the defendant and the bank for defective marble were waived by the defendant and the bank when the architect for the bank permitted the same to be

installed without any objection or protest or claim
to the plaintiff that the same was defective.

XIV.

The Court further finds that there was no objection made to the plaintiff by the defendant or the
bank, or the architect acting for the bank that the
marble was defective in color or in any other respect prior to the completion of the installation of
the marble in the banking quarters of the said
bank building.

XV.

The Court further finds that there was no evidence introduced in this case and no competent
proof that the defendant or the bank sustained any
damages by reason of any claim of defect in the
color or finishing of the marble furnished to defendant by the plaintiff under the contract between
the plaintiff and the defendant.

XVI.

The Court further finds that defendant has
wholly and entirely failed to prove by competent
evidence, or otherwise, each and all of the affirmative defenses set out in the defendant's answer
herein.   [159]

XVII.

The Court further finds that the plaintiff at all
times prior to the bringing of the present action
furnished the defendant all of the marble requested
by the defendant to take the place of any marble
claimed to be defective by the defendant.

XVIII.

The Court further finds that the plaintiff has

never failed or refused to furnish any marble to take the place of any marble claimed to be defective by the defendant, or by the bank, or by the architect acting for the bank.

Done in open court this —— day of October, 1926.

—————————————————,
District Judge.

CONCLUSIONS OF LAW.

The Court finds as conclusions of law:

I.

That the plaintiff is entitled to a judgment in its favor against the defendant in the sum of $7,290.32 with interest thereon at the rate of 6% per annum from April 1st, 1925, to date hereof.

Done in open court this —— day of October, 1926.

—————————————————,
District Judge.

The above and foregoing findings of fact and conclusions of law tendered by the plaintiff are hereby rejected and [160] denied as to each and all of the same, and the plaintiff is allowed an exception to the ruling of the Court in denying each and all of the findings of fact and conclusions of law above set out.

Done in open court this 1st day of November, 1926.

EDWARD E. CUSHMAN,
District Judge." [161]

The defendant presented its findings of fact and conclusions of law, which were over the objections and exceptions of the plaintiff, adopted by the

Court, which said findings of fact and conclusions
of law tendered by the defendant, and the exceptions
reserved by and granted to the plaintiff, are after
title of court and cause, as follows, to wit:

"This cause came on regularly for trial on June
17, 1926, before Honorable Edward E. Cushman,
Judge in the above-entitled court, without a jury,
a jury having been waived by the parties in writing; the plaintiff appearing at the trial by its
President, George W. Eastman, and its attorneys,
Grosscup & Morrow, Charles A. Wallace and J. O.
Davies; the defendant appearing at the trial by its
President, A. P. Robinson and its attorney, H. A. P.
Myers. Thereupon testimony of witnesses and evidence of the parties was introduced. At the close
of the case the same was argued by respective counsel and taken under advisement by the Court.
After the trial, by agreement of both parties, and
in the presence of counsel for both parties, the
Court made an examination of the marble installed
in the Dexter Horton Bank, and thereafter, on the
4th day of October, 1926, the Court filed its memorandum opinion in this cause, and now, on this day,
upon application of defendant in open court, it is
ordered that the prayer of defendant's answer be
amended by adding thereto "and for judgment
against the plaintiff for all counterclaims and setoffs shown by the evidence." And now in accordance with said memorandum opinion, the Court
finds:

I.

That at all times mentioned in the complaint the

plaintiff was and still is a corporation organized and existing under the laws of the State of California, and during all said times the defendant was and is a corporation, organized, existing and doing business under the laws of the State of Washington.

II.

That this action is between the plaintiff, a citizen of the State of California, and the defendant, a citizen of the State of Washington, and the amount involved exceeds the sum of Five Thousand Dollars ($5,000.00), and the Court has jurisdiction of this cause.

III.

That on or about the 30th day of January, 1924, the plaintiff and the defendant entered into a contract in writing whereby the plaintiff agreed to furnish to the defendant certain building marble for use in the Dexter Horton National Bank Building, according to plans and specifications, terms and conditions of the defendant's contract with said bank, dated January 25, 1924, in such a manner as to enable the defendant to faithfully and promptly fulfill its contract with the bank above named, at the agreed consideration of $46,000.00.
[162]

IV.

The Court finds that in addition to furnishing the marble for said building, the plaintiff at the instance and request of the defendant, furnished certain extra marble and material, to the extent and of the value of .$657.92 and no more; and that all other items of extras claimed by the plaintiff have not been established by the evidence and are disallowed.

The original contract for said marble work amounted to the sum of $46,000.00 and, together with the extras herein allowed, amounts to the total sum of $46,657.92, upon which the defendant has been paid the sum of $40,081.24, leaving a balance unpaid upon said original contract and extras in the sum of $6,576.68.

V.

The Court further finds that the plaintiff did not furnish said marble according to said contract but that it greatly delayed the furnishing of said marble and the defendant in performing its said contract with the Dexter Horton National Bank, although defendant urged plaintiff for a speedier delivery of said marble, by numerous letters, telegrams and cablegrams, and personal visits to San Francisco; that a portion of said marble was defective and not according to contract, especially in color, workmanship and material; and the Court also finds that the defendant was required to do certain extra work and labor, hereinafter referred to, in connection with said contract, on account of a portion of said marble not being fabricated according to the plans and specifications.

VI.

The Court further finds that the defendant is entitled to and is allowed the following counterclaims and offsets, as shown by the evidence:

(a) Damages caused by delay on plaintiff's part in furnishing said marble
according to contract............$4,500.00
(b) For defects in said marble........ 2,500.00
(c) For extra work and labor, consisting
of 32 hours of labor in cutting triangles to fit the plans, and 104
hours of labor in drilling holes in
balustrade,—at $1.25 per hour.... 170.00
(d) For wooden hand balustrade on open
stairway to the basement........ 25.00

TOTAL................$7,195.00

and all other items in defendant's counterclaims
and setoffs have not been proven by the witness.
[163]

VIII.

The Court further finds that the defendant has
fully kept and performed all of the covenants and
agreements of its contract with the plaintiff.

As conclusions of law, from the foregoing facts,
the Court concludes :

I.

That the counterclaims and offsets to which the
defendant is entitled exceed the plaintiff's claim
in the sum of Six Hundred Eighteen and 32/100
Dollars ($618.32), and that the defendant is there-
fore entitled to a judgment against the plaintiff in
that sum.

Done in open court this 1st day of November, 1926.

EDWARD E. CUSHMAN, (Sgd.)
Judge.

Plaintiff excepts to Findings numbered 4, 5, 6 and 7, and Conclusions of Law numbered 1, and its exceptions are allowed this 1st day of November, 1926.

EDWARD E. CUSHMAN, (Sgd.)
Judge.''

Thereafter and on said first day of November, 1926, the Court, over the objections and exceptions of the plaintiff, made and entered its judgment herein, in accordance with the findings of fact and conclusions of law, adopted by the Court, which said judgment is found in the judgment-roll.

Thereafter, and on said first day of November, 1926, the Court made and entered its order herein, on good cause shown, granting and allowing the plaintiff up to and including the first day of December, 1926, in which to prepare, serve and file its bill of exceptions herein; and now within the time allowed by law [164] and the order of the Court, the plaintiff presents this, its bill of exceptions, and asks that the same be settled, allowed and filed herein as full, true and correct.

GROSSCUP & MORROW,
Attorney for the Plaintiff.
CHAS. A. WALLACE,
J. O. DAVIES,
Attorneys for the Plaintiff.

Service of the above and foregoing bill of exceptions, acknowledged and a copy received this 29th day of November, 1926. The same is true, full and correct and may be signed and allowed by the Court as a true, full and correct bill of exceptions.

H. A. P. MYERS,

Attorney for the Defendant.   [165]

The parties agreeing to the foregoing the same is settled as the bill of exceptions in the above-entitled cause. Done in open court at Tacoma this 30th day of November, 1926.

EDWARD E. CUSHMAN,

Judge of the United States District Court for the Western District of Washington, Who Presided at the Trial of the Above-entitled Cause.

[Endorsed]: Filed Nov. 30, 1926.   [166]

———

[Title of Court and Cause.]

PRAECIPE FOR TRANSCRIPT OF RECORD.

To the Clerk of the District Court of the United States for the Western District of Washington, Northern Division:

Please prepare a record for the purpose of a writ of error to the United States Circuit Court of Appeals for the Ninth Circuit, and include the following:

(1) Complaint of plaintiff.

(2) Answer of defendant.

(3) Reply of plaintiff.

(4) Judgment.

(5) Order of Court extending time in which to prepare and serve bill of exceptions.

(6) Plaintiff's bill of exceptions.

(7) Order of Court settling bill of exceptions.

(8) Plaintiff's Exhibit No. 3 for identification, same being a sample of marble.

(9) Petition for writ of error.

(10) Assignment of errors.

(11) Bond on writ of error.

(12) Order allowing writ of error.

(13) Original writ of error.

(14) Original citation.

(15) Praecipe.

(16) Stipulation as to bond on writ of error.

Title of Court and Cause on all papers except complaint and all verifications, may be omitted.

GROSSCUP & MORROW,

CHAS. A. WALLACE,

J. O. DAVIES,

Attorneys for Plaintiff in Error.

[Endorsed]: Filed Dec. 9, 1926. [167]

CERTIFICATE OF CLERK U. S. DISTRICT COURT TO TRANSCRIPT OF RECORD.

United States of America,

Western District of Washington,—ss.

I, Ed. M. Lakin, Clerk of the United States District Court for the Western District of Washington, do hereby certify this typewritten transcript of record, consisting of pages numbered from 1 to 167, inclusive, to be a full, true, correct and complete copy of so much of the record, papers and other proceedings in the foregoing entitled cause as is required by praecipe of counsel filed and shown herein, as the same remain of record and on file in the office of the Clerk of said District Court, and that the same constitute the record on return to writ of error herein, from the judgment of said United States District Court for the Western District of Washington to the United States Circuit Court of Appeals for the Ninth Circuit.

I further certify the following to be a full, true, and correct statement of all expenses, costs, fees and charges incurred and paid in my office by or on behalf of the plaintiff in error for making record, certificate or return to the United States Circuit Court of Appeals for the Ninth Circuit in the above-entitled cause, to wit: [167a]

Clerk's fees (Act of February 11, 1925) for
making record, certificate or return, 509
folios, at 15¢ $76.35
Certificate of Clerk to transcript of record,
with seal .... .50
Certificate of Clerk to original exhibits, with
seal .... .50
                                        ———
                    Total............ $77.35

I hereby certify that the above cost for preparing and certifying record, amounting to $77.35 has been paid to me by attorneys for plaintiff in error.

I further certify that I hereto attach and herewith transmit the original writ of error and citation on writ of error issued in this cause.

IN WITNESS WHEREOF I have hereunto set my hand and affixed the seal of said District Court, at Seattle, in said District, this 28th day of December, 1926.

[Seal] ED. M. LAKIN,

Clerk United States District Court, Western District of Washington.

By S. E. Leitch,
Deputy Clerk. [167b]

[Title of Court and Cause.]

WRIT OF ERROR.

United States of America,

Ninth Judicial Circuit,—ss.

The President of the United States, CALVIN COOLIDGE, to the Honorable Judge of the District Court for the Western District of Washington, Northern Division, GREETINGS:

Because in the record and proceedings, as also in the rendition of a judgment of a plea which was in the said District Court before you, between The Mission Marble Works, a corporation, plaintiff, plaintiff in error herein, and the Robinson Tile & Marble Company, a corporation, defendant, defendant in error herein, a manifest error happened to the damage of The Mission Marble Works, a corporation, plaintiff in error, as by said complaint appears, and we being willing that error, if any hath been, should be corrected and full and speedy justice be done to the parties as aforesaid in this behalf, do command you, if judgment be therein given, that then under your seal you send the record and proceedings aforesaid and all things concerning the same to the United States Circuit Court of Appeals for the Ninth Circuit, together with this writ, so that you have the same at the city of San Francisco, in [168] the State of California, where said Court is sitting within thirty (30) days from the date hereof, in the said Circuit Court of Appeals to be then and there held, that the record and proceedings aforesaid being inspected, said Circuit Court of Appeals may cause further to be done therein to correct the error, what of right and according to the laws and customs of the United States should be done.

WITNESS, the Honorable WILLIAM HOWARD TAFT, Chief Justice of the United States, this 9th day of December, A. D. 1926.

[Seal]   Attest: ED M. LAKIN,

Clerk of the United States District Court for the
    Western District of Washington, Northern
    Division.

By ————————,

Deputy Clerk of the United States District Court,
    Western District of Washington, Northern
    Division.   [168a]

[Endorsed]: Filed Dec. 9, 1926.   [169]

CITATION ON WRIT OF ERROR.

United States of America,

Ninth Judicial Circuit,—ss.

To Robinson Tile & Marble Company, a Corporation, and to H. A. P. Myers, Esq., Its Attorney,
GREETINGS:

You are hereby cited and admonished to be and appear at a session of the United States Circuit Court of Appeals for the Ninth Circuit to be holden at the city of San Francisco, State of California, within thirty (30) days from the date hereof, pursuant to a writ of error filed in the Clerk's office of the United States District Court for the Western District of Washington, Northern Division, wherein The Mission Marble Works, a corporation, is plaintiff in error, and Robinson Tile & Marble Company, a corporation, is defendant in error, to show cause, if any there be, why the judgment rendered against the plaintiff in error, as in said [170] writ of error mentioned, should not be corrected and why speedy justice should not be done to the parties in that behalf.

WITNESS the Honorable EDWARD E. CUSHMAN, Judge of the United States District Court for the Western District of Washington, Northern Division, this 9th day of December, A. D. 1926.

[Seal]   EDWARD E. CUSHMAN,

United States District Judge.

Service of the above and foregoing citation ac-

knowledged and copy thereof received this 10th
day of December, A. D. 1926.

H. A. P. MYERS,

Attorneys for Defendant in Error.   [171]

[Endorsed]: Filed Dec. 10, 1926.   [172]

[Endorsed]: No. 5037. United States Circuit
Court of Appeals for the Ninth Circuit. The Mis-
sion Marble Works, a Corporation, Plaintiff in
Error, vs. Robinson Tile & Marble Company, a Cor-
poration, Defendant in Error. Transcript of
Record. Upon Writ of Error to the United States
District Court of the Western District of Wash-
ington, Northern Division.

Filed January 3, 1927.

F. D. MONCKTON,

Clerk of the United States Circuit Court of Appeals for the Ninth Circuit.

By Paul P. O'Brien,

Deputy Clerk.

No. 5037

United States
Circuit Court of Appeals
For The Ninth Circuit

THE MISSION MARBLE WORKS, a Corporation,
Plaintiff in Error,

vs.

ROBINSON TILE & MARBLE COMPANY, a Corporation,
Defendant in Error.

ON WRIT OF ERROR TO THE DISTRICT COURT OF THE
UNITED STATES FOR THE WESTERN DISTRICT
OF WASHINGTON

BRIEF OF PLAINTIFF IN ERROR

Appearances for Plaintiff in Error:

GEORGE CLARK SARGENT,
Hobart Building,
San Francisco, California

GROSSCUP & MORROW,

CHAS. A. WALLACE,

J. O. DAVIES.

2600 L. C. Smith Building, Seattle, Washington.

THE ARGUS PRESS, SEATTLE

No. 5037

United States
Circuit Court of Appeals
For The Ninth Circuit

THE MISSION MARBLE WORKS, a Corporation,
Plaintiff in Error,

vs.

ROBINSON TILE & MARBLE COMPANY, a Corporation,

Defendant in Error.

ON WRIT OF ERROR TO THE DISTRICT COURT OF THE
UNITED STATES FOR THE WESTERN DISTRICT
OF WASHINGTON

BRIEF OF PLAINTIFF IN ERROR

Appearances for Plaintiff in Error:
GEORGE CLARK SARGENT,
Hobart Building,
San Francisco, California
GROSSCUP & MORROW,
CHAS. A. WALLACE,
J. O. DAVIES.
2600 L. C. Smith Building, Seattle, Washington.

THE ARGUS PRESS, SEATTLE

INDEX

CASES CITED

CASES CITED (Continued)

TEXT BOOKS CITED

United States
Circuit Court of Appeals
For The Ninth Circuit

THE MISSION MARBLE WORKS, a Corporation,

Plaintiff in Error,

vs.

No. 5037

ROBINSON TILE & MARBLE COMPANY, a Corporation,

Defendant in Error.

ON WRIT OF ERROR TO THE DISTRICT COURT OF THE UNITED STATES FOR THE WESTERN DISTRICT OF WASHINGTON

BRIEF OF PLAINTIFF IN ERROR

STATEMENT OF THE CASE

The plaintiff in error herein called plaintiff is seeking by these proceedings to review a judgment of the trial court in favor of the defendant in error herein called defendant in the sum of $618.34, which judgment is found at 11 R.

The plaintiff in its complaint, which is found beginning at 2 R. sought to recover from the defendant the sum of $7,290.32 upon a building contract, which contract is found at 30 R. 4. By this contract, known as a sub-contract, the plaintiff was to furnish certain marble to be installed in the Dexter Horton National

Bank in the City of Seattle, Washington, for the sum
of $46,000.00. The defendant had the original con-
tract with the bank for the installation of this marble.
The original contract is found at 56 R. 14, which contract called for the sum of $60,101.00 for the furn-
ishing and installation of the marble in the banking
quarters of the said Dexter Horton National Bank.
Under the two contracts the plaintiff was to furnish
the marble for the banking quarters for the sum of
$46,000.00 and the defendant was to install it for
the sum of about $14,000.00.

The complaint of the plaintiff, after alleging facts
giving jurisdiction to the court below on the ground
that it was a citizen of the State of California and
that the defendant was a citizen of the State of
Washington, alleges that the plaintiff furnish all the
marble required under the contract and in addition
thereto furnished at the instance and request of the
defendant additional marble and performed other
work, labor and service of the reasonable value of
$1,371.56, plaintiff claiming a total amount due under
the contract and for extras in the sum of $7,290.32.

The defendant by its answer, which is found at
4 R. 10, in substance admits that the plaintiff furnished the marble as alleged in its complaint and
seeks to defeat the plaintiff's right of action and
right to recover by setting up certain counter claims
which are found in Paragraph 4 of defendant's answer
at 6 R. 17, and particularly by Subdivision "F"
thereof found at 8 R. 14.

In this paragraph the defendant seeks by its
counter claim to recover against the plaintiff the sum

of $6,000.00, which the architect for the bank withheld from the defendant on the claim that some of the marble furnished was defective and on the claim that the marble was not furnished within the time provided by the contract between the defendant and the bank.

This alleged item of damages in the sum of $6,000.00 for defective marble and damages for claimed delay, formed the chief bone of contention between the parties in the Trial Court and is the chief ground of contention in this error proceeding.

Notwithstanding the fact that the bank withheld from the defendant on its contract with the bank the sum of only $6,000.00, and the defendant in its answer only demanded damages on these two items in the sum of $6,000.00, yet the trial court allowed the defendant damages on these two items in the sum of $7,000.00 (200 R. 1).

The plaintiff, by its reply (9 R. 15) put in issue all the questions raised by the defendant in its answer and by its counter claim contained therein. It is the evidence that was introduced by the parties on this $6,000.00 counter claim of the defendant, that is, damages claimed for delay in furnishing marble and damages claimed for defective marble, that we desire particularly to call to the court's attention.

The cause came regularly on for trial before the trial court on June 16, 1926 (28 R. 15). A jury was waived by written stipulation of the parties and the case was tried by the court without a jury (29 R. 5).

George B. Eastman was called as a witness on

behalf of the plaintiff. His testimony is found beginning 29 R. 10. He testified that he was president of the plaintiff, The Mission Marble Works, and that its place of business was 209 Mississippi Street, San Francisco, and that the plaintiff entered into a contract with the defendant for furnishing marble for the Dexter Horton National Bank of Seattle. The contract was introduced in evidence as plaintiff's Exhibit No. 1. It is found at 30 R. 4. He further testified that all of the marble required by this contract had been furnished by the plaintiff to the defendant and had been actually installed in the banking quarters of the bank at Seattle and was in the bank and being used by the bank at the time of the trial. He further testified that at the request of the defendant the plaintiff furnished additional marble for the said banking quarters and did additional work and labor not covered by the contract, amounting in all to the sum of $1,371.76, which additional marble was also shipped and received by the defendant and installed in the bank and was being used in the bank at the time of the trial.

Mr. Eastman also testified that the marble required by the contract consisted of some eight or nine thousand pieces (40 R. 3). That he understood that under the contract between the defendant and the bank the defendant was given a period of three months in which to install the marble (42 R. 9).

Mr. Eastman also testified that the only claim of defective marble presented or made to the plaintiff was made to him by Mr. Grant after the bank building was completed and that consisted of a claim that one panel of marble was off on account of color and that one check desk had holes in it (43 R. 31).

He testified as to the dates of the shipments of marble, "The first car was shipped about April 26, 1924. The second car was shipped June 4, 1924. The third car was shipped July 28, 1924, and the fourth car was shipped October 11, 1924, and the L. C. L. shipments started there" (46 R. 6). Plaintiff then rested.

The defendant called as its first witness, Francis W. Grant, whose testimony is found beginning at 47 R. 25. He testified that he was an architect and had been such for 25 or 30 years. That he was associated with John Graham, who was architect for the Dexter Horton Building. He testified that there was some difficulty in the work of installing the marble by the failure of the defendant to furnish shop drawings. That complete shop drawings never were furnished. That "When this contract was made there was a sample of marble that was broken in two, one-half going to Mr. Robinson and we kept the other" (50 R. 20). That some of the marble was off in color. And concerning this marble that was off in color, he testified as follows:

"We told Mr. Robinson that we would not accept it. In view of the fact that the bank was very anxious to get in their premises, a compromise was made as to some of it and there was put in some with the understanding and promise of Mr. Robinson to take it out and put in other corresponding with the contract sample.

Some of that has been done and some has not."
(50 R. 27)

He also claimed that there were defects in the check
desks making the claim that they were broken. His
testimony in this respect is as follows:

"They were broken across. Each one was
broken, due to, as I was told, while I was in San
Francisco, because of an accident while on the
saws. We have not accepted them yet. They
were put together again with wax, reinforced on
the bottom with steel and had the waxing been
sufficient to deceive the ordinary eye we probably
would have accepted them." (51 R. 26)

No claim was made that these alleged defects were
called to plaintiff's attention until long after the installation of the marble was completed. He testified
that it would cost $2,500.00 to replace the defective
marble as follows:

"Yes, sir, Mr. Graham and I attempted to
compute that, and in our judgment it would cost
about $2,500.00 to replace and correct all of the
defects." (52 R. 30)

The certificate of the architect withholding from
the defendant the sum of $6,000.00 on account of de-
fective marble and damages sustained by the bank on
account of delay in completing the contract was introduced in evidence as defendant's Exhibit "A-1" and
is found 52 R. 15.

The witness was then asked to estimate the damages due to the bank on account of delay in finishing work. This question was objected to by the plaintiff upon the ground that the witness was incom-

petent to give such testimony. The defendant claimed it was not trying to prove damages suffered by the bank but was offering the testimony to show the good faith of the architect in holding out the sum of $6,000.00. The court overruled the plaintiff's objection and allowed an exception. The witness then estimated the sum of $4,500.00 as damages to the bank for delay (54 R. 15). The witness further testified that the Robinson Title & Marble Company gave a bond for the faithful performance of its contract. "We notified the bonding company that there was evidence of default. Q. Of what did this consist? A. Delay that is all." (60 R. 16)

The contract between the defendant and the bank was introduced as defendant's Exhibit "A-2", and is found at 56 R. 14. This contract provides for a period of three months' time in which to install the marble and is made a part of the contract between plaintiff and defendant by reference, the plaintiff thereby being given a period of three months to furnish the same, which contract also clearly provides that the bank quarters shall be provided by the bank, or the owner, ready for the installation of the marble.

The witness testified that the bank moved into its banking quarters December 20, 1924 (55 R. 26). The witness further testified that the work of installing the marble started on September 22, 1924 (48 R. 12). On cross examination the witness testified that the defendant began the installation of marble in the banking quarters of the bank on the first day that the banking quarters were in a condition to begin the work of installing the marble (67 R. 25).

Over the objection of the plaintiff the witness was allowed to testify as to the time allowed under the contract for the installation of the marble and gave it as his opinion that the time for installing the marble was six months. "Six months after the time they submitted and had their shop drawings approved" (62 R. 30).

The plaintiff contends that, under the said contract and the evidence, September 22, 1924, is the date on which the three months' period to install the marble began to run. The court held that the three months' period for the installation of the marble began to run on April 26, 1924 (183 R. 4) the date of the first shipment of marble.

Plaintiff produced a sample of marble which was marked "Plaintiff's Exhibit No. 3" for identification. The witness refused to identify this as a part of the sample kept by the architect as the criterion for the marble to be furnished under the contract (68 R. 19). The plaintiff demanded of the architect that he produce the sample of marble mentioned in the contract. He testified that he had it and could produce it (69 R. 3). This sample, however, was never produced by the witness or by the defendant. Witness further testified concerning the general character of the marble on pages 69 and 70 of the record. He further testified as follows:

"I stated yesterday that there was some pieces of marble that were off color in the bank.

Q. Those, I believe, are how many? Are there three or four of them that you say that you would not accept?

A. Oh, twenty or thirty.

Q. When you mention twenty or thirty, those are some you had some dispute about, but did you let them go in?

A. Let them go in under protest, with the understanding that they would be taken out and others put in their places.

Q. Twenty or thirty, would you say?

A. I never counted them, but one or the other, twenty or thirty rather than three or four, as you suggested." (70 R. 25)

and that

"Mr. Robinson started to install this marble just as soon as the building was ready for him." (75 R. 1)

The witness further testified beginning 76 R. 23 that as the architect for the bank he inspected each and all and every piece of marble before it was installed in the banking quarters, that not a single piece of marble went into the banking quarters that he rejected although he claimed he let some go in that were defective with the understanding with the defendant that it would later substitute other marble for that which he claimed was defective. There is no claim made by the defendant or the architect that this agreement between the architect and the defendant was ever called to the plaintiff's attention or that the plaintiff ever consented thereto. That the bank moved into the banking quarters on the 20th day of December, 1924 (81 R. 9). That the only portion of the banking quarters that was not completed at that time pertained to the stairs leading from the bank proper down to the basement and those stairs were in condition so that people could use them (81 R. 16).

The next witness called by the defendant was John C. Juengst, whose testimony is found beginning at 82 R. 1. He testified concerning certain work which he claims that he did upon the Dexter Horton National Bank as superintendent for the defendant. The defendant endeavored by this witnesss to show that it was required to do some work in setting the marble and fitting the same which should have been done at plaintiff's plant. He also testified concerning check desks and tried to make it appear that the check desks were broken crosswise of the seams rather than came apart at the seams. He admitted, however (95 R. 2), that in this particular kind of marble the seams run in every direction.

John Graham was the next witness called by the defendant. His testimony is found at 96 R. 1. He was the architect for the bank. He explained the shop drawings and said the shop drawings were the same as setting plans. He testified that the original intention was to have the building, in which the bank was to be installed and have its quarters, completed on Labor Day but that owing to delay the building was not ready at that time. The installation of marble in the bank building for the bank was also delayed and that he made various efforts to hurry up the installation of the marble. He claimed also that some of the marble was defective and not up to standard; that he made an estimate, together with Mr. Grant, as to what it would cost to replace the defective marble with marble that would come up to the standard, and he estimated this to be the sum of $2,500.00. A letter was introduced written by the witness or authorized by him to the United States Fidelity & Guaranty Company advising that company that the defendant was not showing due diligence in the performance of its contract. The exhibit is found at 104 R. 1. No mention is made in the exhibit of any default with respect to defective marble. After the witness had testified to the fact that in his judgment it would cost the sum of $2,500.00 to replace defective marble with marble that conformed to sample, the following proceedings were had:

"Q. What was the other item, besides the item of defective marble, and on account of adjustment of damages sustained by the bank on account of that delay?

MR. DAVIES—That is objected to for the reason he is not competent to estimate the damages to the bank.

THE COURT—I do not see how the Court can go into the bank's damage, or why it should go into it.

MR. MYERS—No more than to justify his refusal and in making payment.

THE COURT—I overrule the objection.

MR. DAVIES—Note an exception." (104 R. 30)

Witness then testified that they withheld from the defendant one month's rent. He did not state, however, how much one month's rent amounted to, nor how much the bank was paying per month for rent, if any thing at all.

On cross examination the witness testified as follows:

> "Mr. Grant was the one who had active charge and supervision of inspecting the marble and passing upon it. Mr. Grant was superintending all that work for me and had all the duties." (106 R. 24)

Albert P. Robinson was the next witness called on behalf of defendant. His testimony is found beginning at 109 R. 10. He testified that he was president of the defendant. He further testified concerning some work which he claimed was done in the setting of marble which he contended was caused by fault of the plaintiff. He further testified concerning delays in the shipment of the marble and that he spent certain sums of money in an endeavor to expedite shipment of the same. He further testified that certain items found in the plaintiff's final statement and invoiced to him were incorrect. Those that he claimed were incorrect are as follows: Item of $7.85, $81.94, $16.09 and $63.20, and the item of $544.78 for boxing and cartage.

In respect to the item of boxing and cartage he testified as follows: "I do not deny that I told them to ship the marble by boat, and we had a shipment coming on practically every boat." He claimed, however, by shipping by boat that there would be less delay in finishing the building and therefore he ought not to be charged with this item. All parties admitted, however, it was necessary to box and crate the marble when shipped by boat and it would not have been necessary if shipped by rail in carload lots as specified in the contract.

A mass of letters and telegrams were introduced in evidence sent from the defendant to the plaintiff complaining of delay in the shipment of marble and urging speedy shipment. No complaint whatever was made in these letters or telegrams concerning defective marble, except as to some few pieces and as to these few pieces all parties agreed that the plaintiff furnish others to take the place of those mentioned in the letters and telegrams.

The witness testified on cross examination that the last payment defendant made to the plaintiff on the contract was made on March 16, 1925, and that this payment was made by check and was in the sum of $4,000.00. This payment was made after the installation of the marble had been completed (146 R. 18).

A letter was then introduced in evidence written by the plaintiff to the defendant as plaintiff's "Exhibit No. 5." The letter is found at 147 R. 1. In this letter defendant admits and states it to be a fact that the defendant at the time of writing was indebted to the plaintiff in the sum of $6,576.64, leaving out and not taking into account any disputed item between them. This letter was written May 25, 1925, more than two months after the last payment was made by the defendant to the plaintiff on the contract and more than two months after its books were closed.

The witness Grant was then re-called for further cross examination by the plaintiff and another demand was made of him by the plaintiff to produce the sample of marble which was required to be kept by the architect under the contract between the defendant and the bank. This sample the witness failed to produce (152 R. 20).

The defendant rested, and the plaintiff in rebuttal called Roland E. Boucher, whose testimony is found beginning at 153 R. 15. He testified that he was familiar with the kind of marble used in finishing the bank in question. That it was not marble of vein formation but was of mass formation. That whatever veins or seams there was in this marble ran in every direction and that there was no regular or uniform direction in which the seams ran. He testified that he was at the plant of the plaintiff at the time the check desks were fabricated and that the check desk tops were not broken during fabrication, that they came apart at seams which ran through the block from which the check desk tops were fabricated and that he never told the witness Grant that they were broken during fabrication. That the check desk tops which came apart, three in number, were put together and finished in a good workmanlike manner. That he was superintendent on the job in the installation of the marble for the defendant and that no piece of marble was installed that was rejected by the architect. His testimony in this respect is as follows:

"Q. Did you at any time put in any pieces of marble after they had been rejected? If they did reject any, and tell you not to put it in?

A. No, sir, no, by no means." (157 R. 11)

That he was working for the defendant at the time he testified.

Beginning at 160 R. 25 the testimony which appears

from the record to have been given by the witness Boucher was not given by said witness. The testimony as found in record from said line 25 to 161 R. 23 was given by the witness Eastman, and it appearing to be given by Mr. Boucher is an error in the making up of the record in the lower court.

Mr. Eastman was the next witness called by the plaintiff in the rebuttal. His evidence is found beginning 161 R. 26. He testified that he was in Seattle at the time the defendant made its last payment to the plaintiff which was March 16, 1925. At that time he and Mr. Robinson and Mr. Grant went down into the banking quarters of the bank in question and made an inspection of the marble installed. That the only defects claimed by the architect Grant at that time were that five or six pieces of marble were off color and one check desk top needed to be re-waxed, and at that time he offered to replace any pieces of marble that were claimed to be defective if the architect or Mr. Robinson would send him the size of the marble which was claimed to be defective and that neither the architect nor Mr. Robinson had ever sent him the size or dimensions of any marble claimed to be defective since the visit and inspection of the marble in the bank. He testified at the trial that he was still ready and willing to do that at any time. He further testified as follows:

> "So far as supplying material or anything that was my portion to do, I have been at all times ready and willing to do anything in my power to furnish any marble or fix up any defects which have ever been pointed out to me. No one has

asked me to furnish anything at any time for that bank that I have not furnished. I have no knowledge of anybody asking me to supply anything since the time I was there to get that check. At all times while we were manufacturing the marble, all the way through, we furnished other pieces of marble to take the place of any that was claimed to be defective. We had items of replacement many times." (165 R. 3)

This testimony is not only uncontradicted but it is also in perfect harmony with the evidence given by all other witnesses who testified at the trial. He further testified that when he visited the defendant on March 16, 1924, and received the $4,000.00 check he went over his accounts with the defendant and defendant made no claim at that time of any single item set out in its counter claim. That the only dispute that arose at that time concerned certain items contained in plaintiff's final invoice to the defendant which were enumerated by the witness and are found at 166 R. 12. Plaintiff's final invoice to the defendant is found beginning at 171 R.

After the close of all the evidence the case was argued to the court by the respective parties and submitted for decision. The court, with the acquiescence of both parties, in rendering its decision, in adopting findings of fact and conclusions of law, and entering judgment, followed the local practice. The court thereafter rendered its written decision in the case, which decision is found beginning at 179 R. 5. Thereafter in due course the court set this cause down for hearing for the adoption of the findings of fact and conclusions of law for November 1, 1926, at which time the plaintiff appeared and tendered its findings of fact and conclusions of law, which findings of fact and conclusions of law are found beginning 190 R. 18. The findings of fact and conclusions of law tendered by the plaintiff were each and all rejected by the court. The plaintiff excepted to this action of the court and plaintiff's exceptions were allowed by the court (196 R. 16).

The defendant presented its findings of fact and conclusions of law, which findings of fact and conclusions of law are found at 197 R. 5. The court adopted each and all of the findings of fact and conclusions of law presented by the defendant.

Plaintiff excepted to the findings of fact Nos. 4, 5, 6 and 7 presented by the defendant and adopted by the court, and the plaintiff also excepted to the conclusions of law No. 1 presented by the defendant and adopted by the court. The exceptions of the plaintiff were allowed by the court (201 R. 5).

The court thereafter rendered judgment in accordance with its findings of fact and conclusions of law over the objections and exceptions of the plaintiff (201 R. 10). The judgment is found at 11 R. 1.

Thereafter and on the 1st day of November, 1926, the court made and entered an order granting the plaintiff until the 1st day of December, 1926, in which to prepare, serve and file its bill of exceptions (201 R. 17), and thereafter and within the time allowed by the court the plaintiff prepared and served its bill of exceptions and the same was allowed by the court and the bill was filed in the cause on November 30, 1926 (202 R. 8).

It is true that the installation of the marble was not completed on the day the bank moved into its quarters. There was some marble yet to be installed to complete the balustrade to the stairway leading from the bank proper to the basement, and this was not completed until March 3, 1925. There is, however, no allegation in the pleadings, or proof in the evidence, or claim made by the parties that the bank suffered any damages by reason of the delay in installing the marble after the bank moved into its quarters.

The action of the trial court in assessing the plaintiff one month's rent to the bank in the sum of $4,500.00 was clearly erroneous.

Thereafter in due course the plaintiff sued out a writ of error to this court seeking a review of the said judgment herein. Original writ of error is found at 205 R., original citation at 207 R. and plaintiff's assignment of errors are found at 13 R.

ASSIGNMENT OF ERRORS

I.

The court erred in overruling the plaintiff's objection to the following question asked the witness, Francis W. Grant, on direct examination by the defendant, and in allowing said witness to testify that as architect on the building he withheld $4,500.00 from the defendant as damages for delay in furnishing marble.

"Q. Referring to the two items upon which the architect declined to make payment on this job until they are corrected, one being for defective marble, and that has never been corrected and that is the item which you have estimated to the court would cost $2,500.00 and the other is on account of delay in the performance of the contract; what was the item of damages under that?" (54 R. 15)

II.

The court erred in overruling the plaintiff's objection to the following question asked the witness, John Graham, by the defendant on direct examination, and in allowing said witness to testify that as architect on the building he withheld from defendant one month's rent as damages received by the bank on account of delay in installing marble.

"Q. What was the other item, besides the item of defective marble, and on account of adjustment of damages sustained by the bank on account of that delay?" (104 R. 30)

III.

The court erred in refusing to adopt the following Finding of Fact tendered by the plaintiff:

"That on or about the 30th day of January, 1924, the plaintiff and defendant entered into a contract in writing wherein the plaintiff agreed to furnish and sell to defendant, and defendant agreed to purchase from the plaintiff, building and finishing marble which was to be used by the defendant in the construction and finishing of the banking quarters of that certain building in the City of Seattle known as the Dexter Horton National Bank Building at the agreed price and value of $46,000.00. That the plaintiff has fully kept and performed all of the covenants and conditions of said contract on its part to be kept and performed, and has furnished all of the marble called for in said contract, and the said marble has been installed in the banking quarters of said building, and is now in the same and being used by the Dexter Horton National Bank." (191 R. 11)

IV.

The court erred in refusing to adopt the following Finding of Fact tendered by the plaintiff:

"The court further finds that under and by virtue of said contract and the terms and provisions thereof plaintiff at the special instance and request of the defendant also furnished to the defendant for use by defendant in the construction and finishing of said banking quarters of the said Dexter Horton National Bank Building, other building marble, in addition to that required to be furnished by the contract above mentioned for the contract price of $46,000.00, of the fair and reasonable price and value of the sum of $826.78." (191 R. 30)

V.

The court erred in refusing to adopt the following Finding of Fact tendered by the plaintiff:

"The court further finds that under the contract entered into between plaintiff and defendant the marble was to be shipped by plaintiff from the city of San Francisco to the city of Seattle, Washington, in carload lots f. o. b. San Francisco, California. That at the special instance and request of the plaintiff forty tons of the marble shipped by the plaintiff to the defendant under said contract was shipped by water in pieces as manufactured by the plaintiff. That under this arrangement it was necessary for the plaintiff to crate said marble. That the reasonable cost of crating said marble was and is the sum of $484.78." (192 R. 10)

VI.

The court erred in refusing to adopt the following Finding of Fact tendered by the plaintiff:

"The court further finds that the change in shipping, as above mentioned, necessitated the carting of said marble from plaintiff's plant to a dock in San Francisco. That the reasonable cost and charge for carting said marble is the sum of $60.00. That the plaintiff paid out the said sum of $484.78 above mentioned, and the sum of $60.00 for the crating and cartage of said marble, and the same is a reasonable and just charge and should be paid by the defendant." (192 R. 24)

VII.

The court erred in refusing to adopt the following Finding of Fact tendered by the plaintiff:

"The court further finds that the total sum which became due and owing to plaintiff for marble furnished the defendant for installation and completion of the banking quarters of the said Dexter Horton National Bank Building was the sum of $47,371.56. That of said sum defendant has paid the plaintiff the sum of $40,-081.24. That there is now a balance due and owing the plaintiff from the defendant, of the sum of $7,290.32. That said sum became due on the first day of April, 1925, and plaintiff is entitled to interest thereon at the rate of six per cent per annum from and after the said first day of April, 1925, to date hereof." (193 R. 4)

VIII.

The court erred in refusing to adopt the following Finding of Fact tendered by the plaintiff:

"The court further finds that all of the allegations of plaintiff's complaint were and are true." (193 R. 18)

IX.

The court erred in refusing to adopt the following Finding of Fact tendered by the plaintiff:

"The court further finds that the banking quarters of said Dexter Horton National Bank Building were not completed and ready for the installation of marble until the 22nd day of September, 1924, and that the said banking quarters were not turned over to the defendant for installation of marble until said date. That under the contract existing between defendant and Dexter Horton National Bank for the installation of marble therein, the defendant was to have three months in which to install the marble after the said banking quarters were turned over to the defendant for the installation of the same. That the Dexter Horton National Bank moved into said banking quarters on the 20th day of December, 1924, and before the time had expired for the installation of said marble. That there was no delay on the part of the plaintiff in furnishing the marble. That neither the defendant nor the said bank suffered any damages by reason of any claim of delay on the part of the plaintiff in furnishing marble for the banking quarters of the said building." (193 R. 21)

X.

The court erred in refusing to adopt the following Finding of Fact tendered by the plaintiff:

"The court further finds that all of the marble furnished to defendant by the plaintiff for installation in said bank building was inspected by the architect acting for the bank before the same was installed as required by the contract between the defendant and the bank and the plaintiff and defendant and was accepted by said architect and permitted to be installed." (194 R. 13)

XI.

The court erred in refusing to adopt the following Finding of Fact tendered by the plaintiff:

"The court further finds that all of the marble furnished by the plaintiff to the defendant for installation in said banking quarters conformed to the requirements of the plaintiff's contract with the defendant in all respects." (194 R. 22)

XII.

The court erred in refusing to adopt the following Finding of Fact tendered by the plaintiff: .

"The court further finds that all claims of the defendant and the bank for defective marble were waived by the defendant and the bank when the architect for the bank permitted the same to be installed without any objection or protest or claim to the plaintiff that the same was defective." (194 R. 28)

XIII.

The court erred in refusing to adopt the following Finding of Fact tendered by the plaintiff:

"The court further finds that there was no objection made to the plaintiff by the defendant or the bank, or the architect acting for the bank that the marble was defective in color or in any other respect prior to the completion of the installation of the marble in the banking quarters of the said bank building." (195 R. 4)

XIV.

The court erred in refusing to adopt the following Finding of Fact tendered by the plaintiff:

"The court further finds that there was no evidence introduced in this case and no competent proof that the defendant or the bank sustained any damages by reason of any claim of defect in the color or finishing of the marble furnished to defendant by the plaintiff under the contract between the plaintiff and the defendant." (195 R. 12)

XV.

The court erred in refusing to adopt the following Finding of Fact tendered by the plaintiff:

"The court further finds that defendant has wholly and entirely failed to prove by competent evidence, or otherwise, each and all of the affirmative defenses set out in the defendant's answer herein." (195 R. 20)

XVI.

The court erred in refusing to adopt the following Finding of Fact tendered by the plaintiff:

"The court further finds that the plaintiff at all times prior to the bringing of the present action furnished the defendant all of the marble requested by the defendant to take the place of any marble claimed to be defective by the defendant." (195 R. 26)

XVII.

The court erred in refusing to adopt the following Finding of Fact tendered by the plaintiff:

"The court further finds that the plaintiff has

never failed or refused to furnish any marble to take the place of any marble claimed to be defective by the defendant, or by the bank, or by the architect acting for the bank." (195 R. 32)

XVIII.

The court erred in refusing to adopt the following Conclusions of Law tendered by the plaintiff:

"That the plaintiff is entitled to a judgment in its favor against the defendant in the sum of $7,290.32 with interest thereon at the rate of 6 per cent per annum from April 1st, 1925, to date hereof." (196 R. 11)

XIX.

The court erred in making and adopting the following Findings of Fact:

"The court finds that in addition to furnishing the marble for said building, the plaintiff at the instance and request of the defendant, furnished certain extra marble and material, to the extent and of the value of $657.92 and no more; and that all other items of extras claimed by the plaintiff have not been established by the evidence and are disallowed.

"The original contract for said marble work amounted to the sum of $46,000.00 and, together with the extras herein allowed, amounts to the total sum of $46,657.92, upon which the defendant has been paid the sum of $40,081.24, leaving a balance unpaid upon said original contract and extras in the sum of $6,576.68." (198 R. 27)

XX.

The court erred in making and adopting the following Finding of Fact:

"The court further finds that the plaintiff did not furnish said marble according to said contract but that it greatly delayed the furnishing of said marble and the defendant in performing its said contract with the Dexter Horton National Bank, although defendant urged plaintiff for a speedier delivery of said marble, by numerous letters, telegrams and cablegrams, and personal visits to San Francisco; that a portion of said marble was defective and not according to contract, especially in color, workmanship and material; and the court also finds that the defendant was required to do certain extra work and labor, hereinafter referred to, in connection with said contract, on account of a portion of said marble not being fabricated according to the plans and specifications." (199 R. 11)

XXI.

The court erred in making and adopting the following Finding of Fact:

"The court further finds that the defendant is entitled to and is allowed the following counterclaims and offsets, as shown by the evidence:

(a) Damages caused by delay on plaintiff's part in furnishing said marble according to contract $4,500.00

(b) For defects in said marble............ 2,500.00

(c) For extra work and labor, consisting of 32 hours of labor in cutting

triangles to fit the plans, and 104
hours of labor in drilling holes in
balustrade—at $1.25 per hour........   170.00
(d) For wooden hand balustrade on
open stairway to the basement........   25.00

TOTAL $7,195.00
and all other items in defendant's counterclaims
and setoffs have not been proven by the witness."
(199 R. 28)

XXII.

The court erred in making and adopting the following Finding of Fact:

"The court further finds that the defendant
has fully kept and performed all of the covenants
and agreements of its contract with the plaintiff."
(200 R. 18)

XXIII.

The court erred in making and adopting the following Conclusion of Law:

"That the counterclaims and offsets to which
the defendant is entitled exceed the plaintiff's
claim in the sum of Six Hundred Eighteen and
32 /100 Dollars ($618.32), and that the defend-
ant is therefore entitled to a judgment against
the plaintiff in that sum." (200 R. 24)

XXIV.

The court erred in adopting each and all of the
above specified Findings of Fact and Conclusions
of Law for the reason that the same are contrary to
the evidence and without evidence to support the

same and are contrary to the law applicable to the case.

XXV.

The court erred in entering a judgment upon said Findings of Fact and Conclusions of Law in favor of the defendant and against the plaintiff over the objection of the plaintiff (11 R. 1).

ARGUMENT

The plaintiff brought this action to recover the balance due on the purchase price of certain marble which the plaintiff agreed to furnish the defendant under written contract of the date of January 30th, 1924, which contract was introduced in evidence as plaintiff's Exhibit No. 1 and is found at 30 R. 4. The contract provided that plaintiff should furnish the defendant with certain specified marble for the Dexter Horton National Bank Building in Seattle, at the agreed price and value of $46,000.00.

Plaintiff also in its complaint alleged that in addition to the marble specified by the contract and furnished to the defendant the plaintiff furnished other marble to the defendant of the reasonable value of $1,371.56. Plaintiff alleged that there is a balance due and owing it of $7,290.32.

At the trial of the cause George B. Eastman, president and manager of the plaintiff, was called and testified on behalf of the plaintiff. His testimony is found beginning at 29 R. 10. His testimony in substance is to the effect that the plaintiff furnished all of the marble required by the contract to the defendant and that the marble was received by the defendant and installed by the defendant in the Dexter Horton National Bank Building in Seattle. That in addition to the marble covered by the contract plaintiff furnished to the defendant other marble and did other work and performed other services for the defendant of the reasonable value of $1,371.56. That all of the marble covered by the contract, together with the additional marble ordered by the defendant, was installed in said building and at the time of the trial was in said building and being used by the Dexter Horton National Bank.

This evidence clearly entitled the plaintiff to recover for any balance due on the contract price of the marble covered by the contract and for the reasonable value of the extra marble furnished.

> *Phillips and Colby Construction Company v. Seymour, et al.*, 91 U. S. 646;
>
> *City of St. Charles v. Stookey*, 154 Fed. 772;
>
> *Omaha Water Co. v. City of Omaha*, 156 Fed. 922.

The evidence of the plaintiff was not contradicted at the trial except as to some minor items of the extras. The plaintiff, therefore, under the well established principles of law in the Federal Courts was entitled to recover the balance due it for the purchase price of the marble, unless the defendant succeeded in defeating the plaintiff's right of recovery by establishing by a preponderance of the evidence some of the counterclaims set out in its answer. The defendant in its answer set out several counterclaims for damages in paragraph IV thereof (6 R. 17).

In one of these counterclaims defendant claims judgment in the sum of $6,000.00 for defective marble and damages for delay in furnishing the marble, which counterclaim is found in subdivision "F" of paragraph IV at 8 R. 13. In this particular counterclaim the defendant alleges briefly that the architect refused to approve the final payment due the defendant from the bank in the sum of $6,000.00 until the defendant should replace defective marble and settle with the bank for its damages for delay.

At the outset, and before we take up the argument in detail on this counterclaim, we desire to call the court's attention to the fact that in this counterclaim the defendant is seeking to offset the plaintiff's claim by damages, which it claims the bank suffered by reason of delay in furnishing marble and for defective marble. In other words, the defendant is not claiming that it suffered any damages by reason of any default on the part of the plaintiff in respect to delay for defective marble. It is only claiming damages on these two items by reason of the fact that the bank withheld from it the sum of $6,000.00. In other words, it is damages suffered by the bank, and not by the defendant, that forms the basis of this counterclaim.

Notwithstanding that only $6,000.00 on these two items of damages was withheld from the defendant, the court by its judgment allowed the defendant on these two items of damages the sum of $7,000.00. See Finding of Fact No. VI (199 R. 30).

It is the action of the trial court in allowing the defendant the sum of $4,500.00 damages for delay and $2,500.00 for defective marble that the plaintiff particularly complains of and seeks review in this proceedings.

Is the Defendant Entitled to Damages for Delay in Furnishing Marble?

Under this head the plaintiff presents to this court for review Assignment of Errors Nos. 1, 2, 9, 15, 20, 21, 23 and 24.

Under this heading we wish first to take up the question as to whether or not there was any delay on the part of the plaintiff in furnishing marble to the defendant for installation in the bank building. In other words, did the plaintiff furnish the marble within the time specified in its contract?

In order to determine this question we believe it to be advisable to first examine the contract and determine, if possible, the period of time granted the plaintiff thereby to furnish the marble. The provision of the contract bearing upon this question reads as follows:

"The Mission Marble Works (hereinafter called the Company) proposes to furnish as required by the specifications prepared by John Graham, Architect (acting for the purposes of this agreement as the Agents of the Purchaser), except as hereinafter modified, for the Dexter Horton National Bank Building at Seattle, Wash., the following building marble, viz.:

Per Sheets Nos. 2, 2A, 5, 6, 8 and 10.

According to the plans, specifications, terms, and conditions of same, and as covered in purchasers contract with said Bank in order to enable the purchaser to faithfully and promptly fulfill their contract with the above bank." (30 R. 17)

This provision of the contract between the plaintiff and the defendant given its broadest construction and the one most favorable to the defendant, simply makes the contract between the defendant and the bank as to time for installing the marble a part of the contract between the plaintiff and defendant.

The contract between the defendant and the bank was introduced in evidence as defendant's Exhibit "A2", and is found at 56 R. 14.

The provisions of this contract with reference to the time for installing marble are as follows:

"It is agreed that the work under this contract shall be commenced immediately after the date hereof and that the first material suitable for immediate installation shall be delivered at the premises not later than three months after setting plans have been approved by the architect and that the remainder of the marble shall be delivered at such subsequent date or dates as shall not interfere with continuous progress with the installation of the marble work, and it is further agreed that the entire contract shall be completed within three months after the first delivery of material as above provided. It is agreed, however, that the Contractor shall not be held liable for delay or failure to perform by acts of the owner or of any other contractor or subcontractor upon the building, or by fires, storms, accidents, strikes, boycotts, lockouts, riots, lack of transportation facilities, delays in transportation, government regulations, acts of God or by any other causes beyond the contractor's reasonable control.

"It is agreed that the contractor shall have possession of the space where marble is to be installed for necessary storage of material after the time for commencement of the work as herein provided, * * *"

It will be noticed immediately on reading the provisions of this contract that there is no definite time set for the beginning of the work and no definite date set for the completion of the work. The contract is dated January 25, 1924. It provides that work under the contract shall commence immediately after the date hereof. It also provides that the first material suitable for immediate installation shall be delivered at the premises not later than three months after the setting plans have been approved by the architect.

The witness, Grant, on behalf of the defendant testified concerning shop drawings or setting plans as follows:

"Q. When did you get shop drawings?
A. I never did get them." (49 R. 17)

"We never received shop drawings or setting plans so we could not approve them. They were never submitted—only partially." (95 R. 14)

Since the setting or shop plan drawings never were

approved by the architect the time for the commencement of the work under the contract cannot be placed from the date of the approval of the shop drawings or setting plans. The contract also provides that the entire contract shall be completed within three months after the first delivery of material as above provided. The first material was delivered on April 26, 1924 (46 R. 6; 167 R. 4). Did the three months period provided in the contract for the installation begin to run from this date? The trial court was of the opinion that it did and so held (183 R. 10), and based its judgment allowing damages upon this conclusion. The question as to whether or not the contract was completed in time depends upon whether or not the trial court was right in so holding.

The evidence, without dispute, shows that the building in which the marble was to be installed was not completed or ready for the installation of the marble or storage of the same until the 22nd day of September, 1924.

"The Robinson Tile & Marble Company began the work of setting the marble on September 22, 1924." (48 R. 13)

"The bank building was in condition to start the work, as far as installing the marble in the bank proper was concerned, on the day the work was started and it was not in condition before that day.

Q. The banking-room itself was not in condition either, prior to the time you started work there for storing marble in there, was it?

A. It was all being used right along.

Q. There was no place for storing marble in there, was there?

A. When?

Q. Until the time you started work, September, 1922?

A. No marble was stored in there; no one wanted to store marble in there.

Q. It was not in condition to store?

A. It was not; it was not built for that purpose.

Q. It was not in condition to be used for storing marble, was it?

A. No." (67 R. 26)

It is plaintiff's contention that the time for the installation of the marble did not begin to run until the bank building was in condition to receive the marble for storage and installation. In other words, that under the contract between the defendant and the bank it was the clear duty of the bank to turn over the building where the marble was to be installed in a condition so that the marble could be installed and in a condition so that the marble could be stored therein before the time for the installation of the marble began to run and the evidence conclusively shows that this date was the 22nd day of September, 1924.

As bearing upon this contention we wish to call the Court's attention to another provision of the contract which is as follows:

"It is agreed that all structural steel supports for marble work will be furnished and set in place by the owner, but that the Contractor shall furnish all required masonry backing and all necessary copper anchors for securing marble to structural steel and masonry backing."

This provision clearly provides that the place where the marble was to be installed was to be provided by the owner ready for installation and not by the contractor. The contract, after providing that the entire contract shall be completed within three months after the first delivery of the material, further provides:

"It is agreed, however, that the contractor shall not be held liable for delay nor failure to perform by acts of the owner nor of any other contractor or subcontractor upon the building . . . nor by any other cause beyond the contractor's reasonable control."

Here we find a specific and definite provision of the contract which relieves the contractor from all liability for any delay which is caused by the owner or any other contractor or subcontractor or any other cause beyond the contractor's reasonable control. The contract makes it the duty of the owner to provide the building ready for the installation of the marble, also definitely provides that the contractor shall not be liable for damages for any delay on the part of the owner in providing the building ready for installation of the marble.

The contract also clearly provides for a period of three months for the installation of the marble after the time for installation of the same begins to run. Our contention that the three months' period provided by the contract for the installation of the marble does not begin to run under the contract taken as a whole as well as the particular provisions thereof with reference to time, seems so obvious particularly when the surrounding facts and the object to be accomplished by the contract are considered, that we would submit this question without further argument were it not for the decision of the distinguished and learned trial court to the contrary.

The trial court held that the three months' provision for the installation of the marble began to run from the date of the first delivery of marble which was as heretofore seen on the 26th day of April, 1924. If this decision of the trial court is correct the three months' period for the installation of the marble ended on the 26th day of July, 1924, and there was approximately two months default before the building was turned over to the defendant or was in a condition to receive the marble or in a condition for installation of the same.

For the trial court to take the position that there was a default in the installation of the marble within the time provided by the contract before the building was turned over for installation of the same, appears to us to be exceedingly harsh, unjust and unreasonable, and we do not believe that any authority can be found supporting such a position. We believe we are justified in saying that the authorities are unanimous in holding that in a building or construction contract the time for the completion of the same does not begin to run until the owner has placed the builder in possession of the premises or in a position where he could begin the work contemplated by the contract.

See Elliott on Contracts, Vol. IV, Sec. 3720, wherein the author says:

> "Mutual assistance in a sense is always understood in building contracts where the work is to be partly performed by the owner or other contractors and whether the contract so stipulates or not the stipulation as to completion by a specified time will be waived by the owner or his agents in not keeping the other portions of the work sufficiently advanced to enable the other party to perform his part."

See

Strobel Steel Construction Co. v. Sanitary District of Chicago, 160 Ill. App. 554-60.

In this case the plaintiff contracted to build certain dams for the defendant. The defendant was to furnish the piers on which the masonry work of the dams was to rest and the defendant was also to do certain other work before the plaintiff could perform the work contemplated by the contract. The contract provided that the work to be done by the plaintiff should be completed by a certain day. And it also provided $100.00 liquidated damages per day for every day's delay in the completion of the work. The work was to begin on a certain day named in the contract. When the day arrived on which the work was to begin the defendant had not performed the preliminary work which the contract provided it should perform and did not do so for a considerable time thereafter. After the defendant had performed the preliminary work required by the contract plaintiff went ahead and finished the work which it was required to perform under the contract but did not complete the work within the time specified by the contract. The defendant held out from the final payment to the plaintiff the sum of $100.00 per day as liquidated damages for delay on the part of the plaintiff. Plaintiff sued to recover this sum.

In deciding the question as to whether or not the defendant was entitled to hold out this sum of money as damages for delay the court said:

"We do not think that a construction should be given to the contract between the parties to this action which would enable the Sanitary District to retard the prosecution of the work by the plaintiff or prevent it from employing all the agencies which, in its judgment, could wisely and advantageously be used in the performance of the work called for by the contract from the date of which the plaintiff was, by the terms of the contract, compelled to commence work to a much later date, and, nevertheless, enforce the clause providing for $100 liquidated damages per day for every day after the date fixed by the contract for the completion of the work. The plainest principles of justice would forbid such a construction of the contract. It is unreasonable to suppose that the parties intended that the contract should be interpreted in that manner, and that it should be left optional with the Sanitary District to retard the work contemplated by the contract, at its pleasure, discretion or convenience, accept the work, and at the same time enforce the heavy penalty for the non-performance by the plaintiff."

Nelson v. The Pickwick Associated Co., **30 Ill.**
App. 333.

This case involves a sub-contractor. The contract in question provided for $25.00 per day liquidated damages for delay in completion of work. The contract among other things provides:

"Should delay be caused by other contractors to the positive hindrance of the contractor hereto a just and proper amount of extra time shall be allowed by architects, provided they shall have given notice to the said architect at the time of such hindrance or delay."

Sub-contractor did not complete his work within the time provided by his contract. He claimed the cause of delay was by the owner and other contractors. The sub-contractor did not apply to the architect for additional time. The question before the court was whether or not not having applied and gotten an extension of time from the architect sub-contractor could excuse the delay and be relieved from the $25.00 per day penalty. The court said:

"The principal question in this case is whether this provision for the allowance for extra time by the architect is the only remedy of the appellant when delayed by other contractors. * * * There is a preponderance of the evidence that such a promise was made and the probability that there was as well as the justice of the claim are strongly supported by the concurrent circumstances. Without reciting that testimony and those circumstances, however, the case may be safely put upon the ground 'that the contract necessarily presupposes and implies on the part of appellees an obligation to supply' that upon which appellant was to do his work. Brooms Legal Maxims, 667. It was the legal duty of the appellees to keep the work in such a state of forwardness as to enable the appellant to perform his contract within the limited time."

Taylor v. Renn, 79 Ill. 181.

In this case the plaintiff contractor was to do the joiner and carpenter work in a brick building. The plaintiff did not complete his work in the time specified in the contract. The owner for the delay held out part of the contract price. Evidence showed that the plaintiff was delayed in doing the carpenter work by reason of the delay of the masons and plasterers. Plaintiff sued to recover the amount held out by the defendant. In deciding the question as to whether or not plaintiff could recover the court said:

"But it is said, the contract was not completed at the time specified. There was no claim set up for the stipulated damages on that account. But was there any delay for which appellees are justly chargeable. The contractor was to do the joiner and carpenter work on a brick building. It needs no argument to show, even if the contract for such work does not so provide, that it is always made with the understanding that the owner of the building shall keep the masons work so advanced as to enable the carpenters to do their work."

Erickson v. United States, 107 Fed. 204.

In this case the government withheld from the plaintiff the sum of $2100.00 as damages for delay in completing a contract for the erection of a lighthouse. The plaintiff claimed that the delay was due to failure of the government to perform its part of the contract in surveying the ground and so forth. The plaintiff sued to recover. In disposing of the question the court, among other things, said:

"In all business transactions the government as a contractor should be treated with the same fairness as private individuals and should also be required to deal fairly on its part. The evidence as a whole shows that in spite of all obstruction and annoyances the plaintiff completed the work substantially and well and that it was accepted as fulfillment of its contract by the government's officers. To punish him by withholding part of the contract price for his work as a forfeiture for delay and by refusing to pay his bill for extras when the delay was entirely caused by the incompetency of the government's agent and the failure of the other contractors, over whom he had no control, to deliver materials on time, and the bad weather in which he had to work in consequence of such delays, would be rank injustice."

In 9 C. J. 783, Sec. 123, it is said:

"The failure of the builder to perform the contract in the time stipulated therefor will be excused, and the owner cannot take advantage thereof, where such failure is caused by the wrongful acts of the owner or by his failure to perform his part of the contract, or by the fault of persons for whose conduct the owner is responsible; and the fact that the contract contains express provisions excusing delay from certain causes does not deprive the builder of the right to excuse delay caused by the wrongful acts or omissions of the owner.

"Applications of Rule. Thus there is a valid excuse for nonperformance within the stipulated time, where the delay is caused by the owner's refusal to permit the builder to complete the work or by his direction of a suspension of the work; or by the owner's failure to deliver to the builder at the proper time the possession of the premises on which the work is to be done."

See also

> *Champlain Construction Co. v. O'Brien*, 117 Fed. 271;
>
> *Larcey Manufacturing Co. v. Los Angeles Gas Co.*, 106 Pac. 413;
>
> *Salisbury v. King*, 119 S. W. 160;
>
> *Herbert v. Weil*, 39 So. 389;
>
> *Morse Dry Dock & Repair Company v. Seaboard Trans. Co.*, 161 Fed. 99.

We believe that this question of law is fully and finally determined beyond question or cavil in favor of the plaintiff by the Supreme Court of the United States in the leading case of

> *McGowan v. American Pressed Tan Bark Co.*, 121 U. S. 575.

In this case the defendant entered into a contract

with the plaintiff on the 23rd day of June, 1881, wherein the defendant undertook to install certain machinery in a boat to be furnished by the plaintiff within sixty days after the date of the contract. The boat was not furnished by the plaintiff until the 10th day of November, 1881. There was delay in the installing of the machinery and also plaintiff claimed the machinery installed was defective. Plaintiff sued the defendant for damages. One of the main questions in the case was whether or not the plaintiff could recover damages for the delay in installing the machinery brought about by the delay of the plaintiff in furnishing the boat.

The Supreme Court held that the defendant was entitled to the sixty days provided by the contract in which to install the machinery after the boat was delivered to it by the plaintiff and in disposing of this question, at page 600, the court said:

"The petition contains an allegation of special damage, from the loss of tan bark occasioned by the delay in not erecting the machinery within sixty days from June 23, 1881; but the bill of exceptions does not show that there was any evidence tending to establish this special damage, except as it may be inferred, from the general charge of the court, that such testimony was offered. But the court, in its general charge, instructed the jury as follows: 'The contract bound the defendants to complete the machinery and set it up on the boat within sixty days. It is too plain for argument, that the failure of the plaintiff to have the boat ready would excuse the defendants from strict compliance with this part of the contract, and that all delay which occurred before the boat was ready is out of the case. The plaintiff was as much responsible for that as the defendants, or sufficiently so to preclude him from complaint on that score.'

"It is, therefore, claimed by the plaintiff that no damages were included in the verdict on account of the delay in not erecting the machinery within sixty days from June 23, 1881. This appears to be a sound proposition. We see no error in the charge of the court, that, if the defendants proceeded under the contract, they were bound to complete the work within the length of time contemplated by the original agreement, and such additional time as was lost by the delay in the construction of the boat. There is nothing in the bill of exceptions to show that the machinery could not have been erected within sixty days after the boat was ready to receive it. The parties treated the contract as in full force, except as to the time in which it was to be performed, and the work was done and the payments were made under the contract as thus extended in time. The defendants made no claim before the suit was brought, that the contract was rescinded by reason of the non-readiness of the boat until the 10th of November, 1881, or that there was any reason in that fact which prevented them from complying with their part of the contract within the sixty days after the delivery of the boat. No such defense is set up by them in their answer, and they introduced no evidence to that effect, so far as the bill of exceptions shows. These views are in accordance with the ruling of this court in *Phillips Co. v. Seymour*, 91 U. S. 646. The plaintiff went on paying the defendants on account for the machinery, and the defendants proceeded in erecting it without complaining of the delay in the furnishing of the boat, and without any claim that they were not required to furnish the machinery within the sixty days after the furnishing of the boat. See, also, *Graveson v. Tobey*, 75 Ill. 450."

The action of the trial court in holding that the three months provision of the contract for installing the marble began to run on the date of the delivery of the first marble, to-wit, April 26th, 1924, instead of September 22nd, 1924, the date the building was turned over by the owner to the defendant, for the installation of the marble not only violates the express provision of the contract but also violates one of the cardinal and fundamental principles of law governing the interpretation of contracts.

See Elliott on Contracts, Sec. 1521, wherein the author says:

"In addition the contract is to be given a reasonable construction. It will not be construed so as to render it oppressive or inequitable as to either party or so as to place one of the parties at the mercy of the other unless it is clear that such was their manifest intention at the time the agreement was made."

See also 2 Williston on Contracts, Sec. 620, wherein the same rule is laid down in the following language:

"A construction which makes the contract fair and reasonable will be preferred to one which leads to harsh or unreasonable results."

It would be an extremely unjust, harsh, and iniquitous rule for the interpretation of contracts to hold that the period for installing the marble began to run before the building in which it was to be installed and which the bank was to provide under the express terms of the contract was turned over to the contractor for the installation of the same and to assess him for damages for delay in installing the same when this delay was caused by the owner in failing to turn over the building in which the marble was to be installed when the contract expressly provides that it shall not be assessed any damages for such a delay.

If the three months' period for installing the marble provided in the contract did not begin to run until the building was turned over for installation of the same, that is, on the 22nd day of September, 1924, then the plaintiff certainly cannot be assessed the sum of $4,500.00, one month's rent to the bank, nor in any other sum as damages for delay for the evidence conclusively shows that the bank moved into and occupied its banking quarters on the 20th day of December, 1924, and at least two days before the three months' period ran.

"The bank moved into its present quarters where this marble was put in on December 20, 1924." (55 R. 25)

"The bank moved into its quarters on the 20th of December, 1924." (56 R. 5)

"When the bank opened, it was on the 20th day of December, 1924." (81 R. 9)

The trial court in assessing damages against the plaintiff in the sum of $4,500.00, a month's rent for the bank, committed a manifest error.

There Is No Sufficient Allegation in the Pleadings or Proof in the Evidence Introduced to Justify the Trial Court in Assessing Any Damage Whatever for Delay on the Part of the Plaintiff in Furnishing Marble

The counterclaim of the defendant for damages for delay must be not only supported by a competent allegation in its answer but must also be supported by competent evidence. It requires no citation of authorities to support this cardinal rule of law. The counter claim as pleaded by the defendant nowhere alleges that the bank suffered any damages whatever by reason of the delay in furnishing the marble on the part of the plaintiff. It merely alleges that the architect refused to approve the final payment of $6000.00 due the defendant on its contract until defective marble is replaced and the bank's damages for delay paid. The evidence wholly and entirely fails to show that the bank suffered any damages whatever by reason of any delay in furnishing marble. All of the evidence introduced upon this question of damages for delay is very brief and we will for the convenience of the court set it out herein in full.

The architect, Grant, over the objection and exception of the plaintiff upon this point, testified as follows:

"By using $2,500.00 for the correction of the defects and $4,500.00 for the delay. I think I have a memorandum of what the figures were. Our original computation contained Twenty-five Hundred for defects, Forty-Five Hundred for damages and $327.00 for some plastering and painting that had to be done after the marble workers had finished their work. That is from a memorandum and the slip that I had that on. It does not seem to be here. I think it is about right. At any rate we did not use that figure in making this deduction.

Q. You deducted less than that amount?

A. We deducted $6,000.00, which is Thirteen or Fourteen Hundred Dollars less than the other."
(55 R. 11)

Upon this question John Graham, architect, over the objection and exception of the plaintiff, testified as follows:

"At that time the bank had expended three hundred and twenty-seven dollars in making good defective plaster; that is, plaster damaged—caused by the marble workers. The damages I speak of were damages caused the plaster of the bank by the workmen of the tile and marble company, caused by the necessity of erecting the marble after the plaster had been done, which should have been done beforehand. The other item, the check desks, and marble that was rejected and mussed up, being replaced and the damages for the delay. Now, in order to minimize that as much as possible, while the work was five months behind them, we concluded, that without any shadow of a doubt there could not be any quibble with the bank, that the bank had been delayed one month by this marble contract, to get away from any element of controversy at all, we figured that nobody could dispute the fact that we were delayed one month; therefore we added a month's rent.

"That was the basis of my writing this letter." (105 R. 13)

This is all of the evidence that we have been able to find in the record on this question of damages for delay in furnishing marble. Granting for the sake of argument that there was delay beyond the three months period provided by the contract for the installation of the marble, yet this evidence is wholly insufficient to justify a court in assessing damages for delay in the sum of $4500 or any other sum or amount. The evidence does not show that the bank was paying or did pay any sum or amount whatever for rent for the banking quarters where the marble was installed prior to the time that the bank moved into said banking quarters and if they did not they could not charge for a month's rent for those quarters. The evidence fails to show whether or not the bank was paying rent in the quarters which it occupied prior to the time it moved into the Dexter Horton National Bank Building wherein the marble was installed. As far as the evidence shows the bank may have been paying only $2000.00 rent in the quarters it was occupying at the time the marble was being installed and that it would have to pay $4500.00 or more for the quarters wherein the marble was installed after it moved in and then instead of being damaged in any sum by way of rent for the delay in installing the marble it would be actually saving money and certainly under such circumstances there could be no rental damages to the bank by reason of a month's delay in getting into its new quarters and there is not a word of testimony that can be found anywhere in the record that the bank suffered any other kind of damages of any character whatsoever. Except this bald statement of the architect that they withheld $4,500.00, a month's rent, as damages for delay, the record is silent.

In fact there was no attempt whatever made by the defendant at the trial to show that the bank suffered any damages by reason of the delay, if any, in furnishing marble. The only attempt that the defendant made at the trial as clearly appears from the record was the very weak attempt to show good faith on the part of the architect in withholding from the defendant some sum of money to cover a supposed one month's rent as damages for delay.

The mere fact that the architect withheld from the defendant $4,500.00 or any portion of that sum from the final payment due the defendant under its contract for claimed damages for delay forms no basis whatever for the trial court's action in assessing any sum for damages for delay against the plaintiff.

In the first place, the contract between the defendant and the bank gave the architect no right or power to withhold from the defendant any sum whatever as damages for delay in finishing the installation of the marble. The only portion of the contract with respect to the power of the architect to withhold or approve payments is as follows:

"In consideration of the faithful performance of the covenants herein contained the owner agrees to pay to the contractor the sum of $60,-101.00 in installments on or about the 5th and 20th of each month, equal to 90% of the value of material delivered to and labor performed at the premises, as ascertained and computed by the architect and upon his certificate, and the remainder upon certificate of the architect that the work is satisfactorily completed and upon a satisfactory showing that all bills and claims of whatsoever nature are fully paid." (59 R. 23)

The architect, certainly, under this provision had no power to withhold any money from the contractor as, or for damages for delay; his action in that respect was wholly void. Even though the architect were empowered to withhold a sum of money as damages for delay by the contract between the defendant and the owner, the action of the architect in this respect would form no basis whatever for assessing damages for delay against the plaintiff, for the very obvious reason that the right of the plaintiff to receive pay for the marble furnished does not depend upon any action of the architect. The right of the plaintiff to receive money for the marble furnished to the defendant is governed wholly and entirely by the con-

tract between the plaintiff and the defendant which
contract provides as follows:

"The price for said marble delivered F.O.B.
cars as aforesaid is ($46,000.00).

Forty six thousand and No/100 Dollars, payable in cash as follows:

On the tenth day of each month 90% of the
amount of marble delivered hereunder during
the previous calendar month and the 10 per cent
before retained within thirty-five days after the
full completion of this agreement by the Company." (31 R. 9)

It seems to have been the idea and belief of the
defendant at the time it filed its answer and at the
time of the trial that it was only necessary for it to
show that the archiect had withheld from it the sum
of $6,000.00 as damages for defective marble and for
delay to entitle it to a judgment for the same amount
against the plaintiff. In this, however, the defendant
was clearly in error under the well established principles of law as well as under the express provisions
of the contract.

See 9 C. J. 834, Sec. 172:

"Provisions in subcontracts that the subcontractor shall not be entitled to payment until the
contractor has secured his compensation from the
builder are valid and enforceable; but if the contractor has by his own fault lost the right to
payment from the owner, the subcontractor will
be entitled to his compensation, and of course unless the subcontract clearly and expressly so provides the right of the subcontractor to payment

is not dependent on the receipt of payment by the contractor, but only on the performance of his subcontract."

George A. Fuller Co. v. B. P. Young Co., 126 Fed. 343;

Greenstone Co. v. Bush Co., 135 N. W. 993.

The action of the trial court in allowing the defendant $4500 damages for delay in furnishing marble cannot be sustained under the contract between the parties, under the pleadings, under the evidence introduced or the law applicable to this case.

The trial court therefore committed a manifest error in allowing the witnesses Grant and Graham to testify over the objections of the plaintiff that they withheld one month's rent as damages suffered by the bank by reason of delay in furnishing marble and also committed an error in allowing the defendant damages against the plaintiff in the sum of $4500.00 for damages suffered by the bank for the delay in the installation of the marble.

The three months' period granted by the contract between the defendant and the bank to install the marble became a part of the contract between the plaintiff and the defendant and the plaintiff was granted the said three months' period to furnish the marble after the bank building was in a condition for the marble to be installed, the same as though it was directly incorporated in the contract between the plaintiff and the defendant. This being the case, neither the defendant nor the architect nor the bank could deprive the plaintiff of this three months' period to furnish the marble by one letter or telegram or a million letters or telegrams demanding that the marble be furnished in a shorter period or by threatening damages if it were not furnished in a shorter period. A party can not be deprived of his contract rights by any such method or procedure. The fact that the defendant may have been willing to install the marble in a less period than three months in order to curry favor of the architect or the bank does not deprive the plaintiff of its right to the full three months' period to supply the marble, and particularly should this be true in this case for the evidence clearly shows that the delay by the bank in furnishing the building ready for the installation of the marble and the storage of the same caused the plaintiff great inconvenience, extra work and trouble and additional costs and expenses. The numerous letters and telegrams sent by the defendant and others to the plaintiff, urging speedy delivery of the marble and threatening damages therefore have no bearing on the merits of this case and should not be considered by this court in arriving at its conclusions upon this question of damages for delay.

DAMAGES FOR DEFECTIVE MARBLE

Specification of Errors Nos. 3, 7, 8, 10, 11, 12, 13, 14, 15, 16, 17, 18, 19, 20, 21, 22, 23, 24 and 25. All of these specification of errors may be grouped together and discussed under one head as they all raise the one question, to-wit: "Did the court commit error in allowing the defendant $2500.00 damages for marble that was claimed by architect to be defective."

First, the plaintiff claims that the court erred in

allowing this sum as damages for claimed defective marble for the reason that the marble was all inspected by the architect before it was installed in the building and it was all permitted to be installed by the architect and no complaint was made to the plaintiff by the architect or the defendant or the bank that the marble was defective until long after all of the marble had been installed. Under these circumstances and the contract governing the rights of the parties the law is that neither the defendant nor the bank can complain of the marble installed even though it may be defective. In discussing this question we desire first to call the court's attention to governing provisions of the contracts. The contract between the defendant and the bank provides that the contractor shall furnish marble

"as shown on Drawings 2, 2-A, 5, 6, 8 and 10, prepared by John Graham, architect, and to the entire satisfaction of said John Graham, architect, * * * the quality and finish of all marble installed shall be equal to that represented by samples on file in the office of the architect and identified by signatures of the parties hereto * * *. It is agreed that the architect shall be the sole judge of the quality and quantity of the work performed under this contract and that all operation hereunder shall be under his direction and control."

The contract between plaintiff and defendant contains this provision:

"The Mission Marble Works (hereinafter called the Company) proposes to furnish as re-

quired by the specifications prepared by John
Graham, Architect (acting for the purposes of
this agreement as the agents of the purchaser)
except as hereinafter modified, for the Dexter
Horton National Bank Building at Seattle, Wash.,
the following building marble, * * * According
to the plans, specifications, terms, and conditions
of same, and as covered in purchaser's contract
with said Bank in order to enable the purchaser
to faithfully and promptly fulfill their contract
with the above bank."

We desire now to call the court's attention to the
evidence introduced at the trial. John Graham, the
architect mentioned in the contracts, testified upon
the stand as follows:

"Mr. Grant was the one who had active charge
and supervision of inspecting the marble and
passing upon it. Mr. Grant was superintending
all that work for me and had all the duties."
(106 R. 22)

Mr. Grant also testified that in inspecting marble
he represented the architect.

"Q. You represented the architect, didn't you?
A. I represented the architect, yes, sir." (77
R. 29)

Mr. Grant further testified beginning at 76 R. 24
that each and every piece of marble that was installed
was inspected by him before the same was installed
and not a single piece of marble that was rejected by
him as such architect and acting on behalf of John
Graham was installed in the building.

Mr. Boucher, the superintendent for the defendant

in the installation of the marble upon this question testified as follows:

> "Q. Did you at any time put in any pieces of marble after they had been rejected? If they did reject any, and tell you not to put it in?
>
> A. No, sir, no, by no means." (157 R. 11)

The record is absolutely devoid of any evidence of any kind or character where the defendant, the architect or the bank ever at any time prior to the completion of the installation of the marble made any complaint of any kind or character to the plaintiff that the marble was defective in any way except as to some few pieces and as to these few pieces all parties agree that the plaintiff supplied others to take the place of these that were complained of.

The court will notice that in the contract between the plaintiff and defendant the architect, John Graham, was to act for the purposes of the contract as the agent of the purchaser. In other words, the purchaser, the defendant, was to be bound by the acts of John Graham and John Graham was not to represent or be the agent of the plaintiff in carrying out the contract. The contract between the defendant and the bank provided that the work should be done to the satisfaction of John Graham and that he should be the judge of the quality and quantity of the work performed under the contract and that all the operations under the contract should be under his direction and control.

In other words, the contracts clearly provide that the marble should be inspected and passed upon by the architect before it was installed and that his decision should be final and controlling as to the character of the marble installed and as to whether or not it complied with the specifications. The contract was given this interpretation by the parties and if there could be any doubt about this construction the practical interpretation given by the parties would control. This is an elementary principle of law.

Town of Packwaukee v. American Bridge Co. of New York, 183 Fed. 359.

We believe we are correct in saying that the law is well settled and particularly so in the federal courts to the effect that in building contracts where the contract provides for the inspection and passing upon the work by an architect or a superintendent or engineer, the party having the right to inspect and pass upon the work must do so with reasonable diligence and that he cannot wait until after the work is completed and then complain that the materials furnished do not comply with the specifications.

See

9 C. J. 797.

"So where the work or materials is under the inspection of the owner or his architect during its progress if the builder is not complying with the contract it is the duty of the owner or architect to object to such work or materials as obviously do not comply with the contract and on his failure to do so the owner cannot after the work is completed complain that the work or materials was not in accordance with the contract."

George A. Fuller Co. v. B. P. Young Co., 126
   Fed. 343;

City of St. Charles v. Stookey, 154 Fed. 772;
   Certiorari denied 208 U. S. 617;

Town of Packwaukee v. American Bridge Co.,
   183 Fed. 359;

Greenstone Co. v. Bush Co., 135 N. W. 993
   (Minn.);

Ashland Lime, etc., Co. v. Shores, 81 N. W.
   136 (Wis.).

These are merely a few of the many cases that
might be cited to the court. We have selected these
merely because of their great similarity to this case
upon the facts. The cases are unanimous in holding
that where the work is under the supervision of an
architect neither the architect nor the owner can complain after the work is completed of any defect in
materials where the architect permitted the materials
to be installed. Under such circumstances both the
owner and the architect are estopped to complain of
any defect in either workmanship or materials used.

We desire to call the court's particular attention to
the provisions in the contract before the court in the
case of

George A. Fuller Co. v. B. P. Young Co., supra.
That contract has this provision:

"The sub-contractor * * * shall within 24
hours after receiving written notice from the
contractor to that effect, proceed to remove from
the ground or building all material condemned by
them, whether worked or unworked, and to take
down all portions of the work which the architect

or contractor shall by like written notice condemn as unsound or improper, or as in any way failing to conform to the drawings and specifications."

This provision of the contract in question in that case clearly gave the contractor and architect the right to condemn the marble after it had been installed. The marble was installed and did not meet the architect's approval and he complained about it at the time it was being installed but however, did not condemn it or order it removed until after the installation of the marble was completed. The architect and the contractor then undertook to condemn the marble. The court held that their action in condemning the marble came too late and that the architect, the contractor and the owner of the building. were estopped to allege the fact that the marble installed did not comply with the specifications.

In this case neither the architect nor the contractor nor the owner had any right under the contract to condemn the marble after it was once installed. The contract provides for inspection and acceptance or rejection before installation and the parties to the same put this interpretation upon the contract by their operations thereunder.

Simple justice and common sense require that the architect or person having the right to accept or reject materials to be installed in a building should act seasonably and promptly. To permit the inspector to wait until the materials had actually been installed and then require the removal of the same would lead to great hardship and injustice.

Speaking upon this question the court in the last quoted case said (page 346):

"Moreover, it appeared that the architects and contractor constantly supervised the erection of the building, saw the offending material being put into place, and did not require its removal, as they had ample authority to do under the sub-contract. There would have been no difficulty about removing the material while the work was progressing, for the marble wainscoting and tiling were not indispensable parts of the structure, but were merely affixed to the walls and floors. What has happened, therefore, is this: The contractor, and through him the owner, have the advantage of $60,000 worth of the sub-contractor's material and labor, which they profess to be unsatisfactory, adopting the architect's declaration to this effect, but nevertheless continue to use and enjoy. In other words, they diminish the contract price by $60,000, retaining that sum as a recompense for what is said to be the unsatisfactory character of the work. They do not reject the offending materials, but keep them in the building, while they insist upon paying a smaller price than was agreed upon. These materials were received and are retained under the contract, which fixes a definite sum to be paid therefor, but they are being treated by the owner and contractor as if they had been furnished for whatever price the architects should decide them to be worth. If this result is a necessary consequence of the written agreement, the sub-contractor must endure it. If one of the parties to an agreement is permitted thereby to assert dissatisfaction with the other's work and materials, while he continues to enjoy them at a diminished price, to be fixed by himself without appeal, it may be conceded that the contract, although improvident, is not unlawful. But a conclusion so harsh should rest upon clear and plain language, which a positive rule of law requires to be enforced. In our opinion, no such conclusion can be properly drawn from the contract of February, 1901. Its provisions must be read together, and its true meaning can only be gathered by considering the instrument as a whole. * * * Nothing was required of the architects or of the contractor except that they should express their dissatisfaction in a specified manner, namely, by notice in writing, in order no doubt to avoid disputes. In other respects their power was practically without limit, and was ample to protect the interests of the contractor and the owner to the fullest extent. Having such a power, it was the plain duty of the architects and of the contractor to exercise it as the work went on. * * *"

That this rule of law is sound there can be no question. Any other rule would lead to great injustice as is clearly demonstrated by the facts in this case. The record in this case clearly shows that had the plaintiff been seasonably informed that the marble now claimed to be defective did not come up to the specifications the plaintiff could have furnished other pieces to take the place of those rejected at a cost of not to exceed

$250.00 and could have done so without delay while engaged in the manufacturing of this particular marble, and now by the failure of the architect and the defendant to notify the plaintiff of claimed defects within a reasonable time, the court below has assessed against the plaintiff damages in the sum of $2,500.00.

This claim, delayed by the architect and the defendant until after the building was completed, that the marble furnished by the plaintiff was defective, was as surprising and shocking to the plaintiff as would be a flash of lightning and sharp clap of thunder coming out of a clear sky to a blushing June bride on her exit from the altar. The law does not sanction such unconscionable and shocking conduct on the part of architects.

The architect at the trial endeavored to justify his conduct and the defendant to support his claim of damages against the plaintiff under a secret agreement between the architect and the defendant.

Supervising Architect, Mr. Grant, testified as follows:

> "In view of the fact that the bank was very anxious to get in their premises, a compromise was made as to some of it and there was put in some with the understanding and promise of Mr. Robinson to take it out and put in other corresponding with the contract sample. Some of that has been done and some has not." (50 R. 27)

No claim is made that this secret agreement or understanding between the architect and the defendant was ever called to the plaintiff's attention or that the plaintiff acquiesced therein. It is now claimed that because of this secret agreement and the fact that the defendant has failed to keep its part that the plaintiff should be compelled to pay the costs of taking out and removing from the bank all marble placed therein under this secret agreement, manufacture other marble to take the place of that removed and install the same in the bank, and it not having done so to pay the sum of $2,500.00 damages and the defendant be entirely dissolved from performing any part of the secret agreement made by itself and be wholly relieved from any damages for its failure to perform said secret agreement.

The contract as we have seen between the plaintiff and the defendant required that the architect inspect the marble and either accept or reject the same before it was installed. That this provision of the plaintiff's contract cannot be changed or modified to the plaintiff's detriment by a secret agreement between the architect and the defendant seems to us too obvious to need the citation of authority. The action of the trial court in refusing to make the findings of fact complained of and to sustain the plaintiff's contention that the defendant, the architect and the bank by accepting the marble and installing and using the same without objection or protest to the plaintiff until long after the same had been installed were estopped from claiming the marble to be defective was clearly erroneous.

Second, there is no proof in the record to sustain the findings of the court that the defendant suffered $2,500.00 damages by reason of defective marble. The architects, Graham and Grant, were the only witnesses who testified upon this question. According to the record the architect, Grant, was the person who had the direct supervision and control of the work and looked after all details concerning the same. It was he who wrote the certificate of the architect, withholding the sum of $6,000.00 from the defendant on account of defective marble and damages for delay. His testimony on this question is as follows:

"Q. Have you made a figuration of what it would cost to replace the defective work in the bank, according to the plans and specifications?

A. Yes, sir, Mr. Graham and I attempted to compute that, and in our judgment it would cost about $2,500.00 to replace and correct all of the defects."

The certificate from the architect withholding the $6,000.00 was then introduced in evidence. The witness was then asked as to the damages for delay which was objected to by the plaintiff. The defendant explained that the evidence was being introduced merely to show good faith of the architect. The witness then continued:

"We did not attempt to assess the damages to the bank nor the cost of storing the defective marble. We had arrived at the sum of $7,300.00.

Q. How did you arrive at that sum?

A. By using $2,500.00 for the correction of the defects and $4,500.00 for the delay. I think I have a memorandum of what the figures were. Our original computation contained Twenty-five Hundred for defects, Forty-five Hundred for damages and $327.00 for some plastering and painting that had to be done after the marble workers had finished their work. That is from a memorandum and the slip that I had that on. It does not seem to be here. I think it is about right. At any rate we did not use that figure in making this deduction.

Q. You deducted less than that amount?

A. We deducted $6,000.00, which is Thirteen or Fourteen Hundred Dollars less than the other."
(52 R. 27)

The evidence of the witness, Graham, goes no further than to corroborate the evidence of the witness Grant quoted. This evidence shows the following state of facts. That when it came time to make the final payment to the defendant for installing the marble the two architects got together and made a preliminary estimate of how much they should withhold in which preliminary estimate they figured $2,-500.00 for defective marble, $4,500.00 for delay, and $327.00 for damages to plaster and painting, making a total of $7,327.00. They, however, did not use these figures in making or arriving at the final sum which they deducted, to-wit, $6,000.00. In arriving at the sum of $6,000.00, which is $1,327.00 less than their preliminary estimate as far as the record is concerned the architects may have concluded and did conclude to reduce the sum of $2,500.00 for defective marble $1,-327.00 or $1,000.00. It is a certainty that they did not put the whole sum of $2,500.00 in for defective marble and there is no evidence in the record to show what sum they finally figured in the $6,000.00 deduction for defective marble and the defendant could not possibly be damaged by reason of defective marble for any greater sum than that withheld from it by the architect.

The architects themselves say that they were not attempting to assess the cost of restoring the defective marble but were merely attempting to decide on a sum which should be withheld from the defendant until the defects were corrected. Their testimony therefore, does not form any basis by which the court could assess damages against this plaintiff in the sum of $2,500.00 for defective marble.

When we come to examine the record as to actual defects in the marble we find no evidence to sustain the finding of the trial court in the sum of $2,500.00, or in any sum approaching that amount. The defendant made no effort whatever to show the amount of marble that was claimed to be defective. The plaintiff repeatedly attempted to get the architect to specify the amount or number of pieces of marble that were claimed to be defective. The architect on each occasion refused.

The nearest the plaintiff was able to get the architect to make a definite statement as to the defective marble is found in the evidence of the witness Grant on 70 R. 24, as follows:

> "I stated yesterday that there were some pieces of marble that were off color in the bank.
>
> Q. Those, I believe, are how many? Are there three or four of them that you say that you would not accept?
>
> A. Oh, twenty or thirty.
>
> Q. When you mention twenty or thirty, those

are some you had some dispute about, but did let them go in?

A. Let them go in under protest, with the understanding that they would be taken out and others put in their places.

Q. Twenty or thirty, would you say?

A. I never counted them, but one or the other, twenty or thirty rather than three or four, as you suggested."

There was also some claim by the architect that certain check desk tops furnished by the plaintiff had not been accepted and as to the defects in those and the reason why they were not accepted by the architect the witness Grant on direct examination testified as follows:

"I think there were five check desks.   *   *   *
They were all broken in the process of manufacture. They were broken across. Each one was broken, due, as I was told, while I was in San Francisco, because of an accident while on the saws. We have not accepted them yet. They were put together again with wax, reinforced on the bottom with steel and had the waxing been sufficient to deceive the ordinary eye we probably would have accepted them."   (51 R. 21)

The evidence shows that there was altogether furnished by the plaintiff for installation in the bank some eight or nine thousand pieces (40 R. 2). Weighing in all some 141 tons (167 R. 13).

Granting for the sake of argument that there is 30 pieces of defective marble, the highest number that the architect would even estimate or guess at, the evidence shows that there were at least 8,000 pieces altogether, and the record also is clear that all the marble was to be installed for $60,101.00. The average cost of each piece would be about $7.50 or a total of $225.00 for the thirty pieces. If the check desk tops, as the architects testified, needed only a little more waxing to make them acceptable this cost could not possibly be more than an additional couple hundred dollars. The evidence, however, clearly shows that these check desk tops were not broken at the factory in the process of finishing but merely came apart at the seams which is always liable to happen in this class of marble.

The witness Boucher who was present at the plaintiff's factory at the time the check desk tops were being fabricated and who at the time of the trial was working for the defendant, testified that the check desk tops came apart during fabrication at seams running through the marble. That they did not fall over or break and that they were put together again and that they were finished in good workmanlike manner (154 R.). The defendant was bound to accept these check desk tops under the contract between the plaintiff and defendant under the following express provisions of the contract:

"It is a well known fact that fancy or colored marbles are more or less unsound, and it is hereby agreed and understood that should any of the fancy or colored marbles to be furnished under this proposal require waxing, filling, cementing or backing in order to present a smooth surface, that we will have the right to do such waxing, filling, cementing or repairing as may be necessary, and this marble, when so treated, is to be accepted."

The above quoted evidence constitutes all of the facts in the case in respect to defective marble upon which an expert could give or base an opinion as to the amount of the costs of remedying the same. Those facts form no basis nor foundation for the opinion of the architect that it would cost $2,500.00 to remedy these defects. Consequently the judgment of the court allowing $2,500.00 damages for defective marble is without evidence to support the same.

For illustration, suppose that a physician in a personal injury action went upon the stand and testified for the plaintiff that he attended the plaintiff for the injury received and that in his opinion the plaintiff's earning capacity was completely destroyed for all time in the future by reason of the injury complained of, and the plaintiff should also go upon the stand and testify and the facts should conclusively show that the plaintiff had for a period of two years prior to the trial and subsequent to the injury complained of been working at the same occupation that he pursued prior to the injury and had received the same wages and that he was able to follow that occupation without any handicap by reason of the injuries received, the expert opinion of the physician under such a state of facts would certainly form no basis or foundation for a verdict of a jury allowing the plaintiff damages for complete destruction or loss of earning capacity and such a verdict would not stand in law.

The judgment of the court in this case allowing

$2,500.00 damages for defective marble has no more foundation or basis upon which to stand than would the verdict of the jury in the assumed case.

JUDGMENT FOR $2500.00 DAMAGES FOR DEFECTIVE

MARBLE IS CONTRARY TO THE CONTRACT

The contract between the plaintiff and the defendant places the duty upon the plaintiff to furnish the marble only. The contract between the defendant and the bank requires the defendant to install the marble. The estimate of the architect, if it may be called such, that damages for defective marble amounting to $2500.00 included the cost of removing the defective marble and installing other marble in the place of the marble removed as well as the cost of the marble.

The claimed defective marble was installed by the defendant as heretofore pointed out without the knowledge or consent of the plaintiff. The plaintiff could therefore under no circumstances be assessed the full sum of $2500.00 as estimated by the architect.

FAILURE TO PRODUCE SAMPLE OF MARBLE

The contract between the defendant and bank provided as follows:

> "The quality and finish of all marble installed shall be equal to that represented by samples on file in the office of the Architect and identified by the signature of the parties hereto."

The plaintiff repeatedly demanded the production of this sample of marble (68 R. 31; 152 R. 20). Neither the architect nor the defendant produced the same. With a claim of defective marble being made it was highly important that this sample with which the marble furnished was to be compared should be produced so that the court would have some guide to enable it to determine this question.

The plaintiff asked the court to view the premises. The court did this and for the court to reach a proper conclusion should have had this criterion by which the marble was to be judged before it. It is true that the plaintiff produced a piece of marble which was marked "Plaintiff's Exhibit '1' for identification," but this was never identified as a part of the sample mentioned in the contract (68 R. 17).

It is the belief and contention of the plaintiff that if the official sample had been produced that it would have shown the same identical variation in color which the architect made complaint of and for which reason he says he condemned some of the marble installed.

The architect testified, referring to the official sample:

> "We have the other half that was cut in two where we can produce it."

This sample being demanded by the plaintiff and not being produced the presumption is that if it were produced it would not support the contention made by the defendant that the marble did not comply with the sample.

We desire to call the court's attention to another most peculiar feature of this case. The defendant is claiming damages in the sum of $2500.00 for defective marble not upon the theory that it was damaged by reason of the defective marble but upon the theory that the bank was damaged by reason of the defective marble. No officer or servant of the bank was called to the stand to testify in this case that it considered the marble defective or had any objection or complaint to make against the marble as installed. Although an officer of the bank was placed upon the witness stand, defendant did not ask him any questions which would tend in any way to show whether or not the bank was satisfied with the marble as installed (149 R. 1).

Under these circumstances the presumption is that the bank is satisfied with the marble as it was installed and if the bank, the real party in interest, is satisfied, then the defendant cannot claim damages on this ground.

These two circumstances throw strong suspicion upon the defendant's claim for damages for defective marble.

"The failure of a party to produce evidence which is within his knowledge, which he has power to produce and which he would naturally produce if it were favorable to him, gives rise to an inference that if such evidence were produced it would be unfavorable to him." (22 C. J. 111)

"We are unable to determine the width of the strap from the lines of wear upon the pin, and are in doubt whether the old breaks could have been seen or not. The block with the strap, through which the pin passed, was in the possession of the vessel after the accident; it was not produced at the trial, and no explanation was given of the failure to produce it. The failure unexplained, to produce evidence which is in the possession of a party, raises a presumption, that if produced, it would not be favorable to the contention of the party possessing it. *Kirby v. Tallmadge*, 160 U. S. 379, *Graves v. United States*, 150 U. S. 120, *Runkle v. Burnham*, 153 U. S. 216."

The Bolton Castle, 250 Fed. 403.

"Where it is apparent that a party has the power to produce evidence of a more explicit, direct, and satisfactory character than that which he does introduce and relies on, it may be presumed that if the more satisfactory evidence had been given it would have been detrimental to him and would have laid open deficiencies in, and objections to, his case which the more obscure and uncertain evidence did not disclose.

"Failure of a party to call an available witness possessing peculiar knowledge concerning facts essential to a party's case, direct or rebutting, or to examine such witness as to the facts covered by his special knowledge, especially if the witness would naturally be favorable to the party's contention, relying instead upon the evidence of witnesses less familiar with the matter, gives rise to an inference that the testimony of such uninterrogated witness would not sustain the contention of the party."

22 C. J. 115.

"Failure to produce at the trial an article which would tend to throw light on the issues may raise a presumption adverse to the party in whose possession it is."

22 C. J. 112.

Before the defendant can have an offset for damages he must actually show he has paid the damages or will be absolutely bound to pay the same in the future.

"As a general rule, the right to sue for indemnity for damages resulting from the negligence, misfeasance, or malfeasance of another accrues only when payment has been legally made by the indemnitee."

31 C. J. 452.

See

Murphy v. No. 1 Wall Street Corp., 127 N. Y. S. 735.

This was a case wherein the plaintiff, Murphy, subcontractor, sued the defendant corporation, contractor, for a balance due on contract. Defendant set up counterclaims for damages for delay. The evidence showed that the plaintiff was guilty of delay and that defendant suffered damages by reason thereof. The evidence, however, failed to show what amount defendant would be called upon to pay the owner by reason of plaintiff's delay. The court in its opinion upon the question of claim for damages said:

"Plaintiff's obligation was not to pay anything that the general contractor might be obliged to pay but only such as plaintiffs themselves might cause, and while the evidence tends to show that the plaintiffs did unnecessarily delay their work, it does not appear that they were wholly responsible for the loss to which the general contractors were put, nor does it appear how much of said loss was properly chargeable to them. There was therefore no basis upon which to estimate the amount which should be allowed defendant as damages for plaintiff's delay."

Fisher v. Edgefield & Nashville Mfg. Co.,
62 S. W. 27.

This is a case where the plaintiff, a subcontractor, sued the defendant for money due under its contract. Defendant was a general contractor putting up a building for the University of the South. The defendant set up a claim of damages for delay. Defendant had not paid the damages, which it set up as a counterclaim. The court held that the counterclaim could not be allowed, and disposed of the question, stating:

"Mr. Wonen, manager of the Edgefield & Nashville Manufacturing Co., does not admit when he was examined upon the subject that his contract would be liable for this claim to the University of the South. It is clear in any event, that he has not paid it and that it has not been agreed upon in the settlement between the University of the South and the Edgefield & Nashville Manufacturing Co., so it is not shown at least to be a fixed liability; and further than that it is not shown that it is a claim which is completely fixed upon the Edgefield & Nashville Manufacturing Co. by the University of the South."

In this case there is no evidence that the defendant has ever been called upon by the bank to pay one dollar or any other sum for damages for delay or defective marble, or for any other reason. Nor is there any evidence that the bank will ever call upon the defendant to pay a single dollar damages for delay or for defective marble. The only evidence in the case is that the architect has held up $6,000.00 until claim for damages and for correction of defective marble has been adjusted. There is no evidence in the record that defendant has in good faith, or at all, offered to change or correct defective marble or requested the plaintiff to furnish it with marble so that it could change defective marble. There is no evidence that the defendant ever at any time, or at all, has attempted to make any settlement with the bank for any claim, if the bank has any, for damages for delay, and it is clearly the duty of the defendant under the contract, and under the law, and under the evidence, to change the marble, if any, which he agreed with the architect to change, and until the defendant performs this duty and shows the court that it has been damaged in a specific sum it cannot make any counterclaim by reason of the fact that the architect has withheld the payment of the $6,000.00. For aught the court knows all the defendant would have to do to get his $6,000.00 would be to go to the bank and say, "I am ready to take out any marble that the architect, or you, say is defective and install marble in place thereof acceptable to you and the architect," and this is a legitimate inference to be drawn, and under the evidence in this case the only inference the court can draw is that the marble is entirely satisfactory to the bank and that the bank under no circumstances would allow the marble installed to be taken out and other marble substituted in place thereof.

The defendant, by its action, now says: "I have been fully paid for my services under the contract. I have no more money coming to me. The $6,000.00 withheld by the bank under the contract belongs to the plaintiff. The plaintiff may recover this any way it sees fit. I will not only not assist it by performing my part of the secret agreement to change claimed defective marble, or by making any attempt to adjust the differences between myself and the bank, but I will do everything I can to defeat the plaintiff in its effort to recover." We are of the opinion that the law does not countenance such conduct on the defendant's part.

In concluding our argument upon this question of damages for defective marble, we desire to call the court's attention to the following testimony given by the plaintiff:

"We were ready and willing to do that at any time. So far as supplying material or anything that was my portion to do, I have been at all times ready and willing to do anything in my power to furnish any marble or fix up any defects which have ever been pointed out to me. No one has asked me to furnish anything at any time for that bank that I have not furnished. I have no knowledge of anybody asking me to supply anything since the time I was there to get that check. At all times while we were manufacturing the marble, all the way through, we furnished other pieces of marble to take the place of any that was claimed to be defective. We had items of replacement many times." (165 R. 2)

This testimony of the plaintiff is wholly uncontradicted in the record. In fact, the testimony all the way through the record shows this testimony of the plaintiff to be true. No one at any time during the trial testified or even intimated that the plaintiff had failed at any time to furnish any marble called for by the contract nor to furnish any marble to replace any broken or defective pieces called to its attention. This is particularly true of the marble now claimed to be defective.

The contract between the plaintiff and defendant only required that the plaintiff furnish the marble and that the marble should comply with contract sample. The architect was to have the decision as to whether or not it complied. It is certainly true that the plaintiff might furnish a piece of marble it thought to comply with the sample and yet the architect might be of a different opinion and reject it. But if this decision of the architect in rejecting the piece of marble was not called to the plaintiff's attention by the architect or the defendant and the plaintiff given an opportunity to replace it we do not see how either in fair dealing, common sense, plain justice or the law, the plaintiff could be held liable in damages.

CRATING AND CARTAGE

ASSIGNMENT OF ERRORS NOS. 5, 6, 15, 21 AND 23

The contract between the plaintiff and the defendant contained the following provision:

"Said marble will be delivered F. O. B. cars at our works, San Francisco, California."

On November 26th, 1924, the defendant sent to the plaintiff a telegram requesting that all subsequent shipments of marble be made by water in the Pacific Steamship Company boats (36 R. 8). Mr. Robinson, on behalf of the defendant at the trial testified as follows:

"I do not deny that I told them to ship the marble by boat, and we had a shipment coming on practically every boat." (118 R. 15)

The plaintiff, through its president, testified at the trial as follows:

"The boxing of shipments in less than carload lots, from November 3, 1924, to February 19, 1925, amounted to the sum of $484.78. The cartage on shipments in less than carload lots, amounted to $60.00. * * * The contract calls for shipment of this marble in carload lots, f. o. b. cars, San Francisco. When shipped by boat in pieces it is necessary to crate the same and cart them to the terminals." (35 R. 16)

"The boxing was necessary as to those shipments which were made by water. You cannot ship marble that way unless it is boxed." (166 R. 23)

The court disallowed the plaintiff these sums upon the following grounds:

"Plaintiff claims extra to the amount of $1,-
371.76. Among these is a charge of $544.78
of which $484.78 is for the expense of boxing
marble, and for cartage $60.00. This expense
was incurred by plaintiff in shipping, at defendant's request, marble by boat instead of by carload lots. The shipments by water avoided
further delay. These items are disallowed." (The
court's opinion, 188 R. 8)

The court, as we have already pointed out, was
clearly wrong in its opinion that the time for installing marble began at the date of first delivery of
marble, to-wit, April 26, 1924, instead of September
22, 1924, the date when the building was ready to
receive the marble for installation and was wrong
in its conclusion that the marble was not installed so
that the bank could occupy its quarters within the
time provided by the contract. This being true the
court was in error in refusing to allow the plaintiff
these items of expense made necessary by the order
and direction of the defendant. Even though there
had been delay there is no fact in the record to show
that the shipment of this marble by railroad as called
for by the contract would have caused any additional
damages for delay.

The court allowed the defendant the sum of $25.00
"for wooden hand balustrade on open stairway to the
basement" (200 R. 9).

There may be some evidence in the record to sustain this finding of the court but after some diligent
search we have been unable to find it.

We desire to call the court's particular attention

to the fact that the court allowed in all $7,000.00 damages for defective marble and delay (200 R. 1).

The bank withheld from the defendant from its contract price for installing the marble the sum of only $6,000 for damages for delay and for defective marble (53 R. 18). There was no evidence introduced of any damages suffered by defendant for delay or defective marble except that held out by the architect. The defendant therefore could not possibly be damaged in any greater sum than $6,000.00 on these two items, and that is all it asked for in its answer (8 R. 12).

DEFENDANT'S WRITTEN ADMISSION OF LIABILITY

The marble was installed so that the bank moved in the banking quarters on the 20th day of December, 1924. At that time all of the marble was installed, except some hand railing upon the stairs leading from the bank down to the safety deposit vaults. This marble, due to the failure of the defendant to furnish setting plans or shop drawings was not fabricated and installed until March 3, 1925.

The last marble shipped by the plaintiff was received by the defendant on February 24, 1925 (56 R. 1). The plaintiff on February 25, 1925, made up and sent to the defendant its final invoice (171 R.). This invoice showed a balance due and owing plaintiff from the defendant of $11,290.32. On March 16, 1925, Mr. Eastman on behalf of the plaintiff went to Seattle to go over the plaintiff's account with the defendant and get payment for the marble furnished (165 R. 16; 146 R. 17). At this time plaintiff received from the defendant a check in the sum of $4,000.00 which was the last sum paid by the defendant on the account (146 R. 19). This payment reduced the plaintiff's claim against the defendant to the sum of $7,290.32, the amount sued for in the complaint. More than two months after the plaintiff and the defendant had gone over their accounts, and on May 25, 1925, the defendant wrote to the plaintiff a letter in which it acknowledged that on that day it was indebted to the plaintiff according to its books in the sum of $6,576.64. This letter was introduced in evidence as plaintiff's Exhibit "No. 5", and is found at 147 R. 1.

The letter in addition to admitting that according to its books it owed the plaintiff $6,576.64, admitted that the defendant had not given the plaintiff credit for items on plaintiff's invoice which were checked over with Mr. Eastman while he was in Seattle. The items which the defendant checked over with Mr. Eastman while he was in Seattle and made objection to are as follows: $7.85, $81.94, $16.09, $63.20, and $544.78, amounting in all to the sum of $713.95. These disputed items added to the amount admitted by the defendant to be due equals the sum of $7,289.59, the same amount that the plaintiff claimed due, except a few cents (116 R. 29).

In other words, on May 25, 1925, more than two months after the books of the plaintiff and defendant had been closed, the defendant voluntarily writes to the plaintiff and states it to be a fact that according to its own books it is indebted to the plaintiff in the exact amount claimed by the plaintiff in its final invoice, except that it reserves the right to dispute a few small items in the invoice amounting in all to the sum of $713.95, and admits absolute indebtedness and liability in the sum of $6,576.64.

Plaintiff brings suit for the full amount it claims due. Not a single thing takes place to change the relation of the parties between the date of the writing of the letter and the date of the trial or date of judgment. The court in its decision not only finds that the defendant is not indebted to the plaintiff at all, but that the plaintiff is indebted to the defendant in the sum of $618.32 and directs and signs a judgment to this effect.

By reason of the facts herein set forth, and the law applicable thereto, the plaintiff feels aggrieved by the judgment of the trial court entered herein and respectfully petitions this court to reverse said judgment on the grounds, and for the reasons herein set out, and to grant the plaintiff a new trial.

Respectfully submitted,

GEORGE CLARK SARGENT,
GROSSCUP & MORROW,
CHAS. A. WALLACE,
J. O. DAVIES.
Attorneys for plaintiff in error.

In the
United States Circuit Court of Appeals

For the Ninth Circuit

No. ~~5307~~ *5037*

THE MISSION MARBLE WORKS,
a Corporation,

Plaintiff in Error

vs.

ROBINSON TILE & MARBLE. COMPANY,
a Corporation,

Defendant in Error

ON WRIT OF ERROR TO THE DISTRICT COURT OF THE
UNITED STATES FOR THE WESTERN DISTRICT
OF WASHINGTON

Brief of Defendant in Error

Appearance for Defendant in Error:
H. A. P. MYERS
1160 Empire Building, Seattle, Washington   F I L

INDEX

CASES CITED

TEXT BOOKS CITED

In the
United States Circuit Court of Appeals
For the Ninth Circuit

No. 5307

THE MISSION MARBLE WORKS,
a Corporation,

Plaintiff in Error

vs'.

ROBINSON TILE & MARBLE COMPANY,
a Corporation,

Defendant in Error

ON WRIT OF ERROR TO THE DISTRICT COURT OF THE
UNITED STATES FOR THE WESTERN DISTRICT
OF WASHINGTON

Brief of Defendant in Error

STATEMENT

The statement of the case made by plaintiff at pp. 1-3 is substantially correct and may be considered true for the purpose of this appeal.

For convenience and brevity, plaintiff in error and defendant in error will be referred to as "P" and "D," respectively.

ARGUMENT

I.

FACTS IN THIS CASE NOT REVIEWABLE

A jury was waived in writing by the parties, and the cause tried by Judge Cushman.

Rev. Stat. Sec. 649 (Sec. 1587 C. S.) is as follows:

"Issues of fact in civil cases in any circuit court may be tried and determined by the court, without the intervention of a jury, whenever the parties, or their attorneys of record, file with the clerk a stipulation in writing waiving a jury. The finding of the court upon the facts, which may be either general or special, shall have the same effect as the verdict of a jury."

Rev. Stat. Sec. 700 (Sec. 1668 C. S.):

"When an issue of fact in any civil cause in a circuit court is tried and determined by the court without the intervention of a jury, according to section six hundred and forty-nine, the rulings of the court in the progress of the trial of the cause, if excepted to at the time, and duly presented by a bill of exceptions, may be reviewed by the Supreme Court upon a writ of error or upon appeal; and when the finding is special the review may extend to the determination of the sufficiency of the facts found to support the judgment."

Rev. Stat. Sec. 1011, Amended (Sec. 1672 C. S.):

"There shall be no reversal in the Supreme Court or in a circuit court upon a writ of error, for error in ruling any plea in abatement, other than a plea to the jurisdiction of the court, or for any error in fact."

WHEN AN ACTION IS TRIED WITHOUT A JURY, BY A FEDERAL COURT, THE FACTS FOUND, WHETHER GENERAL OR SPECIAL, ARE NOT REVIEWABLE

> *Wear v. Imperial Window Glass Co.*, 224 Fed. 60;
>
> *U. S. v. Atchison, T. & S. F. Ry. Co.*, 270 Fed. 1;
>
> *Highway Trailer Co. v. City of Des Moines*, 298 Fed. 71;
>
> *U. S. v. U. S. Fidelity & G. Co.*, 236 U. S. 512;
>
> *U. S. Fidelity & G. Co. v. Bd. of Commrs.*, 145 Fed. 144.

In *Wear v. Imperial Window Glass Co.*, 224 Fed. 60, at p. 63, it is said:

"* * * When an action at law is tried without a jury by a federal court, and it makes a general finding, or a special finding of facts, the act of Congress forbids a reversal by the appellate court of that finding, or the judgment thereon, 'for any error of fact' (Revised Statutes, Sec. 1011 (U. S. Comp. Stat.

1913, Sec. 1672, p. 700)), and a finding of fact contrary to the weight of the evidence is an error of fact.

"The question of law whether or not there was any substantial evidence to sustain any such finding is reviewable, as in a trial by jury, only when a request or a motion is made, denied, and excepted to, or some other like action is taken which fairly presents that question to the trial court and secures its ruling thereon during the trial. United States Fidelity & Guaranty Co. v. Board of Com'rs, 145 Fed. 144, 150, 151, 76 C. C. A. 114, 120, 121, and cases there cited; Mercantile Trust Co. v. Wood, 60 Fed. 346, 348, 349, 8 C. C. A. 658, 660, 661; Barnard v. Randle, 110 Fed. 906, 909, 49 C. C. A. 177, 180; Barnsdall v. Waltemeyer, 142 Fed. 415, 417, 73 C. C. A. 515, 517; Bell v. Union Pacific R. Co., 194 Fed. 366, 368, 114 C. C. A. 326, 328; Seep v. Ferris-Haggarty Copper Min. Co., 201 Fed. 893, 894, 895, 896, 120 C. C. A. 191, 192, 193, 194; Pennsylvania Casualty Co. v. Whiteway, 210 Fed. 782, 784, 127 C. C. A. 332, 334.

"There is another reason why no reviewable question of law is presented to this court in this case. A trial court is entitled to a clear specification by exception of any ruling or rulings which a party challenges and desires to review, to the end that the trial court itself may correct them if so advised, and, if. it fails to do so, that there may be a clear record of the rulings and the challenges thereof. For this purpose a rule has been firmly established that an exception to any ruling which counsel desire to review, which sharply calls the attention of the trial court to the specific error alleged, is indispensable to the review of such a ruling. * * *"

In *U. S. v. Atchison, T. & S. F. Ry. Co.*, 270 Fed. 1, at p. 4, the following language appears:

"The making of special findings of facts in an action at law tried by the court on a waiver of a jury is discretionary with the trial court, and its action in making such findings, in refusing to make requested findings, or in refusing to amend findings made, is not subject to exception, or to a subsequent review in a federal appellate court. City of Key West v. Baer, 66 Fed. 440, 444, 13 C. C. A. 572; Berwind-White Coal Min. Co. v. Martin, 124 Fed. 313, 60 C. C. A. 27; Aetna Life Ins. Co. v. Board of County Commissioners of Hamilton County, 79 Fed. 575, 576, 25 C. C. A. 94."

In *Highway Trailer Co. v. City of Des Moines*, 298 Fed. 71, at p. 73, it is said:

"The only findings of fact or rulings of law that might have been reviewable in this court, if proper requests for opposite findings and rulings had been made, denied, and excepted to before the trial below closed, are found in the judgment and one of them is 'that upon the whole record the plaintiff is not entitled to recover against the defendant herein.' Counsel challenge this and other findings of fact in this judgment; * * *

"The trial in this case ended before or on January 31, 1923, when the court filed its opinion on the merits of the case. There was no such request or motion made, denied, or excepted to before the trial ended, and subsequent requests, and ruling thereon, are, like motions for new trials after verdicts, discretionary with the trial court, and not subject to review in the federal appellate courts.

"Counsel cite statements of facts in the opinion of the court on the merits of the case and base arguments for a reversal thereon. But facts stated in the opinion of the court cannot be treated as special findings of facts in the cause, and, if they could have been, none of them, nor of the declarations of law in the opinion, were challenged by requests for opposite findings or rulings, and exceptions to refusals to make them, before the trial ended, and they are therefore not reviewable now. * * *"

In *U. S. v. Fidelity & G. Co.*, 236 U. S. 512, at p. 527, the Supreme Court said:

"* * * The findings have the same effect as the verdict of a jury, and this court does not review them, but merely determines whether they support the judgment. * * *"

In *U. S. Fidelity & G. Co. v. Board of Commissioners*, 145 Fed. 144, at p. 151, Judge Sanborn used this language:

"The verdict of a jury concludes all issues of fact and of mixed law and fact save those questions of law which have been reserved for review by demurrer, motion, request or exception. A finding of the court without a jury has the same effect, with the single exception that when the finding is special the question whether the facts found sustain the judgment is open to review. In the trial of an action by the court without a jury the rulings of the court in the progress of the trial, and those only, are open to review. The true test for determining whether or not a question or ruling in a trial by the court without a jury is reviewable is the answer to the question whether or not it would have been open to review if the trial had been to a jury."

Art. VII, Amendments to the United States Constitution, is as follows:

"In suits at common law, where the value in controversy shall exceed twenty dollars, the right of trial by jury shall be preserved, and no facts tried by a jury shall be otherwise re-examined in any court of the United States than according to the rules of the common law."

Inasmuch as the court's decision has the same effect as the verdict of a jury, we think the above amendment to the Constitution applies to the case at bar.

II.

WHAT RULINGS OF THE COURT COMPLAINED OF WERE MADE DURING "THE PROGRESS OF THE TRIAL OF THE CAUSE?"

The findings made by the court involved in this error proceeding, are portions of Finding No. VI made by the court as follows:

"The court further finds that the defendant is entitled to and is allowed the following counter-claims and offsets, as shown by the evidence (199 R. 30):

(a) Damages caused by delay on plaintiff's
part in furnishing said marble according to contract_____$4,500.00

(b) For defects in said marble_____ 2,500.00

(200 R. 1)   *   *   *"

If the findings above mentioned are general, they
cannot be reviewed in this court at all.

U. S. Comp. Stat., Sec. 1668;

Cooper v. Omohundro, 19 Wallace 65;

Martinton v. Fairbanks, 112 U. S. 670;

In the *Martinton case,* 112 U. S. 670, this language
appears:

"The theory of the plaintiff in error seems to be
that the general finding in this case, like a general
verdict, includes questions of both law and fact, and
that, by excepting to the general finding, he excepts
to such conclusions of law as the general finding implies; but section 649, Rev. St., provides that the finding of the court, whether general or special, shall have
the same effect as the verdict of a jury. The general
verdict of a jury concludes mixed questions of law
and fact, except so far as they may be saved by some
exception which the party has taken to the ruling of
the court upon a question of law. *Norris v. Jackson*
(9 Wall. 125). But the plaintiff in error has taken
no such exception. By excepting to the general finding of the court, it is in the same position as if it had
submitted its case to the jury, and, without any exceptions taken during the course of the trial, had,
upon a return of the general verdict for the plaintiff,

embodied in a bill of exceptions all the evidence, and then excepted to the verdict because the evidence did not support it.

"The provision of the statute, that the finding of the court shall have the same effect as the verdict of a jury, cuts off the right of review in this case; for the seventh amendment to the Constitution of the United States declares that 'no fact tried by a jury shall be otherwise re-examined in any court of the United States than according to the rules of the common law.' The only methods known to the common law for the re-examination of the facts found by a jury are either by a new trial granted by the court in which the issue had been tried, or by the award of a *venire facias de novo* by the appellate court for some error of law.   *   *   *"

If the findings made should be considered by this court as "special," within the meaning of Sec. 1668 of the Compiled Statutes, then P. cannot question the special facts therein found, or deny them.

. In *Dooley v. Pease*, 180 U. S. 126, it is said:

"Where a case is tried by the court, a jury having been waived, its findings upon questions of fact are conclusive in the courts of review, it matters not how convincing the argument that upon the evidence the findings should have been different.   *   *   *"

Its only right, under the express mandate of the statute, is a claim that the facts found do not support the judgment. This claim, however, could not, we think, be made in good faith by eminent counsel for P.

Assignments of Error Nos. I and II (13 R.) clearly cover "rulings of the court in the progress of the trial." All the other twenty-four assignments of error concern the findings, conclusions and judgment signed by the court, and those proposed by P. and refused. These matters, that is, the exceptions to findings and judgment made by the court, and the request for findings and judgment for P., were made after the judgment was signed, and therefore occurred after the trial ended. Moreover, the twenty-four other assignments of error, which concern the facts only, are immaterial here, as this court is not concerned with the findings made by the trial court. The statute forbids this court from interfering with the facts found.

A TRIAL ENDS WHEN THE JUDGMENT IS RENDERED OR THE DECISION FILED

> *U. S. Fidelity & G. Co. v. Com'rs.*, 145 Fed. 144;
>
> *Mercantile Trust Co. v. Wood*, 60 Fed. 346;
>
> *Highway Trailer Co. v. City of Des Moines*, 298 Fed. 71.

In the *Wood* case, 60 Fed. 346, at p. 348, it is said:

"* * * The finding of the court, whether general or special, performs the office of a verdict of a jury. When it is made and filed, the trial is ended.

Exceptions to the finding, or to statements of legal conclusions contained in it, or in an opinion in which it is contained, or in an opinion filed with it, avail nothing. They are as futile as exceptions to the verdict of a jury. When a case comes to this court upon a writ of error, this is a court for the correction of the errors of the court below solely.   *   *   *"

Therefore, these proposed findings and exceptions to findings cannot be considered "rulings of the court in the progress of the trial."

> *Town of Martinton v. Fairbanks,* 112 U. S. 670;
>
> *Cooper v. Omohundro,* 19 Wall. 65;
>
> *Mercantile Co. v. Wood,* 60 Fed. 346.

The trial ended when the court filed its decision, or memorandum opinion. This was on Oct. 4, 1926. (179 R. 1). All that the court did afterwards—on November 1, 1926—when the findings and judgment were signed, and when those presented by P. were rejected, were rulings made after the trial ended, and not "in the progress of the trial of the cause."

It is plain, we think, that the only rulings of the court in the progress of the trial were those mentioned in Assignments of Error Nos. I and II.

III.

ONLY TWO OF THE ASSIGNMENTS OF ERROR AVAILABLE

It is clear, we think, that the only assignments of error available to P. for argument in this court are Assignments Nos. I and II. We will now argue these:

Did the court commit error in permitting the witness Francis W. Grant to answer the following question:

"Q. Referring to the two items upon which the architect declined to make payment on this job until they are corrected, one being for defective marble, and that has never been corrected and that is the item which you have estimated to the court would cost $2,-500.00, and the other is on account of delay in the performance of the contract; what was the item of damages under that?" (54 R. 15.)

P. objected to this question on the ground that it was immaterial. (54 R. 24.)

If the evidence sought to be adduced by the question was simply immaterial, then the evidence is harmless and not prejudicial.

22 Cyc. 161.

But, we think it cannot be contended that the question is immaterial.

The letters and wire sent to P. and introduced in evidence unmistakably show that D. was greatly delayed in getting the marble to Seattle (Ex. A-11, 120 R) ; the evidence is likewise conclusive as to the defective marble (Ex. A-12, 128 R). These two matters evidently caused the writing of the letter from Graham, the architect, to D., withholding the $6,000.00 from the contract price on account of "defective marble" and "delay." (53 R. 14.)

In view of this, we think it was material to show the grounds, if any, for withholding the $6,000.00, and what damages had been sustained on account thereof. In objecting to the question, counsel admitted that the witness might have a right "to give his estimate as to what it would cost to remedy it" (the defective marble). We think this is an admission that the question, insofar as it concerns the $2,500.00 for defective marble, was proper. Nor do we think the answer showing the delay causing damages in the sum of $2,500.00 is immaterial or improper. The witness did not state how the $4,500.00 damages to D. for delay was caused, but P. did not cross-examine the witness on this item, and hence should not now complain about it.

Did the court commit error in allowing witness John Graham to answer the following question?

"Q. What was the other item, besides the item of defective marble, and on account of adjustment of damages sustained by the bank on account of that delay?" (104 R. 30.)

The objection to this question was that the witness was "not competent to estimate the damages to the bank." (105 R. 5.)

We know of no reason why the witness was incompetent to answer this question, and his incompetency has not been pointed out by P. The witness was the architect for the job. Under the main contract he was the sole judge of the "quality of the work" (58 R. 13); and final payment was not due from the bank until the Architect (withness in question) was convinced that the work was "satisfactorily completed." (60 R. 1.)

The immediate effect of this letter was to stop the payment of $6,000.00 still due to D. as the balance of its contract. This was an immediate damage to D., and we think D. had a right to inquire into this matter to determine whether the damages were feigned and trivial or true and substantial.

If these items of delay and defective marble were proper charges against D., it is clear they were caused by P., and it admitted its responsibility in open court under those circumstances.

Eastman, the President of P., testified:

"Without any question, when I made this contract, I understood that it was my duty to fill that contract, to furnish that marble to Mr. Robinson so as to enable him to carry out his contract with the bank, and I certainly felt my obligation to the Dexter-Horton Bank just as much as though I had myself signed the contract. Every effort was made and I had a personal interest in trying to do so." (167 R. 16.)

Besides, the objections made were too general, to base error in the trial court.

> 1 Wigmore on Evidence, Sec. 18;
>
> *N. Y. Electrical Equipment Co. v. Blair*, 79 Fed. 896;
>
> *Sigafus v. Porter*, 84 Fed. 430;
>
> *Noonon v. Mining Co.*, 121 U. S. 393.

In Wigmore on Evidence, supra, that learned author says, at page 57:

"The cardinal principle (no sooner repeated by courts than it is forgotten by counsel) is, that a general objection, if overruled, cannot avail."

IV.

ANSWER TO THE ARGUMENT OF P. AS TO THE FACTS

Without waiving the contention of D. that no discussion of the facts of the case is proper in this court, for the reason that under the statute and decisions the trial judge is the final arbiter of the facts—in reply to the brief of P. on that subject we wish to submit:

P. claims (p. 30) that under the evidence given by George B. Eastman, President of P., that it was entitled to recover the entire balance of the contract price of marble and extras furnished. We admit this, subject to the counter-claims set up by D.

The authorities cited by P. (p. 30) announce the doctrine that where there has been substantial performance of a contract for the delivery of materials for a building or structure, the contractor may recover thereon, subject, however, to the right of the defendant to recoup for damages caused by *defective material* or for *delay* in completion according to the time fixed in the contract. That is precisely what D. in the case at bar is seeking to do.

In *City of St. Charles v. Stookey*, 154 Fed. 772, at p. 775 (cited by P.), which was an action for the unpaid balance for waterworks, Judge Sanborn said:

"* * * But the contractor substantially performed his agreement, the city took and retained the benefits of his performance, and the court below rightly held and charged the jury that in such a case the contractor may recover the agreed price, less the damages which have resulted to the owner from the former's failure to completely perform. If the repairs

made by the city were caused by improper materials
or workmanship, it was entitled to deduct their cost
from the portion of the contract price it owed to the
plaintiff; * * *"

At page 777, same case, it is said:

"* * * But an action for the purchase price or
for the contract price under an averment of sale or
of performance may be maintained in the federal
courts by one who has substantially but not completely
fulfilled his agreement when the purchaser or the
party with whom he has contracted has received and
retained the benefits of his performance, and he is not
driven to an action on the quantum meruit to obtain
relief. The amount of his recovery may indeed be
restricted to the agreed price, *less the damages sustained by the defendant from his failure to completely
perform his contract* (italics ours), but he may,
nevertheless, recover in his action upon the contract. * * *"

While P. has filed 26 separate assignments of error
(13 R.), yet in the brief argument is made under
only 7 separate contentions. We will make answer to
these in the order followed by P.

V.

DEFENDANT ENTITLED TO DAMAGES FOR DELAY IN FURNISHING MARBLE

At page 33 P. says:

"This provision of the contract between the plaintiff and the defendant given its broadest construction

and the most favorable to the defendant, simply makes the contract between the defendant and the bank as to time for installing the marble a part of the contract between the plaintiff and defendant."

But D. does not agree with this construction of that provision of the contract, nor do we agree that the proper construction is that the bank's contract should be considered as attached to the contract (30 R. 17) between P. and D. The only proper construction is the one in plain harmony with the language used; that is, that P. would furnish the marble to D. so that D. could "faithfully and promptly fulfill its contract with the above bank."

At page 35 P. says:

"* * * The contract also provides that the entire contract shall be completed within three months after the first delivery of material as above provided. The first material was delivered on April 26, 1924 (46 R. 6; 167 R. 4). Did the three months period provided in the contract for the installation begin to run from this date? The trial court was of the opinion that it did, and so held (183 R. 10), and based its judgment allowing damages upon this conclusion. * * *"

As admitted in the brief and shown by the contract between the bank and P. the "entire contract shall be completed within three months after the first delivery of material." P. admits that it is bound by this provision. It is also shown by the evidence (46

R. 6; 167 R. 4), and admitted by P. (p. 35) that the first delivery of material was made April 26, 1924. Hence it was the clear contractual duty of P. to deliver all the marble on or before three months thereafter, or July 26, 1924. This provision as to time of furnishing the marble is plain and unambiguous, and cannot be explained away or excused by other clauses in the contract. The marble was not delivered on or before July 26, 1924.

Mr. Eastman testified:

"* * * The first car was shipped north about April 26, 1924. The second car was shipped June 4, 1924. The third car was shipped July 28, 1924, and the fourth car was shipped October 11, 1924. * * *" (46 R. 7.)

Notwithstanding the definite provision of the contract as to the time of delivery of the marble, P. contends that P.'s limit for delivery was enlarged and extended for the alleged reason that the bank was not ready for installation of the marble on April 26th—when the first car of marble was shipped. We do not think this fact, if true, excuses P. from delivering all the marble within three months of the first delivery, as provided in the contract.

In the trial court P. attempted to excuse the violation of the contract—the great and inexcusable delay in delivering the marble—because P. claimed someone had notified P. to "cease making shipments" (39 R. 23). Mr. Eastman testified:

"We had our contract with Mr. Robinson. I don't believe Mr. Robinson ever told me to let up on the work. Mr. Robinson, during the latter part of the job, wired us many, many times, urging us to send the marble on here. That was after he told us again to start. * * *" (41 R. 16.)

From the evidence of Eastman it appears that he admits Robinson never advised P. to let up on the shipments. John Graham, the architect, testified:

"* * * I never gave Mr. Eastman any authority or said anything to him that could in any way be construed as authority, or twisted into authority to stop, slacken or let up in sending marble here. I did not interfere with the contract between Mr. Robinson and Mr. Eastman. I looked upon Mr. Eastman as a subcontractor under Mr. Robinson and all our business details were carried out with Mr. Robinson. I know my assistant, Mr. Grant, went to San Francisco. He was sent to see if he could expedite the work, not being able to get any communication from the Mission Marble Works, no answer to telegrams or letters and no information, and Mr. Robinson suggested that Mr. Grant might go down and satisfy himself as to what was being done. Mr. Robinson was much perturbed about it, because naturally he was the man to whom we looked. He went down at Mr. Robinson's expense." (106 R. 4.)

The trial court, in discussing this claim that there was some agreement to cease or let up on the shipments said, "it is not reasonable to conclude that such an agreement was ever made." The trial court's conclusion is plain and forceful:

"The main contention has been that the plaintiff did not cause the delay, but that the bank with the acquiescence of defendant, after plaintiff began the delivery of the marble, requested plaintiff to ship no more marble until requested. The evidence does not make this matter entirely clear. There evidently was a conversation between Mr. Eastman, president and manager of plaintiff, and Mr. Robinson, president of defendant, and some officer of the bank, in which the dilemma in which the bank and defendant found themselves because of the lack of a place in the building to store the marble in accordance with paragraph 10 of the above contract was discussed. This situation, in part, was no doubt caused by the condition of the concrete floor in the basement, which the 'owner' was to provide, in compliance with Par. 5 of such agreement. The plaintiff has not sustained the burden of showing any modification of the contract as to the time allowed for completion, as now claimed by it. Plaintiff's contract was to deliver marble f. o. b. cars at its work in San Francisco, so it was not concerned with the storage of marble in Seattle, and it must have been obvious at all times that the expense of storage, elsewhere than in the building, would be a small fraction of that caused by any substantial delay in the performance of the contract. Under the circumstances, and in the light of the conduct of the parties, it is not reasonable to conclude that such an agreement was ever made." (184 R.)

P. complains at page 51 that the evidence does not show that the $4,500.00 was rent which the bank was compelled to pay.

The evidence of Graham fairly shows that the bank's rental was $4,500.00; otherwise there would be no occasion for including this item in the damages caused to the bank. Moreover, P. evidently was satisfied what the $4,500.00 meant, as no cross-examination was had upon this item.

Much evidence and persistent claims were adduced in the trial court to show the delay in furnishing marble, caused by orders to stop or let up on shipments, *but these claims are evidently forgotten in this court.*

At page 38 P. says:

"For the trial court to take the position that there was a default in the installation of the marble within the time provided by the contract before the building was turned over for installation of the same, appears to us to be exceedingly harsh, unjust and unreasonable, and we do not believe that any authority can be found supporting such a position. * * *" (38 R. 19.)

We think this conclusion is erroneous; the trial court did not take the position that because the marble was not installed before the building was turned over for installation of the marble—there was a default in installation, but it evidently did hold that when P.

failed to complete delivery of the marble within three months of the first delivery, which was April 26th, P. violated its contract with D.

D. has no quarrel with P. as to the authorities cited at pages 39-48. They state the law correctly. The trouble is, the application sought to be made of these authorities. Of course, if the Dexter Horton National Bank withheld possession of the building from D. and thus prevented installation of the marble, no default could be claimed by the bank against D. during the time such possession was withheld. But the withholding of such possession did not concern P. nor excuse it from furnishing the marble within the time provided in its contract.

P. contends that there can be no damages assessed against P. for rent, because P. had three months from September 22nd in which to complete the entire contract. However, the court allowed $450.00 as damages for "delay on plaintiff's part in furnishing said marble according to contract (200 R. 1)." In its memorandum opinion the court, referring to the damages for delay, said:

"Under written stipulation, this case has been tried to the court without a jury. By Sec. 9 of the contract last above quoted it was agreed the entire contract should be completed within three months after the first delivery of material. This provision was, by ref-

erence, made a part of plaintiff's contract with the
defendant. The first car of marble was shipped by
plaintiff April 26th, 1924. The bank moved into its
rooms December 20th the same year, before the marble had been all installed, a part of which plaintiff did
not furnish until after the latter date. The architect,
Mr. Graham, and his superintendent, Mr. Grant, testified that this delay had damaged the bank to the
extent of $4,500.00 Aside from the rent that the
bank was paying while kept out of its rooms, probably the exact amount of this damage it would be impossible to establish, a considerable item of which
would be the injury to its business and prestige caused
by carrying on the former in unfinished quarters in
the construction of which workmen were still engaged." (183 R. 11.)

In other words, in fixing the damages at $4,500.00
the court did not limit the damages to rental alone.
Moreover, the marble contract was not even completed
within three months of September 22, 1924.

"When the bank opened, it was on the 20th day of
December, 1924, the railing on the stairs leading from
the bank down to the basement was not completed,
and the side walls of the stairs and there was some
work at the base of the stairs; a portion of the balustrade on the stairs on the first floor was also incomplete. The only portion that was not completed pertained to the stairs leading from the bank proper
down to the basement. That includes the drinking
fountain and the steps of the stairs. The marble on
those was in place so people could walk on them." (81
R. 10.)

The marble for completion of the contract was not all furnished prior to January 17th, 1925. On that date the Dexter Horton National Bank wired P. as follows:

"Mission Marble Company,
"San Francisco, Calif.

"We are extremely desirous of knowing how you can excuse yourselves for the treatment you have shown the Dexter Horton National Bank in attempting to fill the order of marble for our banking room. The original order was placed with you over a year ago and we were kept out of the occupancy of our new banking room for a long period due entirely to our inability to get marble to finish the interior of the room. We took possession of our new room on December twentieth and have had to get along the best we could with the center of our lobby all torn up waiting for the balance of the marble to arrive. It is nearly a month since we moved in and still you have not shipped the balance of order required to complete job. Will you not in fairness and decency see that the order is completed and shipped immediately.

"Please answer.

"DEXTER HORTON NATIONAL BANK.

Paid   Day Letter   Defendant's Exhibit 'A-15"
(178 R. 5)

VI.

Pleadings and Proof Justified Damages Against P. for Delay in Furnishing Marble

P. complains at page 49 that the answer nowhere alleges that the bank suffered any damages on account of delay on P.'s part in furnishing the marble. If this alleged defect in the pleading is fatal, it is singular that learned counsel for P. did not peremptorily end this branch of the case by filing a general demurrer thereto. It would seem that if D. set forth that it had been damaged by the delay, it would be immaterial to allege that some third party had also been damaged. D's damages as to delay and defective marble are set forth in subdivision "f" of Par. IV of the Answer. (8 R. 12.)

It is true that under the answer D. only asked for $6,000.00 under the head of defective marble and delay, but when the findings were signed the court permitted the following amendment:

"* * * It is ordered that the prayer of defendant's answer be amended by adding thereto 'and for judgment against the plaintiff for all counterclaims and setoffs shown by the evidence'." (197 R. 25.)

As to proof of damages the record is brief, and is supported by two men, and is uncontradicted. John

Graham, the architect for the Dexter Horton Building, has more than local professional fame; he has been an architect for 30 or 40 years, on some "very large buildings" in Seattle (96 R. 7). Francis W. Grant has been an architect for 25 or 30 years (48 R. 4); had charge of construction of the United States Post Office at Seattle, and he was formerly Superintendent of Buildings for the City of Seattle (81 R. 1). P. contends that all the evidence on the question of damages is set forth on pages 50-51 of the brief, but we think this evidence only concerns the amount of damages.

There was other evidence of damages caused by the delay in furnishing the marble.

Graham testified:

"* * * I never run into anything like that. We run into a standstill. They said, 'We will get it as soon as we can get it, get it up.' In response to telegrams, letters and entreaties of all sorts, we could not even get the common courtesy of a reply, and finally the banker, Mr. Parsons, himself, took it up. * * * Mr. Eastman was in San Francisco, and we were here and we had no marble and he had, or whatever it was, and it was a difficult situation. We could not even get the common courtesies of business. * * *

"* * * The real cause of this trouble was delay in getting the marble." (102 R. 30; 103 R. 2; 27.)

While Robinson was in San Francisco on November 5, 1924, trying to hurry delivery of marble, the architect wired as follows:

"Seattle Wn. Nov 5 1924

"A. T. Robinson
"Care Mission Marble Works
"San Francisco Cal

"Bank directors very much disturbed at not having marble ready to move bank on Thanksgiving day Stop Have agreed to postpone to December sixth but no later Stop If your work is not finished then will hold you and your bondsmen for all damages caused by delay Stop Everyone else will be finished on time Stop It is outrageous that after all this time you have had since January Twenty-fifth that your work should not be ready Stop Bronze workers have had to stop their work today because of missing marble Stop Will you wire immediately definite statement as to when your work will be completed as printed Invitations are being made for bank opening on December sixth and will have to be cancelled if that date cannot be kept.

"JOHN GRAHAM"

(65 R. 20; 66 R. 1)

Grant testified:

"* * *I have followed the business of an architect twenty-five or thirty years, I guess * * * (48 R. 4.)

"* * * There was insufficient marble to proceed with the setting of it continuously at the time they were ready to set—not enough marble in Seattle, delivered by the Mission Marble Works, to permit the continuous work of setting it in place.   *   *   *" (48 R. 15.)

"* * * I went to San Francisco in reference to this bank job, at the request of Mr. Robinson. * * * I was absent about a week. I talked to Mr. Eastman about expediting the work. I visited the shop of the marble concern to (49 R. 25) whom we had sublet a portion of the work in order to get it done faster. * * * I visited that place and saw the work in progress and endeavored to expedite it. I asked them to rush it as much as possible. Mr. Robinson paid my expenses on that trip.   *   *   *" (50 R. 8.)

"I was right there on the job all the time except for the brief period that I was in San Francisco. The work was delayed for lack of marble." (60 R. 23.)

"* * * The delay was not caused by not getting the shop drawings here. The delay was caused by not getting the marble here. We could go ahead without the shop drawings. We did have difficulty and we proceeded without them; that is, without a complete set. We got part of the shop drawings. * * *" (63 R. 17.)

P. understood that D. wanted all the marble on the job in Seattle in June, whether the bank quarters were ready to have the marble installed or not. On June 27th D. wrote P. a letter, ending as follows:

"* * * If you can get all the marble here before the middle of June, so much the better. You will have it out of your way and will get your money that much quicker." (129 R. 9.)

Default of P. in not getting the marble to Seattle on time; D.'s numerous and urgent demands for the shipment of marble and for the delivery of setting plans (shop drawings) and notice to P. that damages would be assessed against it on account of delays, as well as its silence and failure to reply to numerous telegrams and letters, are abundantly shown by the testimony at the trial and particularly by these telegrams and letters. Mr. Eastman admitted in open court that all of these communications were received. His reply thereto is shown by his own evidence:

"(Q) You haven't offered here to the Court any response that you made to those repeated and numerous demands we made upon you; you haven't offered any to the judge.

(A) I am working with my attorney's advice. My attorney represents me in this matter." (168 R. 18.)

At pages 52-53 it is asserted that the contract gave no right to the architect to withhold any sum as damages for delay in finishing installation; but the architect is the final arbiter as to the quality of the work and material, and as to whether the job has been completed. Under the main contract final payment is not due until the architect has certified that the "work is satisfactorily completed." (59 R. 32.)

CERTIFICATE OF THE ARCHITECT FINAL AND CONCLUSIVE IN THE ABSENCE OF FRAUD OR BAD FAITH

Martinburg etc. v. March, 114 U. S. 549;

Sweeney v. U. S., 109 U. S. 618;

Kihlberg v. U. S., 97 U. S. 397;

Bush v. Jones, 144 Fed. 942;

Michaelis v. Wolf, 136 Ill. 68.

9 C. J. 756, Sec. 94.

In the *Michaelis* case it is said:

"Where in a building contract provision is made for the payment of the price or a portion or portions of such price upon the certificate or certificates of the architect in charge of the construction of the building, the obtaining or presentation of such certificate or certificates is a condition precedent to the right to require payment, and such condition must be strictly complied with, or else a good and sufficient excuse shown for not complying therewith."

P. has made some argument and considerable comment on certain facts, referred to in the court's memorandum opinion, particularly the time allowable for the completion of the job, and the $4,500.00 rental for one month. P., however, in discussing the facts, must confine itself to the findings made by the court, excluding the court's memorandum opinion.

VII.

MARBLE WAS DEFECTIVE

Commencing at page 56 P. argues that the marble was not defective; that same was all inspected by the architect before installation, and that therefore the court erred in allowing any damages therefor. It will be noted that under the terms of the bank's contract the "architect shall be sole judge of the quality and quantity of the work." ((58 R. 13.)

We agree with P. that the decision of the architect "should be final and controlling as to the character of the marble installed and as to whether or not it complied with the specifications.' But this fact, under the evidence adduced at the trial, does not bar or estop the architect from showing that some of the marble installed was and is defective, because on account of the urgent need of the bank for its quarters, some of the objectional marble was installed conditionally and on the agreement that same would be replaced. Architect Grant testified:

"* * * In view of the fact that the bank was very anxious to get in their premises, a compromise was made as to some of it and there was put in some with the understanding and promise of Mr. Robinson to take it out and put in other corresponding with the contract sample. Some of that has been done and some has not. This light color in the marble could have been avoided. * * *" (50 R. 27.)

These facts are special and unusual and are not controlled by the general rule of law set out by P. as shown by the cases cited at pages 60-61. At page 62 P. says:

"* * * To permit the inspector to wait until the materials had actually been installed and then require the removal of the same would lead to great hardship and injustice."

But in the case at bar the architect did not wait until the materials had actually been installed and then object—the objection was made before installation. This agreement evidently came to the knowledge of the bank, for after it had moved into the bank quarters and on January 17, it wired to P.:

"* * * It is nearly a month since we moved in and still you have not shipped the balance of order required to complete job. * * *" (178 R. 20.)

This evidently refers to the replacement of the objectionable marble, as there was no marble needed for anything else—and the marble had been installed.

In speaking of defects in the marble and for replacement thereof, the architect, Grant, testified:

"* * * In this particular marble the seams run in every direction; to some extent only there is a direction in which they predominate. The seams do, however, run in every direction, to some extent. They are, however, stronger in one direction than in another. In all of the marble there in the bank, that was supplied by the Robinson Tile & Marble Company, and that out in the hall and vestibule leading to the elevators, it shows frequently where it has been broken, where it has come apart, where the seams are. It shows this all over and this is true of the marble set in the hall as well as in the bank proper.* * *" (70 R. 4.)

"I had no experience in the actual fabrication of the marble. I stated yesterday that there were some pieces of marble that were off color in the bank.

"Q. Those, I believe, are how many? Are there three or four of them that you say that you would not accept? A. Oh, twenty or thirty.

"Q. When you mention twenty or thirty, those are some you had some dispute about, but did let them go in? A. Let them go in under protest, with the understanding that they would be taken out and others put in their places." (70 R. 25.)

"* * * The marble has splotches of yellow in it— some relatively thin—spread out, and it is sawed along the major axis of the splotch, which will make a very large spot in the marble; while, if it is sawed across the splotch it will make a narrow strip of yellow, which is not objectionable in this marble. In some cases the yellow covered three-fourths of the entire piece of marble, showing that they sawed longitudinally, the long way in the splotch of the yellow in the quarry deposit." (75 R. 28.)

And concerning the check desks the same witness testified:

"Q. Those that were rejected by you were not put in at all?

"A. Some were. I will qualify that. I have already attempted to make it clear. We rejected some, but with the understanding that they would be taken out later. Others were rejected because there was a possibility of getting other marble for it. Mr. Robinson got a piece of marble up from San Francisco and made some substitution, but in other cases he was unable to make those substitutions." (78 R. 2.)

The defective coloring is explained in the memorandum opinion:

"The effect of the pieces of marble splotched with yellow when brought in proximity with the grayer pieces of marble produces a clashing effect—not as pleasing to the eye as it would be if more uniform in color. It is not necessary to determine the cause that produced this effect.

"The contract between defendent and the bank, Par. 3, provided that the 'quality and finish' of the marble should be equal to those represented by samples on file in the office of the architect. The Court concludes, in view of the use to which this marble was to be put—the interior finish of such banking-rooms, that 'quality and finish' in this character of marble includes coloring. The sample filed with the architect was not produced in evidence, but a sample of the marble was offered by plaintiff. This sample shows no indication of any such want of uniformity in coloring as appears in the installed marble. The Court finds the defect in the marble, in this respect, damaged the bank, and defendant, in the amount to

which the architects testified, to-wit, $2,500.00."
(187 R. 13.)

P., at page 65, refers to the agreement that some
of the objectionable marble might be installed subject
to the agreement of D. that same would be replaced,
as being "secret." This is the first time we have
heard this word used in this lawsuit. Besides, this
agreement was beneficial both to P. and D.—It permitted the installation to continue, with permission to
remedy, and at the same time reduced the period of
rent to be paid by the bank.

P. also claims that there is no proof of damages as
to defective marble to the extent of $2,500.00. Both
Graham and Grant testified that the damages in that
regard were at least $2,500.00, and there is no evi-
dence to contradict it. Stronger, however, than this,
is the fact that the trial court made a personal view
and inspection of the marble shortly after the trial.

"After the trial by agreement of both parties and
in the presence of counsel for both parties, the Court
made an examination of the marble installed in the
Dexter Horton National Bank." (197 R. 17.)

The main contract provided:

"The quality and finish of all marble installed shall
be equal to that represented by samples on file in the
office of the Architect and identified by the signatures of the parties hereto."

P. introduced a sample, which both parties admit is the same as the identified sample. The court doubtless compared this with the marble which he saw installed in the bank.

It must be assumed that the court, in making his decision, did not shut his eyes as to what he saw or the condition of the marble personally inspected by him. In consideration of the undisputed evidence as to the amount of damages on account of defective marble, and especially the fact that the trial court examined the marble as installed, we think this court would not be justified in interfering with the award of $2,500.00 damages for this branch of the case.

WHERE A VIEW OF PREMISES IS MADE BY A JURY OR BY A TRIAL COURT, WITH CONSENT OF THE PARTIES, THE COURT OR JURY MAY APPLY THEIR OWN JUDGMENT AND KNOWLEDGE IN CONNECTION WITH THE TESTIMONY IN THE CASE.

In re East Spring St., 41 Wash. 366;

Conness v. Indiana etc. Ry. Co., 193 Ill. 464;

Beveridge v. Lewis, 137 Cal. 619;

Groves etc. Ry. Co. v. Herman, 206 Ill. 34;

Hartman v. Reading etc. Ry. Co., 13 Atl. 774;

City of Kansas v. Butterfield, 89 Mo. 646;

Hatton v. Gregg, 88 Pac. 592.

In the *Beveridge* case, *supra*, the Supreme Court of California used this language:

"The jury may be permitted, in weighing the evidence in an eminent domain proceeding, to exercise their individual judgment as to values upon subjects within their knowledge which they have acquired through experience and observation.' (41 Wash. at 371.)

An in the *Hartman* case, *supra*, the Supreme Court of Pennsylvania had this to say:

"When a jury have viewed and examined the premises (in proceedings to recover damages sustained by the construction of a railroad), their own observation * * * is just as good as that of any of the witnesses; and while they are not to disregard the testimony produced on the trial, they are, nevertheless, not required to repudiate the evidence of their own senses." (41 Wash. at 371.)

P. seems shocked at the $2,500.00 item of damages, and at page 65 says:

"This claim, delayed by the architect and the defendant until after the building was completed, that the marble furnished by the plaintiff was defective, was as surprising and shocking to the plaintiff as would be a flash of lightning and sharp clap of thunder coming out of a clear sky to a blushing June bride on her exit from the altar. The law does not sanction such unconscionable and shocking conduct on the part of architects."

In view, however, of the many letters and wires which P. received from D., the President of P. should not be surprised at any reasonable damages.

On October 31 D. wired P.:

"* * * If you do not rush shipments for above will be forced lay off crew and then will certainly be charged with full damages as set forth in contract Stop Do not understand why you shipped open counter work first instead of grill counter work Stop You will be ahead if you run crews nights and Sundays if this program will not finish marble sufficiently soon you better secure assistance other shops Stop We will do everything possible avoid liquidated damages and you must do your part or suffer consequences accordingly Stop * * *" (121 R. 12.)

On November 6th:

"* * * We hereby notify you (Eastman) and Mission Marble Works that according terms contract between us we will hold you liable for all damages assessed and incurred by reason your failure fulfill contract You can ill afford loss reputation Graham greatly disappointed." (122 R. 21.)

On November 11th:

"* * * Is Eastman home If not where is he and when will he be home This delay is bound to cost considerable money as it is can't you see that it will cost considerable more if we don't show some speed your unbusinesslike methods and inactivity in handling the fabrication of this marble costing us exorbitant expense installation If you don't heed our request and continue to ignore our telegrams which should be answered immediately in all cases we will charge you with our said expense in addition to damages assessed against us by the owners." (124 R. 3.)

VIII.

$2,500.00 DAMAGES ALLOWED ARE PROPER AND CORRECT

At page 73 P. makes a brief contention that in no event could the court allow the full amount of $2,500.00 for defects in said marble. The point sought to be made is that the installing of new marble is a part of this $2,500.00 and it being the duty of D. to install, therefore *all* the damages could not be awarded against P. But it must be remembered that the trouble being defective marble (which P. should not have furnished) it is obliged to and should pay all the damages resulting from such defects, and not a portion thereof only.

IX.

EFFECT OF FAILURE TO PRODUCE SAMPLE OF MARBLE

The law cited by P. at pages 75-77 as to the usual presumption against a party who fails to produce evidence in his possession, which he can produce, is elementary and admitted. It surely is not needed in this court. There are several reasons, however, why the usual unfavorable presumption cannot be indulged against D. under the facts of this case. These are:

(a) A sample of marble was marked as "Plaintiff's Ex. 1" (68 R. 31).

This sample, although not identified as one of the two used when the bank contract was made, was "just as good." (68 R. 18.) P. assumed that said marble was a sample of the marble furnished Robinson by the architect (68 R. 32) and Mr. Grant testified that the marble brought into court by P. is the same as the original sample (68 R. 9). Since P. brought in this marble and used it at the trial, it should not now be permitted to say that the sample was not the same as the originals. If it was not, it was near an imposition on the court to bring it in at all. In other words, the marble marked as "Exhibit 3" is admitted by both parties to be the same as the original sample—neither party denies it. Since that used at the trial is the same as the official sample, no one is hurt or injured by the absence of the original.

(b) No legal or other notice or request was ever served upon D. to produce the official sample of marble desired; this evidently because P. knew the sample as an Exhibit was the same as the original, or official, sample. Consequently, the unfavorable presumption cannot be invoked against D., for it is only parties to the suit that are affected by the presumption; it does not affect parties when the notice or subpoena is served upon a witness.

(c) P.'s counsel in open court requested Grant to produce the original sample (69 R. 11) but it was not in the possession of the witness, although witness thought he could produce it. Afterwards the witness testified:

"I have not brought the sample. I failed to find it." (153 R. 9.)

X.

Cratage and Drayage

These two items, $484.78 for boxing a shipment of marble by boat and $60.00 for cartage, were extras included in the complaint. The contract between P. and D. provided that the marble should be delivered on "cars at the works of the P. in San Francisco." (31 R. 6.)

On October 31, D. wired P. as follows:

"Rush shipments by boat. * * * If you do not rush shipments for above, will be forced lay off crew. * * * Have had eight mechanics working but forced to lay off five and the three now on job working great disadvantage account your not sending stock to complete units as we go." (121 R. 10.)

And on November 26 D. wired to P.:

"Make all shipments by Pacific Steamship Boats." (—— R. ——.)

Concerning the items for cratage and drayage, Mr. Robinson, president of D., testified:

"* * * This boxing and cartage, I claim should not have been charged to us, due to the fact that they did not have the marble out to ship it in the usual and customary manner by rail, to load in a box-car at the time planned. * * * It would have been much cheaper for the Mission people to pay this boxing and cartage themselves than to cause so much delay. The two items making up this amount are $484.00 for boxing and $60.00 for cartage. Had this marble been gotten out on time there would have been no occasion for boxing and cartage. Therefore, I claimed that the delay is a damage to me, to that extent, and which I have charged for.

"I have a letter from Mr. Eastman, suggesting that he would forget those items if I would pay him the amount due." (118 R. 11; 24; 119 R. 1.)

The court's disposal of these items is shown in the opinion, as follows:

"Plaintiff claims extra to the amount of $1,371.76. Among these is a charge of $544.78 of which $484.78 is for the expense of boxing marble, and for cartage $60.00. This expense was incurred by plaintiff in shipping, at defendant's request, marble by boat instead of by carload lots. The shipments by water avoided further delay. These items are disallowed. * * *" (188 R. 8.)

At page 83 P. objects to a $25.00 item allowed for wooden hand balustrade and suggests that there is no evidence to sustain it. This item is for the expense of a wooden hand balustrade to the basement of the bank because marble was not ready (113 R. 7). Besides, this item of $25.00 is the last item of D.'s Exhibit A (113 R.); and the parties stipulated that D. paid in cash the sum of $784.24 on account of the items in said exhibit—one of these is $25.00.

We conclude that the trial court's disposition of the items for cratage and drayage was proper.

XI.

LETTER WRITTEN BY D. ON MAY 25, 1925

At page 84 P. that the marble for the hand railing to the stairs leading to the safety deposit vaults was not fabricated and installed until March 3, 1925, because D. failed to furnish setting plans or shop drawings.

The evidence shows that "shop drawings" and "setting plans" are identical (98 R. 31). While it is true that it was the duty of D. to furnish these, so far as the bank is concerned, yet as between P. and D. it was the duty of *P.* to furnish these shop drawings, as the marble was being fabricated in San Francisco.

"* * * Shop drawings are made by the manufacturer, steel manufacturers, terra cotta manufacturers, masonry manufacturers, and all. It is *usage* through-the building trade. It is our rechecking of the instructions, the manufacturer gives to his own workingmen. That is what it amounts to." (Testimony of John Graham; 98 R. 24.)

Surely P. cannot argue that because it failed to perform part of its contract, namely to furnish shop drawings, that this is an excuse for not furnishing the marble in time so that the contract could be completed within the time provided by the contract.

P. contends at page 85 of its brief that D. has already admitted in a letter dated May 25, 1925, that it owes P. $6,576.64. This letter simply explains the condition of the account between P. and D.—or in other words, the balance of account as shown by the invoices, less the payments.

We think in the account between P. and D. the only proper items that should be inserted in the account on the debit side are the shipments of marble, and on the credit side the cash payments. Items that are open to some possible dispute or question, such as the expenses of the trips made to San Francisco by Robinson and Grant; damages for delay and for defective marble are in their nature unliquidated and could not, with property, be placed in D.'s books of account as proper items. If these are not settled, or paid, they are proper items for suit, but cannot be arbitrarily placed in the debit side of P.'s account against D.

In view of these facts, the letter of May 25, 1925, is of no special significance.

XII.

Occupancy of the Bank Building by the Owner Not a Waiver of Defective Materials or Non-Compliance With Contract

Hurley v. School Dist., 124 Wash. 541;

6 R. C. L. 995.

In the Washington case the Supreme Court of that State held that mere occupancy does not waive the right to recover against the contractor for failure to complete the contract.

And in 6 R. C. L. at p. 995, it is said:

"* * * the mere fact that the owner enters into possession and uses a building which has been constructed for him does not ordinarily constitute a waiver of a non-compliance by the contractor with his contract in erecting the building. The occupancy and enjoyment of the structure by the owner does not necessarily preclude him from showing that the contractor's work has been improperly or defectively executed. * * *"

XIII.

CONCLUSION

In conclusion, this cause should be affirmed, because

First: The facts are not reviewable in this court, except to determine if they support the judgment. And in determining what facts the court found, the court's opinion cannot be considered, but only the findings.

Fleischman Co. v. U. S., 270 U. S. 349;

British Queen Min. Co. v. Baker, etc., 139 U. S. 222.

Second: The only other matter open for argument by P. in this court is "rulings of the court in the progress of the trial" if any, which were excepted to. The only rulings during the trial, which are assigned as error, are those involved in Assignments Nos. I and II.

Third: The other twenty-four assignments of error concern the facts in the case, which are not open for review in this court, except as above noted.

Fourth: Even should the court be permitted to look into the facts, the undisputed evidence shows that the balance on the contract, including extras, is $6,576.68 (199 R.); and that the damages on the counterclaim of D. is $7,195.00 (200 R.). Under these facts the judgment was properly signed for $618.32 (200 R. 27).

But the facts as shown by the court's findings must stand.

Fifth: The facts found by the court (200 R. 1) concerning the damage of $4,500.00 for delay and $2,500.00 for defective marble (if true, and they must be so regarded in this court), are sufficient to support the judgment entered.

Sixth: The Architect's refusal to make a certificate of completion of the work is final and conclusive upon D., as no fraud or bad faith is charged against him. And the matters causing said refusal cannot be rectified. $4,500.00 for delay, and $2,500.00 for damaged marble, cannot be eliminated without payment. These damages have been actually suffered by D.; and since these matters—delay and defective marble —prevented D. from faithfully and promptly fulfilling its contract with the bank, the damages are properly chargeable to P.

Respectfully submitted,

H. A. P. MYERS,
Attorney for Defendant in Error.

5037

No. 5307

United States
Circuit Court of Appeals
For The Ninth Circuit

THE MISSION MARBLE WORKS,
a Corporation,

Plaintiff in Error,

VS.

ROBINSON TILE & MARBLE COMPANY,
a Corporation,

Defendant in Error.

ON A WRIT OF ERROR TO THE DISTRICT COURT OF THE
UNITED STATES FOR THE WESTERN DISTRICT
OF WASHINGTON.

REPLY BRIEF FOR PLAINTIFF IN ERROR.

Appearance for Plaintiff in Error:

GEORGE CLARK SARGENT,
San Francisco, California.

GROSSCUP & MORROW,
C. A. WALLACE and
J. O. DAVIES,
Seattle, Washington.

No. 5307

United States Circuit Court of Appeals
For The Ninth Circuit

THE MISSION MARBLE WORKS,
a Corporation,

Plaintiff in Error,

vs.

ROBINSON TILE & MARBLE COMPANY,
a Corporation,

Defendant in Error.

ON A WRIT OF ERROR TO THE DISTRICT COURT OF THE
UNITED STATES FOR THE WESTERN DISTRICT
OF WASHINGTON.

REPLY BRIEF FOR PLAINTIFF IN ERROR.

Appearance for Plaintiff in Error:
GEORGE CLARK SARGENT,
San Francisco, California.

GROSSCUP & MORROW,
C. A. WALLACE and
J. O. DAVIES,
Seattle, Washington.

United States
Circuit Court of Appeals
For The Ninth Circuit

THE MISSION MARBLE WORKS,
a Corporation,

Plaintiff in Error,

vs.

ROBINSON TILE & MARBLE COMPANY,
a Corporation,

Defendant in Error.

No. 5307

ON A WRIT OF ERROR TO THE DISTRICT COURT OF THE
UNITED STATES FOR THE WESTERN DISTRICT
OF WASHINGTON.

REPLY BRIEF FOR PLAINTIFF IN ERROR.

I.

The defendant in error by its brief seeks to lock and
bolt the doors to this court so that the plaintiff in
error may not enter; it seeks to bar this court from
passing upon the merits of this cause; it seeks to
prevent this court from passing upon issues of law
raised by the plaintiff in error, as to whether or not
there was any evidence introduced at the trial to support the Findings of Fact made by the trial court.
For both its lock and key, the defendant relies chiefly

upon certain decisions of Circiut Court of Appeals of the eighth circuit, written by the distinguished jurist, Judge Sanborn, from whose decisions the defendant has quoted at considerable length. Plaintiff will not undertake to discuss in detail each of those cases or the many other cases that might be cited. Plaintiff has carefully read all of the cases cited by defendant and the facts in none of those cases are at all similar to the facts in this case, and when those cases are read together with other cases decided by the same jurist, this court, we believe, will readily agree with the plaintiff that there is no law laid down in those cases that bars this court from the hearing and determination of this case on the merits. We desire to call the court's attention to the case of

Barnsdall v. Waltemeyer, 142 Fed. 415.

In this case the same Judge said:

"The first defense was that the promise was to pay $10,000.00 out of the first net proceeds of the property sold; that this property consisted of a mine; that the expense of operating it had been more than the income from the ore derived from it; and that there never had been any net proceeds. A jury was waived, and this issue was tried by the court. The evidence was oral testimony. There was sufficient to sustain a finding of the issue either way, and the court made a special finding of facts in favor of the plaintiff. When, in action at law, a jury is waived and the court tries an issue of fact and makes a special finding upon which the substantial evidence is conflicting, the losing party may not reverse it by writ of error because it was not sustained by the weight of evidence. *Hughes County v. Livingston,* 104 Fed. 306, 319, 43 C. C. A. 541, 555. The only reviewable questions upon a writ of error to reverse a judgment upon such a trial are the rulings upon the admission and the exclusion of evidence, upon questions of law upon the question whether or not there was any substantial evidence in support of the finding, and upon the question of the sufficiency of the facts to support the judgment."

The court in that case clearly held that where the trial court made a special finding of fact in favor of the plaintiff upon writ of error, the appellate court would review the question whether or not there was any substantial evidence in support of the finding, and in that case, the court did review the evidence and held that the testimony was sufficient to warrant the findings in the trial court in the following language:

"The testimony was sufficient to warrant the findings, and no other issue is presented by the challenge of the trial of the first defense."

The court in that case in support of its declaration of law that where the trial court makes a special finding of fact, the appellate court can review the question as to whether or not there was any substantial evidence in support of the finding cited many authorities, and these authorities clearly support the rule announced by the court. We desire to call the court's attention to the law laid down in a few of the cases cited in the opinion. In the case of

The Francis Wright, 105 U. S. 381;

the court at page 387, said:

"It is undoubtedly true that if the Circuit Court neglects or refuses, on request, to make a finding one way or the other on a question of fact material to the determination of the cause, when evidence has been adduced on the subject, an exception to such refusal taken in time and properly presented by a bill of exceptions may be considered here on appeal. So, too, if the court, against remonstrance, finds a material fact which is not supported by any evidence whatever, an exception is taken, a bill of exceptions may be used to bring up for review the ruling in that particular. In the one case the refusal to find would be equivalent to a ruling that the fact was immaterial; and in the other, that there was some evidence to prove what is found when in truth there was none. Both these are questions of law, and proper subjects for review in an appellate court. But this rule does not apply to mere incidental facts which only amount to evidence bearing upon the ultimate facts of the case. Questions depending on the weight of evidence are, under the law as it now stands, to be conclusively settled below and the fact in respect to which such an exception may be taken must be one of the material and ultimate facts on which the correct determination of the cause depends."

It is true that that case was one in admiralty. This court, however, speaking through Judge Ross, held that the law therein laid down was applicable to a law case tried before the court without a jury. In the case of

>*Eureka County Bank v. Clarke*, 130 Fed. 325;

in that case this court said:

>"If there be no evidence at all to support the findings of fact made by the trial court, such findings would as a matter of law, be erroneous."

The court cited *"The Francis Wright," supra,* in support of this declaration of law. The court then went on to say:

>"It is true that that was a cause in admiralty but no reason is perceived why what was there said in respect to findings of fact is not equally applicable to an action at law tried by the court without the intervention of a jury."

In the case of

>*The City of New York*, 147 U. S. 72;

also cited by Judge Sanborn, the court said:

>"If the court below neglects or refuses to make a finding, one way or the other, as to the existence of a material fact, which has been established by uncontradicted evidence, or if it finds such a fact when not supported by any evidence whatever, and an exception be taken, the question may be brought up for review in that particular. In the one case the refusal to find would be equivalent to finding that the fact

was immaterial; and, in the other, that there was some evidence to prove what is found, when in truth there was none. Both of these are questions of law, and proper subjects for review in an appellate court."

In the case of

Laing v. Rigney, 160 U. S. 531;

also cited by Judge Sanborn, at page 540, the court said:

"The plaintiff duly excepted to the findings and conclusions, and it is well settled that exceptions to alleged findings of fact, because unsupported by evidence, present questions of law reviewable in courts of error."

This was an act tried before the court without a jury. The court reviewed the evidence and determined that one essential finding of fact was wholly without evidence to support it and for that reason, reversed the judgment and sent the case back to the lower court for further proceedings.

The law laid down by Judge Sanborn in *Barsdall v. Waltemeyer, supra,* has never been modified or changed by the Circuit Court of Appeals of the Eighth Circuit, by any later decisions of that court, on the contrary it has been repeatedly cited by that court with approval and was cited by that court with approval, in the case of

Wear v. Imperial Window Glass Co., 224 Fed. 60;

cited and relied upon by defendant. That case is direct authority for, and supports the plaintiff's position that where the trial court makes special findings of fact, an appellate court on review will look into the evidence to see and determine whether or not the special findings of fact have substantial evidence to sustain the same.

Whatever may be the rule of law in the eighth circuit, the plaintiff in this case is entitled to a decision upon the merits under the law as settled in other circuits and particularly in this circuit. There seems to have arisen in the sixth circuit considerable doubt and confusion as to the proper method of procedure to be followed in the Federal courts in cases tried before the court without a jury, in order that the parties might preserve their rights and have their causes reviewed on the merits on appeal to the Court of Appeals of that circuit, and in order to sell the practice and do away with the confusion which existed, the court, speaking through Judge, now Chief Justice, Taft laid down the proper rules in the case of

> *Humphreys v. Third National Bank,* 75 Fed. 852;

for guiding parties in such trials, and upon that question the court said:

> "When a party in the circuit court waives a jury and agrees to submit his case to the court, it must be done in writing, and if he wishes to raise any question of law upon the merits in the court above, he should request special findings of fact by the court framed like a special verdict of the jury, and then reserve

his exceptions to those special findings, if he deems them not to be sustained by any evidence, and if he wishes to except to the conclusions of law drawn by the court from the facts found, he should have them separately stated and excepted to."

No one can contend that the proceedings had in the trial court do not strictly comply with all of the requirements laid down by the court in that decision. The court then went on to point out the distinctions between the trial of a case before the court with a jury, and before the court without a jury. The court pointed out that where the trial of a case was before the court without a jury, as the court did not and could not be asked to instruct itself, the proper method of procedure was to present to the court specific conclusions of law and ask the court to rule upon, and if they were denied, to take an exception. The same is true in respect to motions for a directed verdict. It is impractical to ask a court to take a case from itself and to decide that there is no evidence upon which it could base a decision. The practical way, therefore, is to present to the court for its adoption, specific findings of fact, and if the court refuses to adopt them, to except thereto.

There seems to have existed in this circuit the same confusion with respect to the proper procedure to follow in the trial of a law case before a court without a jury. This court in several cases held that where a law case was tried before the court without a jury, and the court made a general finding in favor

of one of the parties and no special finding was asked
for by either party, that this court would not review
the question of law whether there was any evidence
in the record to support the general finding of the
trial court, to the confusion and embarrassment of
the attorneys and serious injury to the citizens of
this circuit. With this situation existing this court
in the case of

> Societe Nouvelle d'Armement v. Barnaby,
>   246 Fed. 68;

endeavored to lay down the proper line of procedure
and settle the practice in this circuit in cases tried
before the court without a jury. This court in that
case, after reviewing at considerable length the decisions upon the question, particularly the decisions
of the United States Supreme Court, including all or
nearly all of the decisions of that court which are
cited by the defendant in its brief upon this question,
in an endeavor to point out the proper method of
procedure, said:

> "But where a finding is supported by no com-
> petent evidence, a question of law is presented
> which is reviewable on appeal. The inquiry per-
> tains to the practice or the manner by which
> such a question may be raised or brought into
> the record. Judge Taft has lucidly stated the
> practice, in *Humphreys v. Third National Bank*
> (6th Ct.) 75 Fed. 852, 855, 21 C. C. A. 538, 542.
> If a party, having submitted his case to the
> court, wishes to raise any questions of law upon
> the merits, on review, 'he,' says the distinguished

jurist, 'should request special findings of fact by the court, framed like a special verdict of a jury, and then reserve his exceptions to those special findings, if he deems them not to be sustained by any evidence; and, if he wishes to except to the conclusions of law drawn by the court from the facts found, he should have them separately stated and excepted to. In this way, and in this way only, is it possible for him to review completely the action of the court below upon the merits.' "

This court, after laying down the proper rules for raising the question of the sufficiency of the evidence, goes on to say:

"We do not understand that in any event the appellate court will look into the evidence to determine whether the trial court has rightly decided the question of fact, unless it be that there is no sufficient evidence to support the finding. In other words, it will not try the case, as a jury tries it, to determine the weight or preponderance of the evidence, but only to determine whether there is any competent evidence sufficient to support the finding."

"This brings us to an examination of the record to determine what questions the plaintiff in error has here for our consideration. It made no request to the court for any special findings, nor did it make any request for a general finding of any kind. It did except to the general finding made, which exception was allowed, but this was not sufficient under the statute or the authorities to require an examination of the evidence, to determine even whether it was sufficient to support the finding."

This court, in this case, clearly held that if special findings had been tendered to the court and rejected, or if special findings had been made by the court and had been excepted to, that the court would have reviewed the question as to whether or not there was any evidence to support the same. This court in the case of

Garwood v. Scheiber, 246 Fed. 74;

again laid down the same rule of procedure as announced by it in the *Barnaby* case, and again approved the procedure laid down by the Circuit Court of Appeals of the sixth circuit, in the case of

Humphreys v. Third National Bank, supra.

Tested by these rules laid down by this court in the two cases cited, there can be no question but what the plaintiff in this case is entitled to have its case heard upon the merits, and have the court determine whether or not there is any substantial evidence to support the special findings of fact made by the trial court.

The plaintiff presented to the trial court, special findings of fact, twenty-three in number, 190 R. 7. Each of these findings of fact were rejected and denied by the trial court. Plaintiff asked for an exception to the ruling of the court in denying each of these, and plaintiff was allowed an exception, 196 R. 15. The defendant also presented to the court special findings of fact, eight in number, and conclusions of law, one in number, 196 R. 28. The findings of fact and conclusions of law tendered by the defendant were granted by the court. Plaintiff excepted to findings of fact, numbers four, five, six and seven, and to the conclusions of law, number one, and plaintiff's exceptions were allowed by the court, 201 R. 5. The record therefore clearly shows that the plaintiff in this case complied with the rules laid down by this court in the two cases *supra* in both letter and spirit.

The decisions of the other circuit courts are in harmony with the law laid down by this court in those cases.

> *Geo. A. Fuller Co. v. Brown,* 15 Fed. (2d) 672.

This was an action tried before the court without a jury. The court made special findings of fact. The question before the court was whether or not there was sufficient evidence to support certain of the findings of fact. Discussing this question, the court said:

> "Coming next to the assignments of error, which challenge the sufficiency of the evidence to support the findings made by the Judge, we note that a question was raised in the briefs as to the power of the court to review the evidence for the purpose of determining whether it supports the findings of fact, but we think that the court does have that power upon exception to the findings. *Arcade v. A. R. Corporation* (C. C. A. 2nd) 286 F. 809; *U. S. v. Penn. & Lake Erie Dock Co.* (C. C. A. 6th) 272 F. 839;

Chicago Life Ins. Co. v. Tiernan (C. C. A. 8th) 263 F. 325; *Humphreys v. Third National Bank* (C. C. A. 6th) 75 F. 852, 21 C. C. A. 538. Every finding of fact, however, having reasonable support in the evidence and tending to support the judgment, is binding on this court. *Va. & W. Va. Coal Co. v. Charles* (C. C. A. 4th) 254 F. 379, 165 C. C. A. 599. We do not examine the evidence to see whether the trial court has rightly decided the questions of fact, but only to determine whether there is any competent evidence sufficient to support the findings. *Societe Nouvelle d'Armement v. Barnaby* (C. C. 9th) 246 F. 68, 158 C. C. A. 294.

"After a careful review of the evidence, we think that all the findings excepted to are supported by evidence, except the finding that each of the eight vessels built for the Emergency Fleet Corporation other than hull 1446 was built at a profit commensurate with the profit on hull 1446. The evidence establishes that the profits on the second, third, fourth, fifth and sixth vessels were commensurate with the profits on hull 1446, but there is no showing whatever that this was true with respect to the profits on the seventh or eighth vessels. All that the evidence shows with respect to these vessels is that some profit was made on them. We must hold, therefore, that the tenth finding, in so far as it embraced the seventh and eighth vessels, was not supported by any sufficient evidence, and that the

learned district judge erred in making the finding."

In the case of

> *Arcade & A. R. Corporation v. Kann*, 286
> Fed. 809;

the court said:

"The cause was tried before the court without
a jury, and, while not printed in the record on
review, it appears from the decision and the
judgment that the parties filed a stipulation in
writing waiving a jury. The court filed findings of fact and conclusions of law. Notwithstanding the settled rule to which we have re-
cently called attention, argument is made as to
the weight of evidence in certain particulars.
We therefore repeat that we examine the record
on writ of error solely to ascertain whether
there is some evidence to support the findings
of fact."

This question is, however we believe, fully and
finally set at rest and in favor of the plaintiff's contention by the recent decision of the Supreme Court
of the United States, in the case of

> *Fleischmann Co. v. United States*, 270 U. S.
> 349.

This case was tried to the court without a jury. The
court handed down a written opinion in which it
considered the entire case as to facts and law. No
special findings of fact were thereafter made. The
same day in which the opinion was handed down, a
judgment was entered. The case went to the Circuit

Court of Appeals and was affirmed. Writ of error was sued out in the Supreme Court of the United States to review the judgment from the Circuit Court of Appeals. One of the questions before the court was whether or not it could review the evidence to determine whether or not there were any facts to support the judgment of the trial court. In determining this question, the court said:

"The opinion of the trial judge, dealing generally with the issues of law and fact and giving the reasons for his conclusion, is not a special finding of facts within the meaning of the statute. * * * And it is settled by repeated decisions, that in the absence of special findings, the general finding of the court is conclusive upon all matters of fact, and prevents any inquiry into the conclusions of law embodied therein, except in so far as the rulings during the progress of the trial were excepted to and duly preserved by bill of exceptions, as required by the statute. * * * To obtain a review by an appellate court of the conclusions of law a party must either obtain from the trial court special findings which raise the legal propositons or present the propositions of law to the court and obtain a ruling on them. *Norris v. Jackson, supra,* 129; *Martinton v. Fairbanks, supra,* 673. That is, as was said in *Humphreys v. Third National Bank, supra,* 855, he should request special findings of fact by the court, framed like a special verdict of a jury, and

then reserve his exceptions to those special findings, if he deems them not to be sustained by
any evidence; and if he wishes to except to the
conclusions of law drawn by the court from the
facts found he should have them separately stated and excepted to. In this way, and in this
way only, is it possible for him to review completely the action of the court below upon the
merits."

The defendant has not contended and no one can
contend that the plaintiff did not fully and completely
comply with the procedure therein laid down in both
letter and spirit.

The defendant also cites some cases from the Supreme Court of the United States in support of its
contention that this court cannot review this case
upon the merits. All of these cases cited by the
defendant are mentioned in the opinion of the Supreme Court in the case of

Fleischmann Co. v. United States, supra;
and as that case holds where findings of fact and
conclusions of law are requested of the trial court
and denied, or where special findings of fact are
made by the trial court and exceptions taken, a review may be had upon the merits, no further comment upon the cases cited by the defendant from the
Supreme Court is deemed necessary.

II.

WHEN DID THE TRIAL END?

The defendant contends that the trial ended when the court handed down its opinion in the case. This contention of the defendant cannot be sustained. It is ridiculous and contrary to the actions taken by the defendant at the trial. The cases cited by the defendant not only fail to support the defendant's contention in this respect, but hold directly to the contrary.

In the case of

United States Fidelity & G. Co. v. Board of County Commissioners, 145 Fed. 144;

cited by the defendant, the court on page 151, said:

"The trial ends only when the finding is filed, or, if no finding is filed before, when the judgment is rendered."

When the court used the above language, the specific question was before the court which we are now discussing, that is, the court was determining the question when a party should make its request for special findings of fact or take its exceptions to the special findings of fact made by the court or make any motion it deemed proper in order to preserve its rights on writ of error.

The opinion of the trial court is not a finding, and is specifically so held by the Supreme Court in the case of

Fleischmann v. United States, supra.

The trial court in this case did not treat his

opinion as a finding of fact, neither did the defendant, nor the plaintiff.

The court ordered the matter set down for the purpose of making findings of fact and conclusions of law, and the case was in accordance with the directions of the court, set down for hearing for the first day of November, 1926, for the purpose of making and adopting special findings of fact and conclusions of law, 190 R. 4, and at that time the plaintiff presented its special findings of fact and conclusions of law separately stated and numbered. These were rejected by the trial court and exceptions were asked by the plaintiff, and plantiff's exceptions were allowed by the court, and this was done before the findings of fact and conclusions of law were made by the court, and thereafter the defendant presented its special findings of fact and conclusions of law separately stated and numbered and these were adopted by the court. Before the court adopted the same and before the same were filed, plaintiff stated its objections to the same and asked for an exception to the action of the court in allowing its findings of fact, number four, five, six and seven, and conclusions of law, number one, and the exceptions were allowed by the court. These proceedings were taken before the entry of judgment, all of which appear of record.

The Supreme Court of the United States, in the case of *Fleischmann v. United States, supra*, held that only exceptions which were made after the entry of judgment came too late. There is no Federal

statute or rule of the Federal Court, or decision by
any Federal Court stating in what form a party
should make his exceptions to the special findings
of fact, found by a court without a jury. The plaintiff in this case made his exceptions to the special
findings of fact found by the court in strict conformity to local law. See Section 383 Remington's
Compiled Statutes of Washington, 1922, which reads
as follows:

"Exceptions to the report of a referee or commissioner, or to findings of fact or conclusions
of law in a report or decision of a referee or
commissioner, or in a decision of a court or judge
upon a cause or part of a cause, either legal or
equitable, tried without a jury, may be taken
by any party, either by stating to the judge, referee or commissioner when the report or decision
is signed, that such party excepts to the same,
specifying the part or parts excepted to (whereupon the judge, referee or commissioner, shall
note the exceptions in the margin or at the foot
of the report or decision); or by filing like
written exceptions within five days after the
filing of the report or decision, or, where the
report or decision is signed subsequently to the
hearing and in the absence of the party excepting within five days after the service on such
party of a copy of such report or decision or of
written notice of the filing thereof."

Plaintiff's exceptions should, therefore, be considered sufficient by this court. Written exceptions

would only needlessly encumber the record by again repeating in a different form plaintiff's submitted and requested special findings of fact.

No court has ever held that under the proceedings taken in this case, a plaintiff in error is not entitled to a review of the question whether or not there is substantial evidence of record to support the special findings of fact found by the trial court. Proceedings had in this case are in strict accord with both the letter and the spirit of the law applicable to such proceedings as repeatedly announced by this court, and the Supreme Court of the United States. If the proceedings had in the trial court were not in strict conformity to the procedure laid down governing such proceedings, yet that would be no bar to a decision upon the merits in this case in this court. In the case of

Garwood v. Scheiber, 246 Fed. 74;

on page 77, this court after holding that the plaintiff in error had failed to properly raise the question of the sufficiency of the evidence in the court below, said:

> "However, notwithstanding this failure to bring the question properly into the record for the consideration of the court, we have very carefully examined the whole of the testimony presented by the bill of exceptions, and are fully satisfied it is ample to support the finding. Indeed, we believe the weight of the testimony to be that way."

The defendant has nowhere in its brief pointed out to the court wherein it claims that the plaintiff failed to comply with any rule of court, statute or decision. In this respect, it has left the plaintiff and this court in the dark as to its contentions. If, however, it is the contention of the defendant that the plaintiff's tendered findings of fact and conclusions of law, and its exceptions to those made by the court came too late, then it of necessity follows that the defendant's special findings of facts and conclusions of law also came too late and as the court's general opinion under all of the authorities cannot be taken as a special finding of fact, and as the court made no general finding in favor of either party, there has been no decision by the trial court upon which a judgment could be entered and the judgment entered in this case must of necessity fall.

In concluding our argument upon this question, we wish to respectfully and earnestly urge upon this court that as long as there is no statute or rule of court governing the proceedings or conduct of a trial court in cases tried by the court without a jury, this court ought not to lay down a rule of law which would deprive the trial court of its discretion to conduct proceedings in its court in such a manner as it deems best or advisable for the dispatch of the business before the court. It ought not to lay down a rule which would compel the court or compel attorneys to insist that a court rule upon questions of law or questions of fact before the trial court has had an opportunity to thoroughly and carefully consider such questions and thus deprive the parties of the real benefit and object which they always have in mind in submitting cases to the trial court without a jury. Ordinarily trial courts prefer to give both questions of law and fact submitted to it for decisions, in the trial of cases time for consideration before it is required to decide the same. In this particular case, the trial court outlined the proceeding to be followed at every step and both parties consented thereto. The court at no time made any complaint that the steps taken by the plaintiff were not in time or proper manner. It gave due consideration to the plaintiff's request for special findings and its exceptions to the special findings made by the court. The defendant in all respects consented to proceedings had in the case as outlined by the court and made no contention before the trial court that any proceedings had therein were not had in time, and it ought not now be heard to complain. The defendant has not claimed and no one can contend but that every single question which is now urged upon this court of law or fact was not fully and fairly and thoroughly presented to the trial court for its consideration, and the trial court given an opportunity to rule thereon, and that is all that the spirit of the decisions of this Court or the Supreme Court of the United States require.

III.

The defendant, in paragraph III., page twelve of its brief, makes some contention that the plaintiff's objection to the questions asked the witness, Grant, and the witness, Graham, on direct examination, 54

R. 14, 104 R. 30, are insufficient to permit this court to review the action of the trial court, overruling these objections. The witnesses, Grant and Graham were architects for the bank in the finishing of the banking quarters. They were put upon the stand as expert witnesses and were testifying as such. They had been asked to point out defects in marble and to give their estimate as such experts, of the amount of money it would take to replace defective marble, with marble that came up to the contract standard. The defendant was also claiming delay in the finishing of the marble work and claiming that the bank had been damaged by this delay, and the question that the plaintiff was trying to prove, was "How much had the bank been damaged by this delay?" The plaintiff objected that this damage by the bank from delay, if any, was not a subject for expert testimony by an architect, and the architect was incompetent to give expert testimony as to how much the bank had been damaged, by reason of its being kept out of its quarters for a certain or any period of time. Objection made by the plaintiff was amply sufficient to call this situation to the attention of the court. That such damages cannot be proven by expert testimony is too clear to need the citation of authorities. The defendant realized this and informed the court that it was merely trying to show good faith upon the part of the architect. Good faith on the part of the architect is wholly immaterial, as pointed out in plaintiff's brief. The right of the plaintiff to payment is not dependent upon any action of the archi-

tect, and secondly, the architect was given no power
of authority under the contract between the plaintiff
and defendant, or between the defendant and the
bank, to withhold any payment from the defendant
by reason of any delay in completing the contract.

Therefore, the question asked the architect was
wholly incompetent and immaterial from every stand-
point, and the court was clearly in error in overruling
the plaintiff's objection. The evidence given by these
witnesses in answer to those questions is highly
prejudicial to the plaintiff's case.

V.

Plaintiff desires next to note briefly the argument
of defendant under paragraph V. of its brief.

The defendant nowhere in its brief contends that
the marble was not installed within the time provided
for in the contract between defendant and the bank.
It admits that it was installed at least soon enough
to enable the bank to move into its quarters before
the time for the installation thereof provided for in
the contract ran. If this is true, then of course the
defendant could not be held liable for damages suffered by the bank for delay by way of either one
month's rent or any other period of time, yet the
defendant claim that the plaintiff ought to be held
for a month's rent for failure to furnish the marble
within the time provided for in the contract, between
the plaintiff and defendant. The contract between
the plaintiff and defendant, as to the time in which
the marble should be furnished, makes the provisions of the contract between the defendant and the

bank a part of the contract between the plaintiff and defendant. The defendant quotes the following provision of the contract between the defendant and the bank:

"The entire contract shall be completed within three months after the first delivery of material,"

and contends that as all of the marble was not delivered within three months after the first delivery, that there is a default on the part of the plaintiff. This contention on the part of the defendant is obviously unsound. If any of the provisions of the contract between the defendant and the bank in respect to time, are made part of the contract between the plaintiff and defendant, then all of those provisions are made a part of plaintiff's contract, and the plaintiff is entitled to all of the provisions of the same.

Defendant on page twenty-two of its brief, says:

"Much evidence and persistent claims were adduced in the trial court to show the delay in furnishing marble, caused by orders to stop or let up on shipments but these claims are evidently forgotten in this court."

Plaintiff admits that this statement is true. Upon this issue as upon several other issues presented by the trial court, there is a conflict of the evidence. The plaintiff in this court has abandoned all contentions made in the trial court, which were decided against it, where there was and is a conflict of the evidence.

On page twenty-four of defendant's brief, defendant makes this assertion:

"In other words, in fixing the damages at

$4,500.00 (for delay) the court did not limit the damages to rental alone."

Whether this is true or not, we cannot say. If it is true, then the court in this instance, did the very thing for which the plaintiff is complaining in this court. It assessed damages without a particle of evidence upon which to base the same.

VI.

Under paragraph VI., page twenty-six of defend- ant's brief, defendant says:

"It is true that under the answer D. only asked for $6,000.00 under the head of defective marble and delay, but when the findings were signed, the court permitted the following amend- ment:

" 'It is ordered that the prayer of defendant's answer be amended by adding thereto and for judgment against the plaintiff, for all counter claims and set offs shown by the evidence."

The defendant then contends that under this amend- ment it is entitled to recover the $7,000.00 damages which the court allowed for defective marble and delay. The trouble with this contention of the de- fendant is that the defendant introduced in evidence at 53 R. 13 the certificate of the architect showing that $6,000.00 was the full amount that was with- held from the defendant on account of damages for defective marble and for delay. The evidence, there- fore, conclusively shows that this sum is the great- est amount which the defendant could, under any theory of law or fact recover in this case upon these two items, no matter how much the bank may have suffered.

Defendant at page thirty quotes the following from the Transcript:

"(Q) You haven't offered here to the Court any response that you made to those repeated and numerous demands we made upon you; you haven't offered any to the judge.

"(A) I am working with my attorney's advice. My attorney represents me in this matter."

The plaintiff's answer in the trial court and to this court to the letters and telegrams offered by the defendant urging speedy delivery of the marble, was, and is that during the course of the trial and at the time for the plaintiff to put in rebuttal testimony, the defendant by itself and by its witnesses, had definitely testified and established the fact that the building in which the marble was to be installed was not ready for the installation of the same until the twenty-second day of September, 1924, and that the installation of the marble had been sufficiently completed so that the bank moved into its quarters on the twentieth day of December, 1924, less than three months after the building was ready for the installation of the marble, and as the contract provided for three months period for the installation of the marble, that therefore under the law applicable to the contract between the parties, neither the plaintiff nor the defendant could be charged with any loss of rent by the bank, and as that was the only damages claimed for delay, it was useless for the plaintiff to encumber the records by any rebuttal along this line.

Plaintiff has no critcism to offer to the law announced in the cases, cited by defendant on page thirty-one of its brief. The law there laid down in those cases, however, is not applicable to the present case. The contract between the plaintiff and defendant provides for the manner in which payment shall be made, and the right of the plaintiff under the contract to payment for marble furnished, is not dependent upon the certificate of any architect, 31 R. 10.

VII.

The defendant, in paragraph VII. of its brief, on page 32, makes the contention that some of the marble furnished by the plaintiff was defective. At the trial, the defendant contended that some of the marble furnished was defective. Plaintiff contended that none of the marble furnished was defective, that it all complied with the contract. Upon this issue, evidence was introduced by both parties. The court viewed the premises and determined this issue in favor of the defendant. Since this was a disputed question of fact determined by the court in favor of the defendant, the plaintiff has not contended in this court and is not now contending that some of the marble furnished was not defective. The plaintiff's contentions in this matter before this court are:

(1) That since all the marble was inspected by the architect before it was installed, and was permitted to be installed by the architect without any

claim being made to the plaintiff, that it was defective, that the architect and the defendant and the bank are, under the law, estopped to claim any damages against the plaintiff by reason of any defects in the marble installed.

(2) That even though the defendant is not estopped to claim damages, that there is no sufficient or competent evidence in the record to base an award of damages in the sum of $2500.00 for defective marble or in any other sum approaching that amount.

Plaintiff respectfully submits that the judgment heretofore entered herein should be set aside.

Respectfully submitted,

GEORGE CLARK SARGENT,
San Francisco, California.

GROSSCUP & MORROW,
C. A. WALLACE and
J. O. DAVIES,
Seattle, Washington.
Attorneys for Plaintiff in Error.

United States

Circuit Court of Appeals

For the Ninth Circuit.

THE ATCHISON, TOPEKA & SANTA FE
RAILWAY COMPANY, a Corporation,
Plaintiff in Error,

vs.

ROY SPENCER, an Infant by SARAH E. SPEN-
CER, His Guardian *Ad Litem,* and SARAH
E. SPENCER, Individually,
Defendants in Error.

Transcript of Record.

Upon Writ of Error to the United States District Court of the District of Arizona.

Filmer Bros. Co. Print, 330 Jackson St., S. F., Cal.

United States
Circuit Court of Appeals
For the Ninth Circuit.

THE ATCHISON, TOPEKA & SANTA FE
RAILWAY COMPANY, a Corporation,
Plaintiff in Error,

vs.

ROY SPENCER, an Infant by SARAH E. SPEN-
CER, His Guardian *Ad Litem,* and SARAH
E. SPENCER, Individually,
Defendants in Error.

Transcript of Record.

**Upon Writ of Error to the United States District
Court of the District of Arizona.**

Filmer Bros. Co Print, 330 Jackson St , S. F., Cal.

INDEX TO THE PRINTED TRANSCRIPT OF RECORD.

[Clerk's Note: When deemed likely to be of an important nature, errors or doubtful matters appearing in the original certified record are printed literally in italic; and, likewise, cancelled matter appearing in the original certified record is printed and cancelled herein accordingly. When possible, an omission from the text is indicated by printing in italic the two words between which the omission seems to occur.]

Index. Page

Index.   Page

NAMES AND ADDRESSES OF ATTORNEYS OF RECORD.

CHALMERS, STAHL, FENNEMORE & LONGAN, Fleming Bldg., Phoenix, Arizona, Messrs. E. W. CAMP and ROBERT BRENNAN, Kerckhoff Bldg., Los Angeles, Cal., C. B. WILSON, Flagstaff, Ariz., M. W. REED, F. E. FLYNN, and T. J. NORRIS, Prescott, Ariz.,
> Attorneys for the Defendant (Plaintiff in Error).

MARRON & WOOD, Albuquerque, New Mexico, JAMES R. MOORE, Natl. Bank of Ariz. Bldg., Phoenix, Arizona, ROBERT E. MORRISON, Prescott, Arizona, SIDNEY SAPP, Holbrook, Arizona,
> Attorneys for the Plaintiff (Defendant in Error).

———

CLERK'S CERTIFICATE TO RECORD.

State of Arizona,
County of Navajo,—ss.

I, Lloyd C. Henning, Clerk of the Superior Court of the State of Arizona, in and for the county of Navajo, hereby certify that the following twenty-two pages of typewritten and printed matter is a full, true and correct copy of the record and the whole thereof in Civil Cause No. 1717, heretofore pending in said court entitled: Roy Spencer, an infant, by Sarah E. Spencer, his guardian *ad litem,* and Sarah E. Spencer, Individually, Plaintiffs, vs. The Atchison, Topeka & Santa Fe Railway Company, Defendant.   The said record consisting of the following documents: .

Petition, filed April 18, 1924.

Order, filed April 18, 1924.

Complaint, filed April 18, 1924.

Summons with return thereon, filed April 22, 1924.

Petition for removal, filed May 2, 1924.

Notice of removal, filed May 2, 1924.

Order of removal, filed May 7, 1924.

Bond of removal, filed May 2, 1924.

Minute entries.

All of which appears on file and of record in my office.

IN TESTIMONY WHEREOF, I have hereunto affixed the Seal of the aforesaid Court at Holbrook, Arizona, this 15th day of May, A. D. 1924.

[Seal]   LLOYD C. HENNING,

Clerk.   [1*]

[Title of Court and Cause.]

PETITION FOR APPOINTMENT OF GUARDIAN.

To the Superior Court of the State of Arizona Sitting Within and for the County of Navajo:

The petition of Roy Spencer respectfully shows:

1.   That he is a minor under the age of twenty-one years, to wit, of the age of eighteen years; that his father is dead and that he has a claim or cause

------

*Page-number appearing at the foot of page of original certified Transcript of Record.

of action against The Atchison, Topeka & Santa Fe Railway Company for injuries to him as a result of a collision with a train of the said defendant, caused by the negligence of the defendant in the manner set forth in the complaint herewith tendered for filing.

2. That he desires the appointment of Sarah E. Spencer, his mother, as his guardian *ad litem* for the purpose of commencing and prosecuting an action in this Court to recover from the defendant the damages caused by such negligence, and respectfully petitions this Honorable Court to appoint his mother for such purpose.

Dated this 12th day of April, 1924.

ROY SPENCER,
Petitioner. [2]

State of New Mexico,
County of Bernalillo,—ss.

Roy Spencer, being first duly sworn, upon his oath deposes and says that he is the petitioner named in the foregoing petition, and that the facts therein stated are true.

ROY SPENCER.

Subscribed and sworn to before me this 12th day of April, A. D. 1924.

[Notary Seal]   FLORENCE ANDERSON,
Notary Public.

My commission expires November 1st, 1925.

(On Back): No. ——. In the Superior Court of the State of Arizona, in and for the County of Navajo *County*. Roy Spencer, an Infant, by Sarah E.

Spencer, His Guardian *Ad Litem,* and Sarah E. Spencer, Individually, Plaintiffs, vs. The Atchison, Topeka & Santa Fe Railway Company, Defendant. Petition.   Marron & Wood, Attorneys and Counsellors, State National Bank Building, Albuquerque, New Mexico.

Filed Apr. 18, 1924, at 11 o'clock, A. M.   [3]

[Title of Court and Cause.]

ORDER APPOINTING GUARDIAN.

Upon reading and filing the petition of Roy Spencer, a minor of the age of eighteen years, showing that he has a cause of action against The Atchison, Topeka and Santa Fe Railway Company for personal injuries, and praying for the appointment of a guardian *ad litem* to commence and prosecute said action, naming his mother, Sarah E. Spencer, as such guardian, it is considered ORDERED and ADJUDGED that Sarah E. Spencer be and she hereby is appointed guardian *ad litem* for the said Roy Spencer for the purpose of commencing and prosecuting an action against The Atchison, Topeka and Santa Fe Railway Company to recover for personal injuries alleged to have been inflicted upon the said Roy Spencer by the said The Atchison, Topeka & Santa Fe Railway Company through negligence.

Done in open court this 18th day of April, 1924.

J. E. CROSBY,

Judge.

(On Back): No. ——. In the Superior Court of the State of Arizona, in and for the County of Navajo, *County.* Roy Spencer, an Infant, etc., Plaintiffs, vs. The Atchison, Topeka & Santa Fe Ry. Co., Defendant. Order. Marron & Wood, Attorneys and Counsellors, State National Bank Building, Albuquerque, New Mexico.

Filed Apr. 18, 1924, at 11 o'clock A. M. [4]

———

In the Superior Court of the State of Arizona, in and for the County of Navajo.

ROY SPENCER, an Infant, by SARAH E. SPENCER, His Guardian *Ad Litem,* and SARAH E. SPENCER, Individually,

Plaintiffs,

vs.

THE ATCHISON, TOPEKA & SANTA FE RAILWAY COMPANY,

Defendant.

COMPLAINT.

The complaint of the above-named plaintiff Roy Spencer respectfully shows:

1. That the defendant, The Atchison, Topeka & Santa Fe Railway Company is a corporation organized under the laws of the State of Kansas, and owning and operating a line of railway as a common carrier of freight and passengers and operating its line of railway through the States of New Mexico and Arizona, and having offices, stations and agents

in charge thereof within the County of Navajo, and State of Arizona.

2. That the plaintiff, Roy Spencer, is a minor under the age of twenty-one years, to wit, of the age of eighteen years, and is and has been a resident of the State of Colorado, residing at Montrose in said state, and the plaintiff, Sarah E. Spencer, his mother, has been duly appointed guardian *ad litem* by this Court for the purpose of bringing this action on his behalf, the father of the said Roy Spencer being dead, and that the plaintiff, Sarah E. Spencer, is a resident and citizen of the State of New Mexico, residing at Mountainair in said state.

3. That on or about the 11th day of June, 1923, [5] while this plaintiff, Roy Spencer, was riding in an automobile driven by Benjamin B. Spencer, his father, along a public highway which crosses the tracks of the defendant company about one mile west of the village of Mountainair in the county of Torrance and State of New Mexico, the automobile in which this plaintiff, Roy Spencer, was riding was struck by the locomotive of a train of cars then being operated by the defendant company upon its aforesaid tracks, and received serious and permanent wounds, bruises and bodily injuries, from which he still suffers and, as he is advised and verily believes, will continue to suffer during the rest of his natural life.

4. That the said collision and injury to this plaintiff, Roy Spencer, was caused and produced through the negligence and carelessness of the de-

fendant, its officers, agents and servants in the following particulars:

(a)   That the aforesaid crossing was one of peculiar and extraordinary danger in that the view or warning of approaching trains were cut off and obstructed from both directions by cuts, curves and the configuration of the ground, and the crossing so constructed as to make it very difficult to operate and drive a car over and across the same, or to learn of the danger from approaching trains in time to enable a reasonably careful person to protect himself and guard against said danger.

(b)   That the defendant built its line of railroad across the said highway, and it became and was the duty of the defendant, under the laws of the State of New Mexico then existing, to construct the said crossing and to restore the highway to a reasonably safe condition for public traffic.

(c)   That the construction of the said highway was improperly and carelessly and negligently done in that the   [6]   railroad tracks were built upon a grade some eight or nine feet above the level of the said highway at the point of such crossing, and instead of conducting the said highway under the said tracks as could have been done, and in the exercise of reasonable care and prudence should have been done, by means of a bridge or trestle to conduct the railroad over the said highway, the highway crossing was improperly and negligently constructed by building a steep embankment on each side of the said tracks with a turn at approximately right angles on each side of such grade, and of such a character as to make it unreasonably difficult to operate a car over the said tracks.

(d) That view of approaching trains at the point where the said highway was carried across the railroad tracks was very largely concealed from both directions by deep cuts through which the railroad tracks were constructed and laid, concealing the views of approaching trains from persons passing across the said tracks, of such a nature and character that due to the rise in the grade of the highway and the position of the aforesaid cuts and the configuration of the adjacent land the approach of trains cannot be discovered by the exercise of reasonable care and prudence by a person lawfully attempting to use the said highway crossing until in a position of peril, from which it is unreasonably difficult to avoid or escape.

(e) That the condition of the said crossing, as the defendant constructed the same, and under the surrounding circumstances, was and is an improper, careless, unsafe and unreasonable method of carrying the highway across the railroad tracks due to the peculiar dangers and natural topographical features above recited, and it was readily and reasonably possible and practical, and the duty of the said defendant in restoring the said highway to a safe condition to have carried the highway [7] under the railway tracks and thus to have avoided the dangers and difficulties so produced by the grade crossing.

(f) That due to the difficulty of discovering approaching trains because of the nature and condition of the highway and cuts above mentioned, it was the duty of the defendant to establish and maintain automatic signals or warnings of approaching trains so maintained and operated as to give signals and warnings to persons using the highway of the dangerous proximity of such trains, such as an automatic bell or other reasonable signaling device, the physical condition of such crossing requiring and making necessary extraordinary methods and means of warning persons using the highway of such danger.

(g) That the train which collided with the said plaintiff and injured him was then being carelessly and negligently operated in approaching said crossing at a very high rate of speed so as to be, and was, beyond the reasonable control of the servants of the defendant operating the same while approaching said crossing.

(h) That the agents and servants of the defendant operating the said train approached the said crossing without giving any signal by bell or whistle of its approach within a distance reasonably calculated to be heard by or warn the plaintiff or other travelers of its approach, or within the proper, usual and customary distance from the said crossing, and the approach of said train being then concealed by dust stirred up by the breeze.

5. That by reason of the careless and negligent conduct of the defendant, as aforesaid, the said plaintiff, Roy Spencer, was at the time aforesaid, while lawfully crossing the tracks of the said defendant upon said public highway, struck by the

train of the defendant operated by the defendant
and its servants, receiving a compound fracture of
the leg and serious injuries about the head, chest
and body, as a result of [8]   which said plaintiff,
Roy Spencer, was then taken to a hospital and required to procure medical and surgical attendance
and hospital treatment at g*g*eat expense, and ever
since has been and still is badly disabled and incapacitated from performing any and all kinds of
labor to gain a livelihood, and as a result of which,
as he is advised and verily believes, he will be permanently disabled and incapacitated from performing labor or earning a livelihood.

6.   That as the direct and proximate result of the
injuries so received a diseased condition of his
lungs, nerves and body has resulted and been caused
and directly produced, the exact nature of which
this plaintiff, Roy Spencer, is unable to more particularly state, but which said diseases and diseased
conditions, as this plaintiff is informed and verily
believes, will further permanently disable and in-
capacitate him from earning a livelihood and require
large outlay of money in endeavoring to ameliorate
and cure the said condition and restore him to bodily
health.

7.   That the injuries so suffered by this plaintiff,
Roy Spencer, has subjected him to great bodily and
mental pain and distress and, as he is advised, will
continue to cause him great bodily and mental pain
and distress of a permanent nature.

8.   That this plaintiff, Roy Spencer, has incurred
debts for hospital and surg*u*cal attend*ants* in en-

deavoring to cure him, in an amount exceeding One Thousand ($1,000) Dollars, and, as he is advised and verily believes, will be required to lay out and expend large sums of money in the future for such purpose in excess of Twenty-five Hundred ($2500) Dollars.

WHEREFORE plaintiffs pray judgment against the defendant [9] for the sum of Twenty-five Thousand ($25,000) Dollars, besides their costs of this action.

(Signed)   **MARRON AND WOOD.**

MARRON & WOOD,

Albuquerque, N. M.,

(Signed)   **SIDNEY SAPP,**

Holbrook, Ariz.,

Attorneys for Plaintiffs.

(On Back): No. ——. In the Superior Court of the State of Arizona in and for the County of Navajo. *County.* Roy Spencer, an Infant, by Sarah E. Spencer, His Guardian *Ad Litem,* and Sarah E. Spencer, Individually, Plaintiffs, vs. The Atchison, Topeka & Santa Fe Railway Company, Defendant. Complaint. Marron & Wood, Attorneys and Counsellors, State National Bank Building, Albuquerque, New Mexico.

Filed Apr. 18, 1924, at 11 o'clock A. M. [10]

[Title of Court and Cause.]

Action brought in the Superior Court of the State of Arizona, in and for the County of Navajo, and the complaint filed in said County of Navajo, in the office of the Clerk of said Superior Court.

SUMMONS.

In the Name of the State of Arizona, to The Atchison, Topeka & Santa Fe Railway Company, Defendant, GREETING:

You are Hereby Summoned and required to appear in an action brought against you by the above-named plaintiff, in the Superior Court of the State of Arizona, in and for the county of Navajo, and answer the complaint therein filed with the Clerk of this said court, at Holbrook, in said county, within twenty days after the service upon you of this summons, if served in this said county, or in all other cases within thirty days thereafter, the times above mentioned being exclusive of the day of service, or judgment by default will be taken against you.

Given under my hand and the Seal of the Superior Court of the State of Arizona, in and for the County of Navajo, this 18 day of April, 1924.

[Seal]   LLOYD C. HENNING,
                 Clerk of Said Superior Court.
                      By Olive C. Reed,
                      Deputy Clerk.   [11]

Office of the Sheriff,
County of Navajo,
State of Arizona,—ss.

I Hereby Certify that I received the within summons on the 19 day of April, A. D. 1924, at the hour of 8 A. M., and personally served the same on the 19 day of April, 1924, on W. T. Gates, Agent of said Co. in the County of Navajo, a copy of said summons, to which was attached a true copy of the complaint mentioned in said summons.

Dated this 19 day of April, 1924.

L. D. DIVELBESS,

Sheriff.

By O. C. Williams,

Deputy Sheriff.

Fees, Service......$1.50
Copies ....
Travel 1 miles..... .30
Publication ....

Total $1.80
  #318.   F206.   OK L. D. D.

Filed Apr. 22, 1924, at 3:10 o'clock P. M.   [12]

[Title of Court and Cause.]

PETITION FOR REMOVAL.

To the Honorable, the Superior Court of the State of Arizona in and for the County of Navajo:

COMES NOW the Atchison, Topeka and Santa

Fe Railway Company, a corporation, the defendant in the above-entitled action and presents this, its petition and by it respectfully shows:

I.

That The Atchison, Topeka and Santa Fe Railway Company, your petitioner herein, was at the time of the commencement of this action and now is, a corporation of the State of Kansas with its principal place of business in the city of Topeka in said state, and that it is not, and was not at the time of the commencement of this action, and never has been, a corporation organized and existing under and by virtue [13] of the laws of the State of Arizona; that it was at all of said times and now is a resident and citizen of the State of Kansas and a nonresident of the State of Arizona.

II.

That according to the laws of the State of Arizona, the time within which your petitioner is required to appear and answer, demur or otherwise plead in said action has not yet expired, and your petitioner has not demurred, answered, pleaded or otherwise appeared in said action. That Roy Spencer, one of the plaintiffs in said above-entitled action, was, at the time of the commencement of this action, and ever since has been and now is a citizen and resident of the State of Colorado and that said Roy Spencer is not now and was not at the time of the commencement of this action, a resident of the State of Kansas and at all of such times, the said Roy Spencer has been and now is a nonresident of the State of Kansas.

III.

That Sarah E. Spencer, the other plaintiff in the above-entitled action and guardian *ad litem* for the plaintiff, Roy Spencer, was at the time of the commencement of this action and ever since has been, and now is a citizen and resident of the State of New Mexico, and that the said Sarah E. Spencer, one of the plaintiffs in said action and the guardian *ad litem* of the plaintiff, Roy Spencer, is not now and was not at the time of the commencement of this action, a resident or citizen of the State of Kansas and at all of such times, said Sarah E. Spencer has been and now is a nonresident of the State of Kansas. [14]

IV.

That the place of residence of the plaintiffs and defendant respectively is alleged and appears upon the face of the complaint herein.

V.

That the above-entitled action is a suit of a civil nature brought by the plaintiffs to recover a judgment against your petitioner for Twenty-five Thousand (25,000) Dollars damages for alleged personal injuries sustained by Roy Spencer, according to the allegations of the complaint, who was riding in an automobile driven by his father along a highway crossed by the tracks of the defendant company in the County of Torrance, State of New Mexico, and which automobile, according to the allegations of said complaint, was struck by petitioner's locomotive; that the crossing at which said accident is alleged to have occurred is, according to the allegations of the complaint, of peculiar and extraordinary danger on account of the view of approaching trains being cut off by cuts, curves and configurations of the ground and the crossing is so constructed as to make it very difficult to operate and drive an automobile over and across it or to learn of the danger from approaching trains in time to guard against danger and for alleged negligence of the petitioner, according to the allegations of the complaint, in the construction of railroad tracks on a grade above the level of the highway. And further it is alleged that petitioner's train was negligently operated at a high rate of speed without warnings of its approach to the crossing. [15]

VI.

That the amount in controversy in the above-entitled action exceeds, exclusive of interest or costs, the sum or value of Three Thousand (3,000) Dollars.

VII.

That this is an action in which there is a controversy wholly between citizens of different states and which can be fully determined as between them, and that your petitioner is actually interested in such controversy.

VIII.

That your petitioner offers and presents herewith a good and sufficient Bond and Surety as provided by the Statute in such cases, conditioned that it will, within thirty (30) days of the filing of this petition, enter a certified copy of the record in the above-entitled court, in the United States District Court

for the District of Arizona; and that your petitioner
will pay all costs that may be awarded by said District Court if it shall hold that said action was
wrongfully or improperly removed thereto.

WHEREFORE, your petitioner prays that this
Court accept this petition and said bond and surety
and that said action be removed into said District
Court of the United States for the District of Arizona, pursuant to the statutes in such cases made
and provided, and that this Court proceed no further
in this action, except to make an order for the removal as prayed for, accept and approve the bond
presented herewith, and direct the Clerk of this
court to prepare a certified copy of the record in
the above-entitled action for [16] entry in said
District Court of the United States for the District
of Arizona.

And your petitioner will ever pray, etc.

Dated, April 23, 1924.

> THE ATCHISON, TOPEKA AND SANTA
> FE RAILWAY COMPANY, a Corpora-
> tion.
> By CHALMERS, STAHL, FENNEMORE
> & LONGAN,
> > C. B. WILSON,
> > ROBERT BRENNAN,
> > Attorneys for Said Petitioner.

State of California,
County of Los Angeles,—ss.

W. H. Brewer, being by me first duly sworn, deposes and says that he is the assistant to the general

manager of The Atchison, Topeka and Santa Fe Railway Company, a corporation, the defendant in the above-entitled action; that he has read the foregoing petition for removal and knows the contents thereof, and that the same is true of his own knowledge, except, as to the matters which are therein stated upon his information and belief, and as to those matters, he believes it to be true.

W. H. BREWER.

Subscribed and sworn to before me this 23d day of April, 1924.

[Notarial Seal]   NELLIE B. KEMPER, Notary Public in and for Said County and State.

No. ——. Dept. ——. In the Superior Court of the State of Arizona, in and for the County of Navajo. Roy Spencer, etc., et al., Plaintiff, vs. The Atchison, Topeka and Santa Fe Railway Company, Defendant. Petition for Removal. Received copy of the within petition this —— day of April, 1924. ——, Attorney for Plaintiffs. E. W. Camp, Robert Brennan, M. W. Reed, E. T. Lucey, Kerckhoff Building, Los Angeles, California, Telephone, Main 2980, Attorneys for Defendant.

Filed May 2, 1924, at 9 o'clock A. M.   [17]

[Title of Court and Cause.]

NOTICE OF REMOVAL.

To Roy Spencer, an Infant, by Sarah E. Spencer, His Guardian *Ad Litem* and Sarah E. Spencer, Individually, Plaintiffs Above Named, and to Messrs. Marron and Wood and Sidney Sapp, Esq., Attorneys for Plaintiffs:

You and each of you will please take notice that on the 7th day of May, 1924, at eleven o'clock A. M., or as soon thereafter as counsel can be heard, The Atchison, Topeka and Santa Fe Railway Company, a corporation, the defendant in the above-entitled action, will present to the Superior Court of the State of Arizona in and for the County of Navajo, at the courthouse in the city of Holbrook, said county and state, its petition for and bond on removal of the above-entitled action in the above-entitled court to the District Court of the United States for the District of Arizona, pursuant to [18] the statutes in such cases made and provided; and that a copy of said petition and a copy of said bond, together with a copy of proposed order on removal are hereunto annexed and made a part hereof.

Dated, April 23, 1924.

CHALMERS, STAHL, FENNEMORE & LONGAN,

C. B. WILSON,

ROBERT BRENNAN,

Attorneys for the Defendant, The Atchison, Topeka and Santa Fe Railway Company, a Corporation.

Filed May 2, 1924, at 9 o'clock A. M. [19]

[Title of Court and Cause.]

ORDER OF REMOVAL.

The Atchison, Topeka and Santa Fe Railway Company, a corporation, the defendant in the above-entitled action, having within the time provided by law filed its petition in due form for the removal of said action to the District Court of the United States for the District of Arizona, and having at the same time offered a good and sufficient bond as required by law and said bond having been approved, and it appearing to the Court that said defendant is entitled to have said cause removed to said District Court of the United States for the District of Arizona.

NOW, THEREFORE, it is hereby ORDERED that said action be removed into the District Court of the United States for the District of Arizona, and that all further proceedings in this court in said action be, and they are hereby stayed, [20] and the Clerk of this court is hereby directed to make a certified copy of the record in said action for entry in said United States District Court.

Done this 7th day of May, 1924.

J. E. CROSBY,
Judge.

Filed May 7, 1924, at 12:05 o'clock P. M.   [21]

[Title of Court and Cause.]

BOND ON REMOVAL.

KNOW ALL MEN BY THESE PRESENTS:

That The Atchison, Topeka and Santa Fe Railway Company, a corporation, as Principal, and the National Surety Company, a corporation, as Surety, are held and firmly bound unto Roy Spencer, an infant, by Sarah E. Spencer, as his guardian *ad litem,* and Sarah E. Spencer, individually, plaintiffs in the above-entitled action, their heirs, executors and administrators, in the sum of One Thousand Dollars lawful money of the United States of America, for the payment of which, well and truly to be made, they bind themselves, their successors and assigns, as the case may be, jointly and severally, firmly by these presents.

The condition of the above obligation is such that WHEREAS, The Atchison, Topeka and Santa Fe Railway Company, a corporation, the defendant in the above-entitled [22] action has applied by petition to the Superior Court of the State of Arizona in and for the County of Navajo, for the removal of a certain cause therein pending, wherein Roy Spencer, an infant, by Sarah E. Spencer, his guardian *ad litem,* and Sarah E. Spencer, individually, are plaintiffs, and The Atchison, Topeka and Santa Fe Railway Company, a corporation, is defendant, to the District Court of the United States for the District of Arizona for further proceedings on the grounds in said petition set forth, and that all further proceedings in said action be stayed.

NOW, THEREFORE, if the above-named defendant shall, within thirty days from and after the date of the filing of said petition, enter in said District Court of the United States for the District of Arizona, a duly certified copy of the record in the above-entitled action, and shall pay or cause to be paid all costs that may be awarded therein by the said District Court of the United States, if such Court shall hold that such suit was wrongfully or improperly removed thereto, then this obligation shall be void, otherwise to remain in full force and effect.

Dated, April 23, 1924.

[Corporate Seal]

THE ATCHISON, TOPEKA AND SANTA FE RAILWAY COMPANY, a Corporation.

By W. H. BREWER,

Its Asst. to Gen. Mgr.,

Principal.

Attest:

E. W. JONES,

Its Asst. Secretary.

NATIONAL SURETY COMPANY.

By HOLBROOK INSURANCE AGENCY.

By LLOYD C. HENNING, Manager,

Its Agent.

The foregoing bond is hereby approved as to form and sufficiency of surety this 22 day of May, 1924.

J. E. CROSBY,

Judge.

Filed May 2, 1924, at 11:45 A. M.   [23]

[Title of Court.]

Minute entries made in Civil Cause No. 1717, Roy Spencer, an infant, by Sarah E. Spencer, his guardian *ad litem,* and Sarah E. Spencer, individually, Plaintiff, vs. The Atchison, Topeka & Santa Fe Railway Company, a corporation, Defendant. Filed April 18, 1924. J. E. CROSBY, Judge, Presiding.

MINUTES OF COURT—MAY 7, 1924—ORDER DIRECTING REMOVAL OF CASE TO UNITED STATES DISTRICT COURT OF ARIZONA.

May 7, 1924, 11 o'clock A. M.

(Title of Cause.)

Present: Mr. WOOD and SIDNEY SAPP, attorneys for plaintiffs and C. B. WILSON, representing the defendant.

Mr. Wilson at this time presents his petition for removal and no objection being raised by counsel for plaintiffs, this case is ordered removed to the United States District Court of Arizona.

State of Arizona,
County of Navajo,—ss.

I, Lloyd C. Henning, Clerk of the Superior Court of the State of Arizona, in and for the County of Navajo, do hereby certify that the foregoing one-half page of typewritten matter is a full, true and correct copy of all minute entries made in Civil Cause No. 1717, wherein Roy Spencer, an infant, by Sarah E. Spencer, his guardian *ad litem,* and Sarah E. Spencer, individually, are plaintiffs and the Atchison, Topeka and Santa Fe Railway Company, a corporation, is defendant, and that I have compared the same with the original.

IN TESTIMONY WHEREOF, I have hereunto set my hand and affixed the Seal of the aforesaid Court at Holbrook, Arizona, this 15th day of May, 1924.

[Seal]   LLOYD C. HENNING,
                                        Clerk.
                    By Oliver C. Reed,
                                Deputy Clerk.

[Endorsed]: Filed May 17, 1924.   [24]

[Title of Court and Cause.]

ANSWER.

COMES NOW The Atchison, Topeka and Santa Fe Railway Company, a corporation, defendant in the above-entitled action and for its answer to the plaintiffs' complaint by way of answer, ADMITS, DENIES and ALLEGES as follows:

I.

Defendant ADMITS that it is a corporation organized under the laws of the State of Kansas, and owns and operates a line of railway as a common carrier of freight and passengers and operates its line of railway through the states of New Mexico and Arizona, as alleged in Paragraph 1 of said complaint.

II.

That it does not have knowledge or information sufficient to enable it to answer the allegations contained in Paragraph 2 of said complaint, and placing its denial on that ground, [25] DENIES said allegations and all thereof.

III.

For further answer to said complaint, defendant DENIES that at the time and place mentioned in said complaint, or at any time or at any place or at all, the said Roy Spencer was injured or that the plaintiffs, or either of them, sustained any damage on account of any injuries to said Roy Spencer by reason of any act of negligence of this defendant or any of its agents, servants or employees, either in the manner or to the extent set forth in said complaint, or in any manner or to any extent whatever.

IV.

For further answer to said complaint, defendant DENIES that at the time or place mentioned in Paragraph 3 of said complaint or at any other time or place, or at all, or while the said Roy Spencer was riding in an automobile driven by his father or anyone else, that said or any automobile was struck by any locomotive or train of cars operated by the defendant by reason of any negligence of the defendant, either in the operation of said locomotive or train of cars, or otherwise or at all, and DENIES that on account of any negligence of this defendant, either in the manner or to the extent attempted to be alleged in said complaint or otherwise or at all, the said Roy Spencer received serious or permanent wounds, bruises or bodily injuries, or that on account of any wounds, bruises or bodily injuries the said Roy Spencer suffers or will continue to suffer during the rest of his natural life and DENIES that the said Roy Spencer was or is permanently injured on account of any of the matters attempted [26] to be alleged and set forth in said complaint, on account of any negligence of this defendant or any of its agents, servants or employees.

V.

For further answer to said complaint and particularly to paragraph 4 thereof, this defendant DENIES that any collision in which the said Roy Spencer was involved, or any injury that he may have received, was caused or produced through the negligence and carelessness of the defendant, its officers, agents or servants as therein attempted to be alleged, or otherwise, or at all. That it DENIES that the crossing therein described was of peculiar or extraordinary danger or that the view or warning of approaching trains was cut off or obstructed from both or either direction by cuts, curves or configurations of the ground or that by reason of the construction of said crossing, it was difficult to operate or drive an automobile over or across the same or to learn of the danger from approaching trains in time to enable a reasonably careful person to protect himself and guard against said or any danger.

That it DENIES the imputation contained in subdivision (b) of paragraph 4 of said complaint, that its line of railroad was not constructed across said highway in compliance with the laws of the State of New Mexico, and ALLEGES the fact to be that said crossing was in all respects constructed in compliance with the laws of said state and that said highway was restored to a reasonably safe condition for public travel.

That it DENIES that the construction of said or any highway was improperly or carelessly or negligently done, or [27] that the railroad tracks were improperly or negligently built upon a grade above the level of the highway at the point of said crossing, and DENIES that the said highway crossing was improperly or negligently constructed, either in the manner attempted to be alleged in sub-division (c) of said paragraph 4 of said complaint, or otherwise, or at all.

That it DENIES that the view of approaching trains at the place mentioned in subdivision (d) of said paragraph 4 was very badly or otherwise concealed from both directions by deep cuts or other-

wise, or at all, and DENIES that the approach of
trains could not be discovered by persons lawfully
attempting to use said highway crossing who were
in the exercise of reasonable care and prudence,
and ALLEGES that a person in the exercise of reasonable care and prudence would have a view of
approaching trains in time to avoid being struck
thereby.

That it both generally and specifically DENIES
each and every allegation in subdivisions (e), (f),
(g) and (h) of paragraph 4 of said complaint as
fully and completely as if each allegation thereof
were separately traversed and in like manner denied.

And DENIES that said Roy Spencer was involved in any collision with any engine or train of
cars of the defendant while riding in any automobile or otherwise, or at all, because of any negligence
of this defendant, either because of the physical
conditions existing at and in the vicinity of said
crossing or because of any negligence in the operation of defendant's said or any train or because of
any failure to install automatic bells or other signal devices, or otherwise, or at all, and expressly
DENIES that it was guilty of any negligence causing or contributing to any injury [28] or injuries
of the said Roy Spencer.

V.

For further answer to said complaint and particularly to paragraph 5 thereof, it DENIES that by
reason of any careless or negligent conduct of this
defendant, the said Roy Spencer was struck by de-

fendant's train or that by reason of any careless
or negligent conduct of this defendant, the said
Roy Spencer received a compound or other fracture of his leg or serious or other injuries about his
head, chest or body, or that the said Roy Spencer
was required to procure medical, surgical or hospital treatment or incurred any expense because of
any negligence of this defendant or any of its
agents, servants or employees.

VI.

For further answer to said complaint, and particularly to paragraphs 6, 7 and 8 thereof, defendant ALLEGES that it does not have knowledge or
information sufficient to enable it to answer the allegations thereof, and placing its denial on that
ground, DENIES said allegations and all thereof;
but in this connection, expressly DENIES that any
injuries or ailments from which the said Roy Spencer may be suffering or from which he may continue to suffer were in any wise caused by or due
to any negligence of this defendant in any manner
whatsoever and further DENIES that the plaintiffs have been damaged in the sum of Twenty-five
Thousand (25,000) Dollars or in any other sum or
amount whatsoever.

VII.

For further answer to plaintiffs' complaint and
for a further and separate defense herein, defendant ALLEGES that [29] if the said Roy Spencer was injured at the time or place or in the manner attempted to be alleged in said complaint, or
if the said Roy Spencer or his mother, Sarah E.

Spencer suffered any damages on account thereof, then such injuries and damages resulting therefrom, if any, were caused solely by and resulted wholly from the negligent acts and omissions of the said Roy Spencer and his father who was driving or operating said automobile, in that the said Roy Spencer and his father or other person riding with him in said automobile failed to look or listen for the approach of defendant's engine and train of cars to said crossing, and that the said Roy Spencer failed to take any precaution whatsoever to avoid said collision and failed to observe and heed the warning signals given by the sounding of the whistle and the ringing of the bell of the approach of said engine and train of cars to said crossing; that the injuries to the said Roy Spencer and the damages resulting therefrom, if any there were, did not result from and were not caused by any negligent act or omissions on the part of the defendant or any of its agents, servants or employees.

For a further and separate defense, defendant ALLEGES that any injuries incurred by the said Roy Spencer resulting from the accident attempted to be described in said complaint, were caused by his own negligence as aforesaid, and that his said negligence was the direct and proximate cause thereof and were not caused by any negligence or carelessness of this defendant or any of its agents, servants or employees.

For further answer to said complaint, it generally and specifically DENIES each and every material allegation [30] therein contained not here-inbefore in this answer expressly admitted.

WHEREFORE, defendant having duly answered, prays that plaintiffs take nothing by their action and that it have and recover its costs and disbursements herein incurred or expended.

CHALMERS, STAHL, FENNEMORE and LONGAN,

C. B. WILSON and

ROBERT BRENNAN,

Attorneys for Defendant, The Atchison,Topeka and Santa Fe Railway Company.

[Endorsed]: Filed Jun. 13, 1924. [31]

————

[Title of Court and Cause.]

DEFENDANT'S MOTION FOR DIRECTED VERDICT.

NOW COMES The Atchison,Topeka and Santa Fe Railway Company, a corporation, and moves the Court for directed verdict in its favor in said action upon the following grounds:

1. Upon the ground that plaintiff has not offered sufficient evidence to entitle the cause to be submitted to the jury.

2. That there is no evidence of any negligence on the part of this defendant.

3. That the undisputed evidence shows that plaintiff was guilty of negligence himself, which was the sole and proximate cause of the injury complained of.

4. That the undisputed evidence shows that plaintiff was guilty of negligence and want of ordinary care for his own safety, which contributed to the injury complained of. [32]

5. That the evidence considered in the light of physical conditions shows that, if the precautions required by law had been observed by plaintiff, he would have seen the train in time to have avoided the accident.

6. That the undisputed evidence shows that if plaintiff had looked and listened before attempting to cross the railroad track, he would have seen and heard the train in time to have avoided the accident.

7. That the undisputed evidence shows that if plaintiff had looked in the direction of the on-coming train before attempting to cross the railroad track he would have seen the train in time to have avoided the accident.

8. That the undisputed evidence shows that plaintiff and the driver of the automobile in question, to wit, Benjamin Spencer, were engaged in a joint enterprise at the time of said accident and that if the driver of the said automobile and plaintiff had exercised ordinary care and caution before attempting to cross said railroad track, they could have seen or heard the train in time to have avoided the accident.

9. That the undisputed evidence shows that plaintiff and the driver of the automobile in question, to wit, Benjamin Spencer, were at the time of the accident engaged in a joint enterprise and that,

if plaintiff and the driver of said automobile, or either of them, had exercised ordinary care and caution for their own safety before attempting to cross said railroad track, they could have seen or heard the train in time to have avoided the accident. [33]

10. That the undisputed evidence shows that, if plaintiff and/or the driver of said automobile had looked upon the railroad track in the direction from which the train was approaching when they were 30 feet from the railroad track, they could have seen the train for a distance of not less than 1,500 feet, and in time to have avoided the accident.

11. For the reason that the undisputed evidence shows that if plaintiff suffered from damages or injuries whatsoever at the time, or in the manner or because of the facts alleged in plaintiff's complaint therein, such damages were directly, immediately and proximately caused and occasioned by the fault, negligence and carelessness of plaintiff himself, and not by any fault, negligence or carelessness on the part of the defendant.

12. By the allegation of the complaint and the evidence adduced, it is conclusively shown that plaintiff, Roy Spencer, and his mother, are citizens and residents of the State of New Mexico and that the alleged cause of action sued on herein accrued in that state by reason whereof the plaintiff's right to recover judgment against this defendant is controlled by the laws of New Mexico; that under the laws of New Mexico the plaintiff has not only not offered legally sufficient evidence of defendant's

negligence proximately causing his injury or damage, but on the contrary the evidence establishes such negligence of Roy Spencer as to bar his right to a verdict and judgment in his favor.

13.   That Arizona constitutional provision being Section 5, Art. 18, is not applicable in this case because of the residence and citizenship of Roy Spencer and his   [34]   mother and because the alleged cause of action sued on herein accrued in the State of New Mexico.

14.   That the evidence adduced herein showing that the negligence of Roy Spencer proximately contributed to his injury, there is no controverted question of fact to be determined by the jury, and, therefore, the record presents solely a question of law for the Court, which, notwithstanding the provisions of Sec. 5, Art. 18 of the Arizona Constitution it is the right and the duty of this Court, under the law as administered by the Federal and United States Supreme Courts, and under Rule 60 of the Rules of Practice of this Court, to decide; that said Rule 60 reads as follows:

"The defendant in an action at law, tried either with or without a jury, may either at the close of the plaintiff's case or at the close of the case on both *saides*, move for a nonsuit. The procedure on such motion shall be as follows:

The defendant or his counsel, shall state orally in open court that he moves for a nonsuit on certain grounds, which shall be stated specifically. Such a motion shall be deemed and treated as assuming for the purposes of the motion (but for such purposes only) the truth of whatever the evidence tends to prove, to wit: Whatever a jury might properly infer from it. If, upon the facts so assumed to be true as aforesaid, the Court shall be of opinion that the plaintiff has no case, the motion shall be granted and the action dismissed. The party against whom the decision on the motion is rendered may then and there take a general exception, and may have the same, together with such of the proceedings in the case as are material, embodied in a bill of exceptions. If evidence shall be introduced by either party after the decision on the motion has been made, the same shall operate as a superseding of the motion; but such motion may be renewed at the close of all the evidence.''

Dated this 13 day of August, 1926.

M. W. REED,
ROBERT BRENNAN,
Attorneys for Defendant.

[Endorsed]: Filed Aug. 13, 1926. [35]

[Title of Court and Cause.]

VERDICT.

We, the jury, duly empaneled and sworn in the above-entitled action, upon our oaths, do find for the plaintiff and assess his damages in the sum of

Sixteen Thousand Seven Hundred Fifty Dollars ($16,750.00).

JOHN W. BEACH,
Foreman.

[Endorsed on Back]: Filed Aug. 15, 1926.   [36]

In the District Court of the United States in and for the District of Arizona.

No. 170—LAW (PRESCOTT).

ROY SPENCER, an Infant, by SARAH E. SPENCER, His Guardian *Ad Litem,*

Plaintiff,

vs.

THE ATCHISON, TOPEKA & SANTA FE RAILWAY COMPANY, a Corporation,

Defendant.

JUDGMENT ON VERDICT.

This cause came on regularly for trial on the 11th day of August, 1926. The parties were present in person, and represented by their attorneys, Mr. F. E. Wood of Albuquerque, New Mexico, and Mr. Robt. E. Morrison of Prescott, Arizona, counsel for plaintiff, and Mr. Robert Brennan and Mr. M. W. Reed of Los Angeles, California, and Messrs. Norris, Norris & Flynn of Prescott, Arizona, counsel for the defendant. A jury of twelve persons were regularly impaneled and sworn to try said action. Witnesses on the part of plaintiff and defendant were duly examined, and the cause was thereafter continued and tried on the 12th, 13th, and 14th days of August, 1926. After hearing the evidence, the arguments of counsel and instructions of the Court, the jury retired to consider their verdict, and subsequently, on the 15th day of August, 1926, returned into court with their verdict signed by the foreman in accordance with the law, and being called, answer to their names and say:

"We, the jury, duly empaneled and sworn in the above-entitled action upon our oaths, do find for the plaintiff and assess his damage in the sum of Sixteen Thousand Seven Hundred Fifty Dollars ($16,750).

JOHN W. BEACH,
Foreman. [37]

NOW, THEREFORE, IT IS ORDERED that judgment be entered herein in favor of plaintiff Roy Spencer, an infant, by Sarah E. Spencer, his guardian *ad litem,* and against the defendant The Atchison, Topeka & Santa Fe Railway Company, a corporation, in accordance with the verdict in said cause in the sum of Sixteen Thousand Seven Hundred and Fifty Dollars ($16,750),

WHEREFORE, by virtue of the law, and by reason of the premises aforesaid it is

ORDERED, ADJUDGED AND DECREED that plaintiff Roy Spencer, an infant, by Sarah E. Spencer, his guardian *ad litem,* do have and recover of and from the defendant The Atchison, Topeka & Santa Fe Railway Company, a corporation, the sum of Sixteen Thousand Seven Hundred and

Fifty Dollars ($16,750.00) with interest thereon at the rate of six per cent (6%) per annum from date hereof until paid, and for plaintiff's costs incurred in said action, taxed at the sum of Two Hundred Fifty-four and 20/100 and that execution issue.

Dated and entered in open court this 15th day of August, 1926.

F. C. JACOBS.

[Endorsed]: Filed August 15, 1926.   [38]

[Title of Court and Cause.]

CERTIFICATE OF CLERK U. S. DISTRICT COURT TO JUDGMENT–ROLL.

I, C. R. McFall, Clerk of the District Court of the United States for the District of Arizona, do hereby certify that the foregoing papers hereto annexed constitute the judgment-roll in the above-entitled action.

ATTEST my hand and the seal of said District Court this 21st day of August, 1926.

[Seal]   C. R. McFALL,
                              Clerk.
                    By J. Lee Baker,
                        Deputy Clerk.

[Endorsed on Back]: Filed Aug. 21, 1926.   [39]

[Title of Court and Cause.]

NOTICE OF PROPOSED BILL OF EXCEPTIONS.

To Marron & Wood, Esqs., and Robert E. Morrison, Esq., Attorneys for Plaintiff:

YOU ARE HEREBY NOTIFIED that the following is a copy of defendant's proposed bill of exceptions in the above-entitled cause.

Dated, Oct. 23d, 1926.

CHALMERS, STAHL, FENNEMORE and LONGAN,

ROBERT BRENNAN,

FRANK E. FLYNN,

M. W. REED,

Attorneys for Defendant. [40]

[Title of Court and Cause.]

DEFENDANT'S PROPOSED BILL OF EXCEPTIONS.

BE IT REMEMBERED, that the trial of the above-entitled cause came on regularly to be heard before the Honorable, F. C. Jacobs, Judge of the District Court of the United States, in and for the District of Arizona, sitting with a jury, at the courtroom of said court, in the Elk's Building, City of Prescott, State and District of Arizona, on the 11th day of August, 1926, at 10:00 o'clock A. M., the plaintiff being represented by his counsel, Messrs.

Marron & Wood, of Albuquerque, New Mexico, and
Robert E. Morrison, of Prescott, Arizona, and the
defendant being represented by its counsel, M. W.
Reed, Esq., Robert Brennan, Esq., of Los Angeles,
California, and Frank E. Flynn, of Norris & Flynn,
of Prescott, Arizona, and the parties having announced ready for trial.

J. G. A. Martin, was duly sworn as Shorthand
Reporter. [41]

WHEREUPON, the following proceedings and
none other were had:

Mr. WOOD.—If the Court please, I would like
to have the name of Robert E. Morrison associated
in the case as counsel for the plaintiff.

The COURT.—Very well. Let the record so
show.

A jury was thereupon impaneled and sworn to
try the case.

The opening statements to the jury were then
made by counsel for the respective parties.

DEPOSITION OF FRANK SEDILLO, FOR PLAINTIFF.

Whereupon the deposition of FRANK SEDILLO,
produced on behalf of plaintiff, was read as follows:

Direct Examination.

My name is Frank Sedillo. I live right here in
Mountainair. I have lived in Mountainair, I do
not remember how many years. I was living here
before they put the railroad in. About two years
before. Before that, I lived about two miles south

(Deposition of Frank Sedillo.)

from Eastview. Three miles south of the town. I knew Benjamin Spencer in his lifetime; (24) ever since Spencer came here. I knew when he was struck by the train and killed, but I do not remember the date. I know the place where he was struck by the train. Before the railroad was built, I know there was a road used and traveled by the public at that place where Spencer was killed. There was the same road before they put the railroad where the crossing was. I know that road, ever since 1901, was used for public travel. When I first knew it, it was mostly used for hauling salt from the river. The places on that road between here (Mountainair) and the river were settlements. Well, Lucinio [42] Torres' ranch. This was the first ranch that was settled there before Mountainair right around here. The point, or place, on the river to which that road leads was to different places down the river. This road met with the Mananza Road and many other roads in Abo. Abo was a settlement at that time before the railroad. It was an old town. I know whether or not, and to what extent, the people of Abo used that road—the old road. All the people in Mananza and Puenta passed by there to the river all the time. I am referring to this road that crossed the railroad where Spencer was killed. They used it to go to Mananza and Puenta and the Salt Lakes. I knew that road earlier than 1901. I knew it before 1901, but I knew it very well after 1901. Before 1901 I lived at Eastview. I worked a long time with

(Deposition of Frank Sedillo.)
Spencer in the sawmill. That is where I crippled
my hand. Eastview is more or less about nine miles
northwest of the crossing.

Q. Whereabouts, with reference to where the
present road crosses the railroad did that old road
cross the line of the road before it was built?

A. The old road crossed right in the same place,
only it crossed kind of west, the old road; and this
new road crosses right straight south. That is, the
old road crossed at a much sharper angle than this
road. More toward the west where this road
changes and goes due south. There is a mark at
or near the crossing outside the present line to
show where the old road ran; it shows. Runs like
a creek, or something, on the other side of the track.
The line, with reference to the fence of the railroad
right of way at the point where the road now
crosses, well, it shows—like a creek shows—the old
road right inside the fence. It [43] shows
something like a creek. This mark that I have
called "something like a creek"—it shows the
wagon-track inside of the fence of the railroad
right of way on the south side of the crossing; and
that road which shows inside of the fence on the
railroad right of way to the south of the crossing,
and to the west of the road, is the line of the old
road that I have described followed. It is the old
road, only the water, the rain, has made it a creek.
There is a mark on this, the north side of the track,
showing where the old road ran. On both sides.

(Deposition of Frank Sedillo.)

Q. I show you a diagram showing where the road now comes down and crosses those tracks, with west marked "W" and north marked "N" and south marked "S." In the center of this diagram, with the lines across it, represents the railroad track. The single line crossing it represents the fence there now. Will you take this pencil and mark on that where the line of the old road now shows, and where it crosses the line now occupied with the railroad tracks?

We will offer, in this connection, the diagram which was made by the witness at the time, and which was annexed originally to the deposition, and attached when offered on this trial.

The COURT.—It may be admitted in evidence as Plaintiff's Exhibit No. 1.

Cross-examination.

I think that the present crossing is about at the place where the old road crossed the line. The old road made more of a diagonal crossing there than at present. All the people of Abo crossed by that road. That road led to [44] the Salt Lakes.

DEPOSITION OF JUAN CHAVEZ Y CHAVEZ, FOR PLAINTIFF.

The deposition of JUAN CHAVEZ Y CHAVEZ, a witness on behalf of plaintiff, taken on the 2d day of July, 1925, at Mountainair, New Mexico, was then read as follows:

(Deposition of Juan Chavez y Chavez.)

Direct Examination.

I live five miles west from here in a place called Barano, New Mexico. I am 68 years. In the place where I now live I have been since 1900. Ever since 1873, I have lived close enough here to know the roads in this vicinity, this immediate section of the country. I knew Benjamin Spencer in his lifetime. I remember the occasion when he was killed in a railroad collision, or accident. I was not here at that time. I was in Isleta. I know the crossing about one mile west of Mountainair where the train struck him. I have known the road which crosses the railroad at that point ever since 1903 or 1905. Ever since they started there. I mean when they started the railroad building. The road before the railroad—I have known that road to cross at the point where it does now before the railroad was built. I knew that old road ever since 1873. That was the first time that I knew it. Ever since I first saw it then it shows that it has been traveled a long time.

Q. Where did the road—the old road—cross the line that is now occupied by the railroad? Was it at the place it now crosses, or some other place?

A. Don't remember very well whether it was right at the same point. [45]

Q. Was it approximately more or less right where it is (35) now that it crossed that place?

A. I do not remember. I could not testify whether it was right there, or further. One thing

(Deposition of Juan Chavez y Chavez.)

I know, when they put in the railroad there had been a crossing.

Q. A crossing right where?

A. Right where it is.

Q. Do you remember, more or less, whether that road ran up near the railroad crossing before they built the railroad?

A. The valley that goes on the other side of the hill. I know where that road runs, now, from here —the valley here, up to the crossing where Spencer was killed. And from here up to there, it follows, more or less, the line of this old road that I was speaking of. It used to go as far as Abo. But, leading from here up to the crossing where it crosses the railroad where Spencer was killed, I believe it goes from here to there following the same line as the old road. And from the point of that crossing—the old road—it followed downward; whether it crossed the railroad tracks in approximately the same place it does, I do not remember— (36) I do not recall.

Cross-examination.

Last time I was up to this crossing was in March past. I went up that road from here. And on down the valley crossing the railroad and on down. The road is practically the same as it always was. Ever since they built the railroad it is the same road. It is practically the same as it was before they built the railroad. I think it is [46] the same thing. I do not know whether they made any difference when they built up the railroad. (37)

DEPOSITION OF ENTIMIO LUNA, FOR PLAINTIFF.

Thereupon the deposition of **ENTIMIO LUNA**, taken and produced by and on behalf of **plaintiff**, was read as follows:

Direct Examination.

My full name is Entimio Luna. I live in Eastview. I have lived in this section of the state and county all my life. I am about 67 or 68 years old. I knew Benjamin Spencer. I know the crossing at which he was killed. (38)

Q. Was there any other road leading from the Rio Grande over to the Salt Lakes, and to that section of the county, in the early days, except this one described?

A. Around this place only that one. (41) I mean in the lower part of those places. I mean— the road followed the lower part of that section generally as it passed through.

Mr. WOOD.—We desire to read, in that connection, the stipulation entered into by counsel at the time that this deposition was taken: "It is stipulated by and on behalf of the plaintiff and defendant, by the attorneys for the respective parties, that the witnesses Juaquin Sisneras, Merijildo Sisneres, Miguel Garcia, Comito Contreres, and Natividad Salas, named in the notice to take depositions in this case, if produced on behalf of the plaintiff, would testify in substance and to the same ef-

(Deposition of Meriam White.)
fect as the testimony of the witnesses Frank Sedillo, Chavez y Chavez and Entimio Luna, and for
this reason they are not produced, and it is stipulated that this stipulation may be read in evidence
in lieu of their testimony." (42) [47]

DEPOSITION OF MERIAM WHITE, FOR PLAINTIFF.

The deposition of MERIAM WHITE, taken at
the same time, was then produced on behalf of
plaintiff, and read as follows:

Direct Examination.

My name is Meriam White. I live at Mountainair. I have lived at Mountainair about 17
years. I remember the occasion when a collision occurred which resulted in the death of
Benjamin Spencer. I was in the day coach of the
passenger train going west, at the time of the collision and accident. I was a passenger on the train
that struck him. (43) I boarded the train at
Mountainair. The first thing that called my attention to the accident was the shriek of the whistle
and the shock to the train as the air-brakes were
applied. I think I had better begin a few minutes prior to the time of the accident, in order to
make my story connected. (44) The train was
running very fast and rocking the coach, and I
was a little bit nervous, to be frank. I was talking
with Mr. Finney, the conductor, and noticed that
they were running fast, and I remember that I

(Deposition of Meriam White.)

remarked to Mr. Finney that they were running fast, and he replied that they were fifteen minutes late. This Mr. Finney I speak of,—was the conductor (45) in charge of the train. Mr. Finney is an old friend of mine. I was talking with him. This shriek of the whistle and the application of the brakes, was while I was talking with him. It all happened so suddenly. I was hardly through speaking when we heard this whistle, and the train jarred with the brakes, and at that time we heard the whistle and felt the shock, and the train was filled with a cloud of dust, and [48] I felt that we had struck something. I felt the application of the brakes. Yes, indeed. It threw anyone almost out of their seats; that is, anyone who was careless. The brakes were applied suddenly. The application of the brakes was at the same time as the whistle. My attention was on the movement of the train before (46) that time. I have no recollection of the whistle prior to this whistle for air, after the train left Mountainair. From the time we left Mountainair the train was beginning to speed, and by the time the whistle was sounded the train was rocking, going around curves. My attention was fixed on the movement of the train practically from the time it left Mountainair. After going out of the yards. From the time I left Mountainair until the whistle which accompanied the application of the brakes, I have no recollection of any signal, or whistle, of any kind prior to that. (47) I heard the whistle for brakes. I

(Deposition of Meriam White.)

am not able to state at what point, or where the train was, at the time this whistle was blown that I heard. The train came to a stop, after passing the crossing in question. (48) I noticed particularly the curve, because when the train came to a standstill we were beyond the curve; that is, the other side of the crossing, and the engine which was pulling the passenger train was somewhere about the whistling post, or beyond. . . . And followed them outside and down the track to where they were putting the injured men into a baggage-coach. I recognized Benjamin Spencer. I knew Roy, but I could not see his face from the way they were holding him, but Mr. Spencer's face was turned toward me. (49) I have been a telegraph operator for a period of 13 years. I have served as agent and operator [49] for a period of 13 years. I am versed in railroad work so that I am able to judge reasonably the speed of trains if I am on them. The average rate of speed for a passenger train is about 30 miles, and I would judge that that train was going a full 50 miles an hour when this whistle was blown. I have lived at Mountainair since 1909. It is either '08 or '09. The rocking of the train caused me to notice the speed of the train. As I said, I was nervous over the speed of the train. I knew it was running faster (50) than a passenger train usually runs. And then there was the rocking of the car, and when I mentioned it to the conductor he told me that they were late, and I knew that they were making up for lost time.

(Deposition of Meriam White.)

WITNESS.—(Continuing.) I know that train was delayed at Mountainair. (51) I believe that there was a work-train came in from the west before we started. In this case the work-train was a train that was hauling crushed rock. And I noticed that came in from the west just before the passenger train left Mountainair. The passenger train then left immediately. I could not state how long the passenger train remained standing at Mountainair. I do not know what their regular wait is here. I think the delay was fifteen minutes in Mountainair from the arriving to the time we departed.

Cross-examination.

The conductor told me, when we were speaking together (52) that they were late fifteen minutes. I couldn't state exactly how long I was speaking with him. Mr. Finney and I are old friends, and he usually speaks to me. I should [50] judge we were talking several minutes—when he took my ticket. Talking up to the time—the accident happened. The brakes were suddenly applied.

Q. Probably if any whistle was sounded it was probably during the time that you were talking?

A. I do not know when. At the rate the train was going it would not have been very long from the time they sounded the whistle until they would be there. I was talking to the conductor at the time the train reached the whistling-post east of the crossing. Now, after this accident and the train

(Deposition of Meriam White.)

was stopped, I did not make any memorandum of where we were when the train (53) stopped. I did notice—yes, I first—naturally I looked back to see what had caused this trouble, and I do not believe that I could see the crossing, from my coach. I did not pay particular attention to the whistling-post, but I saw the wreck afterwards, and the old wrecked automobile, and noticed the distance *from* the car was from the crossing. Not where they were struck. Where it was carried. The car was on the cow-catcher of the engine when it stopped. I was there yesterday afternoon. I did not go to the whistling-post. I went round the crossing. I believe nine or ten telephone poles. (54) Mr. Woods was with me. I showed him where, to the best of my knowledge, the train stopped. Going there yesterday and seeing the whistling-post, my judgment as to where the engine stopped from the post is—perhaps beyond a telephone post, or two. How I can state now as to where the engine stopped without regard to the whistling-post—is I know about where the day coach was that I was in when I got off, and the reason I remember is the position of the curve and the high bank, and I could remember [51] yesterday afternoon from the position of the bank about where the day coach was. The place in the train of the day coach is dependent on the number of cars in the train. I know that day we had a long train. As to how many cars between me and the engine—I do not know the length of a car. I could not answer that. I was

(Deposition of Meriam White.)
able to estimate where the train stopped from the
curve in the track. My day coach was past (55)
the curve, and I could look up the track and see
where they were taking this car off, and I could
remember yesterday afternoon about how far. I
noticed at the time of the accident in looking out
of the window—the upper part of the train—and I
now make an estimate as to about where it was
with reference to that whistling-post. (56)

TESTIMONY OF ROY SPENCER, IN HIS OWN BEHALF.

ROY SPENCER, called as a witness in his own
behalf, testified as follows:

Direct Examination.

I am the plaintiff in this case. My age is twenty.
My nearest *birth* is the 19th of April. I was twenty
the 19th of April. I live at Mountainair, New
Mexico. I have lived at Mountainair all my life.
I am a son of Benjamin Spencer, who was killed at
the time I was injured. Up to the time of my injury, my business or occupation was farming and
sawmilling. I worked for (57) my father. My
father's occupation was sawmilling. I worked at
home for him. I attended school. I attended at
Eastview and was in the eighth grade. I suffered
my injury in1923. On June 11th. That day we had
started down to Scholle on an errand to get some
furniture. We left home, I think, about nine
o'clock. I went [52] with my father in a Ford

(Testimony of Roy Spencer.)

truck. I went because my father asked me. (58) He asked me to go down there to help him get some furniture. I was living at home with my father. I was then seventeen. We started out about nine o'clock in the morning. We went to Scholle. (59) Coming back, we left Scholle between one and two o'clock, I think, somewhere along there. We were going in a Ford truck. My father was driving. It was my father's car. He had owned that car since January of the same year. It was a second-hand car when he bought it. It did not have a top. Just one seat, then a box on behind, about three feet and a half square, I think, maybe four feet long, or three, something like that. When we started out that day and went there, the car was running good when we left. It run good all the way to Scholle. Scholle is pretty near due south from my home; it is a little southwest or west of south. (60) Scholle is on the line of the railroad between Mountainair and Belen. And my home was to the north of the railroad some nine miles. Scholle is fourteen miles west of Mountainair on the railroad. A little south of west, I think. It is on the line between Mountainair and Belen.

Q. Did the car commence to show poor motion that day or work badly?

A. After we had left Scholle about two miles, I think, it started to missing. Scholle is just a station, a small station. (61) There is one little store there and a postoffice. The car started to missing after we had gone about two miles from

(Testimony of Roy Spencer.)

Scholle.   At the time we had went about four miles
it got so bad we had to run in low most of the time
then until we got pretty near about the crossing
and then had to run in low, you might say, all the
time.   There was   [53]   —several hills before we
got to this crossing that I had to assist the car up
before we got to those places.   I got out and pushed
behind.   My father requested me to do that between
Scholle and Abo, after the hill I could not say.   İ
don't remember.   (62)   I do remember his asking
me at least once to get out and push.   My father
was 68.   I had gone with him, driving with him
before.   That day the spark had been out and he
had been working on the coils to get it to run.   The
lid was off the top of the coils and he worked, done
something, worked the platinums, I don't just know
how he done.   The coils were down in front of the
car, under the wind-shield—right next to the bottom
of the car.

Q. Show the jury what position you saw your
father take in working with those coils when the
car was running badly?   (63)

A. Well, he would stoop over them little plates
that work up and down that takes the spark, he
would work them that way and look at them; İ
don't know just how he worked them.   The effect,
apparently, that that had on the car, whatever he
was doing in adjusting the plates or wires seemed to
cause the spark to heat up there, I don't know what
part it would play.   I do not know how to operate
a car very good.   Just exactly what my father was

(Testimony of Roy Spencer.)

doing, and what was wrong, I am unable to state. I
saw him bend down when his car would go bad and
do something down under the wind-shield several
times. The driving wheel of this Ford, or the steering-wheel is on the left side of the car. My father
would stoop down—he would stoop to the right. He
would stoop his head at those times when he was
fixing the car—probably to where the wind-shield
connects. (64) [54] something like that. I had
to get out and push that car up the hill twice between
Scholle and this place at the crossing. It was about
four o'clock when we approached that crossing at
which I was hurt. And that was on the eleventh
of June. The condition of the road for a mile or
so back as we approached the crossing in the direction that we were going toward Mountainair is, it
goes down a draw, just a kind of draw between
hills and the earth is held on all sides, just a low
draw down through there. By "draw" I mean a
low sort of valley. As it were, between two rises
on either side. And the highway ran up through
that draw toward the crossing. (65) The railroad
runs more or less in the same direction as the highway at that point. So that, in driving along, you
are running in the same general direction of the
railroad. When we approached the railroad crossing that day, I saw a train passing along. When
I was about 500 yards from the crossing, I saw a
freight train of some kind go along the crossing, I
don't know what kind it was. It was going in the
direction of Mountainair. When I saw the train it

(Testimony of Roy Spencer.)

was just coming over the crossing. Mountainair is the village where I go when I go to any village ordinarily, the nearest to any (66) store. I was familiar with Mountainair; I had been there very much.

Q. How familiar were you with the tracks of this railroad between Mountainair Station and this crossing at which you were hurt?

A. Well, I had been there afoot when I was a kid when going to school at Mountainair, us boys used to go up the railroad there around, I knew the road was there, and the railroad, that is all I knew. I was thoroughly familiar [55] with the railroad between the station at Mountainair and this crossing. There was one track between those two places. There are some switches at the station in Mountainair. Those switches extend about 200 yards west of the station, I think. (67) I don't know; just guessing at it. That is my best judgment, 200 yards west of the station. From there on to this crossing there is but the one track. As to the condition of the railroad as to cuts and positions which it takes running from Mountainair up to this crossing at which I was hurt—well, from the time you leave Mountainair, it is in a cut, right at the depot on one side, and continuing from the time you go about 250 yards west of the depot you have got a bridge over the railroad there that extends away above where the train passes under and continues in that way clear on down to pretty close to the crossing. (68) It has got a slight curve through that cut.

(Testimony of Roy Spencer.)

Q. How deep is that cut between the depot and the crossing at the different points as you go along, describe it generally; give us your best recollection of how the cut extends from where it commences at the depot, clear up to the point you say is near the crossing.

A. It will average about from twenty to thirty feet deep, up to the time it goes out from behind this hill. It goes out from behind the hill about 850 feet from the crossing—That is, in the direction of Mountainair from the crossing.

Q. Now, when you approached, you say you saw the freight train go over the crossing east toward Mountainair. Do you know how far it is from that crossing to Mountainair switch? [56]

A. About two miles, something like that. If the record of actual measurement here shows a mile and three-quarter, I would say that is (69) accurate; I was just guessing at two miles. I would judge that that freight train was moving twenty miles an hour as I saw it going east—about like most any other train travels along there.

Q. Now, Roy, what is the condition of the railroad there as to grade at that point, in which direction does the railroad slant, or grade from the crossing going east?

A. It slants to the west. There is a down grade. With reference to the station at Mountainair, that down grade commences about a quarter of a mile, I think, maybe a little over, west from Mountainair. You start up grade as you go west out of Mountain-

(Testimony of Roy Spencer.)

air.   After you go about a quarter of a mile it turns
down, and from there you go down grade.   As I
approached—as the train passed over, that is, this
freight train passed on—   (70)   we were coming up
the crossing all the time.   The road runs more or
less in the same direction as the railroad up to the
crossing.   When you get to the point of the crossing,
the road turns north and goes over the tracks.   Goes
right around the corner of the fence, you are going
northeast when you come up the tracks, when you
turn right north and go up over the tracks.   You
might call it a square turn around the corner, and
up over the tracks.

Q. Did you stop at all before you went over the
tracks that day, and if so, where?

A. Stopped right at the corner of the fence.   The
picture you now show me of a crossing, with an
automobile standing before it, and a road going over,
and a man and   [57]   a sign on the crossing is a
good representation of the appearance of that crossing at the point where I went over.   (71)   The
position occupied by the automobile shown in the
picture looks to be about the same place as that
where we stopped that day.   We stopped about the
same place where that auto stands.   I know where
the road commences to rise there to go over the
track, whether inside or outside the railroad fence.
It is right at the corner.   I would say just inside,
maybe two or three feet.   Then it commences to rise.
From that point to the nearest rail, the road has to
rise about ten feet.   The road has to rise ten feet

(Testimony of Roy Spencer.)

in going from the railroad fence up to the track. (72) The top or grade of the railroad extends out on a level a few feet beyond the rail. I would judge four feet—outside the rail. The rise from just inside the fence to about four feet outside the rail is ten feet. Where the man is standing in the picture—that is on the rail just where the road crosses. And that is the same crossing. It looks substantially in that picture the same as it did at the time I crossed it.

Mr. WOOD.—We will offer it in evidence. Counsel has kindly consented that we may use this without eliminating the memorandum or record thereon, with the further concession (73) of the fact, as we understand it to be, that the fence shown on the picture at the edge of the railroad's right of way, is one hundred feet from the center of the railroad track, that is my understanding of the fact.

Mr. BRENNAN.—Correct.

The COURT.—It may be admitted as Plaintiff's Exhibit No. 2. [58]

WITNESS.—(Continuing.) The picture you now show me was taken from the road, looks like about half way up to the crossing from the corner of the right of way, looking west up the Valley. (74) About halfway from the fence corner up to the railroad track on the crossing. That correctly shows the appearance of the ground looking west from that crossing in the direction in which the road I traveled came. The fence post that shows first in this picture is a post going up to the right

(Testimony of Roy Spencer.)

of way.   One of the fence posts leading up to the rails.

Q. What is the corner, or fence corner that is shown in that picture, the line of posts running up toward the telegraph-pole, and this one post that shows in the picture the other way?

A. This is the line of posts coming east, the corner would be out here (indicating), and this post the one that leads up to the right of way. That is, the long line of posts that appear there is the fence at the railroad right of way. Approaching from the west. To the left of the fence appears, apparently, a road. That is the road we came up. (75) In and across the corner inside of the angle are some other marks or indications on the ground. They are the indications of the old road, apparently an old road. The appearances which make me say it is an old road are two ditches—it is not real ditches, just kind of cuts in the earth. Marks of wheel tracks. They are what show just inside of the corner. To the right of the picture and to the left of the picture appear rises in the ground or hills, and a low point between—That is the draw of which I have spoken, that comes up along the railroad. And it is   [59]   shown in this picture.

(The picture was offered and received in evidence as Plaintiff's Exhibit No. 3.)   (76)

Mr. WOOD.—Q. I show you another picture. Can you tell from where that was taken, and what it shows with reference to this accident? (Handing photograph to witness.)

(Testimony of Roy Spencer.)

A. Yes, sir; that is taken standing west of the crossing, looking to the crossing. The picture you show me shows a cut or excavation in the ground. That cut is west of the crossing. The white boards shown in the forefront of the picture, are those at the crossing. So that this shows the crossing looking east along the railroad track from the cut.

(The picture last mentioned was admitted in evidence as Plaintiff's Exhibit No. 4.)

Mr. BRENNAN.—Then Exhibit No. 4, if I understand it rightly, shows the condition west of the crossing?

Mr. WOOD.—Yes, it is taken west of the crossing, looking toward the crossing and showing the crossing on the railroad-track and from the railroad track. (77)

Q. The picture shows, looking down the track and in front of you, a tree-covered hill beyond the crossing; what is that hill?

A. That is the hill where the curve goes around and goes into it; it is a hill that comes out across this low place. That hill is a hill that the train comes out of. The hill shown in the front of the picture, extends down to the railroad-track. Beyond the crossing—the railroad track is cut through there. It is nine telegraph—six telegraph-poles from the railroad crossing to where the railroad tracks [60] cut into that hill shown in the front of that picture. It would be 900 feet, just counting the telegraph-poles, 150 feet. It is something like 900 feet from the crossing. (78)

(Testimony of Roy Spencer.)

Mr. WOOD.—I show you another picture. (Handing photograph to witness.) Does that correctly show the appearance of the south side of the track from the crossing at which you were struck, down to this hill that the cut is in to the east?

A. It does. That is a fairly correct representation of the appearance. There is a mark on the picture apparently where that hill touches the track. That is 850 feet from the crossing, where the hill touches the track. And the hill shown, to the right of this picture, is the same hill which is shown in the picture, Plaintiff's Exhibit No. 4, as being down the track as you look out of the cut; east from the crossing to that hill 850 (79) feet east.

(Thereupon the picture was received in evidence as Plaintiff's Exhibit No. 5.)

WITNESS.—(Continuing.) The picture, Exhibit 5, which you last showed me, shows an embankment of the railroad running eastwardly from the crossing. I should judge that the approximate height of the embankment above the surface of the ground at its bottom as it runs across the draw is ten feet from bank to bank. About the same as at the road. You show me another picture—that is the same embankment which is shown in the last picture. And the embankment shown there is the railroad embankment. That picture is taken about a hundred yards east of the crossing. This shows a culvert with a man in the culvert. I know the man. (80) [61] He is about five feet eight. I have never accurately measured the height of the

(Testimony of Roy Spencer.)

railroad at that point. I can judge the height. It is six feet to the top of the culvert and looks about four feet above the culvert to the raise. In the front of this picture appears from the culvert, looking in the direction in which the picture was taken, something in the nature of bushes. This shown here is a kind of mesquite brush of some kind, I don't know just what they do call it; that looks like weeds, part of it I suppose is weeds, the run of that is kind of mesquite brush, something like that. I know what is in the bottom of that draw along at that place in the way of stones or indications of a water (81) course of vegetation, or what that is there from that culvert; down the flat it is grass, after you get down a little ways from the railroad and down the draw, there is grass in that low part of the draw. There is no creek bed or bottom or anything there to show the flow of water there at any time. This shows from there as it was about the time I had the injury. This is about a hundred yards from the crossing east. This is in that draw that is between the crossing and the hill into which the railroad cuts toward Mountainair.

Mr. WOOD.—We will offer this in evidence.

The COURT.—It will be admitted as Plaintiff's Exhibit No. 6.

WITNESS.—(Continuing.) You hand me another picture which shows the same point shown in Exhibit 6, but farther removed to show more of the railroad (82) and of the land between the picture and the railroad. Aside from that it shows the

(Testimony of Roy Spencer.)

same view as Exhibit 6. The bank shown there is the railroad embankment. Between the two cuts. And this is about one hundred yards east of the crossing. This also shows weeds [62] and other things between me and the culvert shown in the other picture. There is nothing on the ground there showing the flow of water, or that running water has been there.

(The picture was thereupon admitted in evidence as Plaintiff's Exhibit No. 7.)

WITNESS.—(Continuing.) There is nothing, nor was there at the time anything in the nature of growth along that embankment up near the rails except just weeds of all descriptions. They average about two feet high up on the railroad embankment. (83) They were that way at that time.

Mr. WOOD.—Exhibit No. 8, which I am offering in evidence, and is a picture taken on the crossing about at the line of the railroad right of way looking east into this hill; it shows a freight train going over the crossing, and the engine just in the cut.

Exhibit No. 9, which I am offering, is taken—

Mr. BRENNAN.—Are you able to tell the jury how many cars were in that train?

Mr. WOOD.—I am not, Mr. Brennan.

Mr. BRENNAN.—I think 52 cars, if it may be so stipulated in connection with your offer. (84)

Mr. WOOD.—I think the testimony will show that somebody stated there were 52 cars in the train, but not between the crossing and the cut, so

(Testimony of Roy Spencer.)
that the number of cars would not be helpful to us.
I think that appeared before, if I recall right.

Exhibit No. 9, is taken about one-third of the way
up, about two fence posts up, which I should judge
is about one-third of the way from the right of way
fence up to the crossing, in the same direction, and
shows a pusher engine on the [63] same train,
showing in 8, just disappearing in the cut.

Q. Now, Roy, you have told us that you stopped
in your father's car about the same position that
the car shown in the first picture was in. What,
if anything, did you do at that time and place in
the way of looking for trains?

A. I got out and looked both ways for trains and
could not see any. Then I proceeded to start up
the hill. The car was running badly. It was running very poorly, and had to start and start it up
this hill; of course, we went slow, naturally, (85)
starting and going up the hill it was running poor.
As the car started up the hill, I started pushing
from behind. I started pushing the car up the hill
from behind the car. I was pushing up the hill
with my head down. My hands were in the back
of the car, on the back of the truck bed. I was trying to push it up the hill, didn't seem to want to go
very good. I pushed all I could. It was just going.
That was about all. I did not continue pushing
from behind all the way up the hill. As I went up,
I went around to the side, after we went up a little
ways, around the side of the car. I will tell you
how I went to the side, what I did; (86) I stepped

(Testimony of Roy Spencer.)
around to the side while pushing, so when I got on top I could step in the car.

Q. When you got around to the side, show the jury just the position you occupied with reference to that car, step down here to this table.

(Witness steps down to the table.)

We will assume that this table is the automobile, which is headed up the hill toward the track that is in front. Now, come around behind and show the jury just what you did as you pushed that car up the hill or helped, when you [64] started, and showing just what you did when you got up to the track.

A. I started pushing behind like that (indicating). Of course, I was higher than that on the truck back—the car was going up the hill and naturally I throwed my arms out with my head that way (indicating); when we had gone probably thirty feet I was pushing on one end.

Mr. WOOD.—The witness illustrates by placing both his hands on the edge of the table, with his feet extended out behind.

Q. About how high was your head when you were pushing, and what attitude did you assume in pushing the car when you started?

A. My head was about that high, I should judge (indicating), and my hands about like that (indicating).

Q. With reference to the back of the car, how high was (87) your head—with reference to the

(Testimony of Roy Spencer.)

body of the car upon which your hands rested, how high was your head?

A. About even with my hands.

Q. Your feet, about how far behind the car were your feet as you were pushing? Illustrate and show.

A. (Indicating.) Just as much as it takes a person to push; of course, the feet naturally goes back so to push. Now, when I ceased that position, I will show the jury what I did, how I made the change. There was a handle on this side, I had hold of, when I stepped around I just stepped that way, held that hand like that (indicating), pulled that way and pushed against the back of the seat with this other hand. There was a brace at the right, or rather, corner of the box of the car. I hung to that brace with the left hand. [65] In stepping around from the rear of the car to get hold of the side, I stepped just as quick as I could. I did not stop pushing at all—could not. (88) I could not because the car would stop. It was all I could do with the car itself to get along, much less stop pushing. I say I didn't stop pushing but stepped around as quickly as I could. And held that handle with the left hand. I show the jury likewise on the table what position I assumed after I got around to the side where that hand was; I had hold of that end back here, projected out about four inches, throwed my arm back here (indicating).

Mr. WOOD.—Illustrating a handle attached to

(Testimony of Roy Spencer.)

the left rear end of the box, and his left hand attached to the handle.

WITNESS.—This projected about four inches, had to put my right hand in against the back of the seat on the front here. I was pulling, and that throwed me under, just like that (indicating). I continued in that position as I passed up the hill on the truck, until the front wheels hit the track, and then I was, I suppose about six feet, anyhow, from the front wheels when they got on the track, and the hind wheels were coming off the grade. (89)

Q. Can you illustrate by moving your feet or showing the jury just about the movements you made in pushing the car with your feet as you came up the hill after you came around to the side?

A. I pushed just like that (indicating) of course, I went as slow as I could, pushing as hard as I could. From the time I went around the end of the car, as I was pushing up the hill on the side of the car, I was facing to the west. [66] I expect we were half-way up the hill, or from the place where we started, when I changed my position from the rear of the car to the side, I should judge it took a full minute to get the car to move from the point where it was at rest from the time it started at the bottom (90) of the hill, until I got up on the track. From what I know of that car, and the way it was running, I know what would have happened if I had ceased for a minute or an instant to push; it would stop. From my knowledge of the grade, I

(Testimony of Roy Spencer.)

know whether or not stopping on that grade it could have been started again without going back to the bottom of (91) the grade; it would have to run back to the bottom of the grade, if it stopped, undoubtedly. I did not cease for an instant to push from the time that car started until it got up on the track. From the position that I was in while I was pushing that car at the side of it, I could not have turned my head around and looked behind me in the direction in which that train was coming. (92) In order to have looked in the direction in which the train came at the time—from the time I came to the side of the car, until I came up on the track—I would had to have taken this arm down and look and quit pushing, and the car would have stopped and would have run down and it would have had to. I have said that at one time my father asked me to get out and push; I do not recall whether or not he said anything to me about —at that particular time—about my getting out and pushing. I couldn't say whether he did or not. I don't remember. I do remember that on some part of the journey he asked (93) me to get out and push. From that time on each time the car got into a position where it would not climb, I got out and pushed. [67] I ceased to push, when the front wheels of the car had just went over the first rail. At that time, I was beside the car and I stepped on the rail—on the running-board, rather, and stepped in the car and turned around. My purpose in stepping in the car just when ceasing

(Testimony of Roy Spencer.)

to push, stepping on the fender, was because it was going over the rails down the other hill. The car was in motion. From the time I raised up and ceased to push, it was just a second or so before I got a view of the train; just as quick as I could step in and look, step in and face to the west, when I turned around and throwed my head down the track. (94)

Q. How much time, if any, elapsed from the time you let go of the car, until you looked in the direction of that train?

A. Just as quick as I could jump in the car and turn around.

Q. As quick as you could jump in—I will ask this additional: How long was that?

A. I don't know; could not have been more than a second or so; maybe two seconds from the time I quit pushing until I was in the car.

Q. Illustrate to us again—using the platform here as the car—just what you did when you ceased to push, where you went to, and show us how fast you did it, and how you looked for the train; then we can get an idea of the time.

A. How quick I jumped in the car?

Q. Yes, sir. Come down here again—or, assume, now, that the platform here is the car—face the other way, please, Assume that the platform here is the car and that [68] you are pushing this way (indicating), and stoop over and get in the position you were when you were pushing, using (95) this end of the platform as the back of the

(Testimony of Roy Spencer.)

car. That is, the rear of the car, now, as you were climbing up the hill last—I don't mean at first—when you were behind, I mean, when you got up to the railroad tracks, as *you* hand let go and you stepped inside the car, and state how quickly you did it.

A. I would say it would be just like that (indicating). Up on the running-board and in the car and turned around and sat down. There could not have been more than a couple of seconds.

The COURT.—Suppose you could stipulate as to approximately the length of time of the demonstration? (96)

Mr. WOOD.—Well, the witness stepped up as rapidly as a man could step.

Mr. BRENNAN.—Yes, it may be so stipulated. He said he got up as rapidly as he could, and he undertook to demonstrate how he did it.

WITNESS.—When I looked then first, I saw the train. It was about 50 yards away. I don't know what was the next thing I saw. The next I remember, they was putting me in the baggage coach of the train. The next I knew after looking up and seeing the train some fifty yards off they were picking me off the engine. (97) My father was sitting in the car, at the time, driving. I did not notice at all what my father was doing as I pushed that car up the hill. Whether he was sitting up or leaning over, I don't know. I don't know how far it is from that crossing until the railroad enters the cut west of the crossing; just a few feet, the north

(Testimony of Roy Spencer.)

side [69] of the embankment runs right up to the highway. It goes in there on a curve; that picture showed the west there a while ago, that curve. The picture, Exhibit 4, is the one I refer to, and shows the track in the curve west of the crossing. That curve extends about between two and three telegraph poles, I think, west of the crossing when you are standing on the track, before a train disappears into that cut to the left. (98) The sides of that cut at the point shown in the picture, Exhibit 4, looked to be about ten feet there. I know, from my knowledge of the cut, about how high they are; about 10 feet where the close part of the picture is shown. This freight train which had come along the track and gone toward Mountainair, was running right along there when I first saw it, just about in the place that picture is taken. It had passed east through the cut and gone on to Mountainair. It seemed like six or eight minutes, maybe ten, had elapsed from the time I saw the train pass the cut going toward Mountainair, and the time when I started up the grade. I am talking about the cut west of the crossing. (99)

Mr. BRENNAN.—About two or three telegraph poles west.

Mr. WOOD.—He says that the train disappears about two or three telegraph poles.

Mr. BRENNAN.—You are talking about the time when he said he was about five hundred yards from the crossing, or track, when he first saw the train?

(Testimony of Roy Spencer.)

Mr. WOOD.—When he first saw it, yes.

Q. This cut through which the train from the west passes, about how long is it, how far back from the crossing does that cut reach, if you know, more or less? [70]

A. I don't know. About three hundred yards, I think. And then it emerges on the other side of the hill. The next I knew was when they were picking me off the cowcatcher of the train. I noticed that my father was there, they taking him off at the same time. Then they put us in the baggage-coach and took us (100) to Albuquerque. The train continued its journey to Belen and they took us up to Albuquerque. There they took us to a hospital The first medical attention I got was at Belen. Dr. Wilkinson treated me there. I don't know who called him. The doctor was there when I got there. Took charge of myself and father, and from there they took us on to Albuquerque and then put us in a hospital. I went into St. Joseph's Hospital at Albuquerque. As to medical attention I had there—there was not nothing done then more than just give morphine for a period of about two weeks. (101) Dr. Lovelace attended me at Albuquerque. He was the railroad physician. I did not call any physician myself. Dr. Lovelace continued to take care of me until I left the hospital. I think it was twelve days after the accident, as well as I remember, before they attempted to set the bones in my leg. Then he operated and put a silver plate in around the bones. (102)

(Testimony of Roy Spencer.)

I will tell just what treatment and attention was given me from that time on at the hospital, narrate the story of my treatment and my efforts to get well from that time on until I left the hospital.

Well, he put that in along towards the last of June, then he never done any more than dress it and treat it for about 54 days, only during this time he put a splint on to hold this silver plate he had put in there, to keep it [71] from bending; he put a splint under my arm and down on my foot, and bound me to it so I could not move. After about 54 days, I think, the contraction of the muscles some way doubled the plate up, and throwed the bones out of the incision, then he took the board off and put me in a plaster paris to hold it. I remained in plaster paris until in January before it was taken off, and before he taken the plaster paris off he took the silver plate out, he taken the silver plate out some time in October, I don't remember just the day, along in October, took that silver plate out from around the bone; that was probably two months after he put the cast on. Then he put in an extension form after he put me in the cast; the cast did not hold the bones right, and he put it on my foot, tied it up and left me that way until January, when I took down sick with appendicitis. When he taken me to the operating room, he took the cast off that night, all but a trough the cast run along, where the break was. Then he put steel splints on (103) top of that and bound it. That

(Testimony of Roy Spencer.)

is the first time he taken the cast off in 54 days after I was hurt.

Q. When were you able at all to leave your bed from the time you were injured?

A. About the first of January some time, I began to get up in a wheel-chair. That was in 1924.

Q. Roy, what, if any, other injuries beside that broken leg did you suffer at the time?

A. Suffered a fractured foot, several bruises on the body, but they didn't amount to much; they was just skinned places.

Q. Tell the jury just your physical feelings and [72] symptoms after you were hurt until you got in the hospital, whether or not you were suffering, and how, and what your feelings were.

A. I was suffering death from the time I was hurt until I got to the hospital; did not suffer so bad after I got to Belen, the doctor there gave me a couple of hypodermics and kind of eased me until I got to the hospital, only when they would move me, or something like that.

Q. Now, what were your symptoms or feelings while you were in the hospital, from the time you commenced, describe (104) what pain and suffering you had or endured in that time?

A. I endured intense pain for two months after that, attacks would come and go; only at these periods I am speaking of when he would operate or raise the leg something like that, then of course there would be a period of three or four days that I would have intense pain again. After two months,

(Testimony of Roy Spencer.)

the majority of the time I felt, you might say, pretty good a little part of the time. Of course, every night is when it would get worse. After two months there was times I felt pretty good at times.

Q. These times when the bones were set, were you given anesthetic, put under the influence of ether or chloroform at those times?

A. He did when he operated. He set my leg two or three times and never give me anything but a hypodermic. I did not, at that time, suffer pain in other parts of the body than the broken leg— more than my foot, in my foot I did. I don't know what was the condition of my foot; just [73] intense pain; I asked the doctor what was the matter, and he said a slight dislocation; I didn't never know what it was; never did locate it; they had me bound down. I did not suffer pains in any other parts of my body (105) at first. Not at that time, no. There were several bruises on my left leg and several on my head, lots of places, just little skinned places, that is all. Two places on my left leg, quite a bunch of skin knocked off, nothing to amount to anything, no more than just the hide knocked off. I did not have any soreness or troubles in my body or abdomen—not then. I left the hospital, I think, the 19th day of February. I had been up out of bed then about three weeks. I could not touch my leg to the floor then, I could just get around then on crutches, just barely could get around that way. I will state as near as I can what has happened and what has been done, what has been the condition

(Testimony of Roy Spencer.)

of that leg from that time down until now; I will take it from that time (106) and tell what happened then and what I did, what I endeavored to do:

When I first got out of the hospital, I went to Colorado about two months. The doctor told me when I left the hospital to steam it and use oil on it. Steam my knee and joints, and rub my leg all over. At that time, my knee was just stiff, my knee was sore, and I could not—my joints all over was sore, all over the leg, the muscles, and everything. He said to rub it with oil and steam it. So I went to Colorado and stayed about two months and treated it. When I came back, I came back through Albuquerque and went to see Dr. Lovelace and asked him what to be done, it didn't seem to be improving much more than I could touch my foot to the floor a little better than I could when I first [74] got out. He said the only thing to be done was to steam it and use oil on my knee—my knee gave me trouble all the time—and I done that from then on. It didn't seem to ever get much better, but my leg got strong, didn't hurt now to stand on it. My knee is still stiff. I have very little motion in that knee now. That there is all. (107) (Illustrating.) That is the extreme of motion that I can give my knee.

Q. Just step up there where all the jury can see, and move your knee to show the amount of motion there is in it.

(Whereupon the witness indicates to the jury.)

(Testimony of Roy Spencer.)

A. As I move it forward, there is a catch there
—stops it dead still when it stops. That knee has
been in that condition a year and a half, I guess.
It has not improved in a year and a half. When
I left the hospital, Dr. Lovelace told my mother—
I was not able to take care of it then—and since
I have got able to take care of it, I take care of it
myself now. For a time my mother took care of
it. My mother took care of it for a period of a
year, I guess. Since then, I have been trying to
do it myself. (108) I cannot give or get any
more motion than I have shown. The length of
that leg as compared with the other is—it is just
an inch shorter, just exactly an inch difference. A
wound was made when they put the plate in and
cut down to the bone, and then the bone came out
through that wound later. That wound has not
entirely healed—there is a little ooze and sore there
yet. This wound just oozes out and scabs over and
oozes out again; the scab will come off and just
pus keeps [75] kind of oozing out.

Q. Will you drop your trousers enough so that
the jury can see both of your legs, see the condition of that leg and the wound as it is now?

A. (Witness indicates to the jury as requested.)
Right there is where it oozes over, where the
break was. (109) Before the injury, both of my
legs were of substantially the same physical development as well as I can tell. It is this right knee
that is stiff. And it is swollen around the bottom
in the joint, as you can see. This wound remains

(Testimony of Roy Spencer.)

open and still oozes. The scar on the leg appearing there, that is the scar made from those operations—he cut that open twice. (110) The doctor cut the bone and the leg twice. The first occasion was when they put the plate on, and the next was when they took it out. It was in September when the leg was found bent and the plate showing up. I don't remember the day. (112) Then at that time, he put a board from the arm down to the foot and bandaged down, clear down, so that I could not move. My condition of health and strength before this injury was good. Never had any sickness, more than just little common ailments. Just grippe, something like that, down a few days and up, a little sick, stayed inside with a cold, something like that. I don't remember how many times and to what extent (113) I ever had even those common ailments; a very few times. As to my condition of strength and ability to work before that— I could do anything anybody else could, I think, anyway. Up to the time of my injury I had been cutting logs at the sawmill for, I guess, three months before the accident. And in the course of cutting logs the particular [76] part of the work I did was sawing and trimming. I worked at or about the sawmill in the handling of lumber. The work I had been able to do and did do in the way of handling lumber at my father's sawmill was taking the lumber away from the saw. As to my strength and athletic capacities before that time— I was pretty good; the best foot racer in the coun-

(Testimony of Roy Spencer.)

try, they all know that. By "the country" I mean
that county and town—Mountainair. (114) I had
taken   part   in   these   athletic   contests. Every
Fourth of July and the Fairs, never was beat from
the time I was 12 until I was hurt. Besides foot
racing, I had done wrestling, never was a boxer,
never could hit with my left hand. There was
nothing wrong that I ever found with my strength
and ability to handle myself up to the time of the
accident. There was nothing that I know of affecting my strength. There was never any tuberculosis or consumption, to my knowledge, in my family,
or my father or mother or brothers or sisters—
(115) None that I ever heard of. There were no
troubles of that kind in me down to the time of the
accident, so far as I know. I first learned that I
was tubercular when I got back from Colorado—I
went there on a visit—going through Albuquerque.
I visited in Colorado and came back through Albuquerque. (116) My condition as to strength, bodily strength, since my injury and down to the present time has been nothing like it was before in any
place. I will state to the jury what my condition
of strength and ability to work is, what I am able
to do, and the effect on me of attempting to do any
hard work; it don't seem to bother as long as I use
my arms, something like that; but just as sure as
I go to lift, use my body, [77] it seems like it
hurts across the stomach, when I sit down a little
while it seems to be all right again. To use my
arms don't seem to bother. Just as sure as I use

(Testimony of Roy Spencer.)

my body to lift, or anything like that, it seems like
that is when my stomach hurts, and I have to sit
down. It seems to hurt across the abdominal region; It seems to be my stomach, apparently. I
have never been able to do any hard work at all
since I was injured. I am not able to now. The
work I have done, or attempted to do, in the (117)
last year, since the former trial is no more than
just a few chores; I have to get out and exercise,
make a practice of probably an hour or two a day
using my arms, something like that, just to be exercising and out in the air. When I start to do this
work it would have the effect I have described.
After I start to work it has that effect just as quick
as I go to pant and get warm. Coming back to the
time and occasion when I was hurt; I have stated
to the jury that while I was pushing the car up the
hill I did not look to the left behind me at the
track down toward Mountainair. The reason I did
not was because I could not.

Q. You have given that reason. Now, is there
any other reason?

A. One reason was, when I started up the hill I
had already looked to the east; you can see a distance there, something like 200 yards, a little over.
And this cut that comes out of the track on the
west, it comes right out; when you are standing at
the corner post you cannot (118) see the train
coming from the west more than a telegraph pole
before it gets to the crossing. The reason why I
did not look and see was I was watching toward

(Testimony of Roy Spencer.)

the west, because [78] it is more dangerous than the east of the track, the crossing there.

Q. Was there anything in the position which you had alongside of the car going up there that would obstruct your view of the train from the east, had you looked in that direction? (119)

A. Yes, there was. The right of way fence where these poles and boards set up there for a distance of four feet; a curve in the track down the road throws your head right against this fence on down the track. You also have to pass a distance of about eight feet before you could see the track. Exhibit No. 5, the photograph which shows a fence alongside of the roadway as you pass up, that is the fence which I am speaking of which would obstruct my vision. (120)

Q. Did you ever notice what, if any, precautions your father habitually took in approaching or crossing the railroad track, as to whether or not he was careful or a careless man?

A. I have been with him lots of times when there was times that he would go up to the track and before ever approaching the crossing—I have crossed the crossing very seldom in an automobile; generally when we crossed the tracks it would be under an underground culvert, something like that, not an automobile on the highway. (121)

From my knowledge and observation of my father, I would say he was a very careful man in watching for danger of trains in crossing the tracks, as I knew it at that time. As we went up

(Testimony of Roy Spencer.)

there, and as I was looking for the train to emerge out of this cut to the west, and watching in that direction [79] as I have testified, as I climbed that, I listened as much as I could for the sound of any train. Of course, (122) the car was running, I could not hear very good. I was listening, yes, sir. And as I went up there, I did not hear any bell, any whistle, any sound of a train, until I saw the train. At that time there was some wind blowing. There was quite a breeze blowing towards the northeast. (123) It would be blowing almost direct down the track, a little bit north of down the track. By "down the track," I mean toward Mountainair.

Q. Was there enough wind blowing, or did the wind carry any dust?

A. Well, none to speak of that could hide the hill, more than it was kind of hazy like to see away off—it was hazy like; right close you could see. It might hide anything completely quite a ways off. It is a dry country around there most of the time. (124) I have noticed at other times as to whether or not, standing at that point from which we started the automobile, I could see an engine coming from Mountainair and the direction in which this train came before it emerges from that cut.

Q. What have you noticed from observation in that particular from the point where you started the car?

A. Well, I think the engine first comes in sight just about the time it gets to the whistling post.

(Testimony of Roy Spencer.)

The top, outside, the north side of the engine, the top, comes in sight there. (125) The right-hand corner, the cowcatcher, you can see it first, the smokestack and top of the engine then comes out more until you can see all of it at the mouth of the cut. [80]

There is no place before it actually emerges from that cut that you can see the whole of the front of the engine, until it passes the edge of the hill as I have described. The only sort of warning or signals there at that crossing at the time was the crossing sign. That crossing sign was two boards crossed, nailed up on a post.

Q. Roy, I had you describe before the height of the embankment in this draw from the picture, and you did describe it, as to its height on the south side; now, what was the condition on the north side, as that draw continued (126) on up?

A. Just the same as on the south side. The draw just kind of goes out against the hills and stops on the north.

Q. The road on the north side, the road on the highway leading to Mountainair, does that cross any hill after passing the railroad, or does it continue on in this draw?

A. Well, I think the draw kind of quits after it goes down a little ways from the crossing. At probably two or three hundred yards, just kind of runs up against the hills and stops. Then the road goes alongside of this big cut. But, north of the track, and for a distance, I think, of two or three

(Testimony of Roy Spencer.)
hundred yards, it is the same draw that is on the
south side.

Cross-examination.

My father had owned five automobiles before he
bought this automobile.   He had owned three Fords
and two Buicks.   I lived at home all of the time
from   (127)   my birth up until the time of this
accident.   I was 17 years   [81]   of age when the
accident   occurred.   That   was   June   11,   1923.
When I was 17 years of age—the 19th day of
April.   This truck I was riding in was a Ford
truck.   It was a truck when we got it; I don't
know whether it had been made over or not.   Back
of the seat to the rear end of the bed of the truck
I would guess was four feet.   The height of the
seat was just about like an ordinary car, I think.
(128)   The height of the bottom of the seat from
the bottom of the bed of the truck where the seat
is attached to the bed of the truck was, I should
guess, a foot—just guessing at it.   Maybe not that
much.   There were running-boards on the sides of
the truck.   The running-board was just like any
other running-board.   I think about a foot from
the ground.   The distance from the running-board
up to the top of the bed of the truck was probably
a foot and a half.   I would guess that the top of
the bed was three feet and   (129)   a half from the
ground, just looking at it.   The bed of the truck
had sides, sideboards.   I think the sideboards were
ten inches high.   There was an end-gate that raised
and lowered.   As we approached that crossing, it

(Testimony of Roy Spencer.)

was raised. I am about five feet four in height. As we left the 100-foot point, that is, the south right of way fence, to go over the crossing, I pushed against the hind end gate, at the right-hand corner—the upper part of the gate. I grabbed hold of the upper part of the hind end gate—the rear end gate—with my left hand to hold and brace on the right- (130) hand corner. When I got within fifty feet of the track, I released my hold with the right hand, and I traveled around, stepped around to the right-hand side of the truck. As I proceeded along the right-hand side of the truck, my left hand was not released [82] at all. I had to remove the position of my left hand as I made that change. I took my position before I started around, just stepped around, put my right hand against the back of the seat. I placed my right hand against the back end of the seat. Up to the cushion or up the top of the seat with my hand, over the top. The seat had a back on it. when I changed my position in the manner I have described, I did not (131) look toward the east. And at no time after I left the south right of way fence, which the testimony shows was one hundred feet from the center of the track, did the truck stop. And as the front wheels of the truck passed over the south rail of the railroad track I jumped up into the car. And in so doing I didn't look to the east, that is, the direction from which the train came, until I jumped in the car. Now, this highway that I have described, over which we traveled

(Testimony of Roy Spencer.)

as we approached the crossing, paralleled the railroad for some distance. It had paralleled the railroad I suppose a distance of half a mile from the crossing. The highway was from two to three hundred yards, I (132) think from the railroad. I think the farthest point must be three hunrded yards.

Q. Was the railroad within sight all of that distance?

A. Well, I couldn't say whether it was away up half a mile or not. I don't remember that.

Q. Well, from the time back, a point half a mile west of the crossing, from that point to the crossing, was the railroad within your view?

A. *Part it* at times, of course, it is cutting through hills, and at times you can see it and at times you cannot. As we made the last half a mile we were traveling [83] very slow. I think he was running in low pretty near all the time—low gear all that time. When we got to the point where we changed our direction of travel, that is, when we reached the point 100 feet south of the center line of the track, we stopped. And I got out of the truck. (133) We stopped at that point I expect half a minute. And then and there I looked in both directions. I had no conversation with my father at that time. Then my father started the car. I don't know at what rate of speed he traveled in going up to the track. I showed you the best I could yesterday, just as well as the car would possibly pull in low and we pushing. I estimate that

(Testimony of Roy Spencer.)
to be probably something like one minute. I have
never observed the time that it takes a Ford to
travel a hundred feet when in low gear. So when
I say it took a minute, it is a guess, on my part,
of the way the car was going; a car can go lots
of different speeds in low gear. It at no time
stopped while it was negotiating that one hundred
feet. (134) I have testified that when we were
within five hundred yards of the crossing I saw
a freight train proceeding in an easterly direction
over the railroad track. I first discovered that
freight train when it was first starting over the
crossing, there, at the opening of the hill. At that
time, we were up about five or six hundred yards
from the crossing.

Q. Where were you when the rear car of that
freight train passed over the crossing?

A. Well, the rear car was the engine. The engine was pushing the train. I didn't take particular notice as to the number of cars in the train, but
I would judge about twenty cars, from what I
remember about the train. (135) [84] There
was no locomotive on the head end of the train.

Q. Where were you with reference to the crossing
or the point at which you stopped before proceeding over the crossing, when the engine passed over
the crossing?

A. Oh, we had not went over 75 yards, I don't
think, while the train was going over the crossing.
It was about a twenty-car train, I think, going
about twenty miles an hour. We probably had

(Testimony of Roy Spencer.)

went 75 or 100 yards, something like that. Just guessing at it; it took us about ten minutes to go from that point where I saw the train going over the crossing, to the point where we stopped. I think that is about right. Probably something like that, yes. I don't think it could have been any longer. I think about ten minutes, that is my best (136) estimate. My father's eyesight was good. His hearing was good. My eyesight was good. · My hearing was good.

Q. Have you ever measured or are you competent to measure the height or depth of the embankment over that crossing?   A. At what point?

Q. From the middle of the crossing to the point you have described as the hill or cut east of the crossing?

A. Why, I think so. It is dug off just like a wall; it is straight up and down like a wall. I never did measure it more than just looked at it.

Q. Roy, as a matter of fact, is not the land on the north side of the railroad track, from the crossing easterly, a considerable distance higher than the land on the south side of the track?

A. Yes, it is more hilly. [85]

Mr. WOOD.—Does your question mean immediately at the   (137) embankment, or farther away?

Mr. REED.—At the embankment, along the embankment, the north side of the embankment.

The WITNESS.—You mean the railway?

Q. (By Mr. REED.) Yes, on the north side of the railway.

(Testimony of Roy Spencer.)

A. Well, there is not much difference along there; it is just the same where they have made the fill, just scooped out along on both sides and made the fill just the same along there. The highway on the north side of the track is higher than the highway on the south side of the track. So, in approaching the railroad, we were traveling westerly away from Mountainair, along the highway—you would not have much of an ascent or grade to climb to get up over the track. It is more than a couple of feet. I don't know just what it is, it is just a small hill.

Q. As you got up to the track and hopped   (138) up into the truck, why didn't you look east?

A. I did.

Q. Why didn't you before getting up into the truck?

A. Oh, before? Well, because the car was going over the track.

Q. Well, what was there in that to prevent you looking eastward before you stepped up on to the running-board?

- A. Well, the car was going over the track and down the other hill; naturally so, I stepped up on the running-board before, and then looked down the track. [86] I was not in school at the time of the accident. I had been in school up to that spring. That was in the summer-time. I hadn't finished the eighth grade when I quit school. I was in the eighth grade when I quit school. (139) Yes, sir, I was supposed to be a bright boy, when I was sev-

(Testimony of Roy Spencer.)

enteen years of age. I had sufficient knowledge to know that to get up into a vehicle as it was about to go on to a railroad track and to do so without looking in one direction, at least, was highly dangerous, unless I had already looked.

Q. You have testified you had not already looked in that direction before you got up into the car.

A. I had looked, yes, sir. I looked at the corner of the right of way but not after that. Not until I got into the car and was about to sit down. I mean to tell the jury that my father having owned five cars previous to this one, and I 17 years of age, had never driven an automobile; nor I have not to this day. I have lived in New Mexico all my life, that I have lived. (140) My father and mother were not residents of that state. My father lived in New Mexico at that time. I don't know how long he had lived there. I think he came in '90, something; I don't know just what it was; 1890 —he had lived there most of the time from that time up to the time of his death. That was his residence. My mother lived there also during that period. I had many friends there. I don't remember whether my father shut down the motor as we stopped at this one hundred-foot point below the track, or not. (141) I think he did.

Q. You have testified on direct examination that when you were at that point, 100-foot point south of the track, you could see the whistling post?

[87]

(Testimony of Roy Spencer.)

A. Well, you can in a way, and in a way you cannot. You can stand at the right of way corner and step up on the bank, where the road goes around this corner post the bank is dug off, and if you will step up on that bank you can see the whistling post, but you cannot by standing down in the road or on the right-hand side of the road. The whistling post is supposed to be a quarter of a mile east of the crossing, I think, but the whistling post is to the north of the track, about six or eight feet, I think. The whistling post looks to be about six or seven feet tall. (142) A locomotive, of course, is taller than the whistling post.

Q. When you are at that one hundred-foot point, you would have no difficulty in seeing a locomotive when it got to the whistling post, would you?

A. Well, you might see the top of it, just a little of the smokestack; I don't know. And as you approach the crossing, or approach the rails at the crossing, you can see farther and farther, easterly.

Q. When you are fifty feet from the track how far can you see in an easterly direction along the track?

A. I don't know; you cannot look down the track that way and tell just how far you are looking. You can see beyond the whistling post quite a little ways. As you get along thirty feet from the track, you can see somewhat farther. I think that when you are at a point twenty feet from the center line of the track, you can see easterly beyond the whistling post (143) I never did notice particularly at that

(Testimony of Roy Spencer.)

point. You can see quite a ways. You cannot tell looking down the track just how far you are away unless you measured, you cannot tell exactly. [88]

On August 18, 1925, I testified in a former trial of this case. On that date, and in this courtroom, in the former trial of this case, I testified as follows:

"Q. What change in the direction of the road takes place when you come around the corner and across the road from the general direction of the road as you approach the crossing?

A. The road is practically going east, a little bit north of east, when you get to the crossing you turn around the corner and go over the track north."

I further testified as follows:

"Q. How much of a turn is there there, a square turn, or more or less—

A. About a 45 degree angle."

That testimony was correct. (144)

At the same time and place I testified as follows:

"Q. And you state that the wind was blowing that day? A. Yes, sir."

Q. Was it blowing very hard?

A. No, not hard. Just a stiff gale. Just enough that a car, by running over the road, would pick up dust, but the wind itself would not pick up dust."

I further testified: "Q. But there was not enough to stop you from seeing anything that you were looking at? A. Well, not unless something would kick the dust up."

I further testified: Q. Well, did anything kick

(Testimony of Roy Spencer.)
the dust up to obstruct your vision?   A. No, not
me, no."

I further testified: "Q. When you did look along
the railroad track, it was perfectly clear, then,
wasn't it, so that you could see?   A. Yes, sir."

That was correct.   (145)   [89]

Redirect Examination.

I mean that when I saw the train, when I looked
there was nothing in the dust which would obstruct
my view of the train aproaching.

Q. In the answers you gave there about you not
being able to see, what were you referring to as regards distance, as you understood it?   (146)

A. Well, I was referring from the mouth of the
cut to the crossing; I was not referring to a farther
distance.   That is what I was referring to in that
former testimony.

Q. Can you state to the jury what the
effect of the dust, such as there was, if
any, would have been upon objects in that cut, differently, if at all, from what it would have on objects that had emerged from the cut and come toward you, if you are able to tell?

A. Well, you cannot see a train very plainly
coming out of the cut, and you take down in the
cut, if there was nothing to hide the smoke, anything like that, it would have to be able to be seen.

Q. From your recollection of the condition of the
dust, what, if any, effect on that day would the

(Testimony of Roy Spencer.)

amount of dust being carried in the air have upon the vision as to objects down in the cut? (147)

A. It would probably hide them altogether.

Mr. REED.—What is that?

The WITNESS.—It would hide them altogether.

Mr. REED.—At what distance below the crossing are you referring to?

Mr. WOOD.—In the cut. [90]

Mr. REED.—"In the cut," is indefinite. How far away, a mile or two miles?

Mr. WOOD.—My question referred to any point in the cut; did you so understand it?

The WITNESS.—Yes, sir.

Mr. REED.—That is all. (148)

Recross-examination.

Referring to Plaintiff's Exhibit No. 5, which you hand me, I have examined that. I stated on direct examination that there is a fence that shows in that picture, and that that fence obscured my vision somewhat. That fence is right here. That is just a white board fence there. It is right on the south side of the track.

Q. Is there not a wing fence at the cattle-guard?

A. I think that is what it is. I think it extends out from the track, the south rail of the track, about six feet—something like that. That is the only fence there is along there at all. What I was referring to is the curve that goes around there. But as I approached the crossing, that is, while I trav-

(Testimony of Roy Spencer.)
eled the distance of one hundred feet, I did not see
that fence at all. (149)

The COURT.—Let me ask a question.

Q. How much time elapsed from the time that
freight-train passed the crossing going to Mountainair, before you were struck by this train going
west?

A. About ten minutes, I should judge. [91]

Redirect Examination.

Q. You said that this fence shown in the picture
you thought was about six feet long; which did you
refer to, the top or bottom of the fence, as being
six feet long?

A. Taking on an average of the middle, probably
about that, not taking that spur there. I never
measured that. No, sir; only just looking at this.
The fence is accurately shown in this picture. I am
giving my estimate now from looking at the picture
as to its length. (150)

TESTIMONY OF JESUS BERRERAS, FOR PLAINTIFF.

JESUS BERRERAS, called as a witness on behalf of the plaintiff, testified, through the Interpreter Gregorio Ruiz, as follows:

Direct Examination.

My full name is Jesus Berreras. I live at Mountainair. I have lived at Mountainair about fourteen years. (151) I don't know exactly, but I

(Testimony of Jesus Berreras.)

think I am about 54 years old. I knew Benjamin
Spencer and Roy Spencer. I remember the time
and the crossing at which the train struck the automobile and injured Roy Spencer. With reference
to that crossing and where the train struck them;
from the crossing my house is kind of in the direction of the west, on a hill. My house is more or
less a hundred yards from that crossing—something like that. As to about how high above the
railroad tracks and the crossing it is to my house,
why, from about the level of the track where my
house stands is about 20 feet, something like that,
maybe. That is, I say my house is about 20 feet
higher than the railroad track up on the [92]
hill. (152) My house is to the west of the crossing. As you pass over the crossing, going in the
direction of Mountainair, I am up on the hill at the
northwest corner of the angle that the road makes
with the track. At the time this train struck
Spencer's car, I was just outside of my house.

Q. When did you first see Spencer or his car that
day, and where were they when you first noticed
them?

A. Why, when I first noticed them they were on
the east side of the track at the corner of the fence
of the right of way. The car was stopped at that
time. I noticed that car as it started and as it
climbed up the grade on to the track.

Q. Tell the jury just what movements you saw
that car (153) make, and what you saw Roy
Spencer, the boy, and Benjamin Spencer, the father,

(Testimony of Jesus Berreras.)

do, as their car moved up from the corner on to the track?

A. I seen the boy pushing the car from behind, and the old gentleman had hold of the steering-wheel and they began to go up the hill. The boy pushed the car along for a distance from behind up the hill and then he turned to the side of the car and pushed from the side of the car. I seen them after they got started, when Roy Spencer pushed the car from behind, then went around to the side of the car, about between the railroad track and the corner of the right of way, where they had started, why, the old gentleman had this hand (indicating) on the wheel and was stooping down and fixing something on the car. I don't know just how far up the grade the car had gotten when the boy went around from the rear to the side pushing, but it looked to me like it was about half the distance from where they started to where   [93]   the track was when he went to the side.

Q. About how far up from the corner, if you remember, was it, the car had gotten when you saw the old man reach down and apparently do something, as you say, in the bottom of the car—about where was the car when you saw that?   (154)

A. Why, it seemed to me something like fifteen yards—or feet—that they traveled about fifteen feet when I seen him when the man stooped over as they climbed that hill and after I saw him stoop over, I did not see Mr. Spencer straighten up again after that.

(Testimony of Jesus Berreras.)

Q. Now, you stated that you saw the boy pushing the car up the hill. Will you please come down to this table and show the jury just how you saw the boy pushing, both from behind and as he went to the side, supposing this table to be the car and this the hind end of the car, and up over the track (indicating).

(Witness leaves the stand as directed.)

A. The car began to go, went quite a little distance, and then got to that side, and after the boy pushing along- (155) side of the car when the car just about went over the rail, or something like that, the boy jumped in the car. Just as the boy got in the car the train hit him. The car was not going very fast as I saw the boy pushing up the grade. Going very slow. He was pushing all the time, the boy was pushing alongside of the car.

Q. Show the jury the way the boy was moving his feet as you saw it.

A. When he changed over to that side (indicating) the boy was pushing like that—in that way (indicating). [94] I don't know what the boy noticed, but when the car, it seemed to me, went over the rail, the boy jumped in the car.

Mr. WOOD.—The witness illustrates by putting his hand on the table and moving his right foot constantly ahead and bringing his left foot after it step by step.

Q. Where did you first see the train on that occasion, where was it when you first saw the train?

A. First I seen the automobile. First I noticed

(Testimony of Jesus Berreras.)

them, and I knew it was a crossing. I have a habit
of looking down toward the crossing there. Then
I looked up the track and I seen the smoke inside
of the cut. I had my eye on the people because I
knew they were in danger and I seen the smoke, and
I would look at them, then look towards the train.
I seen the engine first come out of (156) the cut
—the engine come first.

Q. Now, tell us whether or not you were noticing
and listening for any signal of bell or whistle that
the train gave?

A. Yes, sir, I noticed that, because I noticed
there was danger at the crossing. I never heard
nothing.

Q. Did that train blow any whistle or ring any
bell during all that time that you saw it and that
it was approaching the crossing?

A. No, I didn't hear nothing; I didn't hear no
bell or no whistle. I noticed the speed that train
was making as it approached the crossing that day.
(157) I could not say at what rate of speed it was
coming, but I could see that it was coming fast, because you could see the dust flying to the sides of
the engine. (158) I have noticed passenger
trains coming in the same direction in that way,
but I noticed that they were not coming as fast as
this train was coming. (159) [95] I could not
say how much faster this train was coming than
any other train I have seen come through there before, but I could notice that this train was coming
faster than the others. I could not attempt to esti-

(Testimony of Jesus Berreras.)

mate in miles. No, sir, I could not say. I could say it was coming a whole lot faster than the others.

Cross-examination.

My eyesight is good. Surely my sense of hearing is normal. I can hear them talking here, I guess if anybody thought I was sick or hard of hearing I would not be here. I guess anybody could know if I was sick or hard of hearing that I would not be here where I am at. On the day of the accident, in the month of June, 1923, my hearing was normal and good,—just the same as it is now. (160) It is a short time since I have known Roy Spencer—about five or six years, something like that. I have talked about this to nobody before to-day only here in court. I have talked to Mr. Wood, the attorney for Roy Spencer. I have talked to Mr. Wood twice, this time. I have lived in Albuquerque. I was there for a month. I was working there at the sawmill this last summer. I lived there somewhere before this one. I heard the train, of course, before it got to the crossing. The train was making a great deal of noise. (161) It was making considerable noise.

Q. How much smoke was there coming from the stack of the engine or locomotive?

A. When it was in the cut it showed more smoke, but when it got out of the cut the smoke spread. The train or the locomotive was about 500 yards, 600 yards from the crossing at the time when it emerged from the cut—something like [96] that.

(Testimony of Jesus Berreras.)

Q. So, for a distance of 600 yards between the west end of the cut and the crossing the train was in plain view and making much noise?

A. I don't know from down in the track, but I noticed it from up where I live. From where I was I could see the train while it traveled a distance of 600 yards, approximately, from the west end of the cut to the crossing, during which time the train was making a natural noise like all the trains make. I could see the train at all times when it was traveling from the west end of the cut until it reached the crossing.

Q. There was nothing in the air, such as dust or smoke (162) or wind to obscure or obstruct the vision?

A. There was just a little bit of wind, but it did not bother me to see the train, no, sir. No, sir, it did not bother my sight a bit. I could not hear the automobile or the Ford truck that was being operated by Mr. Spencer and his son. I didn't hear the motion of the automobile at all, either from the motor or from the movement of it. I was out there at the house, but I didn't hear the noise of the car. The automobile was moving very slow while it was going up towards the railroad track. I didn't have no watch; I don't know how many miles per hour. I do not drive an automobile. It continued to make about the same rate of speed during all of the time between the time when it was at the railroad fence until it got up to the crossing; it was going very slow. The automobile, or the Ford

(Testimony of Jesus Berreras.)

truck, did not stop at any time that I observed. (163) No, sir, the car never stopped. The boy pushed it from behind, kept it going, and [97] then from the side—it was going. I do not know whether the engine or motor of the truck was operating. I don't know whether there was any load on the truck; I didn't go down there to see. I seen from up at the house down there. That is all. I did not observe whether there was any load on the truck. I couldn't say exactly from where I was the position of the boy's feet, on the ground, when he was pushing the truck, but I could see the bulk of his body pushing. I could not tell just exactly the position he had his hands or arms against the car; I could see he was pushing behind the truck.

Q. How far was the truck which Mr. Spencer and his son were operating when you first observed it; how far was it from the railway track?

A. For the first time that I seen it it was standing (164) there at the corner of the right of way. I don't know how long the truck stood at the corner. I don't know how long it had been there or how long it had been stopped when I observed it from my house. It was standing there, and I looked down towards the crossing. I don't know how long it stood there, but when I first observed it, until I seen it start, it might have been ten minutes; something like that. As to the length of time it took the car or truck after I observed it start in motion, to reach the railroad crossing—why, to my estimation, I don't know, but it seemed to me just

(Testimony of Jesus Berreras.)

about a minute from where it started until it hit
the track. My house is west, kind of west, from
the railroad crossing at which this accident occurred. I think something like a hundred yards
west of the crossing. My house is more in the direction of the west than it is north from the crossing of the track. (165) From the railroad fence
to [98] my house must be about sixty yards. I
don't think it is north. It is west. It looks to me
like the railroad track and the highway run at the
point of the accident—runs north and south. The
railroad at my house looks like it is running north
and south; beyond my house a little bit it turns like
it was going west.

Q. Was this train that you had observed approaching the crossing, approaching from the east
or the west or the north or the south?

A. We call it north, but it was more east to me
than it was north. I don't recollect seeing a freight
train at any time pass over this crossing before this
accident. At the time when I heard the train and
saw it coming out of the cut just before that (166)
—I was fixing my house. I was nailing some
boards on the side of my house.

Q. On which side of your house?,

A. Toward the east side of my house. My house
faces toward the east. I had been there since nine
o'clock in the morning working on my house. This
accident happened about four o'clock in the even-
ing; I don't know; something like that. I had
worked all afternoon on my house. I didn't pay

(Testimony of Jesus Berreras.)

any attention to what other trains passed over this crossing, either from the west or the east or the north or the south. I don't remember of any other trains, but this train I particularly noticed because I could see the people coming into contact with this train. I seen one other vehicle pass by in the highway or road that day. (167) Yes, sir, I seen two wood haulers that was on this just before the accident they came over to my house with a load of wood. [99] I saw them before they came over to my house—when they crossed the crossing, then I hollered at them and told them to come to my house, and they come up to my house. They seen just exactly what I seen, too, happen. I did not hear any whistle or any bell on any train that day. I didn't hear any bells or whistles. I don't remember of seeing any train other than the one that I have testified about pass over this crossing on the Santa Fe track that day I didn't notice that this train that I saw approaching the crossing made any additional noise when the brakes were put in operation. The train came to a stop. The train didn't whistle before it came to a stop. (168) I once filed a claim against the Santa Fe Railway for a horse they killed for me—it blowed away and I never heard nothing more about it. (169) When Roy Spencer left the rear end of the truck and came around on the side, from where I were standing, I could see Roy Spencer at that time. I seen the body when he turned around from behind to go to the side of the car from here up (indicating).

(Testimony of Jesus Berreras.)

After he got around to the side of the car I could still see him. The automobile or truck was between Roy Spencer and me. Roy Spencer's father was on top of the car. (171) On the side of the car the wheel is at. On this side of the car (indicating); on the left-hand side. Roy Spencer's father's position in the car did not obstruct my view of Roy Spencer because Roy Spencer was farther back than the man at the wheel. I seen Roy Spencer get on top of the car. I saw him go up into the front end of the car on the front seat—close to the wheel where they handle the car. During all that time I was looking at Roy Spencer; [100] I seen the danger coming and I noticed him and I changed my view. I noticed all of the time and kept my vision on him all of the time. I would be looking at the boy and then I would see the train, too. I didn't have no watch—I couldn't tell you just exactly how long the train was in reaching the crossing from the time when I first observed it. (172)

Q. Where was the truck in the road at the time when you first observed the train coming out of the cut?

A. I noticed the train plain out of the cut, one part of the truck was on the track already. The train was then something like five or six hundred yards, something like that, from the crossing.

Redirect Examination.

Q. You have given one or two measurements here of yards. Did you ever measure the distance from

(Testimony of Jesus Berreras.)

this crossing down to the cut to know how long it is?

A. No, sir, I did not measure; I am just saying what I think it is.

Q. And you call the distance from the crossing to the cut six hundred yards?

That is what you meant to call six hundred yards, was the distance from the crossing to the cut? (173)

A. It seems to me that it might be that distance. I don't know whether it is that distance or not. (174) I stated that I remember seeing his engine just as it came out of the cut. Just at that time when I saw the engine coming out of the cut the Spencer automobile was at the corner somewhere there. At that time it began to get in motion, the car. (180) [101]

DEPOSITION OF M. G. ROSE, FOR PLAINTIFF.

Whereupon, the deposition of M. G. ROSE, taken on behalf of the plaintiff, at Albuquerque, New Mexico, on the 7th day of August, 1925, was produced and read as follows:

Direct Examination.

My full name is William Wood Rose. I live at 714 West Leed, Albuquerque, New Mexico. I was 42 last birthday. (181) My occupation is that of a signal maintainer. I am employed by and for The A. T. & S. F. Railway Company. That is the

(Deposition of M. G. Rose.)

defendant in these cases. I have been in their employ as a signal maintainer 13 years in Albuquerque—about 16 years all told in the service. I have been employed as signal maintainer for the Santa Fe for 16 years. I have been in the employ of the Santa Fe for 13 years on that work—continuous service. That is, I have been in that service for the Santa Fe 16 years; 13 in Albuquerque. As to what territory is within my jurisdiction as signal maintainer— (182) all I can give you is the present, it has been varied so much in the last few years. I have charge of maintenance of signalling at Belen. I do not have those on the cut-off, east of Belen. I never have had. As far as I go east of Belen, that is Medron, about six miles east of Belen. I am familiar with the use and the nature of signals had and maintained by the Santa Fe at crossings in my immediate vicinity during the period of time I have been employed by the Santa Fe. And I have been familiar, from my knowledge of the business, with the use of signals, of automatic signals generally in use by railroads. (183)

Q. For how long a time have you known of the defendant which I will call the Santa Fe Railroad, making use of [102] automatic signals at crossings in New Mexico?

A. Well, I came to New Mexico in 1912, and that was the first one I ever saw, was up here at Mountain Road, in New Mexico. Before coming to New Mexico, I was located at Colorado, Trinidad. My occupation was the same at

(Deposition of M. G. Rose.)

Trinidad, but wasn't the same title. As to how long I have known of the Santa Fe Railroad Company making use of automatic signals at any place on its line at crossings—that is pretty hard to answer—well, fourteen years, there has been a bell there at Trinidad, but I don't know when it was installed. Probably it was there in 1911, but I am not sure. (184)

Q. Have you known of the Santa Fe making common use of automatic signals at crossings for fourteen years?

Mr. BRENNAN.—Objected to as irrelevant and immaterial.

The COURT.—Objection overruled.

Mr. BRENNAN.—Exception.

The WITNESS.—They haven't had them at all crossings, just a few that they have picked out. I didn't know that they have made a common use of any.

Q. I do not mean to infer by my question that they use them at every crossing, or generally at crossings. I mean to inquire for how long a time you know of their making use of signals at crossings?

A. Anyhow, fourteen years, I have known them— I think it was 1911 they installed the bell.

Q. At what class of crossings, to your knowledge, has the Santa Fe made use of such signals, automatic signals, for fourteen years, at least?

Mr. BRENNAN.—Objected to as irrelevant and immaterial, and not in issue in this case. [103]

(Deposition of M. G. Rose.)

The COURT.—Objection overruled.

Mr. BRENNAN.—Exception.

Mr. WOOD.—You have ruled on it, your Honor, but may I state that the purpose of the evidence is to show common use by the company of a class of special signals for cases of more than ordinary danger for a long period of time; that is the only purpose of the evidence. (185)

WITNESS.—(Continuing.) To my knowledge the Santa Fe has made use of such signals, automatic signals, for fourteen years, at least at crossings where the traffic was heavy and to the best protection of the public. To explain the kind and workings of the automatic signals which have been used by the Santa Fe to my knowledge, for fourteen years, I will say that originally it was a crossing bell controlled by an interlocking relay governed by two-track service at least 1,500 feet from the crossing. This interlocking energized in controlling the bell service or automatic flagman of the present day —train entering the block shunted the track circuit—either one may be used—either direction, thereby closing the bell circuit, or causing the flagman to operate. An automatic flagman is an automatic signal device for crossings; (186) as well as a bell; it has a light and an arm or stick that works back and forth with all approaching trains, all work at the same time. I wave my hand in explanation of my testimony, from which I mean that something like a pendulum located at the crossing,

(Deposition of M. G. Rose.)

swings back and forth to give warning to persons attempting to cross.

Q. What starts the bell ringing at the crossing, or this automatic flagman, as you call it, moving, and how is it started, and how stopped? [104]

A. Well, the operation of the bell—the train shunting the track circuit, sends out the interlocking relay, closes the bell circuit, or flagman circuit, therefore the flag or current, whatever it is, operates the flag through a motor or through the coils of the magnetic trips.

Q. And how is that motion continued, and when and how stopped?

A. It continues until the train gets across the crossing, and then the track circuit, on which it is shunted, closes the circuit gate and picks up the interlocking relay, thereby opening the bell circuit.

Q. If I understand, then, briefly, the system is as follows: The bell or the pendulum, automatic flagman, is operated by electricity, by a motor?

A. Yes, the bell is not worked by a motor; the flagman (187) is operated by a motor. The bells are operated by coils or magnets. But each are operated by an electric current. The current, or the switch, which starts the current, is closed when the train passes over the switch about fifteen hundred feet from the crossing—that or farther. And continues to be given by bell or automatic flagman until the engine or the head end of the train crosses— the rear end of the train passes the crossing, when it stops.

(Deposition of M. G. Rose.)

Q. Now, to what extent has the Santa Fe made use of those signals in New Mexico during the period of time you have been here, and what crossings, to your knowledge, have they been used at during that period of time?

Mr. BRENNAN.—Objected to as incompetent, irrelevant and immaterial.

Mr. WOOD.—It is offered merely upon the same issue as before, as to the common use of this class of signals or on   [105]   especially dangerous crossings, and for no other purpose.   (188)

The COURT.—Objection overruled.

Mr. BRENNAN.—Exception.

(Whereupon the reading of the deposition continued.)

A. I really don't know what you mean. Do you mean how many?

Q. That will partly answer the question, yes.

A. There is one at Mountain Road, here in Albuquerque; the next one is at New York Avenue. We have one below Barelas, on the lower road to the stockyards; one between Belen and Los Lunas, on the Socorro Road, and there is one at Los Lunas, too, now; one at Medron, on the same road; one at San Acacio; Socorro, one at Deming, and one at Fort Bayard. The one at Mountain Road, I believe, is just outside of the city limits of Albuquerque; just one in the city limits—the one at New York Avenue. The one at Mountain Road is just out of the city limits, but in the settled portions of Albuquerque. The one I mention at the stockyards is

(Deposition of M. G. Rose.)

below the city of Albuquerque. The next one I mentioned is at Los Lunas, that is a small hamlet of a few hundred people, about 20 miles south of Albuquerque. (189) The one at San Acacio—it is a new highway and bridge across the Rio Grande. It is very dangerous for people coming across the bridge and on the tracks. The minute they get across the bridge they are on the railroad tracks, almost, and a bad curve. They installed it when they installed the bridge. I believe it was in November, 1923—'23 or '22, I won't say which. The one between Belen and Los Lunas is at a little Mexican town named Los Chavez, there. That is a community of two or three hundred people. That signal at Los [106] Chavez I believe was installed in 1919. It is on the main highway to Socorro or El Paso, and houses are built close to the track so the view of the approaching train is poor. Medron is between Belen and Socorro, on the main highway —there (190) is a cut, the railroad goes through and the highway drops down to go across the tracks. The purpose of installing automatic signals there has been to warn passengers from the approaching trains from the east, on account of the bad curve. The signal at Medron was installed the same year, 1919, as the one at Los Chavez. These automatic signals that I mention, have been continuously used and still in use by the Santa Fe. And for the purposes specified. That is, at crossings extraordinarily dangerous the Santa Fe has within my jurisdiction here established these automatic signals. I

(Deposition of M. G. Rose.)

know where Mountainair is. (191) I do not have
anything to do with Mountainair. It is east of
Belen. Mountainair is east of the district in which
I operate. That, I think, is on what is known as
the Belen cut-off, where the road branches from
Belen through the mountain pass to the east—I
have never been there on the railroad, so I couldn't
say. By looking at the map I know where Mountainair is, but that is all. I can't tell you how far
it is east of Belen; I don't have any idea. Couldn't
say whether it is more or less than a hundred. I
know the Santa Fe Railroad has a line known as the
cut-off; I know it is called the cut-off. The main
or east line of the Santa Fe, before the (192) cut-
off, passed through New Mexico through Belen and
Albuquerque, Las Vegas and out at Raton, or there-
abouts. I believe the cut-off is the new line to constitute what the [107] name implies, a cut-off, or
better route, more level route, extending from Belen
east, joining the other line at a point in Kansas. I
think the Santa Fe has commonly used these automatic signals at other points outside of my district
and at crossings having the same general conditions
as these that I have described. I know there is a
few. I can't tell you how many years those few
have been in use by the Santa Fe how many year.
I believe they have been in use for five years, at
least; that long, some of them. (193)

Cross-examination.

In determining what crossing signal device of the

(Deposition of M. G. Rose.)

kind described by me should be installed on the line covered by myself—the principal factors taken into consideration are the amount of traffic across the crossing, and the obstruction to the view of the railroad to approaching trains.

Q. At all of the crossings which you have named at which wigwag signals have been installed, can you state how extensively the highways at those points are used?

A. Well, they are usually installed on the main highways, where traffic is heavy. There have been none installed at other places on my territory. Such signal devices, so far as I know, are installed only at heavily traveled and dangerous highway crossings. (194) I am not familiar with the crossing at which the accident occurred, the accident involved in this case. I know nothing about the accident of any kind. Other kinds of protection installed at crossings within my jurisdiction are: We have crossing gates. [108] Gates are installed where the traffic is real heavy, like in the city of Albuquerque, and hard to handle, not pay any attention to a flagman. The same protection is given, the same automatic flagman or bells are installed at crossings which are but comparatively slightly used, I will say that we have one crossing, that at San Acacio, is used very slightly. I forgot that when I made my statement a while ago. A wigwag was installed at that point on account of a bad curve and bridge right at the railroad track, the approach of the bridge is right at the track, people coming

(Deposition of M. G. Rose.)

off of the bridge is right on the track, to give these people warning before coming on the track. (195) And there is a very sharp curve right there at the crossing. A badly obstructed view. But ordinarily, at slightly used crossings, the warning is a signal sign. Large boards, reading, ''Railroad Crossing,'' and sometimes pictures placed up for people. I think all crossings on the Santa Fe and Rio Grande Division have been posted with placards of different kinds—whether or not they are used. There is some wording or lettering on such signs. The usual wording is ''Railroad Crossing.'' I don't remember of any other. That is a cross-arm proposition. That is the standard sign—crossing sign. The crossing at San (196) Acacio is used very seldom. I maintain that crossing, and I go down there at least, you might say, twice a month, and very seldom see a car go by. There is generally horses, or a bunch of cattle or sheep being driven over the bridge. The road is used more or less for driving livestock. It was built for a highway, but on account of the sand it is used very slightly. I believe that crossing was installed in 1923. I believe it was a county highway. I think the [109] County Commissioners was the ones that had the bridge put in there—new bridge. The superintendent required the signal device installed when they put in the bridge, I think. The bridge was put in the same time the crossing was installed. (197) I think the superintendent ordered the signal department to place the protection there. I do not know

(Deposition of M. G. Rose.)

whether the County Commissioners requested or ordered that a signal device be installed at that crossing. I think the superintendent demanded of the County Commissioners—this is in Socorro County, and the way I understand it, Mr. West required it to be placed there, and they put the crossing in. I am not sure, but I understood the commissioners of the county had to stand the expense of it, but I am not sure; I won't swear to that. The extent of my territory is 64 miles, I believe. It is from Mile Post 01 to 65.

Q. Can you give us the amount of public highway graded crossings there are in that district? (198)

A. There are so many of them I don't know whether they are public highways or not, in Valencia County and Socorro County—it is hard to give you an estimate of the number there.

Q. Can you approximate it? Well, confine your answer to that. How many highway crossings are there on that district, highways that are commonly used for traffic?

A. Well, that is something I never stopped to think, how many it would be. There is a good many in the 64 miles. I don't believe there is a hundred; probably 25. Of these crossings, I believe I have seven where the signal devices such as I have described have been installed. I believe it is seven on my territory at the present time.

Q. Can you state what percentage of all the highway crossings, public highway crossings in your

(Deposition of M. G. Rose.)

territory, are [110] protected by a signal device such as you have described?

A. Do you mean of the 25 that I gave, or do you mean on the main highways, that is, like the ocean to ocean highway, or the road to El Paso? Do you mean those or do you mean the small country roads. they use?

Q. I mean all public traveled highways of your territory? (199)

A. Well, I gave 25 as the number of crossings— around 30 per cent of them.

Q. And when you answered that to the best of your knowledge and recollection there are about 25 highway crossings on your district, does that include all traveled highway crossings, on your district, or only the main roads?

A. That includes all those but have fenced gates to go through—lots of private crossings, with gates, that I have not included.

TESTIMONY OF MRS. SARAH SPENCER, FOR PLAINTIFF.

Mrs. SARAH SPENCER, a witness on behalf of the plaintiff, testified as follows, on

Direct Examination.

I am the mother of Roy Spencer, the plaintiff here. I live at Mountainair, New Mexico, near Mountainair, New Mexico. (200) My age is 57. I have lived at or near Mountainair about twenty-eight or nine years. Roy is twenty years old. He

(Testimony of Mrs. Sarah Spencer.)

was born right where we are living now. From his birth down to the time he was injured his duties and occupation was doing chores around the farm and helping his father about the mill. His father operated a sawmill, at Eastview, New Mexico. Mountainair is our nearest town. It is eleven miles from Eastview. Eastview is six miles east [111] from the mountains. As to Roy's condition of health down to the time of his injury; he was a stout, active child, and was well most all (201) the time—sick a very little. He had some contagious diseases when he was small, like mumps, measles, whooping-cough; he was really very small; I don't think he remembers very much about it. He said he has not been sick very much. He had them most all while he was a child. The usual child's diseases of the kind I have described; he had them when he was small. Aside from that, when he grew up, he was well most of the time, stout, hardly ever sick. As far as I know, he never had any serious constitutional disease or ailment whatsoever. As to his general health and strength and ability to work, up to the time he was hurt—from the time he was a boy, as he grew up on the farm; (202) I think he was stout; stouter than most boys, and very active, and never was sick hardly any. He was sick a very little. He was a small boy. He was not a large boy. He could do tolerable heavy work before he was hurt, carried lumber in the mill, helped in the log woods; I know he operated in the mill, and that is not very easy, I think. He

(Testimony of Mrs. Sarah Spencer.)

carried pretty heavy lumber there. I have seen him help do it. I have seen him helping about the mill, carrying freshly cut or green lumber from the logs. He carried lumber with the other men, grown men, helped with tolerable heavy lumber; I couldn't say as to the size. He was very active, ran races; wrestled quite a good deal; always seemed to come out pretty good, nearly always first in the races. (203) He went to school until—he didn't quite finish the eighth grade, he missed a few points in the eighth grade, I think three points he missed in the eighth grade. He stopped the [112] year before he was hurt, the winter, I think. I don't believe that he went to school the winter before he was hurt. I think he stopped one winter before he was hurt. Roy's occupation was helping his father and helping around the house. I do not think he had any other relation to my family except that of a boy helping his father at home. And that was the manner and relation in which he was working, as any boy helping his father at home. I remember this day when they started out on the journey which ended so badly. I heard something at the start as to why it was and for what purpose that Roy was accompanying his father that day; I remember the talk they had at the breakfast table, talked at the breakfast table about it. Mr. Spencer (204) said something at the breakfast table about the oldest boy, Floyd, going with him, and talked about going a while, and finally decided to let Roy go. Roy didn't seem to want to go, said he would rather not

(Testimony of Mrs. Sarah Spencer.)

go. They insisted on him going, because they thought the oldest boy could help at the mill better. The purpose of the journey was to bring some furniture and some goods from Scholle. Mr. Spencer was asking Roy to go along and help him with that. He wanted him to go and accompany him and be with him. After they left home, the next I knew of it was when I learned that the accident had occurred. The next I saw of Roy was when he was in the hospital at Albuquerque the first time I saw him after he left. I went to Albuquerque the next morning. (205) I did not get there until along in the afternoon of the next day. At that time he was suffering intensely; he seemed to be in his right mind at that time, but a good deal of the time he was not rational, but he was when I got [113] there at night. I stayed there at night with him lots of the time. At night he was irrational, didn't know anything. That morning he seemed to be rational, but he was suffering awful. I was at the hospital a long time visiting him daily at the hospital; I don't remember just how long a period. I was there off and on until he was taken out. I was there when he was taken out in January the following year. I went home two or three times, but I stayed there nearly all the time. At the start sometimes at night he would be irrational; for the first two weeks he was irrational half the time, nearly, at night, but in the daytime he seemed to be in his right mind. (206)

(Testimony of Mrs. Sarah Spencer.)

Q. What did you notice as to his appearance, as to whether or not he was suffering intensely during that time?

A. It seemed to be mostly his leg and foot.

Q. (By Mr. WOOD.) Describe his actions as indicating his physical feelings during that entire time, and commence at the first and give it in your own way, without any prompting, from the time you first went over there, until the time of the operation—I mean the operation for appendicitis.

A. Well, I don't know, I will do the best I can. I can't remember all of it. But anyway he seemed to lay there a little while before the doctor set his leg and put in the plate. It must have been two weeks before they got the plate; he had to send to Chicago for it. Roy suffered everything before they got the plate, and they put in the plate and he seemed to suffer very intensely; right after they put the plate in I was with him one night and he didn't know anything all night hardly. Then he was in bad shape there for a couple of months. We didn't know whether he would live or die. (207) And he was in a dangerous condition there for two [114] or three months. I knew at that time when what has been described as to when the bones or the plate protruded out of this wound, out of the leg. That must have been a month, about a month after they put it in, I think, that it worked out, as well as I remember. He was very uneasy then and seemed to suffer quite a good deal, kept wanting to take the plate out; didn't seem to want

(Testimony of Mrs. Sarah Spencer.)

to take it out because they said he was too weak.
Therefore they didn't take it out. It was quite a
while after the plate protruded in that manner
before that was corrected, or attempted to be corrected by the doctor. (208) I don't know just
how long; I couldn't hardly say; it was quite a
while, several weeks. Then they had to operate
on it again and take it out, taken and operated and
taken the plate out. There is nothing else particularly that I can think of that I remember in regard
to the condition from that on until the time of the
operation for appendicitis; I can't remember anything particularly, no more than his suffering; that
is all. He suffered most of the time, that is all. I
remember the time he was operated on for appendicitis. Which was sometime in December or
January. They telegraphed to me. I had gone
home, and they telegraphed to me to come back.
I came over. Yes, I went back to Albuquerque. I
didn't hear that it was claimed he had tuberculosis
of the bowels until along toward the last, until he
was—well, he was operated—when he was operated
on, I think when they taken that plate out, I think
that was when they—after that they told me. (209)
It was along toward the last of his sickness. I
know it was after the operation that they told me
that. After he had opened his stomach I was told
that. Yes, I believe it was after the operation for
appendicitis that they told me that. [115] Up to
that time I never knew or heard or had any information to the effect that Roy had or had had tuber-

(Testimony of Mrs. Sarah Spencer.)
culosis in any form.   No member of my family, to
my knowledge or belief, had that disease, or my
husband's family, as far as I know.   After the
operation for appendicitis, he remained in the hospital two or three weeks, I guess, might have been
a month.   He was able to get out of bed and sit up,
or did get out of bed and sit up in about a couple
of weeks, I think   (210)   Might have been a little
less time, before he got up after being operated on
for appendicitis.   That might have been three or
four weeks before he left the hospital.   This hospital that he was at was at St. Joseph's Hospital at
Albuquerque.   I think they gave him good care.
I think they did the best they could by him.   The
physician or surgeon who treated him during all
this time was Dr. Lovelace.   I don't know who
called Dr. Lovelace to him.   They got Dr. Lovelace when they first taken him to town he was
called.   Called him before I got there.   I didn't
call him.   And he continued to treat the boy all
the way through.   When I left the hospital, he had
been up   (211)   about two or three weeks, I think.
I left the hospital the last part of January, I think,
sometime.   As to his condition then, as to his ability to get around, and condition of his leg at that
time; why, he was very weak.   They thought it best
for us to take him out.   Told us if we left him in
there very much longer they were afraid he would
die.   That is what they told me.

Q. Where did you take him then?

(Testimony of Mrs. Sarah Spencer.)

A. Well, my sister came down from Colorado, Montrose, to visit him, and she wanted him to go home with her, so he decided to go up there, take him up there and stay a while until he got stronger. [116]

I went to Colorado with him. He was there two months. Then we come back home; back to Mountainair. When he left the (212) hospital his leg was much stiffer than it is now. It bends a great deal more now than it did then. As to what kind of treatment has been given the leg since he left the hospital; well, he just rubbed it and used oil, bathed it with hot water and cloths. That was the treatment prescribed by Dr. Lovelace. I gave that faithfully and continuously. I bathed it for quite a while with oil, and used hot cloths. I have known of his giving himself that treatment some. For the last two years, you might say, the leg has been substantially the same as it was shown here to the jury. I think there is some improvement, more improvement the first year than there has been in the last two, it seems to me. I can't say I have noticed any material change or change in the leg in the last year. (213)

Q. There appears now to be a scab and a sore, with oozing out at the point of the original injury. What was the condition of that at the time he came from the hospital?

A. Well, it seeped more then than it does now. There was more matter from it then than now. Run pretty badly when he first came out. Take in

(Testimony of Mrs. Sarah Spencer.)
the last year, since the former trial here; I can't
notice any material change in that—I can't notice
it has improved any in the last year.

Q. What have you noticed in regard to his physical condition and his ability to do any work in the
last year or two years?

A. Well, I don't think he is able to do strong
work. He works some, does chores and light work.
I notice when he gets out and exercises extensively,
the next day he is [117] very tired and lays
down lots; he seems to get tired more than he used
to; didn't use to give way to anything. If he goes
out and walks very far, or anything like that, I
notice the next day he always seems tired and rests
nearly all day. I don't think he can stand nothing
like the work he could before he was hurt. The
kind of work he does now is—well, he chores
around the place, sometimes will work a few hours,
an hour or so, doesn't work steady, don't (214)
expect it of him. He never gets up in the morning until later, don't expect it.

Q. What were his characteristics before his injury, as to whether or not he was ambitious, an
energetic boy, and one would work?

A. I don't think Roy is lazy. He is a pretty
good worker generally.

Cross-examination.

Roy has never had scarlet fever at any time.
He had children's diseases such as measles, mumps;
I don't remember whether—I don't remember him

(Testimony of Mrs. Sarah Spencer.)

having chicken-pox, except he did have it when he was little; I think maybe he had the whooping-cough when he was little; had the diseases when he was small; must have been small, for the reason he does not remember being sick much—along three or four or five years old, along there. The other boys had them. He was the youngest one. I don't remember him having croup, or anything of that kind. He had the usual and ordinary children's (215) diseases up until about five or six years of age. And he went to school from the time when he was about six until he was sixteen. He went in the country; attended [118] country school. We had good teachers. It was in that vicinity of our home. And he, like boys of his age, would do the work, say, when he was ten years old, he would do the work that a boy ten years old would do around a place. And as he waxed stronger and older, his duties became a little more heavy; that is what I mean. During the last year or two he has been able to do some work around the place; does light work around the place. Attended to horses, harnessing horses, feeding them. He does not ride the horses; he has not ridden any since he has been hurt, I (216) don't know why. He drives a team once in a while; goes to town. Gets the provisions in town. Brings them home. Does work of that sort.

Q. Do you have a farm of any kind, or a ranch?

A. We have a ranch—we are living on a ranch. Well, there was 800 acres in the ranch.

(Testimony of Mrs. Sarah Spencer.)

Q. Does Roy sometimes go out on the ranch and help a little around there?

A. He helps the boys sometimes a little about things. When he goes out to help, he does just any little things there might be to do. I don't know just what all. He doesn't go out and labor all day, work all day like a man, with a team, or anything like that.

Q. What kind of work are his brothers and the others on the place engaged in during the working season?

A. He has only got one brother on the place, and he has planted some beans this year—he is the oldest boy. (217) As to what work Roy has been able to do within the last couple of years out in the bean fields, he doesn't do a great deal. My oldest son runs the team, he handles the team altogether. They have not put in a big crop. The oldest [119] boy does nearly all the handling of the team and he has the team ready for his brother, then he feeds them and does the work around the place. I have noted within the last year, that is, a little improvement in Roy's physical appearance, his face. Seems to be a little bit more robust now than he was, say a year ago. I don't know. Roy doesn't weigh quite as heavy as he did last year when he was here.

Q. Don't you see in him, so far as his general appearance is concerned, the color in his face, general complexion and all, the eye, and everything, indications of a marked improvement?

(Testimony of Mrs. Sarah Spencer.)

A. Well, he has been out in the sun more; he is tanned more, has a darker color than he had a year ago, but he is not as heavy as he was a year ago in flesh, he is not as heavy. (218) I don't know exactly how much he weighs; he was telling me he did not weigh as much as he did a year ago.

Q. Well, now, how frequently has he hitched up the team, driven the team to town and done work of that kind, chores of that sort?

A. Well, some days he chores some, some days he does not chore at all. He has not been in school since his accident. I wanted him to go to school, and he didn't seem to want to go, no way, couldn't get him to go to school; didn't seem to want to go.

I have never lived in Colorado—only just visited there. My husband did not formerly live in Colorado; I had a sister living there. I had never lived there, gone to Colorado to live there. No member of my family has gone to Colorado to live on account of his or her health. I am quite sure of that. (219) I have a sister living there at [120] Montrose. She is the only one living there. The other members of my family are: I have a sister that lives in Oklahoma and one in Newton, Kansas; one in Wichita, Kansas, and I have a sister that lives in Albuquerque, New Mexico. None of them have lived in Arizona.

Mr. WOOD.—That is all. We desire now to offer the deposition of Dr. Lovelace. In connection with this we desire to use this machine. (Referring to machine for exhibiting X-ray pictures.) (220)

(Deposition of William R. Lovelace.)

We first offer the deposition taken on the 22d day of July, 1925,—there are two depositions.

DEPOSITION OF WILLIAM R. LOVELACE, FOR PLAINTIFF (July 22, 1925).

Direct Examination.

My name is William R. Lovelace. I am a physician and surgeon residing in Albuquerque. I have practiced surgery in New Mexico since 1906. I am regularly licensed and educated and qualified as a physician and surgeon. (221) I am the local surgeon for the Santa Fe Railroad Company, I have been doing work for the Santa Fe since 1906 —since 1908. I remember the occasion when Benjamin B. Spencer and his son Roy A. Spencer were brought to Albuquerque suffering from injuries. That was June 11th, 1923. I was called to attend them at that time. (222) When I examined Roy Spencer, he was suffering from extreme shock, slight internal injuries, and a comminuted fracture of the lower femur. What I mean by comminuted fracture is a bone that is broken in several pieces—in more than one piece. In this case the left femur was fractured at lower third, several fragments of bone extending into the muscle. The muscular, or other tissues at the point of injury were badly contused—badly [121] contused and injured. This fracture was about eight inches from the knee joint. The treatment I gave him at that time when he was first brought

(Deposition of William R. Lovelace.)

in, were stimulants to combat shock, and temporary immobilization. It was 7:30 A. M., June 11, 1923, when I first saw (223) Spencer. He was brought to St. Joseph's Sanitarium in Albuquerque at that time. As to what treatment I gave the boy after the first, or preliminary treatment, and what I found necessary to do, and what I did: On June 23, 1923, external methods were tried to reduce the fracture and to also retrain it in proper position. We were unsuccessful. So, on the same day, open method was advised, incision was made over line of fracture, and a large piece of bone was removed; distal and proximal—I mean by that the upper end and the lower end—brought into close apposition and retained by a Lane's Plate. A Lane Plate is a metallic splint which contains openings for screws. This is placed over the fracture and attached to bone by screws, which holds the fracture, or pieces of bone, in place. I do not recall the number of pieces this bone was broken in, but remember distinctly removing one large piece and a few small pieces. After the reduction of the shocked condition existing when I was first called, I then endeavored by outside splint to put the bone in place (224) for reunion, and the matter continued in that condition from the 11th of June until the 23d. My idea in prolonging the time for open reduction was to first have my patient recover from shock and at the same time to retain the fracture in as good a position as possible by external methods. And then on the 23d of June,

(Deposition of William R. Lovelace.)

I cut down to the bone, performed the operation of cutting down to the bone and removed the   [122] broken pieces of bone, fastening the remainder in the method I have described.   I found the muscles and other tissues of the leg at the point of fracture when I operated badly contused and lacerated. His condition from the 23d of June until January 14, 1924, was uneventful, with the exception of slow progress.   I know from my experience, and I can state the reason for the slow progress.   (225) It was due to the severe injury which lowered his resistence, and also to an undernourished condition. I think that covers it.   Several weeks prior to the operation the screws in one end of the Lane Plate which I stated I fastened with screws to both ends of the fracture in uniting the bone, became loose, and the plate was misplaced, but fracture was sufficiently healed to not interfere with the healing. It is not unusual—in fact, in my experience, at least fifty per cent of the plates that are used have to be removed within a reasonable length of time after operation.   These metal plates serve much the same function that the wood or fibre splints, if placed on the outside of the limb, do, merely to hold the bones in position while knitting, or uniting.   And could be and frequently was removed after the bone was reunited, or knit.   (226)   This plate became loosened or misplaced because of low vitality of bone tissue and by muscular contraction of the limb.   The action of the muscle at about the plate would tend to displace it after it had become

(Deposition of William R. Lovelace.)

loosened. The bone tissue in which the screws were first turned came away, and allowed the screws to loosen. The boy was necessarily kept in bed and the limb rigid, in attempting to heal this fracture, from June 23d to January 14, 1924. I recall definitely the condition of the leg at the last date I have mentioned. [123]

Large calsified deposit was easily felt at the line of fracture. There was no evidence of motion, which led me to believe that he had a firm union. There was a stiffness of the knee joint, at the time, due to lack of use. There was not a great deal of motion at this time. (227) Assuming that at the present date, the knee is still stiff, and reduced in motion, I am unable to say without examining it what will be the future condition of that, as to what is to be expected—whether the existing amount of mobility in that knee will ever be increased.

My charges against the boy for my services are $300.00. The boy was attended at St. Joseph's Sanatorium during all the time that I mentioned. I am familiar with the fair and reasonable value of the services rendered in a sanatorium. The charges as itemized on the bill you show me, rendered by the St. Joseph's Sanatorium for services are fair and reasonable for such services. (228) Very reasonable.

(That said bill of the sanatorium was thereupon offered and admitted as Plaintiff's Exhibit No. 10.)

Mr. WOOD.—The bill offered in evidence as Ex-

(Deposition of William R. Lovelace.)
hibit 10, is itemized as to services, dressings, board,
and so forth, a total of $672.85.

WITNESS.—The boy was suffering when he
came under my observation, and during the time
that he was in the hospital. He claimed that he
was having severe pain in his left limb and unable
to move without increasing his pain. The limb was
held rigid during the process of reuniting the bones
by a plaster cast. That extended from the pelvis
to the ankle [124] on the left side. I do not
remember the exact length of time it was necessary
to keep the limb (229) in the plaster cast, but I
think it was about ten weeks.

Cross-examination.

As to the condition of this limb at the time the
appendicitis developed: He was ready to leave the
hospital. He was detained on account of the appendicitis operation, from January 14, 1924, to
January 20, 1924. That would have a tendency further to reduce his resistance—his resisting power
and make his progress with regard to the healing
of his limb a little slower. I stated that the slow
progress of the healing of the wound was due to
undernourishment, and to the severity of the injury. (230) The ordinary length of time that that
condition should be healed in an ordinary condition
of health, if not undernourished, should be eight
or ten weeks. This undernourished condition I
speak of was pre-existing. This fracture in the
leg was a complete fracture of the femur. The in-

(Deposition of William R. Lovelace.)

dications were, at the time he left, or before this appendicitis matter came on was that there was a firm union. That means that the bones were completely knitted together. At the time that he left the hospital there was the condition of stiffness in the joint. And I think that was due to nonuse. It is my opinion that that would disappear in time with the use of the limb. (231) At the time this plate became loosened at one end, it had done practically the work it was intended to do. The external methods I used at first before I tried this incision were extension, which I mean—by traction, pulling on the limb. At the same time manipulating the limb. [125]

Redirect Examination.

Physical examination and X-ray examination failed to show any evidence of injury to the knee joint itself at the time of the accident.

Q. With the knee joint held rigid for the length of time that this was necessarily held rigid, could that produce the condition of permanent stiffness and loss of motion?

A. Longer than ten weeks, it could—a partial stiffness which should not be permanent. With that length of time, unless the knee were permanently injured, it should have been restored to as good a condition of mobility as it ever will be, in the lapse of one and a half years, (232) providing he has had treatment for it; massage and manipulations. After a year and a half, I doubt

(Deposition of William R. Lovelace.)

whether there would be any improvement in that condition, without surgical intervention. That surgical intervention would be cutting into the joint and severing the ligaments. The majority of cases get good results. The boy was ready to leave the hospital when he was operated on for appendicitis on January 14, 1924. He was still and up to that time confined to his bed most of the time, as a result of the original injury. I do not mean by leaving the hospital that his cure was complete at that time. He had not regained his strength, but he was in a condition to be able to leave the hospital and be (233) moved to his home. And at that time the condition of appendicitis developed, which kept him in the hospital approximately another month.

Recross-examination.

I made an examination of this boy during the time he was there for tubercular condition of the lungs. [126] His lungs were clear. No evidence of tuberculosis, though we suspicioned the tuberculosis, but we never did find it actively. There had been activity but not in his lungs. It was in abdomen. There was activity at the time of the operation—tubercules in the abdomen; also involving intestines. The effect that would have upon his progress and his recovery was to retard it. Advice was given to this young man before he left the hospital about the future treatment of his limb. (234) He was advised to have a leather splint made at Denver, and to have massage for the knee,

(Deposition of William R. Lovelace.)
and take nutritious food and plenty of fresh air
and sunshine. It was my judgment that if there
is a present condition of stiffness in this knee, it
could probably be removed by surgery.

Redirect Examination.

The condition of tubercules I have described—
this renders the person having that condition and
injury more likely to have that injury result more
seriously than a person in good health.

Q. And what would you expect in a person predisposed to tubercular trouble to result where such
an injury as was received by this boy?

A. It would lower his resistance and make him
susceptible to it. More susceptible to put the tubercular condition into activity. It would also
lower his resistance. So that a person in the condition that I found this boy predisposed to some
tubercular activity would be more likely to suffer
more seriously than a healthful (235) person.
[127]

Recross-examination.

I was familiar with his condition of tuberculosis
of the abdominal region up to the time he left the
hospital. It was worse during the time he was
there than when he came. He complained of more
pain in the abdomen. That condition at the time
he left was very much improved. When he first
arrived in the hospital, he complained of a tenderness in the abdomen, but he did not complain of
severe pain until after he developed appendicitis.

(Deposition of William R. Lovelace.)
Redirect Examination.

I discovered the existence of tubercules I have described at the time of the operation. (236) No, that was not a matter of ten days, or two weeks after the injury. You see, we did not operate until—after the injury. That was January 14th. I discovered the existence of the tubercules January 14, 1924. I did not discover them before that time but suspicioned them being there. I first discovered indications of a tubercular condition in the boy when he complained of a general tenderness all over the abdomen, loss of appetite practically during the entire time of his stay in the hospital. The condition gradually grew worse, until he was operated on and the tubercular appendix was found. He complained of the condition of tenderness of the abdomen when I first examined him. He complained of tenderness all over his body. That could have been the result of the accident, such a tenderness produced. That would have been a very natural result to expect from such an accident as he had encountered. (237) [128]

Mr. WOOD.—We next offer a deposition of William R. Lovelace, the same witness, taken later, on the 14th of August, 1925.

DEPOSITION OF WILLIAM R. LOVELACE, FOR PLAINTIFF (August 14, 1925).

Direct Examination.

WITNESS.—My testimony was taken by deposition in a former examination in this same case.

(Deposition of William R. Lovelace.)

I am the same person who appeared and testified at that time as William R. Lovelace. In that examination I stated, in (238) answer to one question, that I determined some of the conditions of Roy Spencer as a result of X-ray examination and pictures taken at the time of his illness. I will state when and where, under what circumstances and how many of those X-ray pictures were taken, to my knowledge:

The first X-ray was taken June 12th, 1923, that was the day following the injury, to ascertain the position of the fractured bone; the second picture was taken August 3d, 1923—shall I go into detail to tell why?—taken at St. Joseph's, August 3, 1923; the third picture was taken at the same place on September 14th, and the fourth picture was taken September 21st, the same year.

The plaintiff, Roy Spencer, was an inmate and under my care at St. Joseph's hospital in Albuquerque during all that time. The standing and class and condition and reputation of that institution is recognized by the American College of Surgery, which gives it the highest classification. The first picture was taken on the 12th of June. I now produce the X-ray film representing that picture. This is the picture which you hand me, and on which (239) the ticket annexed is marked "June 12, 1923." That picture shows a [129] fractured left femur, which is also comminuted. That is a picture of Roy Spencer's leg. That is as the bone in his leg was and appeared on the 12th day

(Deposition of William R. Lovelace.)
of June, the day after his injury. That is an accurate and correct representation of the appearance of that bone, from my knowledge, as it was at that time.

(The picture the witness then had reference to was admitted in evidence as Plaintiff's Exhibit No. 11.)

Mr. REED.—You are now offering all of these pictures now at this time, are you?

Mr. WOOD.—Now, each one as identified by the picture. (240) I offer all four of the pictures referred to by the doctor, and which are annexed to this deposition.

The COURT.—The next three will be Plaintiff's Exhibits Nos. 12, 13 and 14.

(Whereupon, Mr. Wood exhibits to the jury in the X-ray machine the picture, plaintiff's exhibit, referred to.)

Mr. WOOD.—This picture referred to in the deposition is taken on the 12th day of June.

WITNESS.—I now produce the film or negative of the second X-ray picture that was taken August 3d, 1923. This picture marked August 3, 1923, shows correctly the condition of the leg of Roy Spencer, the plaintiff, at that time.

(Whereupon Mr. Wood exhibited to the jury in the X-ray machine the picture, plaintiff's exhibit, referred to.)

WITNESS.—I now produce the third film or X-ray picture of this boy's leg. (241) This correctly shows the condition of the leg and the position of

(Deposition of William R. Lovelace.)
the bones and the splint fastening them, on September 14, 1923.

(Whereupon Mr. Wood exhibited to the jury in the X-ray machine the picture, plaintiff's exhibit, referred to.) [130]

WITNESS.—I now produce the fourth film and X-ray picture of the leg of the plaintiff, Roy Spencer, that I took or had taken. This is the picture which I show you and hand to you. That correctly shows the condition and position of the bone in Roy Spencer's leg as of September 21, 1923.

(Whereupon Mr. Wood exhibited to the jury in the X-ray machine the picture, plaintiff's exhibit, referred to.)

WITNESS.—In the first picture that I produced, the knee is towards the bottom of that picture on the opposite end from which the ticket is pasted. (24) I never had or took any picture other than these four of this boy's injuries, showing the condition of his leg or bone. Whatever examination that I was aided in making by the result of X-ray pictures, are these four pictures which I have here produced and which have been offered in evidence, with the addition of what we call the floriscope. I made a further examination with the aid of the floriscope. I will describe briefly what the floriscope is and how I used it in making these examinations:

It is a plate or screen which can be placed directly or laterally over any object you want to examine, and shows the bone as plainly as does an

(Deposition of William R. Lovelace.)

X-ray plate; that is the way we generally make our reductions of fractures, by external methods—it is the only way of getting the bones in (243) alignment, or telling that they are in alignment; it is as plain as this plate, or plainer than this plate; you can see the bone exactly how it is. In short, the floriscope is the instrument which is used with the naked eye and directs you in examining the bone with the aid of the X-ray which is also used in taking the pictures that have been produced, the same rays. The picture, Exhibit "A–C–H" shows a badly comminuted— [131] I mean by "comminuted," a bone that is fractured in more than one piece—it also shows a lateral misplacement, together with this fragment of bone. The piece of material which shows to the right of the bone, at the point of fracture, in the picture is a piece fracured loose from the femur. In the operation,— at the time when I cut into the leg and put the silver splint or the Lane's plate, I removed that loose piece of bone. I removed all other segregated pieces of bone that I was able to discover at that time. (244) I testified before that I put this plate on the 23d of June. That is correct. At the time when I put that plate on, the bones were then straight, and I put it in proper position for knitting and reuniting. In the picture, Exhibit "B," it shows the bone greatly bent at the point of fracture, apparently; I will explain how that came about and explain generally the meaning and significance of the picture, Exhibit "B."

(Deposition of William R. Lovelace.)

As stated in my previous deposition, owing to the fact that the patient was suffering severely from shock and undernutrition, it was deemed advisable to postpone any surgical intervention until he had recovered sufficiently from the shock to withstand the surgical reduction of fracture, so the operation for reducing a fracture and retaining fracture in the proper apposition was performed June 23, 1923; at that time the Lane plate was applied, the bone being in proper alignment, and screws applied, holding firmly the bone into correct position—the usual aseptic measures were taken,—limb was then put into and supported by a splint extending from axilla or armpit, to bottom of foot, on affected side—this splint was bandaged firmly to limb so as to completely immobilize it; the limb was so (245) badly contused and lacerated that, in order to give it proper care and treatment, [132] and at the same time to keep it immobilized, the external splint was applied—this splint was retained in position, as above stated, until August 3, 1923; August 2d, 1923, patient stated the following morning that he was taken with a severe pain in region of fracture during the night while attempting to shift his position. Upon examination it was found that the fracture had become misplaced on account of improper calsification. He was removed to the X-ray room and August 3d this picture was taken, showing that the fractured bones were out of line, as stated before, caused by undernourishment, as the usual case of fracture of any bone becomes firmly

(Deposition of William R. Lovelace.)

healed so that you can walk upon it with assistance of a crutch or cane within this length of time. Noticing a displacement by physical examination and X-ray findings, we advised the administration of an anesthetic and limb replaced in as near a correct position as possible by external methods. Wound was sufficiently healed at this time to be able to give it the proper care and attention through a small opening, so a plaster cast was applied. After placing limb in as nearly a correct position as possible, patient was removed to his room. Window was cut in cast and wound was treated. (246) By "window" I mean a hole was left in the cast over the wound proper through which I could treat it. The picture indicates that the bone had then bent, as is quite apparent from ordinarily looking at the picture. At that time, on August 3d, 1923, very little calsification or union of the break had taken place.

Q. In the ordinary case, unless interrupted in some manner, what length of time would be sufficient in a boy of seventeen or eighteen, to cause the union or calsification to take place? [133]

A. Well, we usually consider that within two months, in the normal, healthy individual, to get complete union. The splint reaching from the armpit to the leg had become slightly loosened so as to permit the bone to bend in the position indicated by Exhibit "B," but it was not loosened enough to cause this misplacement, without quite a bit of exertion. From my knowledge of the case and the

(Deposition of William R. Lovelace.)

boy, and the (247) condition, I am unable to say how he managed to succeed, while fastened to a splint, in causing the position of the bone shown in Exhibit "B." He was being given at the time the care and the surroundings were as good as I knew how to give him. At the hospital, there are nurses in charge of the patients and in continual attendance. The picture, Exhibit "B," shows a sharp end, and I will call it sharp point of the lower bone projecting sideways. Now, in order to reduce and restore the condition from that shown in Exhibit "B," by external pressure, I forced the leg into such a position that the bone would be as straight in alignment as I could do it. And is the position of the bone into which I forced it into shape shown in Exhibit "C," the picture of September 14th.

(Whereupon Mr. Wood exhibited, in the X-ray machine, to the jury, the plaintiff's exhibit referred to.) (248)

WITNESS.—And is the position of the bone into which I forced it into shape shown in Exhibit "C," the picture of September 14th.

(Whereupon Mr. Wood exhibited to the jury, in the X-ray machine, plaintiff's exhibit referred to.)

Mr. WOOD.—I am putting in the one there as just referred to as Exhibit "B"—to the right—and the next one he refers to— [134]

Mr. MORRISON.—What is the number of the exhibit here, Mr. Wood?

Mr. WOOD.—Exhibit 13.

(Deposition of William R. Lovelace.)

(Whereupon the reading of the deposition continued.)

WITNESS.—I can't be positive about these dates, these two films.

Q. You can't be positive as to the date of the pictures Exhibit "C" and Exhibit "D," as to which of them was on September 14th and which on September 21st, is that what you mean?

A. That is it.

Mr. WOOD.—The first one I am showing is Exhibit "B," marked September 14th, and Exhibit "C"—

Mr. MORRISON.—Numbered what, now?

Mr. WOOD.—Number 14.

Mr. BRENNAN.—That is "C" and "D." "D" is 14.

(Whereupon the reading of the deposition continued.) (249)

WITNESS.—Together they show substantially the condition of the leg at about those dates. I merely mean I am unable definitely to say which was the earlier date, September 14th, and which was the later date, September 21. But, together, they show the condition of the bone to which I was able to restore it, after the picture Exhibit "B" shows it bent out of position. There is some slight improvement made at a later date when the plate was removed. The X-rays do show that in that restoration, there was a lapping of the ends of the bones. I recall there was a slight shortening in the boy's right leg when he left the hospital, but I don't

(Deposition of William R. Lovelace.)

recall the exact length. (250) It would be the
shortening indicated by this picture—the lapping—
the shortening caused by that and the loss of the
[135] fragments of the bone that were removed.
At the time the two pictures of September 14th
and 21st were taken, and at the time of the last
two pictures, calsification was taking place, and
there was partial union. Calsification proceeding
in a much less degree than would be expected in an
ordinary normal person. In my former testimony,
I stated that in the operation for appendicitis which
I performed in January of 1924, that I found a
tubercular infection of the appendix and bowels.
The effect that tubercular condition which I found
would have upon that wound would be to delay
union. The effect it would have in the leg forming
a solid union, or knitting of the bones, beyond
(251) delaying it would be to prevent him from
regaining the use of the muscles—not as soon as it
would in a healthy individual, on account of the
infected condition due to tuberculosis. His chances
for getting permanent union and perfect use of his
limb would be lessened by reason of the fact of his
tubercular condition. When I operated upon him
in January of 1924, I found the entire peritoneal
cavity, I mean by that, the entire lining of the abdomen, thickly covered with tubercules; I also found
numerous tubercules scattering over the intestines.
It was a case of germ infection or scattering of
tubercules over the bowels and peritoneal lining of
the abdomen. In addition to this T. B. infected

(Deposition of William R. Lovelace.)

peritoneal cavity in intestines, the appendix was acutely extended and studded with tubercules. I stated in my former testimony that I examined and made tests for lung infection for tuberculosis, and found none; that is correct. (*25*) I found no active tuberculosis of the lungs. Not being a T. B. specialist, I could have overlooked infection on that part of the chest. Tuberculosis is a germ disease. [136]

Q. And I will ask you to state whether or not in the ordinary person who never was known or suspected of having consumption or tuberculosis, to what degree are the presence of those germs or their activity discovered in post mortem examinations of people generally not supposed to be afflicted with tuberculosis proper?

A. Statistics show that approximately from 60 to 75 per cent of all individuals that have been examined thoroughly after death—show evidence of having had tuberculosis at one time, or had at the time of their death. A tubercular germ may not be likened to a seed, which germinates, sprouts and produces the disease.

Q. In what way does the germ act and how is its activity produced in the human body?

A. An individual has to have, in my opinion, a predisposition to tuberculosis before he ever contracts it—a real healthy individual, without being predisposed or having had some serious illness, will not contract tuberculosis by coming in direct communication or inhaling (253) bacilli.

(Deposition of William R. Lovelace.)

Q. By a predisposition to tuberculosis, do you mean that the body at the time contains germs or tuberculosis in a latent or inactive state?

A. I mean that he is in a weakened general condition. It is my opinion that the post mortems of 60 to 75 per cent of all cases disclose that the patient has had tubercular germs active at some time in his career. As stated before, in a long, protracted illness, if you have tubercular bacilli in your system, that would make you more susceptible to developing an active case of tuberculosis. [137]

Q. What, if any, effect would an injury such as this boy received, have upon developing tuberculosis?

A. If he had had tuberculosis in a quiescent state, and received a serious injury, this, in my opinion, would increase his chances for developing active tuberculosis. I did not find the existence of tubercules in this boy prior to the time he had that operation of January, 1924, neither did I find (254) activity in the chest, up until the time of the acute attack of appendicitis, but did suspicion a chronic intestinal infection of a tubercular nature before he was operated upon, and insisted upon forced nutrition.

Q. Doctor, in the ordinary and normal case, coming within your experience, with a boy with the amount of tubercular infection and condition which you found in him at the time of that operation of January, 1924, what would be the reasonable out-

(Deposition of William R. Lovelace.)

look as to his future, as you know it as a physician, and from your general experience?

A. Not being able to find any active tubercules in any other part of his body, other than the abdominal condition, would say that his chances for getting a completely arrested condition are about 90 per cent in his favor, due to the fact that the abdomen was opened and we know by experience that this causes an arrest or a cure in about that percentage of cases.

Q. You mean that merely opening the abdomen will cure tuberculosis in 90 per cent of the cases?

A. Of this type of tubercular infection, involving the peritoneal cavity and intestinal covering of the bowels, that is my opinion.   [138]

Cross-examination.

At no time did I find   (255)   any active tubercular condition in the lungs of this boy. I do not know how long the tubercular condition had existed prior to the time I performed the operation in January of 1924, but I suspicioned that he had something of this nature at the time he entered the hospital, on account of his undernourished appearance. I could not state, positively, that there was any causal connection between the injury and the tubercular condition.

Redirect Examination.

Given the existence of latent tubercular germs or propensity existing in the body, not active at the

(Deposition of William R. Lovelace.)

time of the injury, the injury and its resultant reducing of the resisting power of the boy would increase his chances for developing tuberculosis. In the absence of other known causes, I would not attribute the tubercular condition that I found in January of 1924 to the reduced resisting power caused and produced (256) by the injury. The injury itself wouldn't produce tuberculosis, even though it did reduce his resistance, but could be a contributing cause by reason of his weakened condition produced by the injury, but the injury itself would not produce the trouble.

Q. Not directly, you mean? A. Indirectly.

Q. But, in the absence of any other cause known to spur the disease into an active form, would you attribute the resultant disease to the causal connection with the injury and that the injury was capable of reducing his resisting power and then the disease developed as a result of that [139] reduced resistance?

A. The injury would, as stated before, lower his resistance and make him more susceptible to an infection. The effect that lowering resistance would have on a condition of latent tubercular germs in the body, if such then existed, would be to increase the chances, as stated before, to any infection, tuberculosis included.

Q. But I am not talking about infection. I am asking you concerning a case where latent—quiescent—tubercular condition or germs existed at the

(Deposition of William R. Lovelace.)
time of the injury; would the depleting effect upon his strength, his physical condition, his resisting power, be a sufficient cause to stir those latent (257) or quiescent germs into activity and thus produce tuberculosis; by quiescent I mean the presence of tubercular germs, not active, in an arrested condition, then you are using the term the same as I use latent; now, using it in that sense, would not the reduced physical condition, reduced resisting power, caused by the injury and its results as you have described them, be sufficient to cause, to reawaken the quiescent tuberculosis, if it already existed, and to send it into activity?

A. It would make him more susceptible to it by lowering his resistance, but as far as it being a direct cause, I do not believe this possible. It can be a contributing cause. (258)

Recross-examination.

The condition I found in this boy's bowels at the time I made the examination, the tubercular condition (259) I found; it was active at that time. I have no opinion as to [140] how long that condition had existed. I have stated that I did suspect that the first time I examined the boy when he was brought to the hospital, that he was suffering from tuberculosis—with some systemic affliction. I found no other indication of tuberculosis at the time, except an apparent undernourishment at the time the boy was first brought in.

DEPOSITION OF DR. W. T. MURPHY, FOR
PLAINTIFF.

Mr. WOOD.—I would now read the deposition of
Dr. W. T. MURPHY.   (260)

Direct Examination.

WITNESS.—My full name is Walter Thomas
Murphy. I am 41 years of age. I reside at Albuquerque, New Mexico. I have resided at Albuquerque 13 years. My profession is that of practicing medicine. I have been practicing medicine
sixteen years. I am a regularly licensed and qualified physician under the laws of the State of New
Mexico. I have been that fifteen years. (261)
I received a degree of medicine from the University
of Illinois, 1909; served also at St. Anne's Hospital
at Chicago. I have made a specialty in the course
of my profession. My special study has been tuberculosis. In the special study of tuberculosis, I
have been connected with the Modern Woodmen
Sanitarium of Colorado Springs, in 1910; the Albuquerque Sanitarium, Albuquerque, New Mexico,
1912; and director of my own institution since then.
When I refer to my own institution, I mean the
Murphy Sanitarium. That is located at Albuquerque, New Mexico. That sanitarium treated all
forms of tuberculosis. I was at the head of that
institution; I founded it. I was treating, and at
the head of that institution, treating tuberculosis
patients, (262) from 1912 to 1922. I have been
continually in the practice of   [141]   my specialty

(Deposition of Dr. W. T. Murphy.)

and profession up to the present time. And still am. As to the extent of the practice that I had in that connection, and the amount of patients, and so forth, as head of the Murphy Sanitarium:

We had a capacity of 75 patients, and we averaged from fifty to sixty the year round. They were all those tubercular patients. I, myself, am a victim of that trouble. I have given it special and careful study. My institution was the local place in which Santa Fe patients were treated while I operated it here, from 1914 to 1917. (263) Tuberculosis is an infectious disease, due to the tubercular bacilli; this germ may be taken in the system and lay there dormant for years before active disease takes place. The classes of tuberculosis are: The incipient, moderately advanced and far advanced. It is not classified with reference to the organs of the body that it may affect; it affects every organ of the body alike; no tissue of the body is exempt from the tubercular destruction. You may find tuberculosis afflicting one organ and not apparently present in another, however.

Q. Doctor, what causes and produces tuberculosis as a practical proposition? You have stated it has its origin in a bacteria or germ; what causes the activity of that germ resulting in what we call tuberculosis of the body?

A. The tubercular germ is taken in the body, and as long as the resistance of the human body is kept up, the germs are inactive or dormant; but when

(Deposition of Dr. W. T. Murphy.)

there is an exciting cause, such as sudden exposure, or other disease or injury, (264) and the resistance is lowered, we have active tuberculosis.

Q. Doctor, assume the condition of a boy, 17 years of age, raised on a New Mexico ranch, just an ordinarily apparently [142] well, healthy boy, who has not been known by his parents and surroundings to have any chronic disease—just an ordinary ranch boy, a little undernourished in appearance, but no history of tubercular activity that people had discovered—assume that such a boy was injured in a railroad accident to the extent that he was struck by an engine while riding in an automobile, his thigh bone broken in a comminuted fracture, accompanied by jamming and contusing and lacerating of the muscles and tissues surrounding the bone, pieces of the bone being broken loose, and that this injury occurred on the 11th day of June, 1923, that the boy was brought to a hospital—St. Joseph's Hospital in Albuquerque, a good, well-sustained and reputable hospital—and was treated by skilled surgeons; that due to the laceration, bruising and the shock accompanying the injury, it was found improper or unwise to attempt to reduce the fracture for a week's time; that, at the end of that time the fracture was reduced, the condition of the leg and the bone broken being as shown in the X-ray film annexed to the testimony of Dr. Lovelace and marked Exhibit "A," in connection with that testimony; that an operation was then performed, cut-

(Deposition of Dr. W. T. Murphy.)

ting down into the leg, removing a section of bone
at the break, and some small pieces of bone,
placing the leg (265) in position and maintaining it by what is known as a silver or Lane's plate;
that the contused and lacerated condition of the
interior of the leg and wound was such as to make
it inadvisable, in the opinion of the surgeon in
charge, to place the leg in a rigid plaster cast, but
instead, due to that condition, the leg was bound
to a board or splint extending from the armpit to
the leg; that the boy remained in the hospital until
the third of August, when it was discovered [143]
in some way that the splint had become loose,
the boy was complaining of pain in and about the
fracture, and that he was then removed and a second X-ray picture taken showing the bone bent
somewhat out of place, as appears in the X-ray
picture annexed to Dr. Lovelace's testimony, and
marked Exhibit "B," which I show you; that, then
the bone was forced back into position as near as
could be, and a plaster cast put over it and continued in that position until the 14th of September,
when an X-ray picture was taken showing the condition of the bone, the Lane's plate and the surroundings, and another X-ray picture taken
on September 21, 1923, also showing the position of the bone, the plate and the sur
roundings, and that but very little calsification or
knitting or uniting is shown, except as in these
pictures which I exhibit to you, at that time; that,

(Deposition of Dr. W. T. Murphy.)

then another operation was performed, in which Lane's plate was unscrewed and removed and the wound again closed; that the boy continued in the hospital until about (266) the 14th of January, 1924, some seven months after the injury, when an operation was performed for appendicitis; assume that at the time the boy was first seen by the surgeon on the 11th of June, that he appeared to be undernourished, but no tubercular activity was then ascertained or discovered by the physician, nor had there been any tubercular history in his family so far as known, and no apparent activity of the disease in himself or in his surroundings in his family; that at the time of this operation on the 14th of January, 1924, or about that date, the surgeon, upon opening the abdomen, the boy then complaining of extreme pains in the abdomen, and showing indications of appendicitis, it was found that the intestines and the appendix showed a [144] very bad condition of tubercules spread all over the intestines; the appendix and the lining of the abdomen, a genuine case of tuberculosis of the bowels and the lining of the abdomen covered by numerous tubercules spreading over them; from the history of this case, as I have given you, are you able to say with reasonable accuracy, with your experience as a specialist in this disease, what was the producing cause of that tubercular activity found in January, 1924? (267)

A. I am. (268) Taking into consideration the history of the case, the boy being undernourished at

(Deposition of Dr. W. T. Murphy.)

the time of the injury, as I understand, and during the time the bone or the fracture was very slow in healing and calsifying, and a few weeks or months later the surgeon operated on him and diagnosis was a tubercular appendix, with general infection of the intestines, am I right about that, tubercular appendix?

Q. Yes, that was in the question.

A. Would make it without a reasonable doubt, that it was due to the injury that brought on the disease. From my knowledge of the disease and my experience as an expert along these lines, I have had occasion to investigate and examine into the question as to what percentage of persons, apparently in normal health, during their life, whose post mortem examinations disclose, at some time they have had tubercular activity. In these large institutions, such as the Cook County Hospital in Chicago, their statistics show that infants, before they reach the age of one year, 20 per cent of the infants react to the test; up to two years, or older, they show a reaction of 30 per cent; and before they reach the age of twenty, at least 60 per cent; and by the time they are fifty, 85 to 90 per cent of all individuals have been infected (269) [145] with the tubercular bacilli. A person may have quiescent or latent tubercular germs in his system, and live and die of old age, and never discover the existence of the disease, and never suffer any apparent injury from it. As to what conditions cause a person, having those latent germs, or quiescent germs

(Deposition of Dr. W. T. Murphy.)
in his system, to become afflicted with the active disease and what produces the active disease:

The tubercular germ is within the system, it may be laying there merely as an infection, but it may never produce a tubercular disease; it has been proven that a germ may lay in the glands and lay inactivity, without breaking down any tissues, and go on for years, or indefinitely, until there is an exciting cause, such as typhoid fever, pneumonia, accident or injury, to lower resistance, when those germs become active tuberculosis. An injury such as was suffered by the plaintiff in this case, the boy, as you stated his condition, would be, in my opinion, an ample, sufficient cause for stirring any latent tubercular germs that may have been in his system, into activity. In the absence of any other known cause, I would say that, in my opinion, the condition of tuberculosis (270) found in him was the result of this injury.

Q. Doctor, assuming that this injury and the broken bone which you have had described to you and been shown in these pictures, was very slow in knitting and calsifying, as the pictures show, and, further, it was slow and had shown little calsification down to the 21st of August,—suppose also that the wound made when the pieces of bone were removed and the plate taken out, as has been described to you, had never closed, still remains, and a wound showing still oozing and active, and that the leg is approximately an inch [146] shorter than normal—suppose that the boy at this time is still un-

(Deposition of Dr. W. T. Murphy.)

able to work steadily, without strength enough to work steadily at hard work; what, in your opinion, causes that wound still to flow and not to heal thoroughly? Are you able to state, Doctor, from the facts given you, with a reasonable degree of accuracy, what is causing that wound to keep open and not heal?   A. Yes.

Q. And what is it, in your opinion?

A. A tubercular infection within the body of the boy.

Q. Doctor, assume the condition of the boy in regard to the extent of tubercular activity, that has been outlined to you in the previous question, are you able to state, from your experience, with reasonable accuracy, what will be the probable future history of that boy, his ability to (271) work and his chances of recovery from the tuberculosis, are you able to state?

A. Yes, sir. In my opinion, at least 75 per cent of all cases of intestinal tuberculosis are fatal at the start; 25 per cent are very liable to relapse and prove or come into the stage of a chronic condition. Applying my reasons to this specific case; in my judgment, this boy will be a chronic invalid, due to the tubercular infection of the bowels, within a reasonable length of time. In my opinion, this boy will never be able to resume and perform hard labor. I am able to state, from my experience as a physician in the charge of this disease, whether or not this boy will probably ever be able to perform labor to any degree. In my opinion, he will never

(Deposition of Dr. W. T. Murphy.)
be able to perform hard labor. (272) Due to the tubercular infection within his [147] system, and the weakened condition of the intestines.

Cross-examination.

I have never seen or examined Roy Spencer. I am not a surgeon. I operated a hospital for the consumptives in the city of Albuquerque ten years. I discontinued operation in 1922. I sold the Murphy Institution to the Episcopal Church. It was a paying institution while I operated it. It is true that many persons who are infected with consumption, tubercular trouble, are, so far as appearances go, perfectly robust, in good health, and without any accident whatever and without contracting any other disease, they suddenly develop tuberculosis. (273) In my judgment, there is necessarily a causal connection between an injury such as has been described here, and tubercular condition in the body. Tuberculosis does not follow in all cases of injury. It is true that in the great majority of cases, injuries of this sort, no tuberculosis develops. In regard to my answer to your previous question, it is not clear. May I make a suggestion: Will you confine your question to a tubercular infection or a tubercular disease? They are two different diseases entirely.

Q. Is there, in your judgment, necessarily, any causal connection between an injury such as has been described here, and a tubercular condition in the body of the injured person?

(Deposition of Dr. W. T. Murphy.)

A. I would have to answer it "Yes," the way you ask the question. A tubercular infection may be within the system   (274)   for years without any tubercular disease producing toxin to weaken the condition; a tubercular disease is the after effects of the tubercular infection, caused by [148] an exciting cause, such as an injury would cause in the history of this case.

TESTIMONY OF ROY SPENCER, IN HIS OWN BEHALF (RECALLED).

ROY SPENCER, recalled as a witness in his own behalf, testified further as follows, on

Direct Examination.

In addition to those already given by me, the reasons why I did not turn around and look down toward Mountainair when I was pushing the car up the hill was because the train had just gone to the east toward Mountainair, and the west part of the crossing is far more dangerous than the east, and I thought it impossible, being only one track there, for any other train to be coming west at that time.

Thereupon, the American Mortality Tables were offered and received in evidence.

Mr. WOOD.—They (American Mortality Tables) show that the life expectancy of a person twenty years of age is 42.20 years.

Thereupon, the plaintiff rested his case. (276)

Mr. REED.—Now comes The Atchison, Topeka and Santa Fe Railway Company, and moves the

Court for a directed verdict in its favor, in said action, upon the following grounds:

First. Upon the ground that the plaintiff has not offered sufficient evidence to entitle the cause to be submitted to the jury. (277) [149]

Second. That there is no evidence of any negligence on the part of this defendant.

Third. That the undisputed evidence shows that the plaintiff was guilty of negligence himself, which was the sole and proximate cause of the injury complained of;

Fourth. That the undisputed evidence shows that the plaintiff was guilty of negligence and want of ordinary care for his own safety, which contributed to the injury complained of.

Fifth. That the evidence, considered in the light of physical conditions, shows that if the precautions required by law had been observed by the plaintiff, he would have seen the train in time to have avoided the accident.

Sixth. That the undisputed evidence shows that if plaintiff had looked and listened before attempting to cross the railroad track, he would have seen and heard the train in time to have avoided the accident.

Seventh. That the undisputed evidence shows, that if plaintiff had looked in the direction of the on-coming train before attempting to cross the railroad track, he would have seen the train in time to have avoided the accident.

Eighth. That the undisputed evidence shows that the plaintiff and the driver of the automobile in question, to wit: Benjamin Spencer, were engaged in a joint enterprise at the time of said accident, and that if the driver of said automobile and the plaintiff had exercised ordinary care and (278) caution before attempting to cross said railroad track, they could have seen or heard the train in time to have avoided the accident. [150]

Ninth. That the undisputed evidence shows that the plaintiff and driver of the automobile in question, to wit: Benjamin Spencer, were at the time of the accident engaged in a joint enterprise, and that if plaintiff and the driver of said automobile, or either of them, had exercised ordinary care and caution for their own safety before attempting to cross the said railroad track, they could have seen or heard the train in time to avoid the accident.

Tenth. That the undisputed evidence shows that if the plaintiff and/or the driver of said automobile had looked up the railroad track in the direction from which the train was approaching when they were thirty feet from the railroad track, they could have seen the train for a distance of not less than 1,500 feet, and in time to have avoided the accident.

Eleventh. For the reason that the undisputed evidence shows that if plaintiff suffered any damage or injuries whatsoever at the time or in the manner or because of the facts alleged in plaintiff's complaint therein, such damages were directly, immediately and proximately caused and occasioned by the fault, neglect and carelessness of the plaintiff himself, and not by any fault, neglect or carelessness on the part of the defendant. (279)

Twelfth. By the allegation of the complaint and the evidence adduced, it is conclusively shown that the plaintiff, Roy Spencer, and his mother, are citizens and residents of the State of New Mexico, and that the alleged cause of action sued on herein accrued in that state; by reason whereof, the plaintiff's right to recover judgment against this defendant is controlled by the laws of New Mexico; [151] that under the laws of New Mexico the plaintiff has not only not offered legally sufficient evidence of defendant's negligence, proximately causing his injury or damage, but, on the contrary, the evidence establishes such negligence on the part of Roy Spencer as to bar his right to verdict and judgment in his favor.

Thirteenth. That the Arizona constitutional provision, being Section 5, Article 18, is not applicable in this case, because of the residence and citizenship of Roy Spencer and his mother, and because the alleged cause of action sued on herein accrued in the State of New Mexico, and because the evidence of the plaintiff himself, viewing it in the most favorable light, shows that he himself was guilty of contributory negligence.

Fourteenth. That the evidence adduced herein showing that the negligence of Roy Spencer proximately contributed to his injury, there is no controverted question of fact to be determined by the jury, and, therefore the record presents solely a question of law for the Court, which, (280) notwithstanding the provisions of Section 5, Article 18, of the Arizona Constitution, it is the right and the duty of this Court, under the law as administered by the Federal and United States Supreme Courts, and under Rule 60 of the rules of practice of this Court to decide. That said Rule 60 reads as follows:

"The defendant in an action at law, tried either with or without a jury, may either at the close of the plaintiff's case or at the close of the case on both sides, move for a nonsuit. The procedure on such motion shall be as follows: [152]

"The defendant or his counsel, shall state orally in open court that he moves for a nonsuit on certain grounds, which shall be stated specifically. Such a motion shall be deemed and treated as assuming for the purposes of the motion (but for such purposes only) the truth of whatever the evidence tends to prove, to wit: Whatever a jury might properly infer from it. If, upon the facts so assumed to be true as aforesaid, the Court shall be of opinion that the plaintiff has no case, the motion shall be granted and the action dismissed. The party against whom the decision on the motion is rendered may then and there take a general exception, and may have the same, together with such of the proceedings in the case as are material, embodied in a bill of exceptions. If evidence shall be introduced by either party after the decision on the. motion has been made, the (281) same shall operate as a superseding of the motion; but such motion may be renewed at the close of all the evidence."

Dated this 13th day of August, 1926—Robert Brennan and M. W. Reed, Attorneys for Defendant.

(Testimony of R. M. Noble.)

Mr. REED.—I desire to file this motion, your Honor.

The COURT.—Motion for an instructed verdict is denied. You may proceed with the defense.

Mr. BRENNAN.—We desire an exception.

The COURT.—Yes, enter an exception.

Mr. BRENNAN.—Will your Honor also grant us the right to renew our motion at the close of the evidence?

The COURT.—Yes, certainly. (281½) [153]

DEFENDANT'S CASE.

TESTIMONY OF R. M. NOBLE, FOR DEFENDANT.

R. M. NOBLE, a witness on behalf of the defendant, testified as follows, on

Direct Examination. (282)

My name is R. M. Noble. I reside at Clovis, New Mexico. I am in the engineering department of the Santa Fe—a civil engineer. I have been in the employ of the Santa Fe in that capacity since 1914. I am acquainted with the defendant's line of railroad in and about Mountainair, New Mexico. I am familiar with the crossing which is located about a mile and three quarters west of Mountainair. The map or blue-print you now hand me—I made that document. (283) This map shows, in a general way, the alignment of the track and the location of this road crossing. It shows the profile grade of the road, shows profile grade of the tracks,—plan. It shows the right of way and track for a distance

(Testimony of R. M. Noble.)
of about 1,500 feet west and about 2,000 feet east
of the crossing in question. It shows the particular crossing involved in this suit at which the plaintiff was injured. The print shows a profile of the
crossing, that is, the highway crossing—this one
right here (indicating). It also shows a profile of
the railroad track at the crossing and for some distance each side of the crossing. The profile does
not show the grade of the crossing (284) on the
west side of the railroad, that is, the highway, itself; the side from which the car approached is
11½ per cent grade.

Q. In what distance does that grade on the highway crossing extend out from the center-line of the
track?

A. That is the steepest place where the 11½ per
cent [154] grade is, runs from around about 20
feet from the center of the track, to 100 feet out to
the property line—making 80 feet in all. The profile shows the highway to be level for a distance of
20 feet out from the center of the track. From a
point 20 feet from the center line of the track, for
a distance of 80 feet along the highway south of the
track, the grade ·is 11.5 per cent, and then it is
lighter from there on where it goes on the old
trail. I was last at the crossing involved in this
suit (285) about a week before we came over here
on this— Yes, sir, we took field-notes. The print
was made the 21st of June, 1923, field-notes taken,
I believe, about the 14th; I am not sure, about three
days after the accident occurred; I don't remember   .

(Testimony of R. M. Noble.)
the exact date. I can get that for you, however;
have the book out there with the date in it. This
print correctly delineates all of the objects about
the crossing shown thereon.

Mr. REED.—I offer this in evidence, this print,
as (286) Defendant's Exhibit "A."

The COURT.—It may be admitted, Defendant's
Exhibit "A."

WITNESS.—(Continuing.) I can describe the
contour of the highway north of the rails of the
track at the crossing involved in this suit, and for
a hundred feet along the highway.

It is a very light grade on the north side. It
is about, probably two per cent. For a distance of
about 35 or 40 feet (287).

Cross-examination.

Q. May I ask you, please, to step down here to
the edge of the table; I wish to refer to this map introduced in evidence. [155]

(Witness does as directed.)

WITNESS.—The line here I am pointing to, says
"Main Road" (indicating); that is the main road
from Belen, in the direction of Belen, to the crossing, and cuts across the crossing at the place of the
accident. (288) It is the main road along which
I understand the car came. It is the best defined
road on that side of the track; after the crossing
there is two or three trails there. And the outer
line along here is the fence or railroad right of way.

Q. That appears on each side, the inside line

(Testimony of R. M. Noble.)

is intended to be the railroad track. Mr. Noble, on
your map appears here an inscription at the point
where I show you, "Hill 30 feet higher than tracks
—growth of cedar trees."   Now, isn't that the same
hill which is shown in the picture, Plaintiff's Exhibit No. 4, that appears as you look down the track
eastwardly across the crossing, with the cedars on
top of it?

A. That is not.   You cannot see that particular
part of the hill at all.   That hill is the one which
projects out to the track and is the one indicated
upon my map.   (289)

Q. Now, how far is it, Mr. Noble, from the crossing, the place where this road crosses the railroad
track, to where the railroad first cuts into the projection of that hill?

A. The first cut east of there, it is about approximately 900 feet, and then it goes on to a little fill
and goes into another cut.   That is the hill which
is indicated on your profile.   I think I testified before, if I recall right, that it was 850 feet from the
crossing to the point where the track first cut into
that hill.   Well, it might be.   I don't remember, but
it might be 850 or 900 feet.   I am not giving it
with absolute accuracy, but approximately.   [156]

Q. The same point of the hill is the one—the
point that is indicated here upon your map—your
profile, as the hill rising here (indicating), level
with the track, being (290)   the straight line that
passes through it,—that is the same point of the
hill that is shown in the picture, Defendant's Ex-

(Testimony of R. M. Noble.)

hibit "A," is it not? Just look at that and notice where the engine is, please. I am referring to the hill upon the south side of the track. Just notice the position of that engine; that is the same hill, the point beyond which that engine on the train shows, —is it not, and that is the one indicated upon your map at the first elevation above the tracks?

A. That is the hill, yes, sir, I believe. In passing up here, a car would climb in starting from the corner—it would climb in going up the level of the tracks about 9.2 feet—from the corner to the level of the track. I don't know whether I testified before, a little over ten feet. But this is indicated here as 11½ per cent grade, and is approximately 20 feet out there before this grade starts down. Well, at a point nearly 20 feet from the track you get a 11½ per cent grade to the property line, you see, in a distance of about, oh, maybe 80 feet or a little (291) farther, and it rises about something over 9 feet, a little over 9 feet from the right of way line to the level of the track.

Q. Mr. Noble, in getting the data for this map, did you measure accurately the elevation—the elevation at the fence corner, comparing with the elevation at the rail?

A. We have that, yes, sir. I can give that elevation to you exactly to the tenth of a foot, if you want the notes on it, in the other room. My present recollection of [157] that elevation is—I believe that grade figures 9.2 feet—11½ per cent grade, and this is between nine and ten feet. That is, from the

(Testimony of R. M. Noble.)

corner of the railroad right of way outside of the fence, from that point up to the nearest—highest rail—is about ten feet,—possibly a little less, between nine and ten. The roadbed extends on a— comparatively the same level as the rail, at about 20 feet from the center of the track. (292)

Q. Now, crossing there, the fence is, I think, here, approximately at right angle to the crossing, is it not?

A. No, that is not a right-angle crossing.

Q. I say, the crossing with the fences—the road cuts in a little, but is not the crossing itself, and fences there, substantially a right-angle crossing?

A. Practically right angle, the fences are, but the road does not cross through the center line of the fences.

Q. The road goes to the corner and then turns up. Do you know how much farther east the point is where the road crosses the railroad track, than the point where the corner fence post is?

A. I can scale that pretty close. (Witness refers to map.) I don't know exactly. From the southwest fence corner—that is about 62 feet. The point where the road crosses is about 62 feet east of the straight line of that corner fence post. For a little distance after passing the fence post, the road runs much more near in the direction of the railroad than when it strikes the grade of the hill. In other words, the road runs a little east of the fence to about one-third of the way up, as shown in your map, and then on   [158]   the last end of that grade

(Testimony of R. M. Noble.)

is more (293) nearly straight up—it is more nearly right angle, the nearer you get to the track the nearer right angles. I believe the grade is about uniform grade from some place near the property line to that point, 20 feet from the rail. After the road gets around the fence, it is a lighter grade coming west from Belen, but right here, the point of track is fairly uniform grade, the profile indicates there. I have marked upon this map at the point here, "Car was taken from off pilot here." That is about 1,680 feet west of the crossing. And that refers to the place where the car, the Spencer car, was taken off the pilot when the engine stopped. (294) That was my understanding of the place where the engine stopped.

Q. Now, counsel, in his question, asked you about the grade of the road north of the track, as it extends down toward Mountainair—and you,—he limited you to a hundred feet. Did you take the grade of that track as it dropped down into the draw, of the road, I mean?

A. I have it therefor 140, 150 feet, only, from the track—that would put you around this corner along there, see? Along about there (indicating on map). From the center line of the track to a point outside of the fence on the right of way, I didn't take the grade as it went farther than that and dropped down into the draw.

Q. You know it does a little—you did drop down into the same draw on the north side of the track as the south side? (295)

(Testimony of R. M. Noble.)

A. I don't believe there is much drop in there on the north side of the track—this draw runs in this direction (indicating on map).   [159]

The draw runs in the direction from which this car came.   That draw which takes the general direction that the road approaches, and which the car approached, continues right on through under the tracks and right up beyond in the direction of Mountainair.   On my profile, just east of the road crossing, is marked "6450.47 Br. 857A. Cone Arch." That represents a concrete arch.   It looks to be the same arch which is shown in Plaintiff's Exhibit No. 6—I can't see the bridge—the arch that side looks that way.   I examined this district enough to know that there is an arch like that there.   The dimensions of that arch are: (296)

Well, this is six feet from this wall to this wall (indicating), and it is four and a half feet from the floor of this arch up to the string line of the curve— that is curved on a 3-foot radius.   The culvert, or, the arch, is six feet wide across the bottom—that is the end view.   The picture being an end view, as you look at the railroad and look at the arch, and the 4½ feet are from the bottom up to the point where it commences to arch.   The top of the arch is 7½ feet from the bottom.   I am able to tell from accurate measurements or from this picture, by comparison, knowing the height of the arch, what the distance is from the top of the arch up to the rail; the floor of the arch is 12.8 below the base of the rail.   (297)   And from the floor of the arch to the

(Testimony of R. M. Noble.)

top of the arch inside measures 7½ feet, leaves a distance of 5.3 from the inside of the arch to the base of the rail. These figures that I am giving here, are taken from actual measurements which I made at the time, on the ground. We made the measurements—these measurements were checked up a week before we came over here—this last time. I know they are [160] accurate. That culvert shown on this picture, Plaintiff's Exhibit No. 9, about where the arrow is, this picture being taken from the highway looking east, that white mark there shown, is the culvert.

Q. How far is that culvert east of the highway? Please, if you will, mark it position on the main part of the map which shows the railroad.

A. Well, it is right there (indicating). This is it right here, you see. (298)

Q. That is sufficient. The witness points to the indication marked "BR. 857–A," upon his map, crossing the line indicating the railroad. The distance from the highway is about 350 feet.

A. It is a mile and three-quarters from this railroad crossing to the Mountainair station, which is a continuation of defendant's road to the east. I don't know just how far west of Mountainair station in the direction of this bridge or crossing, a switch extends, adjoining the main track—I have not got anything here showing the Mountainair station grounds. I am unable to state the distance exactly. (299) I imagine it is about 800 feet from the depot; no, I don't believe it is that far—I mean

(Testimony of R. M. Noble.)

the farthest west switch. Something like 800 feet, if I remember; I don't know just how far that is. From that point where that switch joins, wherever it is, there is but one track that extends from there to this highway.

Q. Mr. Noble, how·far east of the highway, in the direction of Mountainair, does that curve commence, which extends west, if you are able to tell me the accurate measurement? [161]

A. Yes, I can tell—that is the curve west of the road crossing, that you are talking about now, is it not?

Q. That curve which continues to curve until it passes into the cut west of the road—where does it commence—it (300) commences east of the high-way?

A. Yes, that is the P. C. of the curve right here— that dotted line. P. C. means point of curve. The commencement of that curve is a trifle less than 300 feet, 290 some odd feet, I believe, east of the east cattle-guard—toward Mountainair. And as it passes the highway, it is the same curve which is shown in this picture, Plaintiff's Exhibit No. 4. That curve commences about 300 feet east—of the east of the white cattle-guard fences that are shown in this picture. (301)

Redirect Examination.

I can state whether or not the grade of the high-way, as it approaches the crossing from the south, that is, the portion of the highway extending from

(Testimony of R. M. Noble.)

the south right of way fence to the crossing, is on the same, was built on the surface of the earth. There was no grading of the highway at all any more than just to make a smooth running surface for the road. Of course, that is on the side of the hill, possibly it was cut on one side—filled a little on the other. Generally speaking the roadway was not built so there was a fill, not any more than I said. It was not a fill proposition: You see, it is the side of a hill—you borrow some and fill some, having the cut on one side. I think the highway people call it a balanced section, maybe. (302) A balanced section is where they borrow enough dirt to make all the fill necessary, partly in cut, partly in fill. So the contour of the roadway [162] is coincident with the contour of the land.

Q. Is there any ballast at all along under the track at the crossing?   A. Yes.

Q. How much?

A. Well, ten inches is standard ballast section.

Recross-examination.

Taking this picture, Plaintiff's Exhibit No. 2: Now, what I mean is that the road came along this draw that we have spoken of from the westermost portion of it shown on my map in the bottom of the draw, on toward the bottom of the draw until it got to this point where it had reached the side of the hill into which the railroad cuts at that point. (303) And then it turned and climbs around the corner of the hill at the original grade of the hill.

(Testimony of R. M. Noble.)

You see, on the—you have been over there; you remember this side hill; this point of the railroad is on the north side of the hill, going down this canyon—The railroad is located down Abo Canyon and this particular point is on the north side of the canyon. Now, on this hill there will be little projections or noses sticking out from the side of it, and that is the actual contour at this point—indicated right here on the track profile—see? See this little knob there? The road crosses on that knob—this is the actual ground line under the center line of the track, see? This is from the original profile of the railroad—this little knob I am talking about now—that hill, the little knob, runs right down as indicated here, across this point, falls off this way and the road came right around to get an easy (304) grade as possible around the edge of the hill. [163] The road comes up the draw until it gets to the point on this knob which extends down from the north. And then it climbs up that knob to bring itself up to the level of the railroad track, and that is the natural, the natural knob, that is a sort of cut down there, so that it climbs that knob as it is now. And then goes over on the knob on the other side of the track, and then more or less paralleling the track comes down to the level of the draw that is shown in that profile—

Redirect Examination.

I know for what purpose the culvert was installed. It was installed for the purpose of taking care of

(Testimony of R. M. Noble.)

the surface rain water, protect the road against washout. (305) It takes care of all of the water that falls within the drainage area north of the track. That is the natural run-off place for that water.

Recross-examination.

The actual drainage area north of the track, which is tributary to that, is .22 of a square mile, the actual run-off. I measured it with a transit. That would be somewhere about 120 acres. That is all the drainage that is tributary to that on the north, from which water would come. (306) I am familiar with the country over there. They don't have very heavy rainfall there.

Q. In fact, there is not a thing on the ground there on either side of the culvert that shows that water has ever flowed there, except an occasional bit of weeds that float along—there is no stream bed north and south?

A. There is a well-defined channel there. I do not [164] mean a little depression. I mean a well-defined channel. I do not mean there is a creek-bed, or anything like that. I mean there is a well-defined waterway. The picture, Plaintiff's Exhibit No. 9, which you show me, shows the land reaching right up to that. I cannot show you on this picture where the well-defined waterway is, because this is below the—the draw is getting wider all the time— now, as you go up the waterway, up the watercourse towards the hill, gets steeper all the time near the top of the hill. I mean there is a well-defined water-

(Testimony of R. M. Noble.)

way to the north of this somewhere; beyond the railroad. (307) You can see right here on the profile where it crosses the railroad. This picture shows that grade condition. The waterway is right there where the bridge is. (Indicating.) That is where the waterway is (indicating). I mean that this bridge, the arch, is the waterway. There is a well-defined draw that you have been talking about, showing on the land after you pass through the bridge to the south. Of course, when it rains water runs down that draw. And that is the draw up which this road comes. On the point where the crossing on the railroad is, from the point of that hill (308) straight across to the other hill is about maybe half to three-quarters of a mile. Well, it might not be that far.

Q. The railroad cuts it diagonally from this crossing to where it enters the cut, does it not, and it is a little narrower as you draw a line straight across from hill to hill? I mean the bottom of the hill, the draw where you call the draw, the foot of the hill, between those two.

A. Oh, well, that is a hard question to answer, because it depends on what part of the slope, where you want to call the bottom of the hill; the actual draw is about maybe [165] 150, about, to 500 feet wide. And that is comparatively uniform there. And wider at places, as that draw runs up in the direction of Mountainair. (309)

TESTIMONY OF JOHN BLOCK, FOR DEFENDANT.

JOHN BLOCK, a witness on behalf of the defendant, testified as follows, on

Direct Examination.

My full name is John Block. I reside at Estancia, New Mexico. Torrance County. I have lived in that county, 22 years, I believe. I have held a public office, sheriff's office, from '21, '22, '23 and '24. I know where Mountainair is. I have been there often. I lived in that vicinity when the Santa Fe constructed its railroad line through Mountainair. Lived at Estancia. That is 26 miles from Mountainair. (310) I am familiar with the road which crosses the railroad track about one and three quarters miles west of Mountainair. I knew that road since 1904. I have had occasion to travel over that road prior to and since 1904.

Q. How frequently?

A. Not but a short time ago I have been over it —probably it was three weeks ago I was across that road. As to how frequently, beginning with 1904, I have had occasion to travel over that highway, well, along in 1904 and 1905, I was freighting a good deal down through that country; crossed it several times, maybe twice or three times a month, or maybe twice a week. (311) That highway was pretty publicly traveled in June, 1923. [166]

Q. Do you know whether the highway, before the

(Testimony of John Block.)

line of railroad was constructed across it, was in the same location, in the same position, as it is now?

A. It was practically in the same place, yes, sir. It may be a little different variation in the railroad crossing right at present, as the road did before the railroad was established, it may be just a little bit of difference of a curve, but *it practically* in the same place. I would say the exact point of crossing of the highway and the railroad, is in exactly the same place as it was before the railroad was constructed. The highway (312) follows the natural grade of the earth, contour of the land, as it approaches the crossing—the railroad crossing, from the south—no grade—only the natural lay of the ground as it was before there was any railroads constructed there.

Cross-examination.

That road before the railroad was constructed, followed so near the same course at this crossing that it does now, that it would be hard to detect any difference. As it comes up from the direction of Scholle along the line that Mr. Spencer traveled the day he was hurt, it comes up from that draw from the south direction going east.

Q. And from this highway crossing it runs down along the bottom of that draw as it extends west, does it not?

A. After it goes off the slope there it takes into the bottom. (313)

Q. Where it turns just after climbing the rail-

(Testimony of John Block.)

road, it takes substantially the bottom of the draw, does it not, going west? [167]

A. Well, when you come down from the east there and cross the railroad some little ways, it goes down and strikes the draw practically.

Q. Please keep to my question. After it turns at the railroad fence-corner on the south side of the railroad, and extends down, it goes substantially along the bottom of the draw, for some distance, does it not?

A. Well, it is some little distance, if you call that fence before it hits the bottom of the draw. The picture you show me, Plaintiff's Exhibit No. 3, shows that angle or fence corner on the south side of the railroad at the crossing, the first post being a post in the fence that runs up to the railroad track at the crossing, and the line of posts being the railroad right of way in that corner. I recognized that as (314) a view looking west from the crossing. I see the road in question as it runs west, shown in the picture, commencing near the telephone pole and running west. I believe it is the road. It looks to me like it is a little low down. The road is a little lower down than I thought it ought to be; it comes up. But, as a matter of fact, that picture shows the road and substantially the bottom of that draw leading west from the crossing. I have no doubt that the picture is correct. That draw continues right on up north of the railroad track for some distance—that low place. But the draw is farther east than this crossing is.

(Testimony of John Block.)

Q. Yes, but the draw I am talking about, through which (315) this draw comes, continues right on north of the railroad track, and this bank on the railroad track is built across that draw, is it not? [168]

A. Where the railroad and this crossing is, there is no bank built. There is an embankment a little east from the railroad crossing. I remember that culvert. I would think that embankment I am talking about, and that culvert, is just about in the low place of the draw. To the north of the track that draw does not run exactly up the track, but heads away from it. The same general position as shown in these pictures that the draw occupies below the crossing; I would call it the same general direction—(316) in a general direction it would be. Before this railroad was constructed, before there was any embankment there, this road came out of the draw along up on the hillside on a slope to this crossing, where the crossing is, at a point below this draw, right on out around where this culvert is. I understand what you are getting at —this portion of the road that is shown in Plaintiff's Exhibit No. 3, that you have called my attention to, leading from that fence corner west.

Q. Now, is that just about the same place it was when it was first known by you?

A. That is on the south of the track?

Q. Yes, sir, and where it goes down toward Belen, where I am marking with my pen. (Indicating.) Isn't that about the same place the road al-

'(Testimony of John Block.)

ways was before the railroad was constructed?

A. I don't believe it ever went into the bottom of the draw, it went some distance across the railroad before it took to the draw.   (317)

Q. Well, you won't say that the position of the road, as shown in that picture, Exhibit 3, after it passes the railroad right of way, and extends west, is not in substantially [169]   the same place that it was, will you, Mr. Block?

A. Well, I could not locate it exactly by that picture, I could see it on the natural ground and explain it as I seen it on the natural ground, as often as I have used the road. The road came along up the draw down to the edge along the side and crossed up (indicating). It climbed the hill, before the railroad was there, a little south and west from where the crossing is now.

Q. Oh, it crossed west of where the crossing is? That puts you into the cut, doesn't it?   (318)

A. No, I said it climbed the hill a little west where the crossing is, south and west, is what I said.

Q. My question was, that would put you right into the cut which is shown in Plaintiff's Exhibit No. 4, would it not, that picture (indicating) west of that crossing?

A. Well, I didn't say that it was on this railroad track; I said when it crossed the railroad it went a little southwesterly direction from where it is. I said, when it crossed this railroad it ran a little southwesterly direction up this rise to the bottom of the draw.

(Testimony of John Block.)

Q. Was there any bend there in the road where the bend now is, to cross the tracks?

A. Well, it might have drawn a little more straighter than it does right at present.

Q. That is not quite the answer to my question. Was there any bend or curve before that railroad was built, at the place where that road now curves and crosses the track?

A. I couldn't say for sure; I never paid any attention to it out there, just exactly how the curves went. [170]

Q. At least, can we agree on this? The road came up (319) the draw west of this point, and then after it got over the railroad and proceeded toward Mountainair, it again ran in the lower grounds of the draw, did it not?

A. Well, it went up on the side of the hill, climbed out up where it got on more level ground.

Q. Did it not, north of the railroad, again take to substantially the lower part of the draw, as that draw extended, north of the railroad?

A. Well, I couldn't say that it took to the draw at all—kept on—when it crossed there it was on more level, a little backbone of a place, what I call it. That draw extends in a comparatively even grade right on up past the railroad in the direction of Mountainair.

Q. And then what you want this jury to understand, do you not, Mr. Block, that it came along up the draw to somewhere about where this crossing is, and then it turned and climbed up on that hill,

(Testimony of John Block.)

when there was no railroad there, and then after climbing the hill came down again on the other side, now, that is what you want to testify to, is it?

A. No, sir. (320) Those little draws often are rocky and you cannot get along up them; they have to get out on the side.

Q. Just answer the question and stick to that, please. (Whereupon the reporter read the last question.)

A. Well, I don't quite catch you.

Q. That this road, after coming up in the bottom of the draw to about near this present crossing, turned and climbed up that hill that the railroad penetrates, and then having climbed the hill continued in about the same line that the road takes now, on the north of the crossing; is that [171] what you mean to testify to?

A. I couldn't say that this road took back up the draw at all; it is on a side of the hill, like; it does not go to the bottom and take back up the bottom of the canyon.

Q. That picture shows it in the bottom of the draw, don't it, Mr. Block, that I showed you?

A. Well, I couldn't just like to say exactly by the picture. North of the railroad track, say a couple of hundred feet from the crossing, then on toward Mountainair, that road occupies now something near the same position it did before the railroad track was built. It does not run up on any hill, except to go to the top of the draw and then on comparatively (321) level to Mountain-

(Testimony of John Block.)

air. I would not say that it fell back in the bottom of the draw at all; it climbs along the side of a kind of rolling little backbone that is along through there. I have not a bit of interest in this at all. I am here as a totally independent witness. While I was sheriff, I did not hold a pass issued by this railroad company. (322)

Redirect Examination.

You show me Plaintiff's Exhibit No. 2, which shows the crossing, shows the angle at which the road proceeds up to and over the railroad. I have examined the ground south of the railroad to determine where the highway ran before the railroad was constructed. On that examination, I discovered that the road crossed and ran practically in the same position that it did before the railroad was constructed. (323) As that highway from the west proceeded [172] along in an easterly direction toward the crossing, it climbed gradually up the side of the slope of the hill.

Q. And for some little distance west of the crossing it aproached the point where it now crosses the railroad at a somewhat more oblique angle?

Mr. WOOD.—I don't think that is proper cross-examination.

Mr. REED.—Well, Mr. Block has tried to explain where the road existed, where it was located before the railroad was built, and I would like to bring it out clearly if I can.

Q. Will you state, Mr. Block, whether or not—on which side of this corner post of the south right

(Testimony of John Block.)

of way fence, the highway ran before the railroad was constructed?

A. Yes, it was a little east of it. (324) Then it extended kind of westward, or a little southwestward, say, along the hillside down the canyon.

Q. Taking the other direction toward Mountainair, in which direction did it extend, and give the angle, as near as you can, as the road ran easterly?

A. Well, it would be—I would call it a little bit northeast angle along the side of a little slope, some little backbones through there that went over it. It approached the point where it now crosses the railroad at a more acute angle than the present highway crosses. (325)

TESTIMONY OF BYRON M. KARSTETTER, FOR DEFENDANT.

BYRON M. KARSTETTER, a witness on behalf of the defendant, testified as follows, on

Direct Examination.

My full name is Byron M. Karstetter. I live in about the central part of New Mexico. I guess the name of the [173] town in which I live is Belen—Clovis, either place. (326) I have lived at that place about ten years and a half. My occupation is that of a locomotive fireman. I am working as locomotive fireman for The Atchison, Topeka and Santa Fe Railroad Company, the defendant in this action. I have been a locomotive fireman for that railroad company since Septem-

(Testimony of Byron M. Karstetter.)

ber, 1915. I have had experience also as a locomotive engineer. In the month of June, 1923, I
was employed by The Atchison, Topeka and Santa
Fe Railway Company as locomotive fireman. At
that time the fireman's run on that run was between Belen and Vaughn, New Mexico. In the
course of my employment, and on   (327)   or about
the 11th day of June, 1923, I had occasion to run
through Mountainair, New Mexico. I was working as fireman upon train No. 21. Train No. 21
was a passenger train. It was a local passenger,
considered a local passenger train. On that day it
was a stub train, what they call a stub run, from
Kansas City to Los Angeles. From Chicago to
Kansas City and Kansas City to Los Angeles.
That is the train upon which I was working on the
day of the accident. I recall when I left Mountainair, going westerly, on the day of this accident. (328) It was at 3:52. I remember being on
the engine that day. The locomotive engine that
was drawing the cars in that train was equipped with
an engine-bell. The usual and ordinary engine-bell
such as are used on locomotives of that type. The
number or type of locomotive was No. 1316, an Atlantic, Atlantic type engine. That locomotive was
equipped with a steam-whistle. The engine-bell
was operated or rung by an automatic ringer, air
ringer. That bell is started by a valve right by
the engineer's seat, where he turns the air on the
bell. When this air is turned on, it [174] will
ring the bell until it is shut off. And when that

(Testimony of Byron M. Karstetter.)

bell is rung in that manner, it is not at all necessary to use the rope or hand to ring the bell. (329) On the day of this accident, and as the train left Mountainair, New Mexico, in the course of its journey in a westerly direction, the bell upon that locomotive was ringing. The bell started to ring on leaving the station at Mountainair. It continued to ring until after the train was stopped after the accident. I mean by "after the train was stopped," at a point west of the crossing here in controversy. Between Mountainair, New Mexico, and the crossing at which this accident occurred, the steam whistle on the locomotive was sounded. It was sounded at the whistling-post just east of the crossing. I should judge the whistling-post is about 1,500 feet east of the crossing. About 1,500 feet east of the crossing, the whistle (330) upon the locomotive was sounded by the engineer. It sounded two long blasts and two short. That is what is commonly known as a crossing signal. And the bell at that time was ringing upon the locomotive. I should judge that the rate of speed, approximately, at which the train was being operated after the whistle was sounded, between that point and the crossing, was around 30 miles an hour.

Cross-examination.

I am still in the employ of the Santa Fe. I am now fireman and engineer, both. Before the time of the accident, I had had experience as an (331) engineer. I live at Belen or Clovis, either one;

(Testimony of Byron M. Karstetter.)

that is the division; either place. I run between those places. Belen is the junction station where the Rio Grande Division of the Santa Fe crosses [175] this cut-off division, so-called; it goes through there. The main line of the Santa Fe as it runs through Arizona branches off shortly before getting to Albuquerque a few miles farther, crosses the Rio Grande Division at Belen. And then that cut-off division continues on through Belen—from Belen through Mountainair, Vaughn over to Clovis, where I say I stop. And thus on through Texas, joining the main line of the road in Kansas again. (332) The fireman's division at that time was between Vaughn and Belen—the fireman's terminal was between Vaughn and Belen at that time. I got on the train at Vaughn that day. Vaughn is about 64 miles east of Mountainair. Mountainair is 41 miles from Belen, I believe. I say that my train was a stub train that day. By that I mean it was not the regular equipment. It was called "stub 21."

Q. That was due to the fact that the regular transcontinental train had been held up for floods or some similar cause, was it not?

A. There was a bridge washed out. This stub train was taking its place through New Mexico. (333) I do not know where it started from. It started east of Clovis, New Mexico, I know that. The Canadian River bridge was washed out.

Q. The Canadian River, or the bridge, is in Texas, is it not, or do you know?

(Testimony of Byron M. Karstetter.)

A. I don't know whether it is across the line between Texas and Oklahoma or not; I have not been up there. I don't know whereabouts they would make up a train west of that, to pass through as a stub, covering this route. Clovis, that I speak of, is in New Mexico, and a division point or a point where they have railroad shops and equipment. I know the train did not start from Clovis, (334) because the train came from east of Clovis. As far as I know it came from east [176] of Clovis. I don't know of my own knowledge where it came from. It did come from Clovis, I know that. There is not also another passenger division that runs down to Roswell from Clovis, all in New Mexico. There is a division of the railroad, passenger and freight trains—the railroad goes down there, sure. The train that day consisted of five passenger coaches and a baggage coach. (335) Five cars only. We had the baggage coach and smoker and chair-car and two Pullmans. The engineer of that train was W. E. Spade. My run as fireman at that time was from Vaughn to Belen. Some other fireman took the other division, the other distance. Some other fireman stopped at Vaughn. Vaughn is not a passenger division; it was for the firemen at that time. I couldn't say exactly what time we came into Mountainair that day. (336) I didn't notice. I couldn't say how long we stayed at Mountainair. I don't remember. We stayed there just for the loading and unloading of the regular passengers and baggage, if I remember. We were not delayed for

(Testimony of Byron M. Karstetter.)

any cause whatever at Mountainair that I remember
of. I remember we were not. There was a train
in there ahead of us from the west when we got
there. There was a work-train on the passenger
track when we got there. I remember it distinctly.
(337) That train was in there when we arrived at
the station. I don't remember whether we came
into the station on time. I don't remember whether
we were fifteen minutes late reaching the station.
We were not an hour late, I know that. I don't
remember the exact minutes we stayed at the station.
Something near the usual time—probably two or
three minutes. My recollection is we stayed at the
station two or three minutes only. The reason I
happen to remember right now the exact minute at
which we left the station, [177] as I have given
it to you 3.52, I said, is because it gives an hour and
ten minutes to go to (338) Belen—an hour and
eight minutes to go to Belen. The reason I happen
to remember that it was that time, 3.52, when we left
the station is on account of the running time from
Mountainair to Belen. I don't remember the time,
on the time-card at that time, three years ago, the
train was due to leave Mountainair. I am able to
remember that we left there at 3.52 on account of
our running time from Mountainair to Belen, which
was an hour and ten minutes, on that day—an hour
and eight minutes to go. I remember it by looking
at the time. I remember now that I looked at the
time when we left Mountainair. I still remember
that it was 3.52 when I looked. (339) That was

(Testimony of Byron M. Karstetter.)
the time. My watch was right that day according
to the rest of them. I did not compare my time
with the dispatcher's time. I could not say whether
my watch was correct, or whether it was not. I
did not compare it with the dispatcher's time.

Q. Are not you railroad men obliged always to
compare watches and know you are correct? Is not
that an invariable rule?

A. When you are running the engine you are supposed to, but I was not running the engine on that
day.

Q. Then you don't know whether your watch was
correct or not, do you?

A. It was 3.52, the time we left Mountainair. I
don't know whether my watch was correct or not.
I did not compare with the dispatcher's time. I
don't know whether at any other time I found out
if it had been correct. I had no occasion to compare
it. (340) It was a standard watch. I don't know
whether it was correct or not. I did not compare
[178] it with the dispatcher's time. I could not
say whether my watch was correct at all. We got
to Belen at 5.20. We are given an hour and ten
minutes to make that run. (341) As to how many
times that engine whistled from the time it left
Mountainair until it stopped, after striking these
people—I remember it whistling four times. Those
whistles were given at the whistling board east of
the crossing. All four of them were given right by
the whistling board. (342) There were four
whistles, two long and two short blasts, I remember

(Testimony of Byron M. Karstetter.)

those distinctly. I don't remember whether that engine whistled another time from the time it left Mountainair until it stopped. The whistling post is the only place I remember of it whistling.

Q. How do you know they were delivered at that whistling post?

A. There was two short and two longs, sounded at the whistling post for the crossing. Just at that minute, that instant, I was firing the engine. When I was firing the engine, when he whistled, I looked out for the crossing. I looked out and saw the whistling post (343) just as he blew the whistle. I do take particular notice of every whistling post that he passes, as to whether he blows the whistle. I can testify as to other whistling posts that he whistled at that day. He whistled at all of them east of Belen.

Q. You were watching a little after the accident, I suppose? A. I watched before.

Q. You would undertake to say that you could tell this jury every whistling post at which the engineer blew the whistle on the train the day before? [179]

A. I was not on the day before.

Q. Well, the last day before that that you were on, would you undertake to tell them?

A. If he ever passed a crossing that I was on with him without whistling, I don't know.

Q. If I would ask you if he blew the whistle at that (344) same post on the preceding trip, you would say yes, would you not?

(Testimony of Byron M. Karstetter.)

A. If he ever passed one without whistling, I could not recall it.

Q. Then you would say "Yes" if I asked about the trip before?

A. I couldn't say. If he ever passed one that I was on with him, I could not recall it. I generally pay attention to the whistles the engineer gives. If that engine in going from Mountainair up to that point had whistled at three different places that day, or four, I would know every one that he passed on that day. I mean going to Mountainair, where he stopped after hitting these people, if that engine blew two or three or four different places, I would know it. It did not blow at two or three different places—it whistled four times, for that crossing two long and two short. I will tell the jury this, that (345) that engine whistled going from Mountainair to where it stopped, at only one place that day. That is all I can recall. I cannot recall it whistling but one time. These people when I first saw them, these Spencers, that were hit, were sitting on the pilot of the locomotive when I first saw them. As the train went from Mountainair up to that crossing, I was firing the engine. Not all the time. As we leave Mountainair, the [180] train climbs on a grade for some distance. I don't know the exact distance. Somewhere about a quarter of a mile, —half a mile. After we break and start downgrade, the firing job is some easier. (346) And I would not shovel coal all the way to that crossing. Part of the time I was sitting down. After it went

(Testimony of Byron M. Karstetter.)

through the cut, or started to tip over the grade, I got down and put in a fire and we were at the crossing. From the time we left Mountainair until we stopped at that crossing, I was elsewhere except back of the fire-box. That is, I was sitting on the fireman's seat. I got up to sit at the fireman's seat just shortly after leaving Mountainair, and then until he started downgrade. Leaving the station at Mountainair on the upgrade, I was putting in a fire, and then after he started downgrade. Started firing when he pulled out of Mountainair. Then I started firing on the downgrade. (347) I couldn't say how long it was between those two fires. It was not a couple of minutes. I fire as quickly as that at times. At times I do. I don't know whether I did that day. I don't know how long I sat up on my seat; never paid any attention. I never pay attention to time when I am firing. I did not sit up there so very long. I was not up there five minutes. I couldn't say whether I was up there one minute. (348)

Q. What did you go up for and immediately come back, can you say that?

A. Yes, to get a little rest. It was not necessary to be firing all the time. I don't know how long I sat there. I was firing when we were leaving the station. Then when I got through firing I went up and sat down on this seat. Then, after sitting on the seat, I came back again and fired [181] some more. And the time it took the train to go from Mountainair, where I started, until it struck these

(Testimony of Byron M. Karstetter.)

people at the crossing was, I should judge, about
five minutes. (349) I couldn't say exactly. I did
all this moving around in five minutes.

Q. You stated that you were firing as you started
out of Mountainair; was that the time when you
were then firing when he blew the whistle and you
looked out that first time?

A. I was firing leaving Mountainair; after we
tipped over the grade, why, then I was firing again.
It was the second time I was firing when he blew
these four whistles at the whistling post. I could
not say whether that whistling post is about a mile
and a half from Mountainair. I know where it is,
but I don't know how far. (350) The engineer
was sitting in his seat at that whistling post. He
was looking ahead; he was not talking to me. He
was in his seat. From the seat in which the engineer sits at the side of the engine, he is on the
right side going west. On a straight track, from
the engineer's seat, I don't know how far ahead
of the engine he can see the left rail of the train. I
don't know how close to the engine he can see the
rail; you cannot see very far ahead of the engine—
that is, how close (351) I don't know how far he
can see. Somewhere ahead of the engine he can
see that left rail. I don't know about how close to
the engine. I couldn't say whether or not it would
be the length of the engine. I should judge the train
was running about thirty miles an hour. I don't
know how fast it was running; I was not timing it;

(Testimony of Byron M. Karstetter.)

I should judge that was the speed. The speed varies. I say the speed could vary as [182] far as that goes. I should judge it was making about 30 miles an hour. I was noticing it at that particular time, so (352) that I now remember that I was noticing it.

The COURT.—Were you paying any attention to the speed of the train?

The WITNESS.—Nothing no more than it wasn't making no very great speed.

Q. You still have not quite answered the question —were you paying attention—any attention to the speed the train was making at the time?

A. Nothing more than it was not making no fast speed, no great speed at all. I couldn't say whether that bell ever got out of order, or not. (353)

Redirect Examination.

What I mean is that at that time there was no unusual speed about the train that attracted my particular attention. And my best judgment is, as I have stated, as to the speed of the train. There is one, an overhead crossing, between Mountainair and the crossing where this accident occurred.

Recross-examination.

We do not whistle at an overhead crossing. (354)

TESTIMONY OF FRED RICE, FOR DE-FENDANT.

FRED RICE, a witness on behalf of the defendant, testified as follows, on

Direct Examination.

My name is Fred Rice. I live at present at Amarillo, Texas. My age is 37. I am employed by the Atchison, [183] Topeka and Santa Fe Railway Company as a brakeman and conductor. At present I am brakeman on a passenger train between Amarillo and Belen, New Mexico. In the month of June, 1923, I was employed by the defendant company on the same run as brakeman. (355) On a passenger train. On the 11th day of June, 1923, I was performing the duties of passenger brakeman upon train No. 21. Train No. 21 originated in Chicago. And destined to Los Angeles. It is one of the regular transcontinental trains. The consist of that train on the 11th day of June, 1923, was, as I remember, about five cars. There was two Pullman cars, two day coaches, and one combination car, baggage and mail. That was about half the usual and regular consist of that train. I know why it was that we had such light equipment that day. Part of the train was being detoured by way of La Junta, on account of a washout on the Canadian (356) River. This washout had occurred at a point on the Canadian River some distance east of Amarillo, I am not familiar with the track up there, the country. But it was in the State of

(Testimony of Fred Rice.)

Texas. And part of the train was detoured around La Junta, Colorado. And part of the train was made up and went over the Belen cut-off. One car of that train came out of New Orleans. One of the standard sleeping cars had passengers in it for California, who boarded the train at New Orleans. That was the train upon which I was working as a brakeman, that had the New Orleans car. So far as that train was concerned, upon which I was working, it was part of the transcontinental train known as No. 21. (357) I took that train on that day at Amarillo. I was the rear brakeman. I was stationed in the rear car on the train in the performance of my duties. That would be one of these [184] standard sleeping-cars.

Q. What were your duties as rear brakeman, in a general way? What duties did you perform?

A. Well, there is a whole lot of duties, but the most important thing was to protect that train from being hit by other trains. In case the train comes to a stop before reaching a station, I go back at certain stations, or between stations, and flag. I know that this train was equipped with brakes, air-brakes. I don't remember the number of the engine that was drawing the train on that day. I know it had a bell on it and a whistle. (358) I came into Mountainair, New Mexico, on that train that day, from Amarillo, Texas. It was about four o'clock that the train arrived at Mountainair, New Mexico. Nearly four o'clock. I worked on the train as rear brakeman after that train left Moun-

(Testimony of Fred Rice.)

tainair. Prior to June 11th, 1923, I had served as a passenger brakeman on trains over that division something like ten years. I would make my run from Amarillo to Belen about every day—on that run, 20 times a month. I know where the highway crossing is about (359) a mile and three quarters west of Mountainair. And during the course of my employment, I have known of that highway crossing. I am acquainted with the location of a whistling post east of that crossing—know it very well. I couldn't say about the exact distance east of that crossing towards Mountainair the whistling post was located, or is located. It must be about one-quarter of a mile, something like that. I remember approximately the time when my train left Mountainair, proceeding in a westerly direction. It was two minutes after four o'clock. When the train left Mountainair, I was on the last car in the train, the rear car. And as the train [185] approached the crossing, the whistle on that locomotive was sounded for the crossing. (360) Between Mountainair and the crossing, I heard the whistle being sounded at the proper place. That would be at or near the whistling post. Something like approximately a quarter of a mile east of the crossing. The nature or character of the blasts of the whistle was two long and two short whistles, the regular crossing whistle. I don't remember whether I heard any whistle sounded thereafter up to the point of the accident, I remember an accident taking place there at that crossing. The train came

(Testimony of Fred Rice.)

to a stop when the accident occurred. When the train came to a stop I proceeded to go to the rear, back in the direction of Mountainair. And went back there to flag the train and protect the rear end; that was my purpose in going back there. (361) The speed of the train was about 35 miles an hour. At the point around the whistling post there.

Cross-examination.

I put down the time leaving Mountainair on my reports, and I don't remember anything else between Mountainair and the point where the accident happened until this accident took place—we stopped— except to look at my watch. When we stopped, I put down the time on my book; immediately, shortly after, the train started from Mountainair, before the accident. We were almost on time that day, leaving. I am sure about that. (362) I know we were not very late. By "very late," I mean that we were not a half an hour late. I couldn't just say the number of minutes. I don't remember exactly now what our leaving time at Mountainair was. I [186] always put down in that book, if you call it a book —it is a report—the time of arriving and time of leaving every station. I keep a record—a report. I don't know exactly how long we stayed at Mountainair Station. I don't remember whether we spent more than the usual time there. I put down when we arrive at a station on that report. I haven't that report. It is on file somewhere. I don't suppose I have seen that 4.02 in three years,

(Testimony of Fred Rice.)

to know that that was the time. (363) I don't remember of seeing it. I have carried on my mind three years what I put down on that report "4.02." But I have not carried in my mind at all the time when we reached Mountainair. That is on the report all right—that is on file. When the train stopped at the accident I again wrote that figure down on my report. I wrote the arriving time at that stop. And that is the same report that I turned in somewhere. I have not seen that since —not that I remember. I wrote down on that 4.07. I never was on the stand before in my life. I was on the stand here a year ago—this case is the only one. I have never gone over or been shown or asked to see my reports that I made that day, in order to refresh my recollection to testify. (364) I didn't need to on that one. As to why I was so particular to remember the 4.02 leaving and the 4.07 that I put down—what caused me to remember that —I think there was plenty for me to want to remember that.

Q. What did the 4:02 at which you left Mountainair have to do with it at that time?

A. It was so close to the stopping time, five minutes. I didn't try to remember the time I left or arrived at any other station on that line that day,—it is three years. After the train started, I wrote down this 4:02. I [187] don't know how many whistles that train gave between Mountainair and when it stopped. It is possible—he possibly whistled on the curve getting out of the cut,

(Testimony of Fred Rice.)

about a mile before we got to the crossing. I do not remember, so as to testify, whether he whistled on the curve or any other place except the whistling post. I am not sure about any but the crossing whistle. (365) I had reason to remember that crossing whistle. There was something happened that caused me to remember it right then. I had reason to remember the crossing whistle. This curve I speak about, and the crossing whistle, was probably a little more than half a mile apart. I don't know for what reason he would whistle shortly after leaving Mountainair. He would whistle approaching the stations. I have been running on that division over ten years. Leaving Mountainair, the first station is Abo. (366) That is not a regular stopping place for that train.

Q. Doesn't the engineer almost always respond to the conductor's signal just after leaving Mountainair with the whistle, as to whether he is going to stop at Abo or not?

A. That depends on whether he has a stop or not and whether he has a ticket for that point—it might be two miles below there. I will not say that the train did not whistle just after leaving Mountainair and approximately a mile and a half or more from this crossing on that day; I don't remember that. I can only remember one whistle between the two; that is all I can remember. I remember the crossing whistle. (367) When I heard that whistle, I was riding on that train. Well, I was listening and I heard the whistle. I

(Testimony of Fred Rice.)

was not reading. I don't know whether I was standing up or sitting down. I might have been leaning over. I can't [188] tell the jury whether I was standing up or sitting down—I don't think that has anything to do with it at all. I don't care what I was doing. No, I cannot remember. It don't have anything to do with it. I was on the inside of the train. I know that. (368) I cannot tell which side of the train I was on, the right or left—I would hear it either side. I didn't have to see that whistling-post when the engineer blew that whistle. I don't have to. I know that country. I have worked over it from day to day; I know the hills, the trees and everything.

Q. When he blew that whistle did you look out the window and see just where you were?

A. Why, I glanced out the window, of course, every few minutes in the daylight.

Q. I am asking you what you did do this particular time?

A. No, I don't remember just which way I turned my head every few feet that day. I can tell where I am when I glance out of the window to see where I am. I know where we were, very closely. We were very close to the whistling-post. I don't know which side of the train I (369) was on, or which way I looked out. Doesn't matter, either way. There are windows on both sides. I don't remember whether I was standing or sitting. That don't matter—I might have been leaning over.

TESTIMONY OF W. E. SPADE, FOR DEFENDANT.

W. E. SPADE, a witness on behalf of the defendant, testified as follows, on [189]

Direct Examination.

My full name is W. E. Spade. (370) I live at Clovis. I have lived at Clovis, I expect, about eighteen years. I am employed by The Atchison, Topeka and Santa Fe Railway Company. I have been employed by that company 29 years the 19th of this coming January, in the capacity of a locomotive engineer. I have been a locomotive engineer for the defendant company all but three years and three months of that time. That entire service with the company. I was employed as locomotive engineer on the 11th day of June, 1923. My run at that time was from Clovis to Belen, New Mexico. (371) I have operated a locomotive engine over that division ever since the road had been built, I think that is about 18—about 17 years, I think. I took the train on that day at Clovis. I was engineer of train 21. My train on the particular date to which this inquiry relates consisted of a mail-car, baggage-car, smoking-car, chair-car, and Pullman car. So I had five cars in my train. I remember the type of locomotive that was drawing that train that day. It was 1300 class. That locomotive had upon it a steam whistle. (372) It had a bell. The bell upon that locomotive operated by air,—automatic. What I

(Testimony of W. E. Spade.)

term an automatic air ringer. I remember bringing my train into Mountainair, New Mexico, on the 11th of June, 1923. I could not swear to the time when I arrived at Mountainair with my train, but I think—I know the time we left there; I don't know what time we got there. We left 3:52; I don't know what time we got there, I suppose several minutes before that. I could not say how long I was at the station, at Mountainair, with my train, before I departed west. Between the station of [190] Mountainair, New Mexico, and the crossing at which this accident occurred, there is one overhead highway crossing over the tracks of the company. (373) But there are no other grade crossings. There is a whistling-post near or east of the crossing at which this accident occurred. I think that whistling-post is located about a quarter of a mile east of the crossing. Before we started from Mountainair with my train, I rang the bell on the locomotive. I turned on the ringer. I did that just by opening up a small valve. After leaving Mountainair, and as I proceeded west, I shut this bell off, and then when I whistled for the crossing, I think it is a mile and a half from there, I turned the bell on again.

Q. Then what did you do about whistling, if anything, did you sound the whistle?

A. After blowing the whistle, after sounding the whistle, then I turned the bell on. (374) I couldn't tell you exactly how soon after that the accident happened, after the sounding of the whistle and

(Testimony of W. E. Spade.)
turning the bell on again. I brought my train to a
stop after the accident.

Q. And in bringing your train to a stop, just how
did you do it, what means? What breaks, if any,
did you have there by which you could stop your
train?

A. I had air-brakes on the engine and all cars.

Q. The cars and engines were equipped with air-
brakes, were they?

A. Yes, sir. I make what is commonly called a
"service application," I sounded the whistle at the
whistling-post, east of the crossing. I sounded two
long and two short blasts of the whistle. (375)
[191]

I was operating my engine at that time, I should
judge, at not over thirty or thirty-five miles an
hour. I made the service application of the air,
after striking something and the dust began to fly.
I did not see the object which I struck before striking it. At the time I struck the car I was on a
curve. I was upon the right-hand side of the locomotive. That would be on the north side of the locomotive, going west. When I made the service
application of air, I got a response—the brakes
responded (376) to the air application. I
brought my train to a stop as quickly as I could.
Between the time when I left Mountainair with my
train, and up to the time when I made this application, this service application of the air, I had had
occasion to make a service application of the air
before. Always after leaving Mountainair when

(Testimony of W. E. Spade.)

we get the train started, the company requires us to make what we call a running test of the air-brakes, to be sure that we have air going down the mountain. And leaving Mountainair, such a running test of the air-brakes would be made, at about a quarter of a mile from the station. I think the track going west from Mountainair is level for a distance of about a quarter of a mile. Then after a quarter of a mile (377) the track goes downhill. Just before I started downhill about a quarter of a mile west of Mountainair, I made the service application of air. Just before I went over the hill.

Q. Now, of what did that test consist—just tell the Court and jury how you handled the engine and air-brakes thereon at that time?

A. Well, I leave the engine working steam and I slow the train down to be sure that I satisfy myself that the [192] air is all working, and then I let it off. That is my usual custom. On the particular day of this accident, I did that. I tested my air-brakes by making this running test to which I have referred. I made that running test of my air-brakes at or about the place to which I have referred, namely, about a quarter of a mile west of Mountainair. In making this test of my air-brakes, (378) just about the time I went on this descending grade, I slowed down the train, I expect, to about fifteen miles an hour, and then it takes quite a while to pick up again.

Q. Now, for what distance would your train run

(Testimony of W. E. Spade.)

at this reduced speed after making this test of the air-brakes?

A. Well, it just picks up gradually as the brakes go off; I couldn't tell just— In order for me, as an engineer, to know whether the air-brakes measured up to test, and were working and functioning properly, I would not have to go on that grade, over a quarter of a mile. About that, I should judge. So that would place my train about half a mile west of Mountainair. And then the next thing was the sound of the whistle, I say, and the turning on of (379) the bell again. And then shortly thereafter this accident occurred, and then I made a service application.

Q. Now, let me ask you this, whether or not, by reason of the fact that you had made this test, and this running test of your air at the point to which you refer, that that had any effect, or would have any effect upon an emergency application in case you should want to make one immediately thereafter, or soon thereafter?

A. After using the service application, it takes, I think, about one minute to charge the train up, or a minute [193] and a half. Then you can get an emergency application. You cannot get an emergency application after using the service application. If you want to use the emergency application you have got to use it right now. In other words, if I have occasion to make use of the ordinary service application, that, of course, takes out of the air chamber a certain amount of air. Now,

(Testimony of W. E. Spade.)

before I have the full emergency use of the air, I must await the return of the air through the pump to the chamber. And during that process (380) I don't think I could have got an emergency application of my air.

Cross-examination.

Q. Will you tell us how far, how close to the front of your engine, you can see the left rail, as you sit on the engineer's seat looking ahead, the left rail?

A. You want to know how far ahead of the engine I can see—you can see miles ahead, I assume.

Q. I mean, how close to the engine you can see that rail, as you sit in your seat?

A. From 30 to 40 feet. That depends on the curve, too. In straight track, it would be more than 30 or 40 feet—it would be at least 50 or 60 feet, I expect; 60 feet would be the extreme limit. (381) In other words, as I sit in my engine, if the track is straight my line of vision will cross that left rail from say a matter of sixty feet in front of the engine. The track running up to this crossing, bends a little before it reaches the crossing. It curves to the right. That would enable me to see farther to the left up there than if the track continued straight. [194]

Q. Your line of vision crossing the left rail about 60 feet, for how long a distance before you hit that crossing did you have from your seat a full view

(Testimony of W. E. Spade.)

of the crossing, down as far as the right of way, fence on the left, before you hit that curve?

A. I don't just understand how you mean.

Q. Well, the train—the track extends after finishing this curve, which curves to the right at the crossing, it (382) then extends down and into that cut straight for quite a distance, does it not?

A. Yes, sir.

Q. And how long, how far, more or less, approximately?   A. Oh, I couldn't just exactly say.

Q. Well, it is half a mile or more, isn't it, before the track again bends and disappears in that deep cut?

A. I expect it is three-quarters of a mile. For three-quarters of a mile away from that crossing I had a full and complete view of that crossing. I never saw these men until the automobile was upon the front of my cowcatcher. I was looking down the track at that time. I certainly looked at that crossing to see if there was anyone coming. I kept looking at that crossing as I approached it. It was my duty to do it. (383) Required by the rules of the company to watch the crossing. I kept looking at that crossing clear until I got up there, and never saw that automobile or these men.

Q. And you tell the jury that you were looking at that crossing and not at anything else, will you?

A. When I am sitting behind my engine, going around a curve, I cannot see through that boiler. I cannot see a car coming up on the track. [195]

(Testimony of W. E. Spade.)

Q. You were not going around any curve until you were within three hundred feet of this crossing, were you? or near any curve.

A. Yes. But the way this car come up, coming up on the grade, I could not see them for half a mile where I sat on my seat. (384)

Q. Now, you say the curve interfered with your point of view, do you, that curve interfered with your seeing these men. Do you say that?

A. I can say I cannot see a man when I am sitting here around the curve, I cannot see an automobile coming up the bank behind me when I am sitting back here. The curve does not interfere with my seeing a man on the track, but interferes with my seeing a man down behind the bank. I couldn't tell you exactly how high is my seat and my height, as I sit on the seat above the rails—I expect 12 feet—10 feet. I am 12 feet in the air, and I am looking down; but the boiler prevents me seeing a man over there if I look, or an automobile. (385) I was correct when I said that from my seat I can look across the front of the boiler and see the left rail down to 60 feet from the front of the boiler—on straight track. This curve that was made there favored my wider vision at that point.

Q. Brought these men nearer, nearer into your vision, did it not?

A. Where these men came up with the automobile, they came up behind a fill, all along there the

(Testimony of W. E. Spade.)
road comes up along a fill to where they come up on the track.

Q. That is true, but the curve simply brought them nearer into your vision, did it not, until you struck the [196] curve, only 300 feet away from the crossing; is that not true?   A. Yes, sir.

Q. Now, can you give these men a single other reason why, if you were in a position looking ahead at that crossing, you did not see that automobile and those two men until you found them up on the cowcatcher after you hit them? Can you give any reason?

A. They were not on the track when I was looking; I was looking all the time, and they must have come up there (386) during the time when I could not see them.

Q. When was the time that you could not see them, if you had been looking, and where?

A. Why, during the time the engine was on this curve and I was sitting back here.

Q. And that was just 300 feet away from the crossing, wasn't it, and not more?

(No response.)

Q. Is it the inference that you believe that these men came entirely within your vision and climbed up and got in front of that engine while you were traveling 300 feet, is that your impression? They could do that and climb that hill? Did they do that that day?

A. They must have, because I didn't see them.

(Testimony of W. E. Spade.)

Q. Wasn't it because you were not in your seat and were not looking, that you did not see them?

A. I was sitting in my seat looking, yes, sir. And I didn't see them. When I passed the Whistling-post, I was siting on my seat box. Never moved from that until I hit the crossing. Looked ahead every instant, all the time. (387) [197]

Q. At that crossing which was the danger point you had whistled for?

A. I was looking down the track.

Q. You could look down the track without looking directly at the crossing?

A. I was looking at the crossing, too.

I think we left Mountainair at 3:52. The thing that makes me think so is we had an order to await there for a work-train. I remember we had an order to wait for a work-train there. We did wait for the work-train. When we left there it was 3:52. It was not 4.02 when we left there. I know what time it was. I looked at my watch. I did not put it down on a book or record or memorandum or report. (388) One of the duties of an engineer and conductor is to continually compare their watches and see that they are checking all the time. I did that with someone that day. I know that my watch and Conductor Finney's agreed. I am sure of that. I say it was 3:52. I am sure of that, when we left there. We had an order to wait there until 4:00 o'clock. That has been so long ago, I don't remember our schedule

(Testimony of W. E. Spade.)

time, for leaving there. They have changed the time-card several times. I couldn't say how long we remained at Mountainair awaiting for that work-train to come in. If I remember right, this work-train was there when we got there. (389)

Q. Then why did you wait there for it to come in?

Mr. BRENNAN.—He said he had an order to wait.

Q. (By Mr. WOOD.) My recollection is, you told me that you waited—Did you or did you not?

A. Had a "31" order to wait there until—   [198]

Q. (Interrupting.) My recollection is you told me a few minutes ago that you did wait and were late; am I right about that or not?

A. The time we left Mountainair was 3:52. I don't remember the time we got there. If we were late we waited there until 3:52. If we got there and that work-train wasn't there, and the time was not up, we waited; that is as far as I know. I know we waited there until the time was up.

Q. What time?

A. The wait order to wait there until 4:00 o'clock, 4:00 P. M.

Q. You don't know now whether that was your schedule time or whether it was late or early, do you?

A. I know that was late—I know we had here eight minutes to go down there this trip. (390) We had a wait order to wait at Mountainair until 4:00 P. M. for this work-train. When we got to

(Testimony of W. E. Spade.)

Mountainair and got our work done up, the work-train was there and we left at 3:52. I couldn't swear that that work-train didn't delay us a minute. I could not swear that we pulled out of that station immediately that that work-train got in. That is too long ago. I can't remember. I could not testify as to how long we remained at Mountainair station. I could not testify now to what time we got in there. I do not make those matters a matter (391) of written record on my report at every station. The engineer is not supposed to do that. Whenever my train is late, the conductor makes a report of how late, and the cause of it. I do not make such a report.

Q. The conductor makes it. Mr. Spade, how many times did you whistle after leaving Mountainair that day, and before [199] you struck this car? A. Whistled for the road crossing, once.

Q. Just once. Sure about that?

A. If you want to be exact about it, I will tell you—When I whistled for that road crossing, I looked down the track; there was a Mexican over there on a load of wood, he kept looking down the track, and I whistled twice, to be sure there was not, if there was anybody below that bank, that they would not be on the track.

Q. That is not quite what I asked you, is it? Are you sure that you did—I think you said that you did whistle twice? (392)

A. Yes, I whistled twice, once at the whistling post and once between the whistling post and the

(Testimony of W. E. Spade.)

crossing. This Mexican over there on the load of wood called my attention, I thought there must be somebody down there coming.

Q. You testified here last summer, did you not?

A. Yes, sir. You didn't ask that question. I didn't tell to you or to my counsel about whistling twice. You didn't ask me to tell it. I was asked what whistles I gave approaching the crossing, and I told the one at the whistling post, just as I have now. And that is all I told. Now, we have two whistles, one at the whistling post and one somewhere between the whistling post and the crossing. Yes, sir, that is the truth of it, too, because this man on the load of wood kept looking back, and I thought there must be something down there, and I looked, and could not see a thing.

Q. You looked again at the crossing where these two men were in the car and you could not see a thing, the crossing was clear then, wasn't it; there was not a car or a soul approaching? [200]

A. No, sir, nothing on it. Not a thing at the time I gave the second whistle. (393) I didn't give a screech of the whistle just at the time I struck these men. I was not a bit excited at that time. I was not excited when I found this automobile and these injured men on the front of the engine. It would kind of excite a man just a little to walk around there and find two men sitting up there on the pilot. I suppose it did shock me some. I am able to remember, under the circumstances, accurately and definitely, just what I did at the time.

(Testimony of W. E. Spade.)

Q. Now, this automatic bell doesn't ring all the time when you are out, does it, on the road?

A. It was ringing this day, because I shut it off just (394) as soon as we come to a stop. That automatic bell does not ring continuously when the engine is on the road—it rings when you turn it on. I turned it on after I blew the whistle at the whistling post, that is right. It is my duty to do those things at every crossing. And if you were to ask me if I did that same thing after passing that crossing on any day previous to this, I would give the same testimony. And if you would ask me about any day before this one, I would also give the same testimony as to my blowing the whistle and ringing the bell at that whistling post. The reason I happen to remember the speed (395) of 30 miles an hour at that time is that that is all I have done all my life, just ride and read my watch and study speed. Look at my watch and keep the time.

Q. Was there any reason to hold yourself down to 30 or 35 miles an hour at that point?

A. Yes. After making this running test, I could not have been going faster, I don't think. [201]

Q. You testified before that that is your speed that you always make there, and therefore that is why you know that is the speed at that time?

A. Slowing down—you have a certain way of doing a thing. It is just the same old thing, day in and day out. No, I don't run that 30 or 35 miles an hour all the way from Mountainair to Belen.

'(Testimony of W. E. Spade.)

I hit the speed up somewheres between those places. What makes me remember that I held it down to 35 miles an hour at that place is: After making this running test every day when I go (396) by there I ain't going over 35 miles an hour, I go over that road every third day. There is no other reason why I remember I was not making that speed.

Q. You testified before, I think, that there was an order of the superintendent for permitting a higher speed at that point; do you remember that now?

A. Yes, sir. I don't think I said that. I think there is—they have bulletin instructions out to use so much time between Mountainair and Sais. The instructions were that I was not to go faster than 35 miles an hour between those places. I stated before that sometimes I may have exceeded it a little—after leaving Sais—it never showed up on the reports. You have not a speedometer on these locomotives; you cannot tell just exactly to the mile. I can come pretty near telling. (397)

I was not kicking up considerable dust that day until after we hit the car. I hit the car on a plank crossing. And when I picked it up the front wheels were on one side of the pilot, the cowcatcher, and the rear wheels on the other side, and I had picked it right up off the plank crossing. The way I kicked up dust doing that was the front [202] wheel and the real wheel was dragging on the ties, and the body was dragging in the middle of the track.

(Testimony of W. E. Spade.)

Q. Didn't you pick it right up on the cowcatcher with the wheels up off the ground entirely?

A. It was laying edgeways on the pilot like this (indicating with hands); it was not sitting up; the man and the boy was up on there.

Q. Didn't you testify before that the engine picked the car up entirely up on the cowcatcher; isn't that a fact?

A. It was dragged on the ground there, too. I don't know whether I testified before that it was dragging on the ground or not; you didn't ask me that. (398) We drug the car along on the track there. This train I had that day was a much lighter train than I was accustomed to carry. Train No. 21, as ordinarily constructed, contained, I expect, on an average about 11 cars, on a daily average. Sometimes 12, 13, 14. A very heavy transcontinental train ordinarily. I was using the same class of engine that day as we usually employ on that train. On this occasion I did not have five light cars. There was the same equipment as we always have. I mean, our load was light, it was a stub train. (399)

Thereupon, D. A. Little was sworn as reporter.

I remember that you asked me upon the former trial this question, "By the way, this car, the Ford car, was substantially astride of the pilot or cowcatcher, wasn't it?" and I answered, "Yes, sir, it was on the center of the track," and you asked me, "And was right on the center of the track when

(Testimony of W. E. Spade.)
you hit it?" and I answered, "Yes, sir"; That is
correct. (400) [203]

Q. Do you say now that the wheels on the one
side or the other hung down so as to drag up the
dirt on one side or the other of the engine?

A. I say now when this dust commenced to fly
I first thought it was—I was sitting back here and
I *am* sitting just like I am sitting now and when
this dust started to fly I made a service application
of the brake and I took hold of the reverse lever
and leaned out the side window like this (indicating) and I could see one wheel of the automobile
and I brought the train to a stop. (400) The
dust was coming out from underneath the engine
then on all sides. I expect it was parts of this car
that was dragging that was stirring up the dust.
Yes, I know that was what it was. Dragging in
the middle of the track. I think it was part of the
body or pieces of the car.

You were asking me about the train and I said
that it was a heavy—ordinarily a heavy train and
I had the same engine on on this day. I stated
there were five cars. The first two were mail and
baggage. One was mail and baggage and the other
was a baggage-car. We had two of those and then
the smoker and the chair-car, and one Pullman.
(401) I think that was it. I won't swear to it
but I think that was the makeup of the train. I
could be mistaken as to there being two cars devoted to mail or baggage, but I think that was it.
As to which is the easier to stop, a light train or

(Testimony of W. E. Spade.)

a heavy one—which I can stop the quicker depends on the brakes of the cars. Assuming that the brakes of the cars are equally good on both and assuming that the conditions are the same on both, there should not be any difference in which one I can stop the quicker, a light train or heavy train. [204]

Q. Take a train of five cars and a heavy engine, running at thirty-five miles an hour, in what space of time can you stop them by an application of the service brakes?

Mr. BRENNAN.—I am going to object to that question, because it does not call the witness' attention to the grade or the condition of the track.

Mr. WOOD.—Q. I will say on the grade where you were that day and I thought I asked after passing that crossing. How quickly can you stop that train of five cars that you had there that day, if you are running at (402) thirty-five miles an hour?

A. I expect you could have stopped it in a quarter of a mile.

Q. On the former trial, you testified a half a mile, didn't you

A. Well, you asked me how quick a time I could. This day, I made a service application and, then, looking out of the window on the side, seeing it was a car, I made another application.

Q. Didn't you tell me that day on the former trial that it would take a half a mile to stop the train?

(Testimony of W. E. Spade.)

A. That depends—I said at that time it depends on the brakes the cars—there are no two trains the brakes are alike. There aren't two engines that the same brake—don't brake alike.

Q. Didn't you say on that day that with that engine and under those conditions there you expected it would take you a half a mile to stop it— didn't you say that on the former trial?

A. I expect I did but you asked me if everything was [205] —not what—how quick I could stop, this question. I said a quarter of a mile.

Q. And then I called your attention at that time to the fact that you did stop in less than a quarter of a mile, did I not?

A. I expect you did. (403) I could not tell you just exactly in what space I could stop a train if I was going fifty miles an hour of that sort and under those same conditions. I know I have tried to stop and I figure it generally takes about five or six telegraph poles, counting telegraph poles, going at forty or fifty miles an hour. That is, if you go right after the brake. After I seen what had happened, I went right after the brakes that day; I don't know what distance I stopped in. I did not put on all of the brake I felt that I could put on with that engine to stop. After I seen what had happened, I tried to stop as quickly as I could.

Q. Now, can't you stop a light train with that heavy engine by applying all of the brake that you have quicker (404) than you can stop a heavy train?

(Testimony of W. E. Spade.)

A. They do not teach us that and I do not think they can.

Q. Will you say from your experience that it can't be done, a light train can't be stopped in a shorter distance with a heavy engine in front than a heavy train can be?

A. All cars, the braking power is supposed to be adjusted the same.

I do not know what the grade is from the place where it breaks near Mountainair until it goes down past this crossing. It is about level but after tipping over, they tell me. I haven't got no way of telling only what I [206] hear them say that it is sixty-six feet to the mile. Where I went over the grade at the summit is somewhere about a quarter of a mile beyond the station at Mountainair. I suppose that that is about a mile and a half from this crossing to that summit. How long it will take me to get up full speed, with a light train, with steam on, after I break over that grade (405) Depends on how I will break off and what amount of steam I was working in the engine. Supposing I was feeding it pretty generously, how much space it would take me to get up a speed of fifty miles an hour after I broke over that would depend on whether I put the brake valve in full release and kick all of the brakes off at once or whether I let them gradually release off. Assuming that I was trying to pick up speed and doing whatever I could to pick up speed, I expect it would take me a mile and a quarter, in that down grade, to pick up a

(Testimony of W. E. Spade.)

speed of fifty miles an hour, starting with the usual speed that I would have when I went over that ridge. I could do it in a mile, and that would put me a half a mile from this crossing—still to go. I tried my brakes just before we went over that top. (406) I couldn't just tell you whether I could have tried my brakes at that time and went over the top and picked up this fifty miles an hour and still be a half a mile away from the crossing. I couldn't just tell you.

Mr. WOOD.—All right, that is all.

Redirect Examination.

When I talked about bringing my train to a stop within six or seven telegraph poles, I meant I would make an emergency application of the air, with sand on the rail. The [207] heavier the train and the more cars that I have therein, of course, my breaking power is increased in proportion to the number. (407) In answering counsel's question as to the relative distances within which a heavier or a lighter train might be controlled by the air of the—either the emergency airbrake or the service application, I keep those things in mind.

The COURT.—Q. How many feet between these telegraph poles?

A. They are supposed to be thirty-two, I think, to a mile.

Mr. WOOD.—One hundred fifty feet, I think, is the testimony, between them.

The COURT.—Yes (408)

TESTIMONY OF CLAUDE N. LASSETER, FOR DEFENDANT.

CLAUDE N. LASSETER, a witness on behalf of the defendant, testified as follows:

Direct Examination.

My name is Claude N. Lasseter. I live at Clovis, New Mexico. I have lived there hardly two years. In June, 1923, I was living at Mountainair. I have lived at Mountainair about four or five months, I think. I recall being a passenger on a train on the 11th day of June, 1923. That was the train which collided with an (409) automobile at the crossing west of Mountainair. I boarded that train that day at Mountainair. Somebody else was with me on the train. There was a man sitting with me from Mountainair. I recall the collision at the crossing. I recall hearing signals before reaching that crossing before that accident happened. I heard the regular [208] crossing signal given before the accident happened. I remember that was two longs and two shorts. I heard that about a quarter of a mile from the crossing. I did not hear any other signals, between there and the crossing. (410) I was sitting in the smoker on the north side of the train—on the right side.

Cross-examination.

I was working at the crusher at Sais, at the time. The crusher at Sais was working for the

(Testimony of Claude N. Lasseter.)

Santa Fe. So that I was working for the men that operated the crusher crushing stone for the Santa Fe. My folks were in Mountainair at the time. Sometimes I would ride in and out of Mountainair twice a week and sometimes not only once. I do not know whether that train was on time or not that day. I do not know how long it staid at Mountainair. (411) I went directly from my house to the train. I might have been there a few minutes before the train came in. I have no recollection about it. I was on the train going to Sais back to work. I was engaged in conversation with somebody. I was sitting in the smoker talking to a man from Mountainair. I heard four whistles. They were not the only whistles I heard that train make between Mountainair and the time it stopped.

Q. What other whistles did you hear the train make and where?

A. I heard them blow three times just after leaving the station. Probably 200 yards the other side of the bridge. At the time that train whistled, the first time, (412) after it left Mountainair station, it was probably 300 yards or 400 yards from the station. I heard it whistle then. I remember first hearing it blow when it was down there two or [209] three hundred yards from the station. The sort of a whistle it gave down there just starting from the station the three or four hundred yards up was three short blasts. I could be mistaken about that but I don't think I am. Those three short blasts represented a stop at Abo.

(Testimony of Claude N. Lasseter.)

In other words, that is the response of the engineer to the conductor's signal to stop at Abo. I say I heard another whistle. I think that was at the whistling post. (413) I don't think I could be mistaken about that. What was calling my attention to the fact that he whistled at the whistling post that day at the time when I heard the whistle was I would naturally notice it whether he whistled or not, because it had not been very long since I was on the road. At the time when he blew those two whistles, the first to indicate that he was going to stop at Abo and the second, which I say was at the whistling post—I could not say that any particular thing called my attention to them at that particular time of day at the time the whistles were given. I would naturally pay more attention to the whistles when I am riding on a train than anyone that had never worked on the road. I had been riding on that train, back and forth, approximately, (414) twice a week and sometimes only once. I would not ride on the passenger train all the time. But I would be riding back and forth once or twice a week. So that I would be making from two to four trips counting each way. I don't think it true that I would be likely to pay no attention to whistles traveling that much on the train. I can't remember any particular time that that train did not whistle there. I did not hear a whistle just about the time when the car was struck. The first thing that called my [210] attention to the fact that some accident had happened was the

(Testimony of Claude N. Lasseter.)

way the train was stopping. That was, I think, when I was already over the crossing. (415) I could not say for sure where I was at at the time that I noticed it first. I don't know quite where **I was.** I just noticed the brakes were applied harder than usual when just making an ordinary stop. That was what first called my attention to it. It was not hard enough to jar me out of my seat.

Q. You were particularly paying attention to those whistles indicating the stop at Abo because you were going to stop at the next place, weren't you?

A. The third place was where I stopped. And I noticed those whistles of the engineer signalling that he was going to stop at the next station particularly.

Q. Because he gave those same signals again that he would stop at your place, didn't he? (416)

A. I don't know whether he made that signal out of Scholle or not. I think they usually give those same signals as I was on the train, indicating that they would stop at Sais, as it was where I stopped. I think that whistle particularly I would notice but I never noticed it out of Scholle.

Thereupon the defendant offered the deposition of L. A. AINSWORTH, taken at Mountainair July 22, 1925.

DEPOSITION OF L. A. AINSWORTH, FOR DEFENDANT.

Thereupon, the above-mentioned deposition was read to the jury, as follows:

WITNESS.—My name is L. A. Ainsworth. I reside (417) here in town—in Mountainair. I've been living in town since 1921. I was on the train that struck Mr. Spencer here at the crossing about a mile west of here. I was in the smoker. I got on the train at Mountainair. The train sounded a whistle before [211] it reached that crossing. They whistled as they went around this curve out here. I would have to guess as to how far from the crossing was that whistle. My best judgment is about half a mile. I could not say positive. Then there was another whistle at the time the brakes were applied which must have been just at the time they were hitting the car. I was conscious that something was struck at the crossing. The dust fogged everything up. I couldn't say how far the train did run after it struck the car. Not very far; I suppose (418) they stopped as quick as they could. They were going downhill.

Cross-examination.

I am a farmer. I 'was going to Los Angeles, California, that day. I boarded the train here. I

(Deposition of L. A. Ainsworth.)

did not notice that they stopped here quite a long time. I do not know whether they did, or not. I did not notice that they waited for a gravel train to come in; did not pay any attention to that. If they rang the bell at any time between the crossing and here, I didn't hear it. I say the whistle sounded up here, when they went around the curve. As to what called my attention to the whistle: (419) Well, I just noticed the whistle. I don't know as anything in particular caused it. I ride quite a little on trains. I just noticed it whistling. I couldn't say what specially attracted me. I noticed other places between here and Belen that he whistled that day. He whistled at different crossings and stations. I do not assume it whistled because trains usually do. I was on the smoker nearer the engine, you know. I couldn't say that I was just trying to pick out a whistle each time. I just noticed the whistle. There was nothing in particular that [212] was attracting my attention to this whistle. We were in the deep cut when the whistle blew— just as you come around the bend down here. From here to this crossing, I've an idea is about a (420) mile. And it is my idea that somewhere between here and there—maybe about half a mile away.

I did not notice the speed of the train. There was somebody else on that train from here, or this vicinity, that I knew. There were three who were going to California. There was myself, B. Crowley and Lester Barnes. I got off the train after

(Deposition of L. A. Ainsworth.)

the accident. I knew Mr. Spencer. I ascertained that he was the one that was struck. I saw him. I walked down to the engine. The car was picked up and carried along on the pilot. (421) It was right on the front of the engine when it stopped. Well, the rear wheels were on one side of the pilot and the front wheels were on the other. So that the pilot hit it about between the wheels. I never heard Mr. Spencer speak a word. I figured the boy had a leg broke from the way it was all turned. And they picked them up and took them on. I did not notice how far into the next cut the train had gone—where the pilot was when they stopped. I do not know the distance. (422)

TESTIMONY OF JOHN W. FINNEY, FOR DEFENDANT.

JOHN W. FINNEY, a witness on behalf of the defendant, testified as follows:

Direct Examination.

My full name is John W. Finney. I am a conductor of The Atchison, Topeka and Santa Fe Railway Company. I live in Albuquerque at the present time. Just moved a month or so ago. I was conductor in June of 1923. For the same railroad [213] company. (423) The number of the train and the points between which I ran at that time were No. 21–22 between Amarillo and Belen. I was working on my run on the 11th of June, 1923. I was handling that day a passenger

(Testimony of John W. Finney.)

train. The number of that train was twenty-one. There was five cars in the train. There was a combination mail and baggage, smoker, chair-car and two standard Pullmans. I took that train at Amarillo, Texas. And the end of my run was Belen. I recall an accident which befell my train on that day on that run. That accident occurred (424) a mile and three-quarters west of Mountainair. The locomotive attached to my train was equipped with a bell. It was the standard bell. It was operated by an air device. I recall the. time I arrived at Mountainair that day on that trip. It was 3:51. I left Mountainair at 4:02. We are permitted to leave—arrive at a station five minutes in advance of the leaving time. Usually arrive in Mountainair—our endeavor was to arrive about 3:50, in order to have two minutes to do our station work and leave there on schedule time. (425) That day we arrived at Mountainair at 3:51 and left at 4:02. At the time we left Mountainair, I cannot state whether the bell was set in operation. I don't know. I cannot state whether the bell operated at any time between Mountainair and the point of the accident. The accident occurred at 4:07 P. M. I know that by my watch. When a passenger train stops either at a regular stop or —and any other intermediate stop, the first duty of a conductor is to ascertain the time by looking at his watch; I did that on that occasion. That is a habit. I found that time to be 4:07 P. M. I heard the whistle on that locomotive sounded be-,

(Testimony of John W. Finney.)
tween Mountainair and the crossing or the scene
of the accident. (426) The kind of whistle I
heard [214] was two long and two short blasts,
crossing whistle. I could not say as regards to the
distance from the road crossing that that whistle
was sounded but was at a time before we arrived
at the road crossing. I estimated it was about
thirty seconds before we arrived at the road crossing.

Q. How long did you have to make the trip to
Belen?

A. Dispatchers and trainmen and enginemen figure that it is comfortable time about forty-one
miles from Mountainair to Belen in one hour.
There was at no time any occasion to run the train
at a high rate of speed through that section on that
day. It was a very easy trip for everybody concerned. Because as a usual thing, we handle from
eleven to fourteen cars; on this day and date, we
had just a stub, five cars, everything going along
easy. Leaving Mountainair, as I say, at 4:02,
(427) that gave us an hour and eight minutes to
reach Belen; due at Belen at 5:10 P. M.

Q. Where you leave Mountainair a few minutes
late, as you did on that day, on what section of that
track between Mountainair and Belen do you make
up the time?

A. Well, there is no restriction on that piece of
track in regard to speed between Mountainair and
west of Scholle. There is a restriction on account
of cuts west of Scholle up to a station called Sais.

(Testimony of John W. Finney.)

From there down, it is a race-track. It is an open country, rolling, nothing to obstruct the vision, you can see for miles. This best piece of track—well, that would be—let's see—Abo seven, Scholle fourteen, Sais twenty-one—about twenty-one miles west of Mountainair.

Q. How does the track between Mountainair and the point you have last mentioned compare with the track between [215] that point and Belen with respect to curves or tangents?

A. Well, there is quite a few canyons and quite a few curves (428) between Scholle and Sais. just a short distance down there. I would estimate that about four mile and a half of track of that nature. There are other curves on the mountain but that is the worst place. The straight track or tangent comes more below Sais—west of Sais. And that is where we make up time, if any is necessary to make up. At the time the accident occurred, I would estimate, and did in the report at the time, the speed at which the train was being operated as it approached the crossing at which this accident occurred, as thirty-five miles an hour. I would say that that would be the usual and customary speed for that section of track—that would be just about the average.

Cross-examination.

I do not remember anything which calls my attention to the speed at that time. (429) I have stated that the customary and ordinary speed at

(Testimony of John W. Finney.)
that time was thirty-five miles an hour at that time
and place.

Q. Now, does the fact that that was the ordinary
and usual rate of speed have any influence on your
mind in determining that that was the speed that
was being made at that time?

A. There was nothing unusual to call my attention.

I know a Mrs. White. They call her Mrs. Bill.
I don't know her given name. There was a Mrs.
White on the train that day, what we call Mrs.
Bill White. She is the same one. I know her.
I have known her quite a few years. A good
[216] many. Fifteen or eighteen—something like
that. (430) She was agent or telegraph operator
on the Santa Fe there at Mountainair, and up and
down there quite a few stations I remember her,
yes, sir. I remember her being on the train that
day. I remember having a talk with her just about
the time that this accident happened and a little
before. I was standing right in the aisle on the
left side and she was sitting on the right side and
I was conversing with her, about things in general.
About Bill's business and how he was getting along
in the mercantile business and how old friends do
when they meet and have not seen each other for
a long time. That is all I remember we talked
about. Nothing was said about the speed of the
train at that time by Mrs. White to me. I don't
recollect any conversation about the speed with
anybody. I hardly think that she may have said

(Testimony of John W. Finney.)

it and not remember it. (431) Mrs. White did not comment on the amount of dust that was kicked up before that accident. There was many passengers remarked about the dust after we struck the automobile.

Q. Wasn't that remark made before and while you were in the car and before you struck the automobile? Wasn't there some talk with Mrs. White to you about the dust?

A. Well, why, I did not know the automobile was struck. I did not know what was raising the dust. There was a dust outside and a passenger remarked it and I noticed it. I did not know yet that an automobile had been struck.

Q. Well, do you remember anything about Mrs. White mentioning the amount of dust that was being kicked up?

A. Yes, I remember there was several passengers— Now, I don't remember particularly whether Mrs. White mentioned [217] that fact or not but there was a dust kicked up. I would not say that Mrs. White did not talk about the dust the train was raising. The remark was made by passengers. I don't know just who made it. (432) She may have said that. She may have said that and said that as a generalization. Not to me in particular. I don't remember whether she did or not. We were late that day leaving Mountainair. Our schedule time leaving there was :52. We are permitted to arrive five minutes in advance of that. We are due there—we are due to leave there at :52.

(Testimony of John W. Finney.)

We would be permitted to arrive there five minutes in advance of that to do our station work. We have no real schedule time shown for arriving, and that is the leaving time. We did not leave approximately on time. We are due to leave at 3:52. (433) That is the time we were due to leave according to the time card. But we did not leave at 3:52 that day. It is not the rule that I and the engineer keep track of the time that we leave and arrive at the stations; that is the duty of the conductor, not the engineer. I said that 3:52 was the schedule time that one would get from looking at the time card. I put down the time arriving and time leaving. I put it down at the time. That would be ten minutes late leaving Mountainair. (434) We were delayed there nine minutes—extra —904, a ballast train—two minutes station work, usually unloading of express, baggage and passengers. This work-train that came in from the west delayed us there nine minutes. We were waiting for it. We started out as soon as the work-train pulled into the switch, what we call the call, and then we pulled down to the depot and transacted our usual work, express, baggage and passengers that took two minutes. The reason why we did not do that [218] while we were waiting for the train is we are not permitted to, under the rule. And we did not do it on this occasion. We waited nine minutes for them at the switch.

Q. What sort of a train is this, ordinarily, that

(Testimony of John W. Finney.)
you carry? Is it just a train carrying United
States mail, isn't it?

A. Oh, yes, yes, sir, baggage and express and
passengers. (436) It is not a serious matter to
hold up such a train, for an ordinary work-train.
We just make a daily report covering all delays on
a trip to the dispatcher, who answers for those
things. That is not a serious offense. They do
that every day. Gravel trains hold up through
mail trains. Mrs. White was in the chair-car;
that was the third car from the engine.

I am equally responsible with the engineer to see
that these crossing signals and other similar signals
are given. I am not held by the company responsible with him to see that those signals are given.
Not exactly. His failure to perform (436) that
duty, it is necessary for me to report it. I request
him to do those things and make the report if he
does not do it.

Q. How many whistles did you hear from the
time you left Mountainair until the train struck
this automobile?

A. How many whistles, that is, how many times
it was blown or what kind of signals was given?

Q. Give it either way.

A. Well, he whistled for the crossing. That is
the first one that I heard. And then he whistled
again for the same crossing. I heard two distinct
whistles for the road crossing. I was a witness
here a year ago. I was not asked [219]   at that

(Testimony of John W. Finney.)

time whether I remembered that second whistle. (437)

Q. When did you first remember that second whistle?

A. Before making out the mail report to the superintendent of this accident, my question to the engineer was, "Ed, why did you whistle twice for the crossing." When I was here first as a witness before, I knew that the important thing that I was to testify to here on behalf of the company was that signals were given as we approached that crossing.

Q. And you sat there on that stand and never said one word about this other whistle, didn't you?

A. You asked me the question if he whistled for the crossing and I told you he did. He did. If I am not mistaken, I think you asked me that, if he whistled for the crossing. I did not report to the company and the counsel (438) that there were two whistles before that other trial. I did not make any remark about it at all, because I was not asked the question.

Q. Were you concealing everything on that stand that you were not specifically asked about?

A. The question was asked whether he whistled for the crossing and I answered it that way. There was nothing to conceal. (439)

Q. Did I not ask you on that former trial this question, "Did you hear any whistle sounded?" You answered, "Yes, sir." I asked you, "When

(Testimony of John W. Finney.)
and where"? and you said, "At the crossing
board." That is correct, isn't it?

A. Well, I might have not just exactly put the
cross on the T and dotted the I in there but I would
say, generally speaking, about the whistling board.
I don't know exactly that that is where he blew
the whistle but I   [220]   would say the usual place
at the whistling board.

Q. When I asked you when and where, didn't
that call them both?

A. No, I don't know that it particularly did.
No, sir, in my mind, it did not, when you asked the
question.

Q. You knew at that time that he whistled not
only at the crossing board but between the crossing
board and the crossing?   (440)

A. Well, it all happened there right in succession,
one right after the other. There wasn't very much
difference between—I couldn't tell you just exactly how much difference. Three years ago. It
is quite a long time.

Q. Now, this whistle was all right together that
you heard, wasn't it?

A. Well, it sounded just like about one and then
you go just about—it seemed to me just a short
ways—I couldn't tell you the distance and estimate
the time—and then he give it to it again just the
same thing. I am not giving you any reason why
I didn't tell that when I was here a year ago. I
haven't any reason. I just answered your question
in the beginning on that. I have nothing to con-

(Testimony of John W. Finney.)

ceal at all. Not a thing. With all the travel I
had and with my work as conductor looking after
the passengers, as I was at that very instant, I do
not keep notice outside every time I pass a
whistling post to see whether the engineer blows
the whistle or not. (441) I was not taking notice
at that particular time as I was talking to these
passengers as to whether I was passing that whistling post or not at the time.

Q. How do you know he was at the whistling
post? [221]

A. I said about the whistling post. Estimated.

Q. How much leeway would you allow yourself
on about?

A. Well, that would be considerable, to say about.
I am just estimating for you. I am trying my best
to answer your question. Yes, I think I was noticing just where I was when I heard that whistle.
I think I knew just where I was at. I did not look
out at the time he blew that whistle. The fact of
the matter is that the conductor doesn't have his
head out of the window looking for the whistling
posts along. That is the duty of the engineer.
(442) I could have been a half a mile away at the
time that whistle was blown and I not know it and
I could not—a half a mile away before I reached
the whistling post—that is possible. It could not
have been close to a mile—that is too far.

Q. That whistling post is inside the cut, isn't it—
inside the mouth of the cut?

(Testimony of John W. Finney.)

A. No, that stands pretty well in the open, that whistling post there.   On the right-hand side.

Q. It is down past the point where the railroad cuts into the hill going toward Mountainair, isn't it, Mr. Finney?   (443)

A. It is out of the cut, yes, sir.   It is out of the cut—out of the big cut.

Q. It is down past the point where the railroad cuts into the hill in the direction of Mountainair from this crossing, isn't it?

A. I don't know just what you are getting at, Mr. Wood, in regard to where the railroad cuts in. Do you mean that where we enter the cut going into Mountainair?   Is that what you have reference to? [222]

Q. You know what a cut is, of course?

A. Yes, where the cut is where the railroad goes into the cut.   Is that what you mean?   It is just below there west bound.

Q. A cut, I mean, is where an embankment or where a cut is made through a hill in which to lay the railroad.   Now, going from this crossing toward Mountainair, isn't that whistling post some four or five hundred feet east of where that road first cuts into the hill?   (444)

A. I don't understand your question.   I can't answer that.

Q. Let me illustrate that.   The evidence here shows that that hill is 850 feet east of the crossing where the railroad cuts into it.   Now, is that clear?

(Testimony of John W. Finney.)

A. Eight hundred fifty feet east of the crossing?

Q. Yes, sir, where this accident occurred.

A. I don't know. Something like that I would estimate, yes.

Q. The evidence of the engineer shows so. And he says that the whistling post is 1,500 feet east of the crossing. Now, is or is not, according to your recollection, the whistling post further east than where the road cuts into this hill?

A. Well, just as you get out of the cut, there is a whistling post on the right-hand side of the track to govern the engineer of the east-bound movement. Now, what I am trying to get at is what whistling post you want; whether they were going west or whether they were going east. That is what I can't understand. (445) [223]

Q. And that is the one going west where you were going?

A. Well, this whistling post, then, is situated just out of this cut about—well, I would say about 1,500 feet. I don't know what it measures. That is the standard distance, about 1,500 feet, before you get to this railroad crossing.

Q. From the crossing?

A. Yes, from the crossing.

Q. And you say it is west of the place where the railroad cuts into the hill?

A. I don't get that cut into the hill. I don't get that kind of language. Where it cuts into the hill? Where the railroad cut is?

(Testimony of John W. Finney.)

Q. Is there anything doubtful about the language where the railroad cuts into the hill?

A. Where the railroad cut is, this whistling post stands just outside of it. What I am trying to say is, it is a big cut down below and that it is not in that cut. That is what I am trying to say; I am trying my best to answer your question and come to an understanding. (446) We were not in the big cut when that whistle blew; we just emerged out of that big cut, because to tip over—what we call tip over the top there—that is, when you start descending that grade, it is a little ways out of that big cut. My train was not in the cut at all when I heard the whistle; it was out of the cut. It was coming—coming out of the cut when I first heard the whistle, that is, it was out of the big cut—what we call the big cut there. I did not look particularly to see that fact. I am giving you estimates now. You asked me that. At that time, I heard two whistles. One was at the [224] whistling post and one was somewhere a little beyond there but pretty close together. Those are all the whistles I heard.

Q. You did not hear that whistle when the engineer signalled right close down to the station that he was to stop at Abo?

A. No, sir, no such signal. (447) There was no such signal given. I am sure of that. Positive. I am sure because after the accident, it was necessary to file a wire report to the superintendent directly. While I was working on these mail reports,

(Testimony of John W. Finney.)

I spoke to the porter and told him to pull the cord for Abo; that I wanted to leave these accident reports. If we had pulled it for Abo or we had a passenger for there, I would not have requested the porter to pull the signal for Abo. That is why I am so sure. I am not giving you some conclusions. I am giving you facts. I am giving you facts now.

Q. This accident, with these two men on board, did not excite you? You were perfectly clear and cool and just as you are now, are you?

A. No, I don't think I was. I believe I had a little more sympathy than that for them. I cannot be mistaken about any of these whistles, etc. I am positive about that. These statements I am making, I am positive of. I cannot say that that engineer (448) ever passed a whistling post without whistling. No, I could not say that he never passed a whistling post without whistling or ever did. No, sir, I could not answer that question. I could not say that he never passed a whistling post without whistling. I mean I don't know whether he did or not. If I am within hearing distance of the whistle, I hear the whistles that are blown at every whistling post as we pass them. [225]

Q. And you notice each time in handling your train as to whether or not a whistle is blown at a whistling post?

A. I can hear it each time if he blows it, if I am within hearing distance of the whistle. I don't believe I can tell you what car on the train I got

(Testimony of John W. Finney.)

on that day when I left Mountainair. Sometimes I get on the smoker or the head end and sometimes at other car entrances, just according to (449) work. Sometimes one place and sometimes another. Usually the head end of the smoker. I may have gotten on the head end of the smoker that day. I do not remember. I did not come to Mrs. White on the start when I was taking up tickets. I know I did not because I would not stop to converse until I had my work done. I would say a word to a friend—say how do you do, but I was standing there in conversation with her.

Q. Did you work the smoker before you worked the chair-car, in which Mrs. White was that day?

A. Always worked the smoker first.

Q. Then, you got on the smoker, didn't you?

A. Well, not necessarily. I would not necessarily. I worked the smoker first anyway. (450) Whether I got on it first or not—on this particular occasion, the smoker was one car-length back from the engine.

Q. Why didn't you hear that bell, if it was ringing?

A. I said that I did not remember hearing it. I don't remember hearing the bell at all. It might have been ringing.

Q. You said also that as a railroad man you heard all signals if they were given within your hearing.

A. You asked me about whistles. There are very good reasons why I will hear a whistle and not hear a bell, [226] when within hearing distance. Be-

(Testimony of John W. Finney.)

cause a whistle indicates a condition. There is as much difference between whistle signals as there is between red, white and blue to a railroad man. A whistle for a crossing indicates a crossing. It certainly does.

Q. And the bell for a crossing indicates a crossing, don't it?

A. I don't pay as much attention to the bell, admittedly, as I do the whistle signal, because that conveys to me a condition which I must know. I can't hear the bell. It is out of my hearing so many times that it would not enter into it, as far as I am concerned. It is out of my hearing. If I was in the smoker and that bell rang, it was not within my hearing. I could not hear it if I was in the coach. I can't remember that I did hear it. I answered that and said that I did not remember hearing the bell at any time. I did not hear the whistle at the time when this collision took place. There was no whistle blown at the time the collision took place.

Suppose that I was in the car—the chair-car the day before on that train or the last time (452) I was over there, I do not believe I would be able to testify that I heard the whistle at the same time and place.

Q. I am asking you whether or not you remembered the whistle blowing at that crossing the day before and you answered, if I was in the chair-car— if I was there in hearing, I did.

(Testimony of John W. Finney.)

A. Well, if he blew the whistle, I heard it but I am saying I don't know whether he blew the whistle on the previous trip or not. I could not answer that. That is too [227] far removed from my memory.

Q. Do you ever know of an occasion when the train in your charge passed a crossing without blowing the whistle?

A. I can't answer that question.

Q. You were able to answer it on the former trial, were you not?

A. I said, if I was within hearing distance of the whistle and it was blown, I could hear it.

Q. I asked you that same question on the former trial, did I not, and did you not answer, "Not on the train that I have been on?"

A. I don't remember that, no, sir. (453)

Redirect Examination.

It was before I knew that we had struck an automobile I saw this dust. The engineer made a severe service application and came to a stop and, of course, Mr. Reed, you know, when a train stops, it is a conductor's duty to ascertain immediately,— the first thing he does is to look at his watch and ascertain the time and then to see what has happened. (454)

TESTIMONY OF DR. H. T. SOUTHWORTH, FOR DEFENDANT.

Dr. H. T. SOUTHWORTH, a witness on behalf of the defendant, testified as follows:

Direct Examination.

My name is H. T. Southworth. I reside at Prescott, Arizona. I am a physician and surgeon. I have practiced my profession twenty-five years. Since 1901. I am a graduate of a regular school. The Chicago Homeopathic Medical School. [228] I am duly licensed to practice medicine and surgery within the state of Arizona. I have practiced in this city since 1904. (455) In connection with my practice, I handle a great many surgical cases. I do specialize in surgery, giving special attention to it. Not wholly. I have also done work for The Atchison, Topeka and Santa Fe Railway Company here as surgeon. In the course of my professional career, I have treated a great many fractures and injuries of that character. And I have observed the effect of such injuries upon the individual affected.

Q. Now, Doctor, I am going to propound a question to you—a hypothetical question and, for the purposes of the question, you may assume the facts therein to be established and upon which I shall ask you to predicate your answer. Doctor, assume the condition of a boy seventeen years of age, raised on a New Mexico ranch, an ordinarily, apparently

(Testimony of Dr. H. T. Southworth.)

well, healthy boy, who has not been known by his
parents and surroundings to have any chronic disease, just an ordinary ranch boy, although undernourished in appearance, but no (456) history of
tubercular activity that people had discovered. Assume that such a boy was injured in a railroad
accident to the extent that he was struck by an
engine while riding in an automobile truck, his
thigh bone broken in a comminuted fracture, accompanied by jamming and contusing and lacerating the muscles and tissues surrounding the bone,
pieces of bone being broken loose and that this injury occurred on the 11th day of June, 1923; that
the boy was brought to a hospital—St. Joseph's
Hospital in Albuquerque, New Mexico, a good, well
sustained and reputable hospital, and was treated by
skilled surgeons; that due to the laceration, [229]
bruising and the shock accompanying the injury,
it was found improper or unwise to attempt to reduce the fracture for a week's time; that at the end
of that time the fracture was reduced, the condition
of the leg and the bone broken being as shown in
the X-ray pictures that were offered here in evidence; that an operation was performed, cutting
down into the leg, removing a section of bone at
the break and some small pieces of fragments of
bone, placing the leg in position and maintaining it
by what is known as a silver or Lane's plate; that
the contused and lacerated condition in the interior
of the leg and the wound was such as to make it
inadvisable, in the opinion of the surgeon in charge,

(Testimony of Dr. H. T. Southworth.)

to place the leg in   (457)   a rigid plaster cast but,
instead, due to that condition, the leg was bound
to a board or a splint, extending from the armpit to
the leg; that the boy remained in the hospital until
the 3d of August, when it was discovered, in some
way, that the splint had become loose, the boy was
complaining of pain in and about the fracture and
that he was then removed and a second X-ray picture taken, showing the bone bent somewhat out
of place; that then the bone was forced back into
position, as near as could be, and a plaster cast put
over it and continued in that position until the 14th
day of September, 1923, when an X-ray picture
was taken, showing the condition of the bone, the
Lane's plate and the surroundings, and another
X-ray picture was taken on September 21, 1923,
also showing the position of the bone, the plate and
the surroundings and that but very little calsification or knitting or uniting was shown; that then another operation was performed, in which the Lane's
plate was unscrewed and removed and the wound
again closed; that the boy continued in the hospital
until about [230] the 14th of January, 1924,
some seven months after the injury, when an operation was performed for appendicitis. Assume that
at the time the boy was seen by the surgeon on the
11th day of June, 1923, that he appeared to be
undernourished but no tubercular activity was then
(458)   ascertained or discovered by the physician;
nor had there been any tubercular history in his
family, so far as known, and no apparent activity

(Testimony of Dr. H. T. Southworth.)

of the disease in himself or in his surroundings—
in his family; that at the time of this operation on
the 14th of January, 1924, or about that time, the
surgeon, upon opening the abdomen, the boy then
complaining of extreme pains in the abdomen and
showing indications of appendicitis, it was found
that the intestines and the appendix showed a very
bad condition of tubercles spread all over the intestines, appendix and the lining of the abdomen,
a genuine case of tuberculosis of the bowels and
the lining of the abdomen covered by numerous tubercles spread over them. From the history of this
case as I have given it to you through this question, are you able to say, Doctor, within reasonable
accuracy, based upon your experience as a practicing physician and surgeon and based upon the experience derived from cases which you have handled, are you able to say what was the producing
cause of that tubercular activity found in this boy as
described in the month of January, 1924?

A. I am.

Q. What, in your opinion, Doctor, was the producing cause of tubercular activity in this boy that
has been outlined in the question as it existed on
the (459) 14th day of January, 1924? [231]

A. Secondary infection of tuberculosis.

Q. What do you mean by secondary infection,
Doctor, of tuberculosis?

A. I mean by that that he had some place in his
body, presumably within the chest, as is the most
likely place, a primary infection.

(Testimony of Dr. H. T. Southworth.)

Q. How is tuberculosis contracted, ordinarily?

A. Tuberculosis is contracted in childhood most usually either through the respiratory system or the gastro-intestinal tract.

Q. Now, Doctor, before we have you tell us a little more about the causes of tuberculosis, I would like to ask you first this question. In view of the history of this boy, as I have given it to you in this hypothetical question, I will ask you whether, in your opinion, the accident therein described and the injury therein mentioned to his leg and the operation performed thereon had any relation whatsoever to the tubercular condition that was found at the time that the operation for appendicitis was performed?

A. It had not. In my experience in handling cases where individuals have suffered comminuted fractures of the femur and other bones, I never have known of a case of tuberculosis having developed therefrom. (460) I could not say, roughly, about how many cases of bone fracture I have handled. I do not handle them every day—not necessarily every day, no, but, then, it is a very common occurrence. I say that the injury or trauma was wholly disassociated from and wholly unrelated to the tubercular condition; that is my opinion. The cause of tuberculosis is always the infection by the tuberculosis bacilli. [232]

Q. Are there certain conditions that will produce that? What I mean by condition, whether a person was, say, overworked, or whether— What are the

(Testimony of Dr. H. T. Southworth.)
causes that produce a primary infection of tuber-culosis, ordinarily?

A. Well, a primary infection must always depend upon the reception within the body of the germ, re-gardless of whether it is overwork or what not. (461) After you get your primary infection, they are necessarily active at first. Then they may be-come quiescent.

Q. Then, I would like to have you explain what would cause this bacilli or bacteria that is latent or dormant or quiescent to become active?

A. Perhaps, to make that clear, I had better first tell why they become quiescent in a primary infec-tion.

Q. Yes.

A. Assuming, because we must in a case of this kind assume that the primary infection in a case of this kind was not enormous, because in all cases when we are born every one is susceptible to tuber-culosis, if the tuberculosis becomes present in the body. Those enormous infections result invariably in fatality and it is only in the smaller infections, that is, with regard to number, not excessive doses of infection that the baby or the child is able to render the infection quiescent and he or she reach middle—adult, or middle-aged life. Nature picks up these germs and they are filtered out of the lymph stream by usually the first lymph node through which it passes. Wherever it has soft and fibrous connective tissue, an envelope is thrown round and about the tubercular bacilli and, (462) whether

(Testimony of Dr. H. T. Southworth.)

there is ever a second infection from that primary one, will depend upon two things, [233] first, the size and the number and whether or not they are encased within this envelope and, next, whether or not that envelope remains intact or patent. In other words, nature has confined the bacilli to which I have referred within a limited area and now I say that they have remained quiescent or more under the circumstances and conditions outlined by me.

Q. Now, then, Doctor, will you proceed to tell your experience as a physician and surgeon, what you have read in the medical authorities, the causes that may be revived or make the bacilli more active, the known causes, for instance, that produce this secondary to which you have referred?

A. As I stated before, one of two things, that is, with reference to the infection itself, must happen and one of the two must happen in order to produce a secondary infection. If the number of bacilli within this capsule becomes sufficiently great, regardless, then, of the thickness or the continuity of that capsule, it will undoubtedly break just by reason of their (463) increase and the poisons which they exude. The capsule itself might become porous and permit the germs to pass through it. Statistics show that there are four important conditions that give rise to a secondary infection of tuberculosis. First is undernourishment. First in importance, so far as statistics are concerned, undernourishment. Second, acute respiratory infections and measles. Third, dissipation. Fourth,

(Testimony of Dr. H. T. Southworth.)

overwork. By secondary infection I mean that the germs that have been quiescent are brought into activity by and through this second infection.

The COURT.—Q. It does not necessarily mean the introduction of additional germs in the body, does it? [234]   A. Secondary infection?

Q. No.

A. No, I say a secondary is always secondary to a primary.

Q. All children are born with tuberculosis?

A. Oh, none are born with it, your Honor, but all, when they are born, are susceptible to an infection. None of them are immune if they should come in the presence of tuberculosis bacilli. In other words, no child is immune from it. Probably none of them have it when they are born—very, very hard to recognize a (464) possibility where they would.

Q. Heredity?

A. No, sir, there is only, so far as we know, there is only one disease where there may be actual—one is syphilis and the other smallpox that is given directly from the parent to the child.

Mr. BRENNAN.—Q. Doctor, we sometimes hear the expression of lowered resistance. I notice, in your answer to one of my questions, that in enumerating the causes of the second infection, you made no reference to this phrase lowered resistance. I am just wondering, Doctor, if lowered resistance is a phrase or term well known to the medical pro-

(Testimony of Dr. H. T. Southworth.)
fession and does it have any significance—has it any
relation to the secondary infection in tuberculosis?

A. Lowered resistance is a term that has been
very loosely used by the medical profession and the
laity. It is an easy phrase to say but really has
no basis in scientific pathology. [235]

Cross-examination.

By the term primary infection, which I have used,
in tuberculosis, I mean the original taking into the
body of the germ of (465) tuberculosis. That is
what I referred to by primary infection. And that
germ is so much present in various things that a person comes in contact with that there are few, if
any, people who do not take that germ into their
body. I think that is particularly true of people
who live in civilized or so-called civilized community.
And those who live in what you might call a civilized
community, where people are thrown together and
where they live together, those germs are continually present and being taken up into the human
body.

Q. And that taking into the human body, as I
have described it, is what you meant to express to
these men by the term primary infection?

A. Yes, sir. Well, where— May I amend my
answer?

Q. Yes.

A. Because I do not quite get your question. If
(466) you mean germs that pass some portal of

(Testimony of Dr. H. T. Southworth.)
entry within the system itself, yes, that is what I meant.

Q. If you mean that the germ does what?

A. That gains admission into the system through a portal of entry.

Q. Yes, that is correct. And that portal of entry may breathe it in from the air into the lungs or in through the mouth with food or other manner or in some degree it may pass through the pores of the skin, may it not?

A. Well, I rather doubt the latter.   [236]

Q. Well, the first two, at any rate?

A. The first two but that is a question I thought, perhaps, you had in mind and which I wanted to modify, if you did mean that, because one may breathe tuberculosis germs and then be caught in the nose and then suppressed without getting into the system through the portal of entry.

Q. I think you are correct, Doctor, and we understand each other. What we refer to is taking them into the lungs?

A. Even the lungs, they may never get into the system.

Q. I am speaking of the port of entry rather than the final destination.

A. Oh, yes, that is one way.   (467)

Q. And through the—into the stomach through the ordinary—

A. That is another way. Now, after they get into the lung, they may be cast right out again or they may pass from the lung into the body generally

(Testimony of Dr. H. T. Southworth.)
through the circulatory system or similar manner. And the same through the stomach. As they pass into the body through the lung or through the stomach, in the way that I have mentioned, they may go into the circulation and then reach, as I say, the lymph node.

Q. Which is one of the little traps that nature has provided to catch these wicked strangers and throw them out? This trap may catch that little fellow on his way through and shunt him off and out through the kidneys or through the bowels or some other manner that nature has provided?

A. Well, that is encapsulated. The lymph node is a trap which takes poison or a germ out of the circulatory [237] system or out of the body and shunts it into the *elimenentory* channels; (468) it does that too. If into the lungs, it don't throw them off that are taken in there, neither, presumably, does the bowels or stomach, and some of them and a good many of them get by this lymph node or the trap which catches him to throw him out of the system. Some of them get by that and then those that get by that, the little fellow is caged by another effort of nature to protect the human body against its ills that cage taking this wicked germ and putting a shell around him where he can't work.

Q. Just suppose I would take the germ out of a kernel of wheat, if we can imagine it, and then put it in and put the kernel of wheat and the shell around it the germ is in there but the shell and the

(Testimony of Dr. H. T. Southworth.)
wheat keeps the germ from doing anything until
some new occasion arises; have I got that right?

A. Excepting multiplying its practice as to elimi-
nation of toxemia.

Q. I am limiting myself, now, Doctor, to this   (469)
question of the latent, quiescent, inactive condition
of this germ?   A. Yes, sir.

Q. And it would not be improper to liken this
germ, then, to the little germ that is within the
kernel of wheat, surrounded by the protecting layers
outside, which nature has thrown around the germ
to keep him from being active; that is correct, isn't
it?

A. Something like that.   Now, these germs are so
plentiful in the atmosphere that even in young
babies a year or two old, when they die and you
examine their body, you   [238]   find in quite a
large percentage of them that this germ has been
in there and moderately active so as to show its presence.   I don't know as I know the percentage, but
perhaps in twenty or twenty-five per cent of young
babies, say under two years old, the medical profession finds that this germ has actually been in and to
a degree active but not enough to produce what we
call tuberculosis or to indicate the child had the disease to the ordinary observer.   (470)   Now, when
they grow up and reach the age of, say seventeen
or eighteen years old, the same investigation—post
mortem of people who die from something else and
gave no particular indication of tuberculosis in their
lifetime, you would find, upon examining their

(Testimony of Dr. H. T. Southworth.)
body, that in possibly ninety to ninety-five per cent of those that germ had been in there all of the time and doing some activity. By that, I mean that this little germ is in the bodies of very nearly everybody at that age. And almost everybody at that age has what I have termed, then, primary infection from tuberculosis. Then, having that germ, or primary infection in the manner that I have described at that age, it takes something else from the outside to break that shell or from the inside, as you will, and start that germ into the activity which produces tuberculosis as a disease. And, when you have found the efficient cause which broke his shell and made him active, you then   (471)   have found what produced the tuberculosis. I gave you four causes generally recognized which I say produced the tuberculosis. I said one was undernourishment. I mean by that—well, perhaps I might describe it as lack of proper food or lack of proper quantity of food resulting in under weight for the age and height of an individual. I don't know that I offered to give   [239]   an explanation of what that undernourishment does to the individual—the human body which provokes the germ to break his shell and become active. I merely said that statistics showed that that was a condition. Statistics give undernourishment as one of the producing causes. I can tell you, from my knowledge and experience, why the undernourishment makes that germ active in the disease—what the result on the body is—how it acts.   (472)

(Testimony of Dr. H. T. Southworth.)

When one is undernourished, the body must feed upon its own fat and, unless more food is taken within the body to make up for the loss of that fat, we have the condition which you speak of and which we all speak of as undernourishment—undernourished but I did not guarantee to tell you what that particular undernourishment does or how it does it to this encapsulated primary infection to turn it loose.

Q. Well, I think we could all agree, as a common-sense proposition, that it at least does this. It weakens the body and weakens the resistance of the individual to the encroachment of germs and disease?

A. That is a theory and it is rather a loose theory, because, what is resistance? How do you measure it?

Q. All medicine, like some law, is loose theory, isn't it? A. Yes. Not all medicine.

Q. So we will have to do some—I will not quite use the word guessing, but you do have to use a good (473) deal of theories in medicine?

A. In some parts of medicine and other parts are quite scientific. [240]

Q. That is true and that is also true in law. Then the theory is in medicine that the undernourishment, by weakening the body and, as you said, I will use the term again loosely, diminish the resisting power of nature; will make the party more susceptible and spur latent germs or permit latent germs to become active; am I not right?

(Testimony of Dr. H. T. Southworth.)

A. I don't know that you are, because you active—

Q. Do you know that I am not?

A. Well, as I say, you have used a term there that I can't recognize, because it does not mean anything to me.

Q. What term do you object to and I will try to eliminate it?   A. Lowered resistance.

Q. Very well.   I will eliminate lowered resistance. What principles in the body proper is it that makes nature build this shell around the germ and hold the thing present?

Q. Please term that or explain it in terms, as near as you can so that we will all get it?   (474)

A. Well, I don't know as I can put it in language that you will understand.

Q. Well, it is the power of a vigorous nature that the Almighty has wonderfully put into the body that in a vigorous body is intended to protect it against the ravages of germs and disease; isn't that it?

A. Not quite.

Q. Very well, modify it until it will express it quite from your point of view?

A. Because it would be quite immaterial whether the body were vigorous or not, if it were a primary infection—the presence of the tubercular bacilli. [241]

Q. Doctor, isn't it true that the tubercular bacilli in a strong, vigorous, healthy body is far more likely to remain quiescent and inactive than if the body becomes weak, emaciated and injured by disease or any other cause?

(Testimony of Dr. H. T. Southworth.)

A. It does not necessarily follow in tuberculosis. It depends upon the size of the infection and the allergy.

Q. You haven't quite answered my question, Doctor. Please read it and get the point.

(Question read by the reporter.)   (475)

A. I would have to answer the same; that that does not necessarily follow but would depend upon the size of the infection and the allergic reaction which that individual possessed.

Q. And, let me see, have you defined that allergy reaction so that we know what we are talking about?

A. I don't think so. You interrupted me when I started.

Q. I beg your pardon. I am sure I did not get your definition. Please give it to us.

A. It is a condition produced in the tissues of the body by the presence of the tubercular bacilli, which stimulates the proliferation of connective tissue cells within the tissue where the infection is situated, the connective tissue cells being the framework of that structure.

Q. I would like to have that read for my own benefit, to see whether I have it expressed.

(Answer read by the reporter.)

A. In addition to that effect, locally it has the effect of rendering tissues, even in distant parts of the body allergic or, in terms which the lay person would understand, on guard.   (476)   [242]

A. A form of immunity.

(Testimony of Dr. H. T. Southworth.)

Q. Are you through, Doctor, with your definition?

A. I think that will stand.

Q. Can't you put that in simpler common language? Can't it be done? I am not insinuating that it can, but I am trying to get from you whether or not—frankly, I haven't gotten very far tracing that from your description of the—

A. Well, when one gets down into common terms, you are likely to say something that would be interpreted differently and I would be misquoted. Perhaps, if you asked me some questions, I may—

Q. Well, the way is so blind to me, from your definition, that I hardly know where to butt in, but I will try along other lines, Doctor, and maybe you will develop the same thing. I think we have handled the question of undernourishment and you say you can't tell us just how or why the effect on the body where undernourishment can produce the active condition of secondary infection? You can't tell us that?

A. I don't know what happens to this in cases of undernourishment. Undernourishment weakens the body. That is the only thing that happens that we are all sure of. I don't know that that is the thing that stirs the germ into activity. I won't say it is not.

Q. Unless you can tell us something better that does it, wouldn't we be fully justified in assuming that the weakening of the body by undernourishment is the reason why undernourishment spreads the germ—won't we be justified in supposing that?

(Testimony of Dr. H. T. Southworth.)

A. Well, I think that is getting right back to the question of lowered powers of resistance, which is a loose [243] term and, if it suits you, it is all right, but it does not suit me.

Q. I was simply trying to use a term that did suit you, because it expresses pretty clearly to me what it means. Let's pass from undernourishment. The second reason which you give is acute respiratory infection. What do you mean by that?

A. Would you mind completing it for my benefit?

Q. That is, as I take it, it had been. (478)

A. You missed measles, then. I said acute respiratory infections and measles and I would much prefer to use it all, if you don't mind.

Q. I misunderstood you. I thought you gave acute respiratory infections as one and measles as another. Was I wrong?

A. Yes, sir. Because measles is so closely allied. Acute respiratory infection and measles is all one cause. I can't separate them very well, because they are so closely allied. Measles and acute respiratory infections are so closely allied, in their common symptoms is what I am trying to say. The term acute respiratory infection covers a condition without any thought of measles. But I want to put in measles there, too. (479)

Acute respiratory infection may occur without measles. Measles may occur without acute respiratory infection. It is not the combination of the two things together that may produce tuber-

(Testimony of Dr. H. T. Southworth.)

culosis. May I explain? Acute respiratory infections is given as one of the causes of secondary infection of tuberculosis. Measles, because of the fact that it is the second (480) most prominent symptom, the [244] first being the eruption and the second being the terrific coughing, is so closely allied to and yet can't be called an acute respiratory infection that it is classed in with them as the causes but certainly not that you must have an acute respiratory infection and measles in order to produce tuberculosis, no.

Q. Well, now, I guess I get you; that those two things in combination may be a provoking cause?

A. Well, I don't think that I get you. If you mean in combination in the same patient—if an individual patient has both at the same time to provoke it, no. Let me illustrate again. Willie has the measles and it is found, following the measles, he has a secondary infection of tuberculosis. Johnnie has repeated colds and, as a result of his repeated colds, he has secondary infection of tuberculosis. Tommy has pneumonia and, after he is convalescent or about the time he is convalescent from his pneumonia, he has secondary. Jack has bronchitis—acute bronchitis and we find that he has secondary infection. Now, I have named four different conditions but they are all in one group but not all in one patient. I don't know that I have made it clear.

Q. As far as you have gone, you have made it

(Testimony of Dr. H. T. Southworth.)
very clear to me now. I think I am following you
now. Go (481) on.

A. I shall be very glad, if I haven't made it clear,
to try to. I have finished that unless you wish me
to go ahead with other illustrations.

Q. I want you to go ahead, if you can make that
any clearer. You are giving us acute respiratory
infection and measles and explaining the relation
of it and you started in by illustration, which I
thought, perhaps, you had not completed. [245]

A. I think that is quite sufficient. I could not
add to those various things I have given, typhoid
fever. (482) Very strange to say, it is practically unknown and that is just exactly the point,
sir, why I object to having it tacked on to me lowered resistance. Nothing else in the wide, wide
world that will lower resistance of an individual
than typhoid fever and yet secondary infection of
tuberculosis practically never follows it, that is,
immediately or even in the immediate future.

Q. That is your opinion?

A. That is the opinion of the medical profession.

Q. All of it, Doctor?

A. Oh, I suppose you might find some that would
disagree with me. And I dare say some people
who thought they knew as much about it as I did.
I presume it is true there is frequent disagreement
on medical questions such as we are now discussing.
From my point of view, I eliminate typhoid fever
as one of the producing causes. I cannot tell you
why typhoid fever does not produce secondary in-

(Testimony of Dr. H. T. Southworth.)

fection, when these other troubles do, excepting this: (483) That typhoid fever does not produce the condition within the chest, where 95% of all primary infections are quiescent, as happens in measles or the acute respiratory infections. That, I give, as a probable reason.

Q. But, you would not give that as the definite statement of uncontroverted fact, would you, Doctor?

A. Well, I would have to see controversion first before I would accept it. I don't think it has been successfully controverted.

Q. There are other reasons which might account for it, are there not? Let me illustrate. I will try to be brief, [246] if possible, because I do not want to follow this too far. One of the cures practiced for tuberculosis is the injection of tuberculin, isn't it—inoculation?

A. That is a treatment that is followed by some men.

Q. And the principle of that, upon which they claim that it works, is that that tuberculin produces something in the body that kills the germs of the tuberculosis, roughly speaking. I am not giving it, of (484) course, with medical accuracy. Isn't that the fact?

A. I think the man who has done the greatest amount of work is Sir Arbuthnot Wright in England, who calls it the opsonic index and that the effect is that it does one of two things, either renders the tubercular bacilli more readily to at-

(Testimony of Dr. H. T. Southworth.)

tack by the white blood corpuscles or that it stimulates the white blood corpuscles to greater powers of assimilating the germ. In effect, usually killing the germ, but not always.

Q. And that is the same thing that inoculation and vaccination for smallpox does in the same sort of a way—in a general way?

A. Well, that is explained by Ehrlich by his side chain theory.

Q. And it is also understood and talked in a degree in the medical profession that at times one disease renders you immune from the recurrence of that disease or a similar one in the body? (485)

A. Well, is this the smallpox; do I understand your question?

Q. That would be one, yes, sir. [247]

A. Such as that, yes, sir.

Q. Now, is it possible, assuming that typhoid fever does not produce tuberculosis—is it possible in medical theory that typhoid fever is one of the things which possibly operates to immunize or kill or render inactive the tubercular germ?

A. I have no idea. I have never heard that advanced before.

Q. What is it that these other diseases which you have mentioned do to the body which produces that germ to get into activity? (486)

A. It does not produce the germ.

Q. Induce it?

A. I think, in respiratory conditions which we have mentioned, there is no doubt but what the

(Testimony of Dr. H. T. Southworth.)

fact that there is an inflammatory process in the immediate neighborhood of the primary infection which is quiescent tends to break down the capsule and the second is the continuous cough and the surging of the chest wall which produces the same effect. That, in a sense, would be a traumatic or force cause that produces the activity.

Q. These quiescent, inactive germs, is there any part of the body in which they might be expected to be found? Is there any part of the body or any definite spot in which they would not be expected to be (487) found?

A. Oh, yes. I think, many. The most usual place is in the lymph nodes at the base of the lungs is the—very, very much less often but next in importance would be the retroperitoneal or lymph nodes. The next would be the cervical glands. Very, very few in the axillary glands. [248]

Q. It is your opinion that they do not go generally through the body? A. In a primary infection.

Q. In a primary infection, yes.

A. And then what?

Q. They are in various parts of the body?

A. No, sir. If they go generally through the body, your patient dies. Just can't overpower all of this massive infection. If stopped, it must be a small numerical infection.

Q. You misunderstood my question. A single germ getting into the body, is that not likely to stop and become quiescent in most any part of the body?

(Testimony of Dr. H. T. Southworth.)

A. Oh, I think, sir, when you come right down to
(488) actual facts, I doubt if any single germ
ever does lodge in the body.

Q. Let's take two, then. I want to get a small
unit. You said you doubt if a single germ—I mean
one unit of primary infection, if that will be clearer
to you—

A. No, it is not, because it may be two different
things. A primary infection may consist of several
or many or, as you suggest, possibly one and it is
rather hard to conceive of just one single, solitary
germ. It is so infinitely small that unless it were
a secondary infection the allergic reaction might be
great. (489)

Q. What I mean is this. You have not caught
my meaning. My meaning is this. Doctor, you
have spoken about germs in the course of primary
infection—a germ or germs producing primary infection. Now, let's take the minimum unit of such
germs producing a primary infection, whatever we
call that few, may not that minimum unit find an
emplacement [249] in most any part of the
body?

A. If you mean that all things under God are
possible, yes, but I would say that it was very im-
probable.

Now, dissipation, I mentioned as the third
cause. (490) What it is that dissipation does
to the body which stirs these germs into activity is again theory. The theory is that the alcohol

(Testimony of Dr. H. T. Southworth.)
consumes the body fat, weakens the body, if that satisfies you.

Q. Isn't that what you mean?

A. I did not say that, sir. They might be quite strong physically but devoid of fat. I did not say that it is the absence of fat which stirs the germ into activity. That is the first effect of your under-nourishment and your alcoholism, where it reaches —to that extent, consumes the fat—to burn the fat. I don't know whether there is any other reason besides the consuming of the fat that the alcoholic dissipation has upon the body that stirs the germ into activity. I don't know and I would not say that the weakening of the body eventually in its general sense is what stirs the germ. (491) I could not say that, for this reason; you would then ask me what about mailiary tuberculosis. It is the second infection and it more often than not attacks the apparently strong and I would be un-able to answer you and that is true. Now, by dis-sipation, I did not limit myself merely to the use of alcohol. Any kind of dissipation, I think, if carried to sufficient extreme, probably would have some of that effect. Insufficient sleep and rest and general roistering, will have the same effect if car-ried to sufficient point.

Q. Isn't that begun not by diminishing the fat but by weakening the body that produces the result; isn't that true? [250]

A. I don't know what you mean by weakening the body.

(Testimony of Dr. H. T. Southworth.)

Q. Isn't that a sensible common term that any man understands? (492)

A. Well, may be so, but it is not to me. I don't understand it.

Mr. BRENNAN.—You have said overwork. Is that a question that you can't make him understand?

Mr. WOOD.—I can't make him understand my point of view by any other questions along that line.

Q. You have stated overwork. What does overwork do that produces this germ into activity or spurs this germ into activity?

A. I don't know that any of these things spur any germ into activity.

Q. You said that is what caused the—active condition of the tuberculosis, didn't you?

A. I know, but I don't think that I said that it produced it by spurring the germ to activity. I stated that overwork was one of the—causes of secondary infection of tuberculosis.

Q. And, you have explained what secondary infection meant. Now, how does overwork cause the secondary infection?

A. I am not prepared to tell you, sir, just what it does. (493)

Q. Again, isn't it by lowering the resisting powers and weakening the body?

A. The expression means nothing to me.

The collision of a train with this young man, who had never before shown symptoms of tuberculosis,

(Testimony of Dr. H. T. Southworth.)

the breaking of his leg, crushing of a bone in a comminuted fracture, the six months required in the hospital, with the various operations [251] and treatment that have been outlined in the question; these things could have none of the results upon the body, in my judgment, that overwork or dissipation would have that might cause the secondary infection of tuberculosis that has been described. It could only be overwork and dissipation and it could not be the effect on the body from these injuries. (494)

TESTIMONY OF DR. JOHN W. FLYNN, FOR DEFENDANT.

Dr. JOHN W. FLYNN, a witness on behalf of the defendant, testified as follows:

Direct Examination.

My name is John W. Flynn. I reside at Prescott, Arizona. I am a physician. I have practiced my profession a bit more than thirty years. I am a graduate of a regular school,—of the Medical University, Montreal. I have practiced medicine in Arizona, since '98· (495) All of the time here at Prescott, except three years in Kingman. I have specialized in tuberculosis. For the past twenty years, I have devoted all of my time to the treatment of that disease. For three or four years before that, the greater part of my time. In the course of my practice, I have come into professional contact with quite a large number of cases.

(Testimony of John W. Flynn.)

I have made it a special study for the purpose of trying to determine the sources and causes of tuberculosis. And the treatment of that disease, so as to overcome it.

Q. Now, Doctor, I am going to propound a question to you—a hypothetical question but, for the purpose of the question and your answer thereto, you may assume all of the facts therein stated to be as stated and, with that understanding, Doctor, I am going to ask you this. (496)

Assume the condition of a boy seventeen years old, [252] reared on a New Mexico ranch, an ordinarily, apparently, healthy boy, who had not been known by his parents and surroundings to have any chronic disease, just an ordinary ranch boy, although undernourished in appearance, but no history of tubercular activity that people had discovered. Assume that such a boy was injured in a railroad accident; that he was struck by an engine while upon or riding upon an automobile truck, his thigh broken in a comminuted fracture, accompanied by jamming and contusing and lacerating of the muscles and tissues surrounding the bone, pieces of the bone being broken loose and that this injury occurred on the 11th day of June, 1923; the boy was brought to a hospital—St. Joseph's Hospital in Albuquerque, New Mexico, a good, well sustained and reputable hospital, and was treated by skilled surgeons; that due to the laceration, bruising and the shock accompanying the injury, it was found improper or unwise to

(Testimony of John W. Flynn.)

attempt to reduce the fracture for a week's time. At the end of that time, the fracture was reduced, the condition of the leg and the bone broken being as shown by the X-ray; that an operation was then performed, cutting down into the leg and removing a section of bone at the break and of some small pieces of fragments of bone, placing the leg in position (497) and maintaining it by what is known as a silver or Lane's plate; that the contused and lacerated condition in the interior of the leg and the wound was such as to make it advisable, in the opinion of the surgeon in charge, to place the leg in a rigid plaster cast but, instead, due to that condition, the leg was bound to a board or splint, extending from the armpit to the leg; that the boy remained in the hospital until the 3d of August, 1923, when it was discovered in some way that the [253] splint had become loose, the boy was complaining of pain in and about the fracture and that he was then removed and a second X-ray picture taken, showing the bone bent somewhat out of place; that then the bone was forced back into position, as near as could be, and a plaster cast applied and continued in that position until the 14th of September, 1923, when another X-ray picture was taken, showing the condition of the bone, the Lane's plate and the surroundings, and another X-ray picture taken on September 21, 1923, also showing the position of the bone, the plate and the surroundings and that but very little calsification or knitting or uniting of the bone was shown; that an-

(Testimony of John W. Flynn.)

other operation was performed, in which the Lane's plate was unscrewed and removed and the wound again closed; that the boy continued in the hospital until about the 14th of January, 1924, (498) some seven months after the injury, when an operation was performed for appendicitis. Assume that at the time the boy first was seen by the surgeon on the 11th day of June, 1923, that he appeared to be undernourished but no tubercular activity was then ascertained or discovered by the physician, nor had there been any tubercular history in his family, so far as known, and no apparent activity of the disease in himself or in his surroundings in his family; that at the time of his operation on the 14th of January, 1924, or about that date, the surgeon, upon opening the abdomen, the boy then complained of extreme pains in the abdomen and showing indications of appendicitis, it was found that the intestines and the appendix showed a very bad condition of tuberculosis spread all over the intestines, the appendix and the lining of the abdomen, a genuine case of tuberculosis of the bowels, and the [254] lining of the abdomen covered by numerous tubercles spread over them. From the history of this case, Doctor, as I have given it to you, are you able to say with reasonable accuracy, with your experience as a specialist in this disease, what was the producing cause of that tubercular activity found in January, 1924?　A. I am.

Q. What, in your opinion, Doctor, was the producing (499) cause of the tubercular activity in

(Testimony of John W. Flynn.)
this boy that has been outlined in the question as
it existed on the 14th day of January, 1924?

A. There were two causes. One was a large infection with the tubercular bacilli in childhood and
the other was a thin envelope of fibrous tissue
around these bacilli. The accident to which attention has been called in this question and the injury
sustained by the boy described in that question had
nothing to do with the stirring up of these tubercles.

Q. Was there any connection with that accident
of any trauma or injury produced thereby or therefrom that, in your opinion, caused the tubercular
condition described in the hypothetical question?

A. Nothing that appeared in your question. In
my opinion, the effect upon the boy by reason of
the fact that he was in the hospital from the 11th or
12th day of June, 1923, down to January or February of 1924, being confined to his bed as outlined in
the question and receiving the benefit of good hos-
pitalization would be beneficial, so far as tending
(500) to prevent the spread of the tuberculosis
to the bowel and the peritoneum and to the spread
of the tuberculosis in the bowel and the peritoneum.
I will explain just why I say that, [255] in the
first place, it has been definitely proven that more
than 90% of all people in civilized countries are
infected with tuberculosis in their childhood.
What happens, when these tubercular bacilli enter
the body, is that unless they produce death within
a comparatively short time, nature forms an en-

(Testimony of John W. Flynn.)

velope of white fibrous tissue around these bacilli and encapsulates them and they remain encapsulated there and most of them live as long as the person lives. When a person develops active tuberculosis, what happens is this—a person in adult life—what happens is this; it is not that they become reinfected from the outside but the fibrous tissue—the fibrous envelope which is around these bacilli weakens and breaks down and the bacilli are spread to different parts of the lungs or diffent parts of the body. An infection of the bowels and peritoneum, such as was described in this hypothetical question, is necessarily a secondary (501) infection borne to those parts through the blood stream. What happened was a nodule in some part of the body, in which nodule there were a number of tubercular bacilli and the fibrous envelope on the outside—the fibrous envelope weakened, the bacilli got in the blood and I might say carried down to the peritoneum and the bowels and set up the active tuberculosis there. The reason that the resting would tend to prevent this is that rest is one of the most recognized means of thickening up the fibrous tissues. I don't think that that would be any more accentuated from the fact that the patient was a boy say seventeen or eighteen years of age—I would not expect to get better results in a patient of that age than I would in an older person. [256]

Q. But, you would get these beneficial results to which you have alluded?

(Testimony of John W. Flynn.)

A. The tendency would be beneficial.

Q. Now, Doctor, in your opinion, and based upon your expert knowledge on this subject and this disease, what do you say as to whether the accident described in the question and the injury produced thereby was in any wise related to or produced the tubercular condition described in the question?

A. I don't think it was related to or produced it— (502) related to it in any way or had anything to do with producing it.

An expression is commonly used, called lowered resistance. I come upon that term in my experience very frequently. I attach to that phrase no importance whatever, so far as tuberculosis is concerned. The reason why I attach no importance to it is because it is an indefinite term which has never been—which has never been described and, so far as I know, probably means something different to practically everybody who uses it. In other words, so far as I know, it has no definite meaning. In view of the condition of this boy, as outlined and described in the hypothetical question, the tubercular condition probably would have followed, without regard to any railroad accident or other accident or trauma.

Cross-examination.

Counsel has inquired of me about the term "lowered resistance," which I have stated (503) has no definite meaning. That term is not used by medical text writers who are well posted on this

(Testimony of John W. Flynn.)

subject. It is used in [257] medical text-books, yes, sir. And written by men who claim they are sufficiently expert to write books for the profession, but not to have technical knowledge on that particular question. I don't think it is used in a great many books that have been written by medical men on the question of tuberculosis—not a great many. I don't know what those men mean when they use the term "lowered resistance."

Q. Ordinarily, that is a term which the ordinary person quite well understands, isn't it, "lowered resistance"?

A. No, I don't know what the ordinary person does. I am not able to understand. The term "lowered resistance" or weakening of the bodily strength to resist means, technically, nothing at all to me. Regarding tuberculosis (504) I am speaking, now, altogether. I have stated that tuberculosis of the kind shown in this boy is what is known as secondary infection and by that I mean the spreading of germs previously in his body. I stated to the jury that these germs, as they are taken into the body, and they are practically in your body and mine this minute, are closed up in the envelope that nature has built around them. That is correct. And some cause must come along and weaken that envelope before they break out and become active. And, then, the quest that we have before us is to find out, if possible, what that cause was that weakened and tore down the envelope. (505) It is not possible to do it right now,

(Testimony of John W. Flynn.)

for a certainty. It is a matter of opinion, based on experience, of course, and investigation. My own profession, feels that there is very much to be learned still about tuberculosis and the action of the germs. Wise men have, in the last twenty years, developed and demonstrated to some degree that theories in regard to [258] the activity of these germs existed and a few more years have torn down and destroyed very many of those theories. It is impossible for you or I to say that the next five or ten years would not cause you or I to modify the view I have now expressed in regard to the activity, although it is not probable in this case at all. This is pretty well in the realm of established fact.

Q. Well, that is substantially the view of many, many able men who have written on this subject for twenty (506) years that would be—what they thought were established facts have later been found to be erroneous?

A. But that was before the days, very largely of the scientific methods that are now being employed in the study of tuberculosis. And those methods are being increased and improved constantly.

Q. Doctor, let me, without reading it, because you will have it generally in mind, ask you to bear in mind the question put to you by counsel, the hypothetical question illustrating the condition of the boy—let me eliminate entirely from that question any undernourishment in appearance of this boy at any time and leave him a strong, vigorous,

(Testimony of John W. Flynn.)

healthy boy at the time he received this injury—
eliminating, you understand, merely this undernourishment—now, what produced that tuberculosis
to follow this injury—that active condition of infected bowel on the face of what I have given you
there?

A. A breaking down of the fibrous tissue to allow
the escape of the tubercular bacilli. I don't know
what broke it down. (507) [259]

Redirect Examination.

Based on my experience, I do know that it was
not the accident. The accident had no connection
with it.

Recross-examination.

Q. And, yet, the question, Doctor, eliminates
every other known cause, don't it, when you have
eliminated the undernourishment? The question
and the facts stated in it, give you no other known
reason that produced this activity?

A. Oh, yes, there are four probable reasons.

Q. Oh, I know, theory, but, I say the question
gives you not another basis or reason upon which
to base it—the question as asked?

A. Nothing except the general experience that I
have outlined to you. If this boy never had had
the injury at all or had continued in the same degree of health he would probably have developed
all of (508) this tuberculosis sooner. The rest
put it off—development of it.

(Testimony of John W. Flynn.)

Q. Why, Doctor, you won't tell the jury that the injury such as this boy received and that we outlined to you in that question would not more than counteract any resting that he had in the hospital, would you?

A. It certainly would not.

Redirect Examination (Continued).

Q. Doctor, what are the four theoretical things or principals that would account for the stirring up of this bacilli?

A. May I change the word? Not theoretical. Absolutely established in principle. Perhaps theoretical regarding [260] any particular case.

Q. Well, from a scientific examination of the subject and based upon your experience as an expert, together with the literature that you have read upon the subject, you may now state what the four principal reasons that suggest themselves to your mind would account for the stirring up or spreading of the bacilli?

A. The first and most important is undernourishment. The second one is, and probably next in importance, (509) acute diseases of the respiratory tract. The third is dissipation and the fourth is overwork, about in that way.

THEREUPON the defendant rested its case. (510)

THEREUPON, the jury having retired from the courtroom, and counsel being present, the following proceedings were had:

(Testimony of John W. Flynn.)

Mr. REED.—If the Court please, at this time, the defendant, with the consent of the Court, renews the motion heretofore made for a directed verdict, upon all of the grounds heretofore stated in support of the motion made at the close of the plaintiff's case—each and all of the grounds.

The COURT.—The motion will be denied.

Mr. REED.—Am I right in my understanding, your Honor, that the reason or grounds for denying the motion is that it is based on the constitutional provision of the State of Arizona?

The COURT.—I don't care to limit the reason for the ruling. That is one of the reasons.

Mr. REED.—We reserve an exception to the ruling on the motion. (511) [261]

THEREUPON, at the hour of 1:30 P. M., the jury were returned into court.

THEREUPON, the cause was argued to the jury by counsel.

AND THEREUPON, counsel for defendant requested the Court in writing to give to the jury the following instructions:

REQUESTED INSTRUCTIONS OF COURT TO JURY.

(Title of Court and Cause.)

"You are instructed to return a verdict in favor of the defendant, The Atchison, Topeka and Santa Fe Railway Company," which said instruction the Court refused to give to the jury, and to the refusal of the Court to give the same, the defendant, before the jury had retired from the courtroom, duly excepted.

"As the collision between plaintiff and defendant's railroad train occurred in the State of New Mexico, and as Roy Spencer is a resident of that state, the laws and construction thereof by the courts of New Mexico are controlling upon you in deciding this case.

Under the system in New Mexico it is contributory negligence for a person to approach a railroad track without paying careful attention for the approach of trains on such track in both directions," which said instruction the Court refused to give to the jury, and to the refusal of the Court to give the same, the defendant, before the jury had retired from the courtroom, duly excepted.

"You are instructed that under the laws of the State of New Mexico a railroad company has the right to construct its railroad at grade across any highway which its said [262] railway shall intersect, but that such company shall restore such highway to its former state, as near as may be, so as not to unnecessarily impair its use. You are further instructed in this connection that the laws of the said State of New Mexico govern and fix the rights and duties of the defendant in so far as the construction of said crossing at which plaintiff was injured are concerned.

You are, therefore, instructed that if you find from the evidence that in constructing its railroad across the highway on which the plaintiff was traveling when injured, the defendant restored said highway at the place intersected by its said railroad to its former state, as near as may be, so as not unnecessarily to impair its use, you cannot find the defendant negligent in respect to that matter"; which said instruction the Court refused to give to the jury, and to the refusal of the Court to give the same, the defendant, before the jury had retired from the courtroom, duly excepted.

"You are instructed that a railway track is of itself a warning. It is a place of danger. It can never be assumed that cars or trains are not approaching on a track, or that there is no danger therefrom; it is, therefore, the duty of every person who approaches a railroad track to exercise proper vigilance to ascertain and to know whether any trains or cars are approaching thereon before attempting to cross thereover. Proper vigilance requires every such person to look and listen before going upon said track or so close thereto as to be in danger. The exercise of ordinary care required of a traveler upon the highway under such circumstances, also [263] requires such person or persons to stop, if necessary, in order to look and listen; and if for any reason, the view to any extent is obstructed, the person or persons endeavoring to cross such track or tracks or going so close thereto as to be in danger either of their person or property, should exercise a greater precaution than ordinary prudence would dictate in such an exigency; that is, in stopping and looking and listening. The looking and listening and stopping thus required should be exercised at the last moment which ordinary prudence would dictate before passing from a place of safety to one of danger. If the ability either to look or listen be obstructed by natural or artificial objects, then the duty to stop before entering into a zone of danger in order that proper investigation may be made, becomes absolute.

It is also equally the duty of one accompanying the driver of an automobile to exercise care for his own safety, and if you believe and find from a preponderance of the evidence that said Roy Spencer, without objection or protest, permitted the said Benjamin B. Spencer to drive upon the railroad crossing immediately in front of an approaching engine and train of cars without stopping to look or listen or without having the automobile under such control that it might be stopped in time to avoid going on the crossing in front of the approaching train, or without exercising any care on his own part to ascertain whether the crossing was safe, then you are instructed that the plaintiff, Roy Spencer, would be guilty of negligence contributing to his own injuries, and your verdict must be for the defendant,'' which said instruction the Court refused to give to the jury, and to the refusal of the Court to give the same, the defendant, before the jury had been retired from the courtroom, duly excepted.
[264]

''You are instructed that even if you find from the evidence that Roy Spencer looked and listened for the approach of the train in a place of safety, yet your verdict must be for the defendant if you further find from the evidence that said plaintiff was not constantly and actively vigilant in approaching and listening for the approach of the train during the time occupied by him in going from the place where he may have stopped, looked or listened to and upon the track," which said instruction the Court refused to give to the jury, and to the refusal of the Court to give the same, the defendant, before the jury had retired from the courtroom, duly excepted.

"The passenger train involved in the accident described in the complaint and referred to in the evidence was running in interstate commerce between states and it was the right of the defendant to operate said train at a high rate of speed and no presumption of negligence is to be indulged from the mere fact of the speed at which the train was operating. The reasonableness of that rule is apparent. The great purpose to be subserved by railroads is promptness, speed and dispatch in carrying passengers and freight. And under ordinary circumstances, in the open country, there is no duty resting upon the railway company to slacken the pace of its trains at crossings. To hold otherwise would, to a great extent, destroy the advantages derived from modern facilities for transportation. To impose upon the defendant the duty of operating said train at a rate of speed so that it could be stopped at every grade crossing in Arizona in time to prevent accident should any person or vehicle be crossing the   [265]   track on the highway, would be invalid under U. S. Constitution Article I, Section 8, as a direct and unreasonable burden upon interstate commerce when applied to said train and therefore you are instructed that the defendant had the right at the time and place and in the circumstances disclosed by the evidence to operate its said train in approaching the place of the alleged collision at a rate of speed of fifty miles an hour and you should not consider the speed of the train in this case as disclosed by the evidence as establishing the negligence of the defendant,'' which said instruction the Court refused to give to the jury, and to the refusal of the Court to give the same, the defendant, before the jury had retired from the courtroom, duly excepted.

"If you find that the automobile in which plaintiff, Roy Spencer, was traveling with his father on the day in question was in bad repair or deficient, that fact is not to be considered by you as in any way excusing plaintiff from exercising due care and circumspection for his own safety in crossing the railroad track,'' which said instruction the Court refused to give to the jury, and to the refusal of the Court to give the same, the defendant, before the jury had retired from the courtroom, duly excepted.

WHEREUPON, the Court gave to the jury the following instructions:

INSTRUCTIONS OF COURT TO THE JURY.

Gentlemen: It now becomes my duty to charge you with reference to the law that will guide you in your deliberations for the purpose of arriving at a verdict in this case. In the trial of civil cases, it is the exclusive province of the jury to determine the facts. It is the exclusive [266] province of the judge of the court to charge the jury as to the law by which they must be guided in their consideration of the case and this law, as stated to you in these instructions, it is your sworn duty to accept and follow as the law of this case. Any ideas that you may have as to what the law should be, it is your duty to disregard and to be governed by the law as given you by the Court. If the Court commits any error in stating the law to the jury, that error may be corrected in the manner provided by law. The pleadings in this case have been read to you and I will not take the time to read them to you in these instructions.

This is an action by the plaintiff, Roy Spencer, against The Atchison, Topeka and Santa Fe Railway Company, defendant, to recover damages in the sum of $25,000.00 for injuries received from an accident at a railroad crossing about one and one-half miles west of the village of Mountainair, in Torrance County, State of New Mexico. (513)

It is contended by the plaintiff that on or about the 11th day of June, 1923, while the plaintiff and his father, Benjamin Spencer, were riding along the public highway, which crosses the tracks of the defendant company about one mile west of the village of Mountainair, the automobile in which plaintiff and his father were riding was struck by a locomotive of a train of cars then being operated by the defendant company upon its tracks and that the said plaintiff, Roy Spencer, sustained injuries which resulted in serious and permanent wounds and bodily injuries from which he still suffers and will continue to suffer during the rest of his natural life. It is contended by the plaintiff that said collision was caused by the negligence of the defendant in operating its train. [267]

It is contended by the defendant that the collision or accident which resulted in the injuries to the plaintiff, Roy Spencer, were caused solely by and resulted wholly from the negligent acts and omissions of the said Roy Spencer, in that the said Roy Spencer failed to look or listen for the approach of defendant's engine and train of cars to said crossing and failed to take any precaution whatever to avoid said collision and failed to observe and heed the warning signals given by the sounding of the whistle and the ringing of the bell of the approach of said engine and train of (514) cars to said crossing and that said injuries to the said Roy Spencer did not result from an accident caused by any negligent acts or omissions on the part of the defendant or any of its agents, servants or employees. It is incumbent upon the plaintiff, in civil actions, to prove to you by a fair preponderance of evidence every material allegation of the complaint, in order to entitle him to recover. By a preponderance of evidence does not necessarily mean the greater number of witnesses on one side or the other, but it means the weight of the evidence, the degree of proof which is most convincing and persuasive to your minds, which satisfies your minds to that extent that you may act upon it as intelligent jurors. There may be many witnesses testifying to a fact on one side and a less number of witnesses or only one on the other side and yet the testimony of the one witness may be more convincing and per- suasive to your minds than the testimony of several witnesses on the other side, in which event the pre- ponderance of evidence would be with that side pro- ducing the one witness whose evidence was more con- vincing and more persuasive to your minds. [268]

You are instructed, Gentlemen, that negligence is the omission to do something which a reasonably prudent man, guided by those considerations which usually (515) regulate the conduct of human affairs, would do or is the doing of something which a prudent and reasonable man would not do. It is not intrinsic or absolute but is always relative to some circumstance of time, place or person. Neg- ligence is of no consequence unless it was the proxi- mate cause of the injury. The proximate cause of an injury is that cause which in actual and continu- ous sequence, unbroken by any efficient intervening cause, produces the injury and without which the result would not have happened. It is the efficient cause, the one that necessarily sets other causes in operation. This definition of negligence, as it re- lates to this cause of action, applies alike to the de- fendant and to the plaintiff.

When we speak of contributory negligence of the plaintiff, we mean the kind of negligence above de- fined. Contributory negligence is the negligence of the plaintiff, contributing with the defendant's negligence as a cause of the plaintiff's injury. The mere fact that an accident has occurred from which injury has been sustained will not give rise to a presumption that the injury is due to the negligence of one who is made a defendant in an action based thereon. If the evidence does not show any negligence on the part of the defendant, there can be no recovery, no matter how free from negligence the facts show the plaintiff to be. (516) The evidence must point to the fact that the defendant was guilty of the negligence charged in the complaint. However, all that the plaintiff, upon this branch of his case, is required to do is to make it appear by a preponderance of evidence that the [269] injury came in whole or in part from the defendant's negligence and from no other cause.

Negligence, which is the want or absence of ordinary care, is the gist of the action and the burden is upon the plaintiff to prove by a preponderance of evidence that the defendant's negligence was the proximate cause of the injury. The evidence need not be direct and positive. The fact of negligence in any given case is susceptible of proof by evidence of circumstances bearing more or less directly upon the fact. Where the evidence introduced by a plaintiff does not show negligence on his part and all other essential facts are proven to entitle him to recover, the defendant, to defeat recovery by the plaintiff must establish by a preponderance of evidence that the accident was occasioned by negligence of the plaintiff which contributed in whole or in part to cause the injury.

By proximate cause is meant a cause from which the injury is a natural and probable consequence and without which the accident would not have happened. The defendant has alleged as a defense that the negligence (517) of the plaintiff was the proximate cause of the accident and not the negligence of the, defendant. Therefore, the Court instructs you that, if you are satisfied by a preponderance of evidence that the proximate cause of the accident was not the negligence of the defendant, then the plaintiff cannot recover, even though you may be satisfied from the evidence that the defendant was negligent in some or all of the respects alleged in plaintiff's complaint. If, after considering all of the evidence in the case, you further believe that the plaintiff, Roy Spencer, was also negligent to any extent and that his negligence concurred and co-operated with the negligent act or [270] acts of the defendant to the extent that both the negligence of the defendant and the negligence of the plaintiff was the proximate cause of the injury, then the plaintiff cannot recover. You are instructed that the mere happening of the accident raises no presumption that the defendant was negligent but the burden of proof rests with the plaintiff in this as well as all of the issues in the case, except that of contributory negligence, and, as to this last issue, the burden of proof rests upon the defendant, unless contributory negligence appears from the evidence offered on behalf of the plaintiff.

You are instructed that you should not permit

any sympathy for the plaintiff or bias against the defendant (518) to influence you in any manner in arriving at your verdict. Your verdict must be based solely upon the evidence received and the law as given you by these instructions and not upon anything that you have otherwise heard or read. You must, likewise, be uninfluenced by the evidence that the defendant is a corporation engaged in the operation of a railway as a common carrier of freight and passengers. The parties to this litiga-tion are entitled to your calm and dispassionate judgment, the same as if they were both corporations or both individuals, no more and no less.

You are instructed that the degree of care and caution required to be exercised by a person approaching a railway crossing is dependent upon the circumstances and conditions existing at or near such crossing at the time and, the greater the danger existing by reason of such circumstances and conditions, the greater must be the care and caution required to be exercised by such person. [271]

You are instructed that a traveler on a highway approaching a railway crossing must use ordinary care in selecting the time, and place to look and listen for coming trains and he should stop to make observations, when necessary. It is his duty to use all of his faculties and it is not enough that he carefully listen, believing those in charge of the approaching train will (519) ring a bell or sound a whistle or operate the train at a reasonable rate of speed. He must take advantage of every reasonable opportunity to look and listen and he has no right to omit looking and listening because he has heard no appropriate signal of the approach of the train, even though the law requires that signals be given.

You are instructed that even though you should find that the employees of the defendant in charge of the train involved in the accident described in the complaint failed to sound the whistle or ring a bell as a warning of the approach of said train toward the point where the collision is alleged to have happened or operated said train at an excessive rate of speed, still such failure to signal the approach of said train or the operation of such train at such an excessive rate of speed did not relieve the plaintiff from the duty of exercising ordinary care to look and listen. You are instructed that a person traveling upon a street or highway crossed by the tracks of a railroad company have no right to presume that the railway employees engaged in running trains of cars or cars upon such track will signal the approach of all trains and plaintiff had no right to omit the performance of any part of his duty to look and listen for approaching trains, upon the presumption that the trainmen (520) would do their duty. [272]

You are instructed that while the defendant railway company and Roy Spencer each had the right to use the road across which the track of said defendant is and was laid, yet, because such trains and cars are run on a fixed track and with great momentum, it had the right of way or precedence upon or over those parts of the road occupied by the railway track and it was the duty of Roy Spencer to give unobstructed passage to the approaching train.

You are further instructed that nothing said by the Court in these instructions concerning the question or amount of damages is to be taken by the jury as any indication by the Court that the plaintiff is entitled to recover damages.

In the consideration of your verdict, you are not to be influenced by passion, prejudice or sympathy for either party.

The first question to be considered by you is the question of negligence and, in the consideration of that question, you should lay entirely aside the nature and extent of plaintiff's injuries and damage. If you find that the defendant railroad company was free from negligence or that the plaintiff was negligent, however slight, which in any wise contributed to the collision, you need not consider at all the nature (521) and extent of plaintiff's damages but you should, without hesitation, return a verdict for the defendant railway company.

If the railroad grade at any given place where a railroad is being constructed across a highway is so far above or so far below the level of the road or crossing, or so near to extraordinary cuts or curves as to seriously interfere with its use or seriously increase the danger of such use by bringing the road crossing to the level of the tracks and the situation [273] is such that the road can be carried over or under the tracks without unreasonable cost and expenditure and the danger or obstruction cannot otherwise be reasonably removed, then it is the duty of the railroad company to construct a crossing under its rail or carry the road above its tracks by a bridge, as the occasion requires. If it fails to perform such duty and, in an attempt to carry the road over its tracks at grade that is on a level with the rails, it seriously and unreasonably increases the danger and difficulty at the crossing, then the railroad company will be liable for any injury or damage proximately resulting because of such dangerous condition to one rightfully using the road.

This duty of the railroad company to restore roads as near as reasonably may be to their former con- (522) dition of safety and convenience is not limited to highway regularly established and laid out as such but it applies as well to all roads which at the time the public were openly and actually using—and notoriously using as a road or highway, nor will any length of time a crossing has existed relieve the railroad company from that duty. If a railroad crossing is so constructed at grade as to create a condition of extraordinary danger due to existence of adjacent cuts or other physical conditions concealing the approach of trains so as to render it probable that persons using the road would not be warned by the usual and ordinary measures or discover by the exercise of reasonable care the approach of trains in time to avoid a collision, then it is the duty of the railroad company to use extraordinary and additional precautions such as automatic signaling devices or other like methods to give warning of the approach of its trains or to approach such crossing carefully and with the train under such reasonable [274] degree of control as would enable them to discover and avoid injury to persons so placed in danger. It is for you to say from the evidence whether or not the defendant railroad company, in constructing its road at the point where the accident occurred, did restore the crossing as near as was reasonably possible and practical to its former condition of safety and use- (523) fulness and, if you find that they did not and that the plaintiff suffered his injuries as a proximate result of such failure, then the company is liable to respond in damage for the injuries they so caused, unless the plaintiff himself carelessly and negligently by his conduct contributed to produce his own injury. Independent of statutory provisions, it is the duty of a railroad company to cause suitable and reasonably efficient warning to be given by blowing of whistle of the approach of its trains to road crossings commonly used by the public, if danger to persons using such crossings may reasonably be apprehended from it and failure to give such warning. The laws of the State of New Mexico, which apply to this case in that respect, provide that every railroad corporation shall cause a bell of at least twenty pounds weight attached to its locomotive to be rung at a distance of not less than eighty rods from any road-crossing such as the one on which the plaintiff received his injuries and that for failure to comply with the statute the corporation would be liable for all damages which may be sustained by any person because of such failure. If you find that the defendant failed to comply with these requirements, then it would be guilty of negligence under the definition of negligence above given you. Ordinarily the rate of speed at which a train is travelling is not (524) [275] evidence of nor is it to be considered as negligence. However, if a railroad company in completing its line across a traveled road creates a condition of extraordinary danger where the opportunity of persons using the crossing to discover with reasonable care and diligence the approach of a train in time to protect themselves from injury is materially lessened, then the duty of the company and its employees to safeguard and warn persons using such road is increased proportionately to the danger so created and to approach such a crossing at so high a rate of speed as to make it impossible for them to control the train and avoid injuring persons so using the highway after discovering the peril may be taken into consideration in determining whether or not the defendant used reasonable care to avoid this injury. If you find from the evidence that the defendant failed of its duty in any of the particulars above specified and that such failure was the proximate cause of the collision and, if you further find that the plaintiff exercised reasonable care and prudence at the time for his own safety, then your verdict should be for the plaintiff. It was the duty of the plaintiff, Roy Spencer, to use that degree of care that a reasonably prudent person would use under like circumstances surrounding the case. It was his duty to look and listen for approaching trains before venturing (325) on the tracks. It was also his duty to ren-der all necessary and reasonable assistance to his father in operating the automobile. The law does not require a person approaching a crossing to keep his eye fixed on the track during the whole time to the exclusion of all other necessary duties. What would be due care in that respect under one set of surrounding circumstances might be negli-gence under another. The fundamental [276] rule and test is what should a person of ordinary care and prudence, situated as was the plaintiff, surrounded with all the circumstances and duties that he was, have done and would the plaintiff's conduct measure up to that requirement. In de-termining whether or not the plaintiff was negli-gent, it is for the jury to determine, under all of the circumstances surrounding the plaintiff at the time, and you have a right to take them into con-sideration, and it is for you to say whether, under all of those circumstances, a reasonably prudent person would have acted as did the plaintiff. If you find that such a reasonably prudent person, under those circumstances, would not have so acted but would and should have discovered the ap-proaching train in time to have avoided the danger, then your verdict should be for the defendant. If, on the other hand, you find that, considering all of the circumstances surrounding this plaintiff, he used as high a degree of care as a (526) rea-

sonably prudent person would have used and conducted himself as such a reasonably prudent person would under all of the circumstances in which
he was placed at the time, he was not chargeable
with any negligence which should deprive him of
his right to recover. Any carelessness or negligence on the part of Benjamin B. Spencer, the
father of plaintiff, in failing to observe the approach of a train in time to avoid the collision, if
he was so negligent, cannot be imputed to our prejudice the rights of the plaintiff, Roy Spencer, in this
case.

You are instructed that a traveler on a highway
approaching a railway crossing must use ordinary
care in selecting a time and place to look and listen for coming trains and he should stop to make
observations, when necessary. It is his duty to
use all of his faculties and it is not [277] enough
that he carefully listened, believing that those in
charge of the aproaching train will ring a bell or
sound a whistle or operate the train at a reasonable rate of speed. He must take advantage of
every reasonable opportunity to look and listen
and he has no right to omit to look or listen because
he has heard no appropriate signal of the approach
of the train, even though the law requires such signals to be given.

You are instructed that even if you should find
that (527) the employees of the defendant in
charge of the train involved in the accident described in the complaint failed to sound the whistle
or ring a bell as a warning of the approach of said

train toward the point where the collision is alleged to have happened or operated said train at an excessive rate of speed, still, such failure to signal the approach of said train or the operation of said train at such excessive rate of speed did not relieve the plaintiff of the duty of exercising ordinary care to look and listen.

If you find for the plaintiff under the instructions already given you, then you will award him such sum, not exceeding $25,000.00, which is the amount sued for in the complaint, as you find from the evidence will fully and fairly compensate him for all the damages he has suffered as a proximate result of the injury. He cannot bring any other suit against the defendant for this injury and is entitled to be compensated in this suit for the damages, whether past, present or future, that reasonably and proximately have been or will be caused by the injury. As elements of this amount, he would be entitled to be compensated for the pain [278] and suffering he has endured in the past and also for such pain and suffering he must endure in the future, if any, as you find will result from the injuries received. (528) He should also be compensated for the physical disabilities and deformities that have resulted and for the mental pain and suffering, as well as the physical, in the future that you find he will suffer and endure in the future because of such deformities and disabilities. He is entitled to recover such additional sum as will compensate him for the debt and liabilities he has incurred to pay for the medical and surgical care and nursing that he has received. He is further entitled to be compensated for such loss of earning power and capacity to earn a livelihood resulting from the injury, if any, that you find he will suffer after attaining his majority at the age of twenty-one years and during his reasonable expectancy of life and so long as you find such disabilities will exist. In determining this reasonable expectancy of life, you are entitled to take into consideration the mortality tables that have been introduced in evidence. While these tables are not controlling on you, you may consider them as an aid to guide you in determining that question. If the plaintiff is entitled to recover at your hands, under the instructions given you, as already stated, he should be compensated for all the damages and injuries that are the proximate result of the injuries which he has received in the collision.

If the plaintiff, at the time of his injury, was in (529) such physical condition because of any inactive, quiescent or latent germs of a disease and the injuries he received, by lessening his vitality or otherwise, aroused or spurred these latent and quiescent germs or tendency into [279] activity and thus produced or caused the disease to develop and become active, such disease, if so produced, created or brought into activity through causes and means stirred up and set in motion by an unbroken chain of cause and effect by the injury plaintiff received, then such disease would be the proximate result of the injury and the plaintiff would be entitled to recover for all the damages resulting to him or that will hereafter result to him because of such disease. The fact that one person, because of his physical condition, is subject to and suffers more serious results than another from a given injury and thus suffers unusual and unexpected results does not relieve the party causing the injury from the legal duty to make full compensation for all the damages proximately resulting to the injured party. It is for the jury to say, from all the evidence, whether any disease which the plaintiff was found to be afflicted with and suffering from after the injury or is now suffering from or afflicted with was the result of causes set in action by the injury itself and forming an unbroken chain of cause and effect reaching back to the injury without (*53*) interruption and, if you so find, then you should award the plaintiff full compensation for all such injuries and diseases and for whatever further disability, if any you find from the evidence, he will suffer in the future from such disease and its results. You should not in any manner consider or speculate on the reasons why the plaintiff brought this case in the State of Arizona instead of the courts in New Mexico, where the plaintiff lived and the accident happened. The plaintiff had a legal right to sue the defendant in the Arizona courts and the defendant, being sued in the state courts of Arizona, had a right to [280] remove this cause to the Federal Court and has done so and you should not consider these questions in any manner nor let them have the slightest influence or effect upon your verdict.

You gentlemen are the sole judges of the facts proven; also, the credibility of each and every witness who has testified before you and the weight that you will give to his testimony. In determining the credibility of any witness, you have a right to take into consideration his manner and appearance while giving his testimony, his means of knowledge of the facts to which he has testified, any interest or motive he may have, if shown, and the probability or improbability of the truth of his statements, when measured in (531) connection with all other evidence in the case. If you believe that any witness has wilfully sworn falsely as to any material fact, then you have the right to wholly disregard the testimony of such witness, except in so far as the same may be corroborated by other credible evidence or facts and circumstances proven in the case.

I charge you, Gentlemen, that your verdict must be based solely upon the evidence in this case. You must not render what is known in law as a quotient verdict. A quotient verdict is arrived at by each juror writing on a piece of paper the amount or sum to which he thinks the plaintiff is entitled, adding these sums together and dividing the total by twelve, the quotient being the amount of the verdict. A verdict of this kind is unlawful and cannot stand and I caution you against it.

Evidence has been introduced by the plaintiff, showing the use by the defendant of automatic signals at [281] certain classes of crossings. This evidence was received solely for the purpose of showing that signals of that kind are known, available and in use by railroads to protect crossings presenting conditions of extraordinary danger, requiring something more than the ordinary signboard or the customary signals to protect and warn persons using such highways and you should not consider that evidence for any other purpose.

You have heard statements during this trial that (532) the case was tried once before and you may have seen references in the daily press or otherwise as to what was the result or what verdict was rendered. You are instructed, Gentlemen, that the Court, for reasons it thought right, and it had nothing to do with the merits as now presented, ordered this case to be tried over again and set aside entirely the verdict and all proceedings had on a former trial and you are not to take any notice of that former trial or the verdict or consider or speculate on why the new trial was ordered. Those questions should not be considered by you but you should treat this case just as though it had never been tried before and that this was the first trial and you should consider nothing but the evidence as you have heard it here and the law as the Court now gives it to you.

Two forms of verdict have been prepared for you. One, "We, the jury duly impaneled and sworn in the above-entitled action, upon our oaths do find for the defendant." The other, "We, the jury duly impaneled and sworn in the above-entitled action, upon our oaths do find for the plaintiff and assess his damages in the sum of blank dollars." In the

event you find for the plaintiff, you will insert the
sum to which you think the plaintiff is entitled in
the blank space left in the verdict for that purpose.
When you have (533) [282] retired to the jury-
room, you will elect one of your number as foreman
of the jury, and, when you have agreed upon a
verdict, you will cause your foreman to sign that
verdict which represents your conclusion and return it into open court. Your verdict must be
unanimous. You may swear the bailiffs.

And the Court gave to the jury no further or
other charge or instruction.

And thereupon before the jury had been directed
to retire, and consider their verdict, and while the
jury were present in court, the defendant excepted
to those portions of said instructions of the Court
as charged as follows:

"If the railroad grade at any given place where
a railroad is being constructed across a highway is
so far above or so far below the level of the road
or crossing, or so near to extraordinary cuts or
curves as to seriously interfere with its use or seriously increase the danger of such use by bringing
the road crossing to the level of the tracks and
the situation is such that the road can be carried
over or under the tracks without unreasonable cost
and expenditure and the danger or obstruction cannot otherwise be reasonably removed, then it is the
duty of the railroad company to construct a crossing under its rails or carry the road above its tracks
by a bridge, as the occasion requires. If it fails to
perform such duty, and in an attempt to carry the

road over its tracks at grade, that is, on a level with the rails, it seriously and unreasonably increases the danger and difficulty at the crossing, then the railroad company will be liable for any injury or damage proximately resulting because of such dangerous condition to one rightfully using the road." (522) [283]

"If a railroad crossing is so constructed at grade as to create a condition of extraordinary danger due to existence of adjacent cuts or other physical conditions concealing approach of trains so as to render it probable that persons using the road would not be warned by the usual and ordinary measures or discover by the exercise of reasonable care the approach of trains in time to avoid a collision, then it is the duty of the railroad company to use extraordinary and additional precautions such as automatic signalling devices or other like methods to give warning of the approach of its trains or to approach such crossing carefully and with the train under such reasonable degree of control as would enable them to discover and avoid injury to persons so placed in danger." (523)

THEREUPON, the jury retired. (535)

THEREUPON, at the hour of 8:10 P. M., Sunday, August 15, 1926, all parties present and represented by counsel, the jury returned into open court with their verdict as follows:

LAW–170—PCT.

"ROY SPENCER, an Infant, by SARAH E.
SPENCER, His Guardian *Ad Litem,*
Plaintiff,

vs.

THE ATCHISON, TOPEKA AND SANTA FE
RAILWAY COMPANY, a Corporation,
Defendant.

VERDICT.

We, the Jury, duly empaneled and sworn in the
above-entitled action, upon our oaths, do find for
the Plaintiff and assess his damages in the sum of
Sixteen thousand seven hundred fifty Dollars
($16,750.00).

JOHN W. BEACH,
Foreman." [284]

THEREUPON, and on the same date, to wit,
August 15, 1926, the Clerk of said court entered
judgment in favor of the plaintiff and against the
defendant in accordance with the verdict, as follows:

"In the District Court of the United States in and for the District of Arizona.

No. L.—170.

ROY SPENCER, an Infant, by SARAH E. SPENCER, His Guardian *Ad Litem,*

Plaintiff,

vs.

THE ATCHISON, TOPEKA & SANTA FE RAILWAY COMPANY, a Corporation,

Defendant.

JUDGMENT ON VERDICT.

This cause came on regularly for trial on the 11th day of August, 1926. The parties were present in person, and represented by their attorneys, Mr. F. E. Wood of Albuquerque, New Mexico, and Mr. Robt. E. Morrison of Prescott, Arizona, counsel for plaintiff, and Mr. Robert Brennan and Mr. M. W. Reed of Los Angeles, California, and Messrs. Norris, Norris & Flynn of Prescott, Arizona, counsel for the defendant. A jury of twelve persons was regularly impaneled and sworn to try said action. Witnesses on the part of plaintiff and defendant were duly examined, and the cause was thereafter continued and tried on the 12th, 13th, and 14h days of August, 1926. After hearing the evidence, the arguments of counsel and instructions of the Court, the jury retired to consider [285] their verdict, and subsequently, on the 15th day of August, 1926, returned into court with their verdict signed by the foreman in accordance with the law, and being called, answer to their names and say:

'We, the Jury, duly impaneled and sworn in the above entitled action upon our oaths, do find for the plaintiff and assess his damages in the sum of sixteen thousand seven hundred fifty Dollars ($16,750.)

JOHN W. BEACH,

Foreman.'

NOW, THEREFORE, IT IS ORDERED that judgment be entered herein in favor of plaintiff Roy Spencer, an infant, by Sarah E. Spencer, his guardian *ad litem,* and against the defendant, The Atchison, Topeka & Santa Fe Railway Company, a corporation, in accordance with the verdict in said cause in the sum of Sixteen Thousand Seven Hundred and Fifty Dollars ($16,750.00),

WHEREFORE, by virtue of the law, and by reason of the premises aforesaid, it is

ORDERED, ADJUDGED AND DECREED that plaintiff Roy Spencer, an infant, by Sarah E. Spencer, his guardian *ad litem,* do have and recover of and from the defendant The Atchison, Topeka and Santa Fe Railway Company, a corporation, the sum of Sixteen Thousand Seven Hundred and Fifty Dollars ($16,750.00) with interest thereon at the rate of six per cent (6%) per annum from date hereof until paid, and for plaintiff's costs incurred and expended in said action, taxed at the sum of Two Hundred Fifty-four & 20/100 and that execution issue.

Dated and entered in open court this 15th day of August, 1926.

F. C. JACOBS.'' [286]

THEREUPON, to wit, on the 15th day of August, 1926, on motion of defendant, orders were made by said Court, and entered by the Clerk of said court in the minutes thereof extending defendant's time within which to prepare and file its bill of exceptions, and granting defendant a stay of execution of judgment as follows:

No. L.—170.

ROY SPENCER, an Infant, by SARAH E. SPENCER, His Guardian *Ad Litem,*

Plaintiff,

vs.

THE ATCHISON, TOPEKA & SANTA FE RAILWAY COMPANY, a Corporation,

Defendant.

ORDER EXTENDING DEFENDANT'S TIME TO FILE BILL OF EXCEPTIONS, AND ORDER STAYING EXECUTION.

On motion of the defendant it is now ORDERED that the time of the defendant within which to prepare and file its bill of exceptions herein be and the same is hereby extended to and including sixty days from this 15th day of August, 1926, and it is further

ORDERED that the defendant be and it is hereby granted stay of execution of judgment for the period of sixty days from and after this 15th day of August, 1926." [287]

THEREAFTER, to wit, on the 28th day of August, 1926, a special order was made by said Court, and entered by the Clerk of said court in the minutes thereof, continuing all further proceedings in said cause to the September term of the Prescott Division of the Court for the purpose in said order mentioned, said order being in the words and figures following:

"Regular March, 1926, Term—At Prescott.

In the United States District Court, in and for the District of Arizona.

Honorable F. C. JACOBS, United States District Judge, Presiding.

(Minute Entry of Saturday, August 28th, 1926.)

No. L.–170—PRESCOTT.

ROY SPENCER, an Infant, by SARAH E. SPENCER, His Guardian *Ad Litem,*
                                        Plaintiff,

vs.

THE ATCHISON, TOPEKA AND SANTA FE RAILWAY COMPANY, a Corporation,
                                        Defendant.

SPECIAL ORDER.

IT IS ORDERED that this case and all further proceedings in said case be continued to the September Term of the Prescott Division of this Court, for the purpose of enabling the defendant to prepare, serve and file a bill of exceptions within sixty (60) days from date of the verdict and to prepare, serve and file a motion for new trial, and to enable the Court to retain control over this case for the purpose of entertaining further proceedings herein.'' [288]

THEREAFTER, to wit, on the 30th day of August, 1926, an order was made by said Court, and entered by the Clerk in the minutes thereof, further extending defendant's time to prepare, serve and file its bill of exceptions, said order being in the words and figures following:

"Regular March, 1926, Term—At Prescott.

(Minute Entry of Monday, August 30th, 1926.)

Honorable F. C. JACOBS, U. S. District Judge, Presiding.

No. L.-170—PRESCOTT.

ORDER EXTENDING TERM TO SERVE AND FILE BILL OF EXCEPTIONS.

(Court and Cause.)

It appearing to the Court that the business of the Reporter preceding the trial of this case is such that he will be unable to file a transcript of the testimony in time to enable the defendant to prepare the bill of exceptions in the 60 days heretofore granted to the defendant, and that it is necessary to extend the time 15 days,

IT IS ORDERED that the time be extended 75 days from and after the verdict in which the defendant may prepare, serve and file bill of exceptions.

F. E. Flynn, Esq., of Norris, Norris & Flynn, is present for the defendant." [289]

THEREAFTER, to wit, on the 22d day of September, 1926, and on the 23d day of September, 1926, the defendant served upon counsel of record for plaintiff by United States Mail and personally, respectively, and, on the 25th day of September, 1926, filed with the Clerk of said court its petition for new trial, in the words and figures following:

"In the District Court of the United States in and for the District of Arizona.

No. L–170—Prescott.

ROY SPENCER, an Infant, by SARAH E. SPENCER, His Guardian *Ad Litem*,

Plaintiff,

vs.

THE ATCHISON, TOPEKA AND SANTA FE RAILWAY COMPANY, a Corporation,

Defendant.

PETITION FOR NEW TRIAL.

To the Honorable the United States District Court in and for the District of Arizona:

COMES NOW The Atchison, Topeka and Santa Fe Railway Company, a corporation, the defendant in the above-entitled action, and petitions this Honorable Court to grant the defendant a new trial in this action and to vacate and set aside the verdict of the jury rendered herein on the 15th day of August, 1926, and the judgment entered thereon, in favor of the plaintiff, Roy Spencer, and against the defendant, for the following causes, materially affecting the substantial rights of the defendant herein, to wit: [290]

(1) Excessive damages appearing to have been given under the influence of passion or prejudice.

(2) Insufficiency of the evidence to justify the verdict.

(3) That the verdict is against the law.

(4) Error in law occurring at the trial.

(5) Misconduct of the jury.

Insufficiency of the Evidence:

Your petitioner specifies the following particulars wherein the evidence is claimed to be insufficient:

The following evidence introduced on behalf of the plaintiff shows affirmatively and conclusively that the plaintiff was guilty of contributory negligence which bars recovery.

Roy Spencer testified as follows:

My age is 20. I have lived in Mountainair all my life. I am a son of Benjamin Spencer who was killed at the time I was injured. Up to the time of my injury my business or occupation was farming and sawmilling. I worked for my father. I attended school. I suffered my injury in June, 1923. That day we started down to Scholle on an errand to get some furniture. I went with my father in a Ford truck. He asked me to go down there to help him get some furniture. I was living at home with my father. I was then 17. Coming back we left Scholle between 1 and 2 o'clock, I think, somewhere along there. We were going in a Ford truck. My father was driving. It was my father's car. He had owned that car [291] since January of the same year. It was a second-hand car when he bought it. I did not have a top. Just one seat, then a box on behind, about three feet and a half square, I think, maybe 4 feet long, or 3, something like that. When we started out that day, and went there, the car was running good when we left. It run good all the way to Scholle, Scholle is on the line of railroad between Mountainair and Belen, and my home was to the north of the railroad some 9 miles.

Q. Did the car commence to show poor motion that day, or work badly?

A. After we had left Scholle about two miles, I think, it started to missing. Scholle is just a station—a small station. There is one little store there and a postoffice. The car started to missing after we had gone about 2 miles from Scholle. At the time we had went about 4 miles it got so bad we had to run in low most of the time then until we got pretty near about the crossing and then had to run in low, you might say, all the time. There were several hills before we got to this crossing that I had to assist the car up before we got to those places. I got out and pushed behind. My father requested me to do that between Scholle and

Abo; after the hill I could not say, I do not remember. I do remember his asking me at least once to get out and push. My father was 68. I had gone with him driving with him before. That day the spark had been out and he had been working on the coils to get it to run. The lid was off the top of the coils and he worked, done something, worked the platinums, I don't know just how he done it. The coils were down in front of the car, under the wind-shield—right next to the bottom of the car. [292]

Q. Show the jury what position you saw your father take in working at those coils when the car was running badly.

A. Well, he would stoop over them little plates that work up and down and takes the spark, he would work them that way, look at them; I don't know just how he worked them. The effect, apparently, that that had on the car, whatever he was doing in adjusting the plates or wires, seemed to cause the spark to hit up there, I don't know what part it would play. I don't know how to operate a car very good. Just exactly what my father was doing, and what was wrong, I am unable to state. I saw him bend down when his car would go bad and do something down under the wind-shield several times. The driving-wheel of this Ford or the steering-wheel is on the left side of the car. As my father would stoop down he would stoop to the right. He would stoop his head at those times when he was fixing the car—probably to where the wind-shield comes, something like that. I had had to get out and push that car up the hill twice between Scholle and this place at the crossing. It was about 4 o'clock when we approached that crossing at which I was hurt. The condition of the road for a mile or so back as we approached the crossing in the direction that we were going toward Mountainair is, it goes down a draw, just a kind of draw between hills and the earth is held on all sides, just a little draw down through there. By "draw" I mean a low sort of valley, as it were, between two rises on either side. And the highway ran up through that draw toward the crossing. The railroad runs more or less in the same direction as the highway at that point. So that, in driving along, you are running in the same general direction of the railroad. When we approached the railroad crossing [293] that day I saw a train passing along. When I was about 500 yards from the crossing, I saw a freight-train of some kind go along the crossing. I don't know what kind it was. It was going in the direction of Mountainair. When I saw the train it was going just over the crossing. Mountainair is the village where I go when I go to any village ordinarily, the nearest to any store. I was familiar with Mountainair. I had not been there very much.

Q. How familiar were you with the tracks of this railroad between Mountainair station and this crossing at which you were hurt?

A. Well, I have been there afoot when I was a kid; when going to school in Mountainair, us boys used to go up the railroad there around, I knew the road was there, and the railroad, that is all I knew. I was thoroughly familiar with the railroad between the station at Mountainair and this crossing. There was one track between those two pieces. There are some switches at the station in Mountainair. Those switches extend about 200 yards west of the station, I think. I do not know; just guessing at it. That is my best judgment, 200 yards west of the station. From there on to this crossing there is but one track. As to the condition of the railroad as to cuts and positions which it takes running from Mountainair up to this crossing at which I was hurt—well, from the time you leave Mountainair it is in a cut right at the depot on one side, and continuing from the time you go about 250 yards west of the depot you have got a bridge over the railroad there that extends away above where the train passes under and continuing in that way clear on down to pretty close to the crossing. It has got a slight curve through that cut. [294]

Q. How deep is that cut between the depot and the crossing at the different points as you go along, describe it generally; give us your best recollection of how the cut extends from where it commences at the depot, clear up to the point you say is near the crossing?

A. It will average about 20 to 30 feet deep, up to the time it goes out from behind this hill. It goes out from behind the hill about 850 feet from the crossing—that is, in the direction of Mountainair from the crossing.

Q. Now, when you approached, you say you saw
the freight-train go over the crossing east towards
Mountainair. Do you know how far it is from that
crossing to the Mountainair switch?

A. About 2 miles, something like that. If the
record of actual measurement here shows a mile
and three-quarters I would say that is accurate;
I was just guessing at two miles. I would judge
that that freight-train was moving 20 miles an hour
as I saw it going east—about like most any other
train travels along there.

Q. Now, Roy, what is the condition of the railroad there as to grade at that point, in which direction does the railroad slant, or grade from the
crossing going east?

A. It slants to the west. There is a downgrade.
With respect to the station at Mountainair that
downgrade commences about a quarter of a mile, I
think, maybe a little over, west from Mountainair.
You start up grade as you go west out of
Mountainair. After you go about a quarter of
a mile it turns down, and from there you go downgrade. As I approached—as the train passed over—
that is, this freight-train passed on—we were coming up the crossing [295] all the time. The road
runs more or less in the same direction as the railroad, up to the crossing. When you get to the
point of the crossing the road turns north and goes
over the track. Goes right around the corner of
the fence, you are going northeast when you come
up to the tracks, when you turn right north and

go up over the tracks. You might call it a square turn around the corner, and up over the tracks.

Q. Did you stop at all before you went over the tracks that day, and, if so, where?

A. Stopped right at the corner of the fence. The picture you now show me of a crossing, with an automobile standing before it, and the road going over, and a man and a sign on the crossing, is a good representation of the appearance of that crossing at the point where I went over. The position occupied by the automobile shown in the picture looks to be about the same place as that where we stopped that day. We stopped about the same place where that auto stands. I know where the road commences to rise there to go over the track, whether inside or outside the railroad fence; it is right at the corner. I would say just inside, maybe two or three feet. Then it commences to rise. From that point to the nearest rail the road has to rise about 10 feet. The road has to rise 10 feet in going from the railroad fence up to the track. The top or grade of the railroad extends out on a level a few feet beyond the rail; I would judge 4 feet outside of the rail. The rise from just inside the fence to about 4 feet outside the rail is 10 feet. Where the man is standing in the picture—that is on the rail [296] just where the road crosses. And that is the same crossing. It looks substantially in that picture the same as it did at the time I crossed it.

Mr. WOOD.—We will offer it in evidence. Counsel has kindly consented that we may use this without eliminating the memorandum or record thereon, with the further concession of the fact, as we understand it to be, that the fence shown on the picture at the edge of the railroad's right of way, is 100 feet from the center of the railroad track, that is my understanding of the fact.

Mr. BRENNAN.—Correct.

The COURT.—It may be admitted as Plaintiff's Exhibit No. 2.

WITNESS.—(Continuing.) The picture you now hand me was taken from the road, looks like about halfway up to the crossing, from the corner of the right of way, looking west up the valley. About halfway from the fence corner up to the railroad track on the crossing. That correctly shows the appearance of the ground looking west from that crossing in the direction in which the road I traveled came. The fence post that shows first in this picture is a post going up to the right of way—one of the fence posts leading up to the rails.

Q. What is the corner, or fence corner that is shown in that picture, the line of posts running up toward the telegraph pole, and this one post that shows in the picture the other way?

A. This is the line of posts coming east, the corner would be out here (indicating), and this post the one that leads up to the right of way. That is, the long line [297] of posts that appear there is the fence at the railroad right of way. Approaching from the west, to the left of the fence, appears, apparently, a road. That is the road we came up. In and across the corner inside of the angle are some other marks or indications on the ground; they are the indications of the old road, apparently an old road. The appearances which make me say it is an old road are two ditches— it is not real ditches, just kind of cuts in the earth —marks of wheel tracks. They are what show just inside of the corner. To the right of the picture and to the left of the picture appear rises in the ground or hills, and a low point between; that is the draw of which I have spoken, that comes up along the railroad. And it is shown in this picture.

The picture was offered and received in evidence as Plaintiff's Exhibit No. 3.

Mr. WOOD.—Q. I show you another picture. Can you tell me where that was taken, and what it shows with reference to this accident?

A. Yes, sir; that is taken standing west of the crossing, looking to the crossing. The picture you show me shows a cut or excavation in the ground. That cut is west of the crossing. The white boards shown in the forefront of the picture, are those at the crossing. So that this shows the crossing look- ing east along the railroad track from the cut.

The picture last mentioned was admitted in evi- dence as Plaintiff's Exhibit No. 4.

Mr. BRENNAN.—Then Exhibit No. 4, if I un- derstand it rightly, shows the condition west of the crossing? [298]

Mr. WOOD.—Yes, it is taken west of the cross- ing, looking toward the crossing and showing the crossing on the railroad track, and from the railroad track.

Q. The picture shows, looking down the track and in front of you, a tree-covered hill beyond the crossing; what is that hill?

A. That is the hill where the curve goes around and goes into it; it is a hill that cuts out across this low place. That hill is a hill that the train comes out of. The hill shown in the front of the picture extends down to the railroad track. Beyond the crossing—the railroad track is cut through there. It is 9 telegraph—6 telegraph poles from the railroad crossing to where the railroad tracks cut into that hill shown in the front of that picture. It would be 900 feet, just counting the telegraph poles 150 feet. It is something like 900 feet from the crossing.

Mr. WOOD.—I show you another picture. Does that correctly show the appearance of the south side of the track from the crossing at which you were struck, down to this hill that the cut is in to the east?

A. It does. That is a fairly correct representation of the appearance. There is a mark on the picture apparently where the hill touches the track. It is 850 feet from the crossing, where the hill touches the track. And the hill shown, to the right of this picture is the same hill which is shown in the picture, Plaintiff's Exhibit No. 4, as being down the track as you look out of the cut; east from the crossing to that hill, 850 feet east.

(Thereupon the picture was received in evidence as Plaintiff's Exhibit No. 5.) [299]

WITNESS.—(Continuing.) The picture, Exhibit No. 5, which you last showed me, shows an embankment in the railroad running eastwardly from the crossing. I should judge that the approximate height of the embankment above the surface of the ground at its bottom as it runs across the draw is 10 feet from bank to bank. About the same as at the road.

Mr. WOOD.—Exhibit No. 8, which I am now offering in evidence, and is a picture taken on the crossing about at the line of the railroad right of way, looking east into this hill; it shows a freight train going over the crossing, and the engine just in the cut.

Exhibit No. 9, which I am now offering, is taken—

Mr. BRENNAN.—Are you able to tell the jury how many cars were in that train?

Mr. WOOD.—I am not—

Mr. BRENNAN.—I think 52 cars; it may be so stipulated in connection with your print.

Mr. WOOD.—I think the testimony will show that somebody stated there were 52 cars in the train, but not between the crossing and the cut, showing that the number of cars would not be helpful to us. I think that appeared before, if I recall right.

Exhibit No. 9 is taken about one-third of the way up, about 2 fence posts up, which I should judge is about one-third of the way from the right of way fence up to the crossing, in the same direction,

and shows a pusher engine on the same train, showing in 8, just disappearing in the cut.

Q. Now, Roy, you have told us that you stepped
in your father's car about the same position that
the car shown [300] in the first picture was in.
What, if anything, did you do at that time or place
in the way of looking for trains?

A. I got out and looked both ways for trains
and could not see any. Then I proceeded to start
up the hill. The car was running badly. It was
running very poorly, and had to start and started
up this hill; of course we went slow naturally,
starting and going up the hill it was running poor.
As the car started up the hill I started pushing
from behind. I started pushing the car up the hill
from behind the car. I was pushing up the hill
with my head down. My hands were in the back
of the car, on the back of the truck bed. I was
trying to push it up the hill, didn't seem to want
to go very good. I pushed all I could. It was just
going. That was about all. I did not continue
pushing from behind all the way up the hill. As I
went up I went around to the side, after we went
up a little ways, around the side of the car. I will
tell you just how I went to the side, what I did;
I stepped around to the side while pushing, so when
I got on top I could step in the car.

Q. When you got around to the side, show the
jury just the position you occupied with reference
to that car; step down here to this table.

(Witness steps down to the table.)

We will assume that this table is the automobile, which is headed up the hill toward the track that is in front. Now, get around behind and show the jury just what you did as you pushed that car up the hill or road, when you started, and show just what you did when you got up to the track.

A. I started pushing behind like that (indicating). [301] Of course, I was higher than that on the truck back—the car was going up the hill and naturally I throwed my arms out with my head that way (indicating); when we had got up probably 30 feet I was pushing on one end.

Mr. WOOD.—The witness illustrates by placing both his hands on the rear of the table with his feet extended out behind.

Q. About how high was your head when you were pushing and what attitude did you assume in pushing the car when you started?

A. My head was about that high, I should judge (indicating) and my hands about like that (indicating.)

Q. With reference to the back of the car, how high was your head—with reference to the body of the car behind which your hands rested, how high was your head?

A. About even with my hands.

Q. Your feet, about how far behind the car were your feet as you were pushing?

A. (Indicating.) Just as much as it takes a person to push; of course, the feet naturally goes back so to push. Now when I ceased that position, I will show the jury just what I did, how I made the change. There was a handle on this side, I had hold of. When I stepped around I just stepped that way, held that hand like that (indicating), pulled that way and pushed against the back of the seat with this other hand. There was a brace at the right, or, rather, corner of the box on the car. I hung to that brace with the left hand. In stepping around from the rear of the car to get hold of the seat I stepped just [302] as quick as I could. I did not stop pushing at all—could not. I could not because the car would stop. It was all I could do with the car itself to get along, much less stop pushing. I said I did not stop pushing but stepped around as quickly as I could, and held that handle with the left hand. I show the jury likewise on the table what position I assumed after I got around to the side where that hand was; I had hold of that end back there, projected out about 4 inches, throwed my arm back here (indicating).

Mr. WOOD.—Illustrating a handle attached to the left rear end of the box, and his left hand attached to the handle.

WITNESS. — (Continuing.) This projected about 4 inches, had to put my right hand in against the back of the seat on the front here. I was pulling, that throwed me under just like that (indicating). I continued in that position as I passed up the hill on the truck until the front wheels hit the track and then I was, I suppose 6 feet, anyhow from the front wheels when they got on to the track, and the hind wheels were coming off the grade.

Q. Can you illustrate by moving your feet or showing the jury just about the movements you made in pushing the car with your feet as you came up the hill after you came around to the side?

A. I pushed just like that (indicating). Of course, I went as slow as I could, pushing as hard as I could. From the time I went around the end of the car, as I was pushing up the hill on the side of the car, I was facing to the west. We were half way up the hill or from the place where we started when I changed my position from the rear [303] of the car to the side. I should judge it took a full minute to get the car to move from the point where it was at rest from the time it started at the bottom of the hill until I got up on the track. From what I know of that car, and the way it was running, I know what would have happened if I had ceased for a minute or instant to push; it would stop. From my knowledge of the car, I know whether or not stopping on that grade it could have been started again without going back to the bottom of the grade; it would have run back to the bottom of the grade if I stopped, undoubtedly. I did not cease for an instant to push from the time that car started until it got on the track. From the position that I was in while I was pushing that car at the side of it, I could not have turned my head around to look behind me in the direction in which that train was coming. In order to have looked in the direction in which the train came at the time—from the time I came to the side of the car, until I came on the track—I would had to have taken this arm down and looked and quit pushing, and the car would have stopped and would have run down and it would have had to. I have said that at one time my father asked me to get out and push; I do not recall whether or not he said anything to me about— at that particular time, my getting out and pushing. I could not say whether he did or not. I do not remember. I do remember that on some part of the journey, he asked me to get out and push. From that time on, each time the car got into a position where it would not climb, I got out and pushed. I ceased to push when the front wheels of the car had just went over the first rail. [304] At that time I was beside the car and I stepped on the rail—on the running-board, rear, and stepped in the car and turned around. My purpose in stepping in the car just when ceasing to push, stepping on the fender, was because it was going over the rails and down the other hill. The car was in motion. From the time I raised up and ceased to push it was just a second or so before I got a view of the train; just as quick as I could step in and look. Step in and face to the west, when I turned around and throwed my head down the track.

Q. How much time, if any, elapsed from the time you let go of the car until you looked in the direction of the train?

A. Just as quick as I could jump in the car and turn around.

Q. As quick as you could jump in? I will ask this additional; how long was that?

A. I do not know; could not have been more than a second or so; maybe two seconds from the time I quit pushing until I was in the car.

Q. Illustrate to us again—using the platform here as the car—just what you did when you ceased to push, where you went to, showing how fast you did it, and how you looked for the train; then we would get an idea of the time.

A. How quick I jumped in the car?

Q. Yes. Come down here again—assume, now, that the platform here is the car—face the other way. Assume that the platform here is the car and that you are pushing this way (indicating), and stoop over and get in the position you were when you were pushing, using this end of the [305] platform as the back of the car. That is, the rear of the car, now as you were climbing up the hill last—I don't mean at first—when you were behind, I mean, when you got up to the railroad tracks, as your hand let go and you stepped inside the car, and state how quickly you did it.

A. I would say it would be just like that (indicating). Upon the running-board and in the car and turned around and sat down. There could not have been more than a couple of seconds.

The COURT.—Suppose you could stipulate as to approximately the length of time of the demonstration?

Mr. WOOD.—Well, the witness stepped up as rapidly as a man could step.

Mr. BRENNAN.—Yes, it may be so stipulated.

He said he got up as rapidly as he could and he undertook to demonstrate how he did it.

WITNESS.—When I looked then first, I saw the train; it was about 50 yards away. I don't know what was the next thing I seen. The next I remember, they were putting me in the baggage coach of the train. The next I knew after looking up and seeing the train some 50 yards off they were picking me off the engine. My father was sitting in the car at the time driving. I did not notice at all what my father was doing as I pushed that car up the hill. Whether he was sitting up or leaning over, I do not know. I do not know how far it is from that crossing until the railroad enters the cut, west of the crossing; just a few feet, the north side of the embankment runs right up to the highway. It goes in there on a curve; that picture showed the west there a while ago, shows that curve. The picture, Exhibit No. 4, is the one I refer to, as showing the track in the curve west of the crossing. [306] The curve extends about between two and three telegraph poles, I think, west of the crossing when you are standing on the track, before a train disappears into the cut to the left. The size of that cut at the point shown in the picture, Exhibit No. 4, looked to be about 10 feet there. I know, from my own knowledge of the cut, about how high they are; about 10 feet where the closed part of the picture is shown. This freight train which had come along the track and gone toward Mountainair was running right along there when I first saw it, just about in the place that picture is taken. It had passed east through the cut and gone on to Mountainair. It seemed like 6 or 8 minutes, maybe 10, had elapsed from the time I saw the train pass the cut going toward Mountainair, and the time when I started up the grade. I am talking about the cut west of the crossing.

Mr. BRENNAN.—About two or three telegraph poles west?

Mr. WOOD.—He says that the train disappears about two or three telegraph poles.

Mr. BRENNAN.—You are talking about the time when he said he was about 500 yards from the crossing, the track, when he first saw the train?

Mr. WOOD.—When he first saw it, yes, sir. This cut through which the train from the west passes, about how long is it, how far back from the crossing does that cut reach, if you know, more or less?

A. I don't know. About 300 yards, I think. And then it emerges on the other side of the hill.

Coming back to the time and occasion when I was hurt; I have stated to the jury that while I was pushing the [307] car up the hill, I did not look to the left behind me at the track down toward Mountainair. The reason I did not was because I could not.

Q. Is there any other reason?

A. One reason was, when I started up the hill, I had already looked to the east; you can see a distance there, something like 200 yards, a little over. And this cut that comes out of the track on

the west, it comes right out; when you are standing at the corner post you cannot see the train
coming from the west more than a telegraph pole
before it comes to the crossing. The reason why
I did not look and see was I was watching toward
the west, because it is more dangerous than the
east of the track, the crossing there.

Q. Was there anything in the position which
you had alongside of the car going up there that
would obstruct your view of the train from the
east, had you looked in that direction?

A. Yes, sir, there was. The right of way fence
where these poles and boards set up there for a
distance of 4 feet; a curve in the track down the
road throws your head right against this fence on
down the track. You also have to pass a distance
of 8 feet before you could see the track. Exhibit
No. 5, the photograph which shows a fence along
side of the roadway as you pass up, that is the
fence which I am speaking of which would obstruct my vision.

Q. Did you ever notice what, if any, precautions
your father habitually took in approaching or
crossing the railroad track, as to whether or not
he was careful or a careless man? [308]

A. I have been with him lots of times when there
was times that he would go up to the track and
before ever approaching the crossing—I have
crossed the crossing very seldom in an automobile;
generally when we crossed the tracks it would be
under an underground culvert, something like that,
not an automobile on the highway. From my

knowledge and observation of my father, I will
say he was a very careful man in watching for
danger of trains in crossing the tracks, as I knew it
at that time. As we went up there, and as I was
looking for the train to emerge out of this cut to
the west, and watching in that direction as I have
testified, as I climbed that, I listened as much as
I could for the sound of any train. Of course the
car was running; I could not hear very good. And
as I went up there, I did not hear any bell, any
whistle, any sound of a train until I saw the train.
At that time there was some wind blowing. There
was quite a breeze blowing towards the northeast.
It would be blowing almost direct down the track,
a little north of down the track. By "down the
track" I mean toward Mountainair.

Q. Was there enough wind blowing, or did the
wind carry any dust?

A. Well, none to speak of that could hide the
hill more than it was kind of hazy like to see way
off—it was hazy like; right close you could see. It
might hide anything completely quite a ways off.
It is a dry country around there most of the time.
I have noticed at other times as to whether or
not, standing at that point from which we started
the automobile, I could see an engine coming from
Mountainair and the direction in which this train
came before   [309]   it emerges from that cut.

Q. What have you noticed from observation in
that particular from the point where you started
the car?

A. Well, I think the engine first comes in sight

just about the time it gets to the whistling post. The top, outside, the north side of the engine, the top, comes in sight there. The right-hand corner or cowcatcher, you can see it first, the smokestack and the top of the engine then comes out more until you can see all of it at the mouth of the cut. There is no place before it actually emerges from that cut that you can see the whole of the front of the engine, until it passes the edge of the hill, as I have described. The only sort of warnings or signals there at that crossing at the time was the crossing sign. That crossing sign was two boards crossed and nailed up on a post.

Cross-examination.

My father had owned five automobiles before he bought this automobile. He had owned three Fords and two Buicks. I lived at home all of the time from my birth up and until the time of this accident. I was 17 years of age when the accident occurred. That was June 11, 1923. Back of the seat to the rear end of the bed of the truck I would guess was 4 feet. The height of the seat was just about as high as the ordinary car, I think. The height of the bottom of the seat from the bottom of the bed of the truck—where the seat is attached to the bed of the truck—was, I should guess a foot— just guessing at it. Maybe not that much. There were running-boards on the sides of [310] the truck. The running-board was just like any other running-board. I think about a foot from the ground. The distance from the running-board up to the top of the bed of the truck was probably a foot and a half. I would guess that the top of the bed was three feet and a half from the ground, just looking at it. I think the side boards were 10 inches high. There was an end-gate that raised and lowered. As we approached that crossing it was raised. I am about 5 feet 4 inches in height. As we left the 100 foot point, that is, the south right of way fence, to go over the crossing, I pushed against the hind end-gate at the right-hand corner—the upper part of the gate. I grabbed hold of the upper part of the hind end-gate—the rear end—with my left hand to hold the brace on the right-hand corner. When I got within 50 feet of the track, I released my hold with the right hand, and I traveled around, stepped around to the right-hand side of the truck. As I proceeded along the right-hand side of the truck, my left hand was not released at all. I had to remove the position of my left hand as I made that change. I took my position before I started around, just stepped around, put my right hand against the back of the seat. I placed my right hand against the back of the end of the seat up to the cushion or up to the top of the seat with my hand over the top. The seat had a back on it. When I changed my position in the manner I have described I did not look toward the east. At no time after I left the south right of way fence, which the testimony shows was 100 feet from the center of the track, did the truck stop. And as the front wheels of the truck passed over the south rail [311] of the railroad track I jumped up into the car. And in so doing, I did not look to the east, that is, the direction from which the train came, until I jumped into the car. Now, this highway that I have described, over which we traveled as we approached the crossing, paralleled the railroad for some distance. It had paralleled the railroad I suppose a distance of half a mile from the crossing. The highway was from 2 to 300 yards, I think, from the railroad. I think the farthest point must be 300 yards.

Q. Was the railroad within sight all that distance?

A. Well, I could not say whether it was that way up to half a mile or not. I do not remember that.

Q. Well, from the time it took, a point half a mile west of the crossing, from that point to the crossing was the railroad within your view?

A. Part of it at times, of course it is cut in through hills, and at times you can see it and at times you cannot. As we made the last half mile, we were travelling very slow. I think he was running in low pretty near all the time. Low gear all that time. When we got to the point where we changed our direction of travel, that is, when we reached the point 100 feet south of the center line of the track, we stopped. And I got out of the truck. We stopped at that point I expect half a minute. And then and there I looked in both directions. I had no conversation with my father at that time. Then my father started the car. I do not know at what rate of speed he travelled in

going up to the track. I showed you the best I could
yesterday just as slow as the car would possibly pull
in low and my pushing. I figured that to be probably something like one minute. I [312] have
never observed the time that it takes a Ford to
travel a 100 feet when in low gear. So, that when I
say it took a minute, it is a guess on my part of the
way the car was going; a car can go lots of different
speeds in low gear. It at no time stopped while I
was negotiating that 100 feet. I have testified that
when we were within 500 yards of the crossing, I
saw a freight train proceeding in an easterly direction over the railroad tracks. I first discovered
that freight train when I was first starting over the
crossing, there, at the opening of the hill. At that
time we were up about 5 or 600 yards from the
crossing.

Q. Where were you when the rear car of that
freight train passed over the crossing?

A. Well, the rear car was the engine—the engine
was pushing the train. I did not take particular
notice as to the number of cars in the train but I
would judge about 20 cars, from what I remember
about the train. There was no locomotive on the
head end of the train.

Q. Where were you with reference to the crossing or the point at which you stopped before proceeding over the crossing when the engine passed
over the crossing?

A. Oh, we had not gone over 75 yards, I do not
think, while the train was going over the crossing.
It was about a 20-car train, I think, going about 20

miles an hour. We probably had gone 75 to 100
yards, something like that. Just guessing at it; it
took us about ten minutes to go from that point
where I saw the train going over the crossing, to
the point where we stopped. I think that is about
right. Probably something like that, yes, sir. I
do not think it [313] could have been any longer.
I think about 10 minutes, that is my best estimate.
My father's eyesight was good. His hearing was
good. My eyesight was good. My hearing was
good.

Q. As you got up to the track and hopped up into
the truck, why didn't you look east?   A. I did.

Q. Why didn't you before getting up into the
truck?

A. Oh, before? Well, because the car was going
over the track.

Q. Well, what was there in that to prevent you
looking eastward before you stepped up onto the
running-board?

A. Well, the car was going over the track and
down the other hill; naturally so, I stepped up
on to the running-board before, and then looked
down the track.

Q. You have testified you had not already looked
in this direction before you got up into the car?

A. I had looked, yes, sir. I had looked at the
corner of the right of way, but not after that. Not
until I got up into the car and was about to sit
down. I mean to tell the jury that my father
owned 5 cars previous to this one, and I, 17 years of
age, had never driven an automobile. I do not re-

member whether my father shut down the motor as we stopped at this 100-foot point below the track or not. I think he did.

Q. You have testified on direct examination that when you were at that point, 100-foot point south of the track, you could see the whistling post?

A. Well, you can in a way, and in a way you cannot. You can stand at the right of way corner and step up on the [314] bank, where the road goes around this corner post, the bank is dug up, and if you will step up on that bank you can see the whistling post, but you cannot by standing down in the road or on the right hand side of the road. The whistling post is supposed to be a quarter of a mile, I think, east of the crossing, but the whistling post is to the north of the track about 6 or 8 feet, I think. The whistling post looks to be about 6 or 7 feet tall. A locomotive, of course, is taller than the whistling post.

Q. When you are at that 100-foot point, you would have no difficulty in seeing a locomotive when it got to the whistling post, would you?

A. Well, you might see the top of it, just a little of the smokestack, I don't know. But as you approach the crossing, or approach the rails at the crossing, you can see farther and farther easterly.

Q. When you are 50 feet from the track, how far can you see in an easterly direction along the track?

A. I don't know; you cannot look down the track that way, and tell just how far you are looking. You can see beyond the whistling post quite a little ways. As you get along 30 feet from the track you can see farther. I think that when you are at a point 20 feet from the center line of the track, you can see easterly beyond the whistling post. I never did notice particularly at that point but you can see quite a ways. You cannot tell looking down the track just how far you are away. Unless you measured you cannot tell exactly.

On August 18th, 1925, I testified in a former trial of this case. On that day, and in this court-room, in [315] the former trial of this case I testified as follows:

"Q. What change in the direction of the road takes place when you come around the corner and cross the road from the general direction of the road as you approach the crossing?

A. The road is practically going east, a little bit north of east, when you get to the crossing and turn around the corner and go over the track north."

I further testified as follows:

"Q. How much of a turn is there there, a square turn, or more or less?

A. About a 45 degree angle."

That testimony was correct. At the same time and place I testified as follows:

"Q. And you state that the wind was blowing that day?   A. Yes, sir.

Q. Was it blowing very hard?

A. No, not hard. Just a stiff gale. Just enough that a car, by running over the road, would pick up dust, but the wind itself would not pick up dust."

I further testified:

"Q. But there was not enough to stop you from seeing anything that you were looking at?

A. Well, not unless something would kick the dust up."

I further testified:

"Q. Well, did anything kick the dust up to obstruct your vision? A. No, not me, no." [316]

I further testified:

"Q. When you did look along the railroad track, it was perfectly clear, then, was it not, so that you could see? A. Yes, sir."

That was correct.

Redirect Examination.

I mean that when I saw the train, when I looked there was nothing in the dust which would obstruct my view of the train approaching.

Q. In the answers you gave there about you not being able to see, what were you referring to as regards distances, as you stated it?

A. Well, I was referring from the mouth of the cut to the crossing; I was not referring to a farther distance. That was what I was referring to in that former testimony.

Recross-examination.

Referring to Plaintiff's Exhibit No. 5, which you hand me; I have examined that. I stated on direct examination that there is a fence that shows in that picture, and that that fence obstructed my vision somewhat. That fence is right here. That is just a white board fence there. It is right on the south side of the track.

Q. Is there not a wing fence at the cattle-guard?

A. I think that is what it is. I think it extends out from the track, the south rail of the track, about 6 feet—something like that. That is the only fence there is along there at all. What I was referring to is the curve [317] that goes around there. But as I approached the crossing, that is, while I travelled the distance of 100 feet, I did not see that fence at all.

The COURT.—Let me ask a question.

Q. How much time elapsed from the time the freight train passed the crossing going to Mountain-air, before you were struck by that train going west? A. About 10 minutes, I should judge.

Redirect Examination.

Q. You said that this fence shown in the picture you thought was about 6 feet long, which did you refer to, the top or bottom of the fence, as being 6 feet long?

A. Taking on an average of the middle, probably about that, not taking that spur there. I never measured that. No, sir,—only just looking at this. The fence is accurately shown in this picture. I am giving my estimate now from looking at the picture as to its length.

JESUS BERRARAS testified on behalf of plaintiff as follows:

Cross-examination.

Q. How far was the truck which Mr. Spencer

and his son were operating when you first observed; how far was it from the railway track?

A. For the first time that I seen it it was standing there at the corner of the right of way. I don't know how long the truck stood at the corner. I don't know how long it had been there or how long it had been stopped, when I observed it, from my house. It was standing there, and I [318] looked down toward the crossing. I do not know how long it stood there, but when I first observed it, until I seen it start, it might have been 10 minutes; something like that. As to the length of time it took the car or truck after I observed it start in motion, to reach the railroad crossing—why, to my estimation, I don't know, but it seemed to me just about a minute from where it started until it hit the track.

The Verdict is Against the Law.

The verdict is against the law for the reasons given as to the insufficiency of evidence to support the verdict and for the following reasons:

1. The verdict is contrary to the instruction of the Court reading as follows:

"If, after considering all of the evidence in the case, you further believe that the plaintiff, Roy Spencer, was also negligent to any extent and that his negligence concurred and co-operated with the negligent act or acts of the defendant to the extent that both the negligence of the defendant and the negligence of the plaintiff was the proximate cause of the injury, then the plaintiff cannot recover."

2. The verdict is contrary to the instruction of the Court reading as follows:

"You are instructed that a traveler on a highway approaching a railway crossing must use ordinary care in selecting the time, and place to look and listen for coming trains and he should stop to make observations, when necessary. It is his duty to use all of his faculties and it is not enough if he carefully listen, believing those in charge [319] of the approaching train will ring a bell or sound a whistle or operate the train at a reasonable rate of speed. He must take advantage of every reasonable opportunity to look and listen and he has no right to omit looking and listening because he has heard no appropriate signal of the approach of the train, even though the law requires that signals be given."

3.   The verdict is contrary to the instruction of the Court reading as follows:

"You are instructed that even though you should find that the employes of the defendant in charge of the train involved in the accident described in the complaint failed to sound the whistle or ring a bell as a warning of the approach of said train toward the point where the collision is alleged to have happened or operated said train at an excessive rate of speed, still such failure to signal the approach of said train or the operation of such train at such an excessive rate of speed did not relieve the plaintiff from the duty of exercising ordinary care to look and listen."

4.   The verdict is contrary to the instruction of the Court reading as follows:

"You are instructed that a person traveling upon a street or highway crossed by the tracks of a rail-

road company have no right to presume that the
railway employes engaged in running trains or
cars upon such track will signal the approach of all
trains, and plaintiff had no right to omit [320]
the performance of any part of his duty to look
and listen for approaching trains, upon the presumption that the train men would do their duty.''

5. The verdict is contrary to the instruction of
the Court reading as follows:

''You are instructed that while the defendant
railway company and Roy Spencer each had the
right to use the road across which the track of said
defendant is and was laid, yet, because such trains
and cars are run on a fixed track and with great
momentum it had the right of way or precedence
upon or over those parts of the road occupied by the
railway track, and it was the duty of Roy Spencer
to give unobstructed passage to the approaching
train.''

6. The verdict is contrary to the instruction of
the Court reading as follows:

''If you find that the defendant railway company
was free from negligence or that the plaintiff was
negligent, however slight, which in anywise contributed to the collision, you need not consider at all
the nature and extent of plaintiff's damages but
you should, without hesitation, return a verdict
for the defendant railway company.''

7. The verdict is contrary to the instruction of
the Court reading as follows:

''It was the duty of the plaintiff, Roy Spencer, to
use that degree of care that a reasonably prudent

person would use under like circumstances surrounding the case. [321] It was his duty to look and listen for approaching trains before venturing on the tracks.''

8. The verdict is contrary to the instruction of 'the Court reading as follows:

''You are instructed that a traveler on a highway approaching a railway crossing must use ordinary care in selecting a time and place to look and listen for coming trains and he should stop to make observations, when necessary. It is his duty to use all of his faculties and it is not enough that he carefully listened, believing that those in charge of the approaching train will ring a bell or sound a whistle or operate the train at a reasonable rate of speed. He must take advantage of every reasonable opportunity to look and listen and he has no right to omit to look or listen because he has heard no appropriate signal of the approach of the train, even though the law requires such signals to be given.''

Errors of Law Occurring at the Trial.

1. The Court erred in denying defendant's motion for a directed verdict, made at the close of plaintiff's testimony as follows:

''NOW COMES The Atchison, Topeka and Santa Fe Railway Company, a corporation, and moves the Court for a directed verdict in its favor in said action upon the following grounds:

1. Upon the ground that plaintiff has not offered sufficient evidence to entitle the cause to be submitted to the jury. [322]

2. That there is no evidence of any negligence on the part of this defendant.

3. That the undisputed evidence shows that plaintiff was guilty of negligence himself, which was the sole and proximate cause of the injury complained of.

4. That the undisputed evidence shows that plaintiff was guilty of negligence and want of ordinary care for his own safety, which contributed to the injury complained of.

5. That the evidence considered in the light of physical conditions shows that, if the precautions required by law had been observed by plaintiff, he would have seen the train in time to have avoided the accident.

6. That the undisputed evidence shows that if plaintiff had looked and listened before attempting to cross the railroad track, he would have seen and heard the train in time to have avoided the accident.

7. That the undisputed evidence shows that if plaintiff had looked in the direction of the on-coming train before attempting to cross the railroad track he would have seen the train in time to have avoided the accident.

8. That the undisputed evidence shows that plaintiff and the driver of the automobile in question, to wit, Benjamin Spencer, were engaged in a joint enterprise at the time of said accident and that if the driver of the said automobile and plaintiff had exercised ordinary care and caution before attempting to cross said railroad track, they   [323]

could have seen or heard the train in time to have avoided the accident.

9. That the undisputed evidence shows that plaintiff and the driver of the automobile in question, to wit, Benjamin Spencer, were at the time of the accident engaged in a joint enterprise and that, if plaintiff and the driver of said automobile, or either of them, had exercised ordinary care and caution for their own safety before attempting to cross said railroad track, they could have seen or heard the train in time to have avoided the accident.

10. That the undisputed evidence shows that, if plaintiff and/or the driver of said automobile had looked up the railroad track in the direction from which the train was approaching when they were 30 feet from the railroad track, they could have seen the train for a distance of not less than 1,500 feet, and in time to have avoided the accident.

11. For the reason that the undisputed evidence shows that if plaintiff suffered any damages or injuries whatsoever at the time, or in the manner or because of the facts alleged in plaintiff's complaint therein, such damages were directly, immediately and proximately caused and occasioned by the fault, negligence and carelessness of plaintiff himself, and not by any fault, negligence or carelessness on the part of the defendant.

12. By the allegation of the complaint and the evidence adduced, it is conclusively shown that plaintiff, Roy Spencer, and his mother, are citizens and residents of the State of New Mexico and that the

alleged cause of action [324] sued on herein accrued in that state by reason whereof the plaintiff's right to recover judgment against this defendant is controlled by the laws of New Mexico; that under the laws of New Mexico the plaintiff has not only not offered legally sufficient evidence of defendant's negligence proximately causing his injury or damage, but on the contrary the evidence establishes such negligence of Roy Spencer as to bar his right to a verdict and judgment in his favor.

13. That Arizona Constitutional provision being Section 5, Art. 18, is not applicable in this case because of the residence and citizenship of Roy Spencer and his mother and because the alleged cause of action sued on herein accrued in the State of New Mexico, and because the evidence of the plaintiff, viewing it in the most favorable light, shows that he himself was guilty of contributory negligence.

14. That the evidence adduced herein showing that the negligence of Roy Spencer proximately contributed to his injury, there is no controverted question of fact to be determined by the jury, and, therefore, the record presents solely a question of law for the court, which, notwithstanding the provisions of Sec. 5, Art. 18, of the Arizona Constitution it is the right and the duty of this Court, under the law as administered by the Federal and United States Supreme Courts, and under Rule 60 of the Rules of Practice of this court, to decide; that said Rule 60 reads as follows:

"The defendant in an action at law, tried either with or without a jury, may either at the close of the plaintiff's case or at the close of the case on both sides, move for a nonsuit. The procedure on such motion shall be as follows:

The defendant or his counsel, shall state orally in open court that he moves for a nonsuit on certain grounds, which shall be stated specifically. Such a motion shall be deemed and treated as assuming [325] for the purposes of the motion (but for such purposes only) the truth of whatever the evidence tends to prove, to wit: whatever a jury might properly infer from it. If, upon the facts so assumed to be true as aforesaid, the Court shall be of opinion that the plaintiff has no case, the motion shall be granted and the action dismissed. The party against whom the decision on the motion is rendered may then and there take a general exception, and may have the same, together with such of the proceedings in the case as are material, embodied in a bill of exceptions. If evidence shall be introduced by either party after the decision on the motion has been made, the same shall operate as a superseding of the motion; but such motion may be renewed at the close of all the evidence."

2. The Court erred in denying defendant's motion, made at the close of all the testimony, to direct the jury to return a verdict for the defendant, said motion being based upon each and all the grounds

theretofore stated in support of the said motion for directed verdict, made at the close of the plaintiff's case.

3. The Court erred in refusing to instruct the jury as set forth in defendant's requested written instruction numbered D–I, as follows:

"You are instructed to return a verdict in favor of the defendant, The Atchison, Topeka and Santa Fe Railway Company."

4. The Court erred in refusing to instruct the jury as set forth in defendant's requested written instruction numbered D–VI, as follows:

"As the collision between plaintiff and defendant's railroad train occurred in the State of New Mexico, and as Roy Spencer is a resident of that state, the laws and construction thereof by the courts of New Mexico are controlling upon you in deciding this case. [326]

Under the system in New Mexico it is contributory negligence for a person to approach a railroad track without paying careful attention for the approach of trains on such track in both directions."

5. The Court erred in refusing to instruct the jury as set forth in defendant's requested written instruction numbered D–VIII, as follows:

"You are instructed that under the laws of the State of New Mexico a railroad company has the right to construct its railroad at grade across any highway which its said railway shall intersect, but that such company shall restore such highway to its former state, as near as may be, so as not to unnecessarily impair its use. You are further instructed in this connection that the laws of the said State of New Mexico govern and fix the rights and duties of the defendant in so far as the construction of said crossing at which plaintiff was injured are concerned.

You are therefore, instructed that if you find from the evidence that in constructing its railroad across the highway on which the plaintiff was traveling when injured, the defendant restored said highway at the place intersected by its said railroad to its former state, as near as may be, so as not unnecessarily to impair its use, you cannot find the defendant negligent in respect to that matter."

6. The Court erred in refusing to instruct the jury as set forth in defendant's requested written instruction numbered D–XI, as follows: [327]

"You are instructed that a railway track is of itself a warning. It is a place of danger. It can never be assumed that cars or trains are not approaching on a track, or that there is no danger therefrom; it is therefore, the duty of every person who approaches a railroad track to exercise proper vigilance to ascertain and to know whether any trains or cars are approaching thereon before attempting to cross thereover. Proper vigilance requires every such person to look and listen before going upon said track or so close thereto as to be in danger. The exercise of ordinary care required of a traveler upon the highway under such circumstances, also requires such person or persons to stop, if necessary, in order to look and listen; and if for any reason, the view to any extent is obstructed, the person or persons endeavoring to cross such track or tracks, or going so close thereto as to be in danger either of their person or property, should exercise a greater precaution than ordinary prudence would dictate in such an exigency; that is, in stopping and looking and listening. The looking and listening and stopping thus required should be exercised at the last moment which ordinarily prudence would dictate before passing from a place of safety to one of danger. If the ability either to look or listen be obstructed by natural or artificial objects, then the duty to stop before entering into a zone of danger in order that proper investigation may be made, becomes absolute.

It is also equally the duty of one accompanying the driver of an automobile to exercise care for his own safety, and if you believe and find from a preponderance of the evidence that said Roy Spencer, without objection or protest, permitted the said Benjamin B. Spencer to drive upon the [328] railroad crossing immediately in front of an approaching engine and train of cars without stopping to look or listen or without having the automobile under such control that it might be stopped in time to avoid going on the crossing in front of the approaching train, or without exercising any care on his own part to ascertain whether the crossing was safe, then you are instructed that the plaintiff, Roy Spencer, would be guilty of negligence contributing to his own injuries, and your verdict must be for the defendant.''

7. The Court erred in refusing to instruct the jury as set forth in defendant's requested written instruction numbered D–XX, as follows:

''You are instructed that even if you find from the evidence that Roy Spencer looked and listened for the approach of the train in a place of safety, yet your verdict must be for the defendant if you further find from the evidence that said plaintiff was not constantly and actively vigilant in approaching and listening for the aproach of the train during the time occupied by him in going from the place where he may have stopped, looked or listened to and upon the track.''

8· The Court erred in refusing to instruct the jury as set forth in defendant's requested written instruction numbered D–XXIII, as follows:

''The passenger train involved in the accident described in the complaint and referred to in the evidence was running in interstate commerce between states and it was the right of the defendant to operate said train at a high [329] rate of speed and no presumption of negligence is to be indulged from the mere fact of the speed at which the train was operating. The reasonableness of that rule is apparent. The great purpose to be subserved by railroads is promptness, speed and dispatch in carrying passengers and freight. And under ordinary circumstances, in the open country, there is no duty resting upon the railway company to slacken the pace of its train at crossings. To hold otherwise would, to a great extent, destroy the advantages derived from modern facilities for transportation. To impose upon the defendant the duty of operating said train at a rate of speed so that it could be stopped at every grade crossing in Arizona in time to prevent accident should any person or vehicle be crossing the track on the highway, would be invalid under U. S. Constitution Article I, Section 8, as a direct and unreasonable burden upon interstate commerce when applied to said train and therefore you are instructed that the defendant had the right at the time and place and in the circumstances disclosed by the evidence to operate its said train in approaching the place of the alleged collision at a rate of speed of fifty miles an hour and you should not consider the speed of the train in this case as disclosed by the evidence as establishing the negligence of the defendant.''

9. The Court erred in refusing to instruct the jury as set forth in defendant's requested written instruction numbered D–XXV, as follows:

"If you find that the automobile in which plaintiff, Roy Spencer, was traveling with his father on the day in question was in bad repair or deficient, that fact is not [330] to be considered by you as in any way excusing plaintiff from exercising due care and circumspection for his own safety in crossing the railroad track."

10. The Court erred in instructing the jury as follows:

"If the railroad grade at any given place where a railroad is being constructed across a highway is so far above or so far below the level of the road or crossing, or so near to extraordinary cuts or curves as to seriously interfere with its use or seriously increase the danger of such use by bringing the road-crossing to the level of the tracks and the situation is such that the road can be carried over or under the tracks without unreasonable cost and expenditure and the danger or obstruction cannot otherwise be reasonably removed, then it is the duty of the railroad company to construct a crossing under its rails or carry the road above its tracks by a bridge, as the occasion requires. If it fails to perform such duty, and in an attempt to carry the road over its tracks at grade, that is, on a level with the rails, it seriously and unreasonably increases the danger and difficulty at the crossing, then the railroad company will be liable for any injury or damage proximately resulting because of such dangerous condition to one rightfully using the road," for the reason that such instruction is contrary to the laws of the State of New Mexico, which are controlling in this case, and for the further reason that the plaintiff failed to offer any evidence to prove that the condition or character of the crossing contributed in any way to the collision between him and defendant's train. [331]

11. The Court erred in instructing the jury as follows:

"If a railroad crossing is so constructed at grade as to create a condition of extraordinary danger due to existence of adjacent cuts or other physical conditions concealing approach of trains so as to render it probable that persons using the road would not be warned by the usual and ordinary measures or discover by the exercise of reasonable care the approach of trains in time to avoid a collision, then it is the duty of the railroad company to use extraordinary and additional precautions such as automatic signalling devices or other like methods to give warning of the approach of its trains or to approach such crossing carefully and with the train under such reasonable degree of control as would enable them to discover and avoid injury to persons so placed in danger,'' for the reason that no evidence was offered to substantiate the allegations that the railroad crossing at which plaintiff was injured created a condition of extraordinary danger or that persons using the road would not be warned by the usual and ordinary measures or discover by the exercise of reasonable care the approach of trains in time to avoid a collision.

The foregoing petition is made and based on the matters hereinabove set forth, upon the evidence, records, files, minutes and stenographers' transcript of the proceedings in said action taken, had and made.

Respectfully submitted,
CHALMERS, STAHL, FENNEMORE & LONGAN,

ROBERT BRENNAN,
FRANK E. FLYNN,
M. W. REED,
Attorneys for Defendant.'' [332]

THEREUPON, and on the same day, to wit: September 25th, 1926, defendant filed with the Clerk of said court an affidavit showing service by mail on counsel for plaintiff the said petition for new trial as follows:

"In the District Court of the United States in and for the District of Arizona.

No. L.–170—PRESCOTT.

ROY SPENCER, an Infant, by SARAH E. SPEN- CER, His Guardian *Ad Litem,*

Plaintiff,

vs.

THE ATCHISON, TOPEKA AND SANTA FE RAILWAY COMPANY, a Corporation,

Defendant.

AFFIDAVIT OF SERVICE BY MAIL.

State of California,
County of Los Angeles,—ss.

M. W. Reed, being first duly sworn, deposes and says that he is one of the attorneys of record for the defendant in the above-entitled action; that he resides in the city of Los Angeles, County of Los Angeles, State of California, and has his office in said city; that Messrs. Marron and Wood are the attorneys of record for the above-named plaintiff and they reside and have their offices at the city of Albuquerque, County of Bernalillo, State of New Mexico; that between the said city of Los Angeles and Albuquerque, there is a daily communication by United States mail; that on September 22d, 1926, affiant served defendant's petition for new trial, hereto [333] annexed, in the above-entitled cause, on the above-named plaintiff by enclosing a true copy thereof in a sealed envelope addressed "Messrs. Marron and Wood, Attorneys and Counsellors at Law, State National Bank Building, Albuquerque, New Mexico," and deposited the same in the United States Post Office in said City of Los Angeles, California, with postage thereon prepaid.

M. W. REED.

Subscribed and sworn to before me this 22d day of September, 1926.

[Seal]   S. A. FORRESTER,
Notary Public in and for Said County and State.
        [334]

THEREAFTER, to wit, on the 12th day of October, 1926, on motion of attorneys for defendant, an order was made by the Court as to certification by the Clerk of said court of certain exhibits to the United States Circuit Court of Appeals for the Ninth Circuit and filed with the Clerk of the District Court, in the words and figures following, to wit:

"In the District Court of the United States of America, in and for the District of Arizona.

No. L.–170—PRESCOTT.

ROY SPENCER, an Infant, by SARAH E. SPENCER, His Guardian, *Ad Litem,*

Plaintiff,

vs.

THE ATCHISON, TOPEKA AND SANTA FE RAILWAY COMPANY, a Corporation,

Defendant.

ORDER.

On Motion of Chalmers, Stahl, Fennemore & Longan, attorneys for defendant:

IT IS ORDERED, that if and when the Clerk of this court certifies and transmits to the Clerk of the United States Circuit Court of Appeals, for the Ninth Circuit, a transcript of the record and proceedings on writ of error in the above-entitled cause, the said Clerk of this court shall in addition to such transcript, certify and transmit to the said Clerk of the United States Circuit Court of Appeals for the Ninth Circuit, at San Francisco, California, for the inspection and consideration by said [335] Circuit Court of Appeals, in connection with said transcript, the following original papers in said action, to be by him safely kept and returned to this court upon the final determination of this cause in said Circuit Court of Appeals, namely:

Plaintiff's Exhibits, as follows:

Plaintiff's Exhibit #1—diagram.

Plaintiff's Exhibit II—photograph.

Plaintiff's Exhibit III—photograph.

Plaintiff's Exhibit IV—photograph.

Plaintiff's Exhibit V—photograph.

Plaintiff's Exhibit VI—photograph.

Plaintiff's Exhibit VII—photograph.

Plaintiff's Exhibit VIII—photograph.

Plaintiff's Exhibit IX—photograph.

Plaintiff's Exhibit X—Sanitorium Bill.

Plaintiff's Exhibit XI—X-ray Film.

Plaintiff's Exhibit XII—X-ray Film.

Plaintiff's Exhibit XIII—X-ray Film.

Plaintiff's Exhibit XIV—X-ray Film.

Plaintiff's Exhibit   Mortality Tables.

Defendant's Exhibit, as follows:

Defendant's Exhibit "A"—Blue-print.

IT IS FURTHER ORDERED that defendant need not inset in its bill of exceptions, nor in the transcript of record on writ of error, any of the hereinabove mentioned exhibits.

Dated October 12th, 1926.

F. C. JACOBS,
United States District Judge." [336]

THEREAFTER, to wit, on the 13th day of October, 1926, on application of the defendant, an order was made by the Court, and filed in the office of the Clerk of said court, further enlarging defendant's time to serve and present its proposed bill of exceptions, as follows:

"In the District Court of the United States in and for the District of Arizona.

No. L.–170—PRESCOTT.

ROY SPENCER, an Infant, by SARAH E. SPENCER, His Guardian *Ad Litem*,

Plaintiff,

vs.

THE ATCHISON, TOPEKA AND SANTA FE RAILWAY COMPANY, a Corporation,

Defendant.

ORDER.

ON APPLICATION of the defendant, it is ordered that defendant's time to serve and present its proposed bill of exceptions in the above-entitled cause be and it hereby is enlarged until and including November 13th, 1926.

Dated at Phoenix, Arizona, October 13th, 1926.

(Signed) F. C. JACOBS,

United States District Judge." [337]

THEREAFTER, to wit, on the 19th day of October, 1926, an order denying defendant's petition for a new trial was made by said Court, and by the Clerk entered in the minutes thereof, said order being as follows:

No. L.–170—PRESCOTT.

ᵈROY SPENCER, an Infant, by SARAH E. SPENCER, His Guardian *Ad Litem,*
Plaintiff,

vs.

THE ATCHISON, TOPEKA AND SANTA FE RAILWAY COMPANY, a Corporation,
Defendant.

ORDER DENYING MOTION FOR NEW TRIAL.

It is ordered that defendant's motion for a new trial herein be, and hereby is denied. Exception is saved to the defendant.''   [338]

The foregoing, containing all the evidence offered at the trial of said cause, with the defendant's exceptions thereto, and containing all the proceedings on the trial of said cause, and all the proceedings taken in said cause subsequent to the entry of judgment herein on the 15th day of August, 1926, is hereby offered as the defendant's proposed bill of exceptions.

CHALMERS, STAHL, FENNEMORE & LONGAN,

ROBERT BRENNAN,
FRANK E. FLYNN,
E. T. LUCEY,
M. W. REED,
Attorneys for Defendant.

IT IS HEREBY STIPULATED that the foregoing bill of exceptions is a true and correct bill of exceptions, and that the same may be settled, and allowed by the Court.

Dated, this —— day of ——, 1926.

—————————————.

—————————————.

—————————————.

The foregoing bill of exceptions has been settled and allowed, and is correct in all respects, and is hereby approved, allowed and settled and made a part of the record herein.

Dated this 10th day of November, 1926.

F. C. JACOBS,

Judge. [339]

———

[Title of Court and Cause.]

CERTIFICATE AND ORDER SETTLING AND ALLOWING BILL OF EXCEPTIONS.

THE UNDERSIGNED, Judge of the above-entitled court, hereby certifies that the foregoing bill of exceptions, containing all of the evidence offered at the trial of said cause, with the defendant's exceptions thereto, and containing all of the proceedings on the trial of said cause and all the proceedings taken in said cause subsequent to the entry of judgment herein on the 15th day of August, 1926, is a true and correct bill of exceptions, and the same is hereby settled and allowed and ordered to be filed.

Dated, this 10th day of November, 1926.

F. C. JACOBS,

Judge. [340]

[Endorsed on Back]: Received Copy of the Within Proposed Bill of Exceptions This 27 Day of October, 1926. Robt. E. Morrison, Attorney for Plaintiff.

[Endorsed on Back]: Filed Nov. 10, 1926. [341]

[Title of Court and Cause.]

PETITION FOR WRIT OF ERROR AND SUPERSEDEAS.

The Atchison, Topeka and Santa Fe Railway Company, a corporation, defendant in the above-entitled cause, feeling itself aggrieved by the verdict of the jury and judgment entered thereon on the 15th day of August, 1926, comes now by Chalmers, Stahl, Fennemore & Longan, Robert Brennan, Frank E. Flynn, E. T. Lucey and M. W. Reed, its attorneys, and files herewith an assignment of errors, and petitions said court for an order allowing said defendant to procure a writ of error to the Honorable the United States Circuit Court of Appeals for the Ninth Circuit, under and according to the laws of the United States in that behalf made and provided, and also that an order be made fixing the amount of security which the defendant shall give and furnish upon said writ of error, and that upon the giving of such security all further proceedings in this court be suspended and stayed until the determination of said writ of error by the United [342] States Circuit Court of Appeals for the Ninth Circuit.

Dated, November 8th, 1926.

> CHALMERS, STAHL, FENNEMORE
> & LONGAN,
> > ROBERT BRENNAN,
> > FRANK E. FLYNN,
> > E. T. LUCEY and
> > M. W. REED,
> > Attorneys for Defendant.

[Endorsed]: Filed Nov. 10, 1926.   [343]

[Title of Court and Cause.]

ASSIGNMENT OF ERRORS.

NOW COMES The Atchison, Topeka and Santa Fe Railway Company, a corporation, the defendant in the above-entitled action, by Chalmers, Stahl, Fennemore & Longan, Robert Brennan, Frank E. Flynn, E. T. Lucey and M. W. Reed, its attorneys, and files the following assignment of errors, upon which it will rely upon its prosecution of the writ of error in the above-entitled cause, petition for which writ of error is filed at the same time as this assignment of errors.

I.

The Court erred in denying the defendant's motion for a directed verdict, made by it at the close of plaintiff's case.   [344]

II.

The Court erred in denying the defendant's motion, made by it at the close of all the testimony in the case, to direct the jury to return a verdict for the defendant.

III.

The Court erred in refusing to instruct the jury as follows:

"You are instructed to return a verdict in favor of the defendant, The Atchison, Topeka and Santa Fe Railway Company."

IV.

The Court erred in refusing to instruct the jury as follows:

"As the collision between plaintiff and defendant's railroad train occurred in the State of New Mexico, and as Roy Spencer is a resident of that state, the laws and construction thereof by the courts of New Mexico are controlling upon you in deciding this case.

"Under the system in New Mexico it is contributory negligence for a person to approach a railroad track without paying careful attention for the approach of trains on such track in both directions."

V.

The Court erred in refusing to instruct the jury as follows:

"You are instructed that under the laws of the State of New Mexico a railroad company has the right to construct its railroad at grade across any highway which its said railway shall intersect, but that such company shall restore such highway to its former state, as near as may be, so as not to unnecessarily impair its use. You are further instructed in this connection that the laws of the said State of New Mexico govern and fix the rights and duties of the defendant in so far as the construction of said crossing at which plaintiff was injured are concerned. [345] You are, therefore, instructed that if you find from the evidence that in constructing its railroad across the highway on which the plaintiff was traveling when injured, the defendant restored said highway at the place intersected by its said railroad to its former state, as near as may be, so as not unnecessarily to impair its use, you cannot find the defendant negligent in respect to that matter.''

VI.

The Court erred in refusing to instruct the jury as follows:

"You are instructed that a railway track is of itself a warning. It is a place of danger. It can never be assumed that cars or trains are not approaching on a track, or that there is no danger therefrom; it is, therefore, the duty of every person who approaches a railroad track to exercise proper vigilance to ascertain and to know whether any trains or cars are approaching thereon before attempting to cross thereover. Proper vigilance requires every such person to look and listen before going upon said track or so close thereto as to be in danger. The exercise of ordinary care required of a traveler upon the highway under such circumstances, also requires such person or persons to stop, if necessary, in order to look and listen; and if for any reason, the view to any extent is obstructed, the person or persons endeavoring to cross such track or tracks, or going so close thereto as to be in danger either of their person or property, should exercise a greater precaution than ordinary prudence would dictate in such an exigency; that is, in stopping and looking and listening. The looking and listening and stopping thus required should be exercised at the last moment which ordinary prudence would dictate before passing from a place of safety to one of danger. If the ability either to look or listen be obstructed by natural or artificial objects, then the duty to stop before entering into a zone of danger in order that proper investigation may be made, becomes absolute.

"It is also equally the duty of one accompanying the driver of an automobile to exercise care for his own safety, and if you believe and find from a preponderance of the evidence that said Roy Spencer, without objection or protest, permitted the said Benjamin B. Spencer to drive upon the railroad crossing immediately in front of an approaching engine and train of cars without stopping to look or listen or without having the automobile under such control that it might be stopped in time to avoid going on the crossing in front of the approaching train, or without exercising any care on his own part to ascertain whether the crossing was safe, then you are instructed that the plaintiff, Roy Spencer, would be guilty of negligence contributing to his own injuries, and your verdict must be for the defendant.'' [346]

VII.

The Court erred in refusing to instruct the jury as follows:

''You are instructed that even if you find from the evidence that Roy Spencer looked and listened for the approach of the train in a place of safety, yet your verdict must be for the defendant if you further find from the evidence that said plaintiff was not constantly and actively vigilant in approaching and listening for the approach of the train during the time occupied by him in going from the place where he may have stopped, looked or listened to and upon the track.''

VIII.

The Court erred in refusing to instruct the jury as follows:

''The passenger train involved in the accident described in the complaint and referred to in the evidence was running in interstate commerce between states and it was the right of the defendant to operate said train at a high rate of speed and no presumption of negligence is to be indulged from the mere fact of the speed at which the train was operating. The reasonableness of that rule is apparent. The great purpose to be subserved by railroads is promptness, speed and dispatch in carrying passengers and freight. And under ordinary circumstances, in the open country, there is no duty resting upon the railway company to slacken the pace of its trains at crossings. To hold otherwise would, to a great extent, destroy the advantages derived from modern facilities for transportation. To impose upon the defendant the duty of operating said train at a rate of speed so that it could be stopped at every grade crossing in Arizona in time to prevent accident should any person or vehicle be crossing the track on the highway, would be invalid under U. S. Constitution Article I, Section 8, as a direct and unreasonable burden upon interstate commerce when applied to said train and therefore you are instructed that the defendant had the right at the time and place and in the circumstances disclosed by the evidence to operate its said train in approaching the place of the alleged collision at a rate of speed of fifty miles an hour and you should not consider the speed of the train in this case as disclosed by the evidence as establishing the negligence of the defendant." [347]

IX.

The Court erred in refusing to instruct the jury as follows:

"If you find that the automobile in which plaintiff, Roy Spencer, was travelling with his father on the day in question was in bad repair or deficient, that fact is not to be considered by you as in any way excusing plaintiff from exercising due care and circumspection for his own safety in crossing the railroad track."

X.

The Court erred in instructing the jury as follows:

"If the railroad grade at any given place where a railroad is being constructed across a highway is so far above or so far below the level of the road or crossing, or so near to extraordinary cuts or curves as to seriously interfere with its use or seriously increase the danger of such use by bringing the road crossing to the level of the tracks and the situation is such that the road can be carried over or under the tracks without unreasonable cost and expenditure and the danger or obstruction cannot otherwise be reasonably removed, then it is the duty of the railroad company to construct a crossing under its rails or carry the road above its tracks by a bridge, as the occasion requires. If it fails to perform such duty, and in an attempt to carry the road over its tracks at grade, that is, on a level with the rails, it seriously and unreasonably increases the danger and difficulty at the crossing, then the railroad company will be liable for any injury or damage proximately resulting be-

cause of such dangerous condition to one rightfully using the road.''

XI.

The Court erred in instructing the jury as follows:

> ''If a railroad crossing is so constructed at
> grade as to create a condition of extraordinary
> danger due to existence of adjacent cuts or
> other physical conditions concealing approach
> of trains so as to render it probable that per-
> sons using the road would not be warned by the
> usual and ordinary measures or discover by
> the exercise of reasonable care the approach
> of trains in time to avoid a collision, then
> [348] it is the duty of the railroad company
> to use extraordinary and additional precau-
> tions such as automatic signalling devices or
> other like methods to give warning of the ap-
> proach of its trains or to approach such cross-
> ing carefully and with the train under such
> reasonable degree of control as would enable
> them to discover and avoid injury to persons
> so placed in danger.''

XII.

The Court erred in accepting the verdict of the
jury and entering judgment thereon for the reason
that the evidence shows affirmatively and conclusively that the plaintiff was guilty of negligence
proximately contributing to his injury.

XIII.

The Court erred in accepting the verdict of the

jury and entering judgment thereon for the reason that the jury failed to follow the instruction of the Court reading as follows:

"If, after considering all of the evidence in the case, you further believe that the plaintiff, Roy Spencer, was also negligent to any extent and that his negligence concurred and co-operated with the negligent act or acts of the defendant to the extent that both the negligence of the defendant and the negligence of the plaintiff was the proximate cause of the injury, then the plaintiff cannot recover."

XIV.

The Court erred in accepting the verdict of the jury and entering judgment thereon for the reason that the jury failed to follow the instruction of the Court reading as follows:

"You are instructed that a traveler on a highway approaching a railway crossing must use ordinary care in selecting the time, and place to look and listen for coming trains and he should stop to make [349] observations, when necessary. It is his duty to use all of his faculties and it is not enough if he carefully listen, believing those in charge of the approaching train will ring a bell or sound a whistle or operate the train at a reasonable rate of speed. He must take advantage of every reasonable opportunity to look and listen and he has no right to omit looking and listening because he has heard no appropriate signal

of the approach of the train, even though the
law requires that signals be given.''

XV.

The Court erred in accepting the verdict of the
jury and entering judgment thereon for the reason
that the jury failed to follow the instruction of the
Court reading as follows:

"You are instructed that even though you
should find that the employes of the defendant
in charge of the train involved in the accident
described in the complaint failed to sound the
whistle or ring a bell as a warning of the approach of said train toward the point where
the collision is alleged to have happened or
operated said train at an excessive rate of
speed, still such failure to signal the approach
of said train or the operation of such train at
such an excessive rate of speed did not relieve
the plaintiff from the duty of exercising ordinary care to look and listen."

XVI.

The Court erred in accepting the verdict of the
jury and entering judgment thereon for the reason
that the jury failed to follow the instruction of the
Court reading as follows:

"You are instructed that a person traveling
upon a street or highway crossed by the tracks
of a railroad company have no right to presume that the railway employes engaged in
running trains or cars upon such track will
signal the approach of all trains, and plaintiff
had no right to omit the performance of any

part of his duty to look and listen for approaching trains, upon the presumption that the train men would do their duty." [350]

XVII.

The Court erred in accepting the verdict of the jury and entering judgment thereon for the reason that the jury failed to follow the instruction of the Court reading as follows :

"You are instructed that while the defendant railway company and Roy Spencer each had the right to use the road across which the track of said defendant is and was laid, yet, because such trains and cars are run on a fixed track and with great momentum, it had the right of way or precedence upon or over those parts of the road occupied by the railway track, and it was the duty of Roy Spencer to give unobstructed passage to the approaching train."

XVIII.

The Court erred in accepting the verdict of the jury and entering judgment thereon for the reason that the jury failed to follow the instruction of the Court reading as follows :

"If you find that the defendant railroad company was free from negligence or that the plaintiff was negligent, however slight, which in anywise contributed to the collision, you need not consider at all the nature and extent of plaintiff's damages but you should, without hesitation, return a verdict for the defendant railway company."

XIX.

The Court erred in accepting the verdict of the jury and entering judgment thereon for the reason that the jury failed to follow the instruction of the Court reading as follows:

> "It was the duty of the plaintiff, Roy Spencer to use that degree of care that a reasonably prudent person would use under like circumstances surrounding the case. It was his duty to look and listen for approaching trains before venturing on the tracks." [351]

XX.

The Court erred in accepting the verdict of the jury and entering judgment thereon for the reason that the jury failed to follow the instruction of the Court reading as follows:

> "You are instructed that a traveler on a highway approaching a railway crossing must use ordinary care in selecting a time and place to look and listen for coming trains and he should stop to make observations, when necessary. It is his duty to use all of his faculties and it is not enough that he carefully listened, believing that those in charge of the approaching train will ring a bell or sound a whistle or operate the train at a reasonable rate of speed. He must take advantage of every reasonable opportunity to look and listen and he has no right to omit to look or listen because he has heard no appropriate signal of the approach of the train, even though the law requires such signals to be given."

XXI.

The Court erred in denying defendant's motion for a new trial.

And upon the foregoing assignment of errors, and upon the record in said cause, the defendant prays that the judgment entered in said cause on the 15th day of August, 1926, be reversed.

Dated, November 8th, 1926.

> CHALMERS, STAHL, FENNEMORE & LONGAN,
>
> > ROBERT BRENNAN,
> > FRANK E. FLYNN,
> > E. T. LUCEY,
> > M. W. REED,
> > Attorneys for Defendant.

[Endorsed]: Filed Nov. 10, 1926. [352]

[Title of Court and Cause.]

ORDER ALLOWING WRIT OF ERROR.

UPON MOTION of Chalmers, Stahl, Fennemore & Longan, Robert Brennan, Frank E. Flynn, E. T. Lucey and N. W. Reed, attorneys for defendant, and upon filing a petition for a writ of error and an assignment of errors,

IT IS ORDERED that a writ of error be, and hereby is, allowed to have reviewed in the United States Circuit Court of Appeals for the Ninth Circuit, the verdict and judgment heretofore rendered and entered herein on the 15th day of August, 1926.

DONE this 10 day of November, 1926.

F. C. JACOBS,

Judge.

[Endorsed]: Filed Nov. 10, 1926.   [353]

———

[Title of Court and Cause.]

ORDER STAYING PROCEEDINGS.

The defendant, The Atchison, Topeka and Santa Fe Railway Company, a corporation, having this day filed its petition for writ of error from the verdict and judgment made and entered herein on the 15th day of August, 1926, to the United States Circuit Court of Appeals for the Ninth Circuit, together with an assignment of errors, within due time, and also praying that an order be made fixing the amount of security which defendant should give and furnish upon said writ of error, and that upon the giving of said security all further proceedings of this court be suspended and stayed until the determination of said writ of error by the said United States Circuit Court of Appeals, and said petition having this day been duly allowed:

NOW THEREFORE, it is ORDERED that upon the said defendant filing with the Clerk of this court a good and sufficient bond in the sum of Twenty Thousand Dollars ($20,000), to the effect that if the said defendant and plaintiff in error shall prosecute said writ of error with effect,   [354]   and answer all damages and costs, if it fails to make its plea good, then the said obligation to be void, else to remain in full force and virtue, the said bond to be approved by this Court, that all further proceedings in this court be and they are hereby suspended and stayed until the determination of said writ of error, by said United States Circuit Court of Appeals.

Dated, November 10th, 1926.

F. C. JACOBS,
Judge.

[Endorsed]: Filed Nov. 10, 1926.   [355]

[Title of Court and Cause.]

BOND ON WRIT OF ERROR.

KNOW ALL MEN BY THESE PRESENTS:

That we, The Atchison, Topeka and Santa Fe Railway Company, a corporation, as principal, and National Surety Company, a corporation, as surety, are held and firmly bound unto Roy Spencer, an infant, the plaintiff above named, in the sum of Twenty Thousand Dollars ($20,000.00), to be paid to said Roy Spencer, to which payment, well and truly to be made, we bind ourselves, and each of us, jointly and severally, and our and each of our successors and assigns, firmly by these presents.

Sealed with our seals and dated this 10th day of November, 1926.

WHEREAS the above-named defendant, The Atchison, Topeka and Santa Fe Railway Company, a corporation, has sued out a writ of error to the United States Circuit Court of Appeals for the

Ninth Circuit, to reverse the judgment in the above-entitled cause by the District Court of the United States [356] of America in and for the District of Arizona, rendered and entered in the said cause on the 15th day of August, 1926:

NOW, THEREFORE, the condition of this obligation is such that if the above named, The Atchison, Topeka and Santa Fe Railway Company, a corporation, shall prosecute the said writ with effect, and answer all costs and damages if it shall fail to make good its plea, then this obligation to be void, otherwise to remain in full force and effect.

THE ATCHISON, TOPEKA AND SANTA
FE RAILWAY COMPANY.
By W. K. ETTER,
Its General Manager,
Principal.
[Seal]   Attest: C. W. JONES,
Its Assistant Secretary.
NATIONAL SURETY COMPANY.
By W. B. BARR,
Its Attorney-in-fact,
Surety.
[Seal]   Attest: E. W. WILSON,
Its Resident Asst. Secretary.

Approved as to form and surety this 10th day of November, 1926.

F. C. JACOBS,
Judge U. S. District Court, for District of Arizona.

[Endorsed]: Filed Nov. 10, 1926.   [357]

[Title of Court and Cause.]

ORDER DIRECTING THE CLERK TO TRANSMIT ORIGINAL EXHIBITS TO CIRCUIT COURT OF APPEALS.

On motion of Chalmers, Stahl, Fennemore & Longan, Attorneys for Defendant:

IT IS ORDERED, that if and when the Clerk of this Court certifies and transmits to the Clerk of the United States Circuit Court of Appeals, for the Ninth Circuit, a transcript of the record and proceedings on writ of error in the above-entitled cause, the said Clerk of this court shall, in addition to such transcript, certify and transmit to the said Clerk of the United States Circuit Court of Appeals for the Ninth Circuit, at San Francisco, California, for the inspection and consideration by said Circuit Court of Appeals, in connection with said transcript, the following original papers in said action, to be by him safely kept and returned to this court upon the final determination of this cause in said Circuit Court of Appeals, namely:

Plaintiff's Exhibits, as follows:

Plaintiff's Exhibit #I—diagram.

Plaintiff's Exhibit #II—photograph. [360]

Plaintiff's Exhibit #III—Photograph.

Plaintiff's Exhibit #IV—Photograph.

Plaintiff's Exhibit #V—Photograph.

Plaintiff's Exhibit #VI—Photograph.

Plaintiff's Exhibit #VII—Photograph.

Plaintiff's Exhibit #VIII—Photograph.

Plaintiff's Exhibit #IX—Photograph.

Plaintiff's Exhibit #X—Sanitorium Bill.

Plaintiff's Exhibit #XI X-ray Film.

Plaintiff's Exhibit #XII—X-ray Film.

Plaintiff's Exhibit #XIII—X-ray Film.

Plaintiff's Exhibit #XIV—X-ray Film.

Plaintiff's Exhibit #   Mortality Tables.

Defendant's Exhibit, as follows:

Defendant's Exhibit "A"—Blue-print.

IT IS FURTHER ORDERED that defendant need not insert in its bill of exceptions, nor in the transcript of record on writ of error, any of the hereinabove-mentioned exhibits.

Dated October 12th, 1926.

F. C. JACOBS,
United States District Judge.

[Endorsed]: Filed Oct. 12, 1926. [361]

[Title of Court and Cause.]

AFFIDAVIT OF SERVICE.

State of Arizona,

County of Maricopa,—ss.

John M. Longan, being duly sworn, on oath deposes and says: That on the 10th day of November, 1926, he served on Robert E. Morrison, one of the attorneys for plaintiff in the above-entitled cause,

Petition for writ of error and supersedeas,

Assignment of errors,

Order allowing writ of error,

Order staying proceedings,

Bond on writ of error,

Writ of error, and

Citation on writ of error,

by depositing, on said date, in the United States Postoffice, at Phoenix, Maricopa County, Arizona, true and correct copies of said petition for writ of error and supersedeas, assignment of errors, order allowing writ of error, order staying proceedings, bond on writ of error, writ of error, and citation on writ of error, which said copies so deposited were inclosed by affiant in an envelope addressed: "Mr. Robert E. Morrison, Attorney at Law, Prescott, Arizona," and which said envelope containing said copies was sealed by affiant, the postage thereon prepaid, and deposited by affiant as aforesaid. [362]

That affiant is a member of the law firm of Chalmers, Stahl, Fennemore & Longan, the members of which law firm are the attorneys for the above-named defendant in the above-entitled action; that the members of said law firm are: L. H. Chalmers, Floyd M. Stahl, H. M. Fennemore, Thomas G. Nairn, and this affiant; that said members of said law firm, and each of them, and Thomas G. Nairn, at the time of said service, resided and had their office, and ever since then and now reside and have their office in Phoenix, Maricopa County, Arizona.

That Robert E. Morrison, at the time of said service upon him, resided and had his office, and now resides and has his office at Prescott, Yavapai

County, Arizona; that there was, at the time of said service upon the said Robert E. Morrison, and now is, between Phoenix, Arizona, and Prescott, Arizona, regular communication by mail.

That affiant was, at the time of said service, and is now, a male citizen of the United States, and of the State of Arizona, over the age of 21 years, and fully competent to be a witness in the above-entitled cause.

JOHN M. LONGAN.

Subscribed and sworn to before me this 10th day of November, 1926.

[Seal]   H. W. ALLEN,
                                   Notary Public.

My commission expires Sept. 5, 1928.

[Endorsed]: Filed Nov. 10, 1926. [363]

[Title of Court and Cause.]

PRAECIPE FOR TRANSCRIPT OF RECORD.

To the Clerk of Said Court:
Sir:

Please issue and certify a transcript for plaintiff in error upon writ of error to the Circuit Court of Appeals for the Ninth Judicial Circuit, of the record in the above-entitled cause, and include therein:

(1) The judgment-roll.
(2) Petition for removal.
(3) Notice of removal.

(4)   Order on removal

(5)   Bond on removal.

(6)   Defendant's bill of exceptions.

(7)   Petition for writ of error and supersedeas.

(8)   Assignment of errors.

(9)   Order allowing writ of error.

(10) Order staying proceedings.

(11) Bond on writ of error.

(12) Writ of error.

(13) Citation.

(14) Order directing Clerk to certify up original exhibits.

(15) Affidavit of service by mail of copies of petition for writ of error, and supersedeas, assignment of errors, order allowing writ of error, order staying proceedings, bond on writ of error, writ of error, and citation.

(16) This praecipe.

(17) Certificate of Clerk authenticating the record.

Dated: November 10, 1926.

CHALMERS, STAHL, FENNEMORE & LONGAN,

ROBERT BRENNAN,
FRANK E. FLYNN,
E. T. LUCEY and
M. W. REED,

Attorneys for Defendant and Plaintiff in Error.

[Endorsed]: Filed Nov. 13, 1926.   [364]

[Title of Court and Cause.]

ORDER EXTENDING TIME TO AND IN-CLUDING JANUARY 15, 1927, TO FILE WRIT OF ERROR AND DOCKET CAUSE.

ON MOTION of Chalmers, Stahl, Fennemore & Longan, attorneys for defendant in the above-entitled action, and for good cause shown, it is hereby ordered that said defendant's time within which to file the record thereof on its writ of error and docket the above-entitled cause with the Clerk of the United States Circuit Court of Appeals for the 9th Circuit, at San Francisco, California, is hereby enlarged and extended from December 8, 1926, to January 15, 1927.

F. C. JACOBS,
Judge.

Dated, November 30th, 1926.

[Endorsed]: Filed Nov. 30, 1926. Original sent to C. C. A. [365]

[Title of Court and Cause.]

CERTIFICATE OF CLERK U. S. DISTRICT COURT TO TRANSCRIPT OF RECORD.

United States of America,
District of Arizona,—ss.

I, C. R. McFall, Clerk of the District Court of the United States for the District of Arizona, do hereby certify that I am the custodian of the records, papers and files of the said United States

District Court for the District of Arizona, including the records, papers and files in the case of Roy Spencer, an infant, by Sarah E. Spencer, his guardian *ad litem,* and Sarah E. Spencer, individually, plaintiffs, versus The Atchison, Topeka & Santa Fe Railway Company, a corporation, defendant, said case. being numbered 170 on the Law Docket of the Prescott Division of the said court.

I further certify that the foregoing 365 pages, numbered from 1 to 365, inclusive, constitute a full, true and correct copy of the record, and of the assignment of errors and all proceedings in the above-entitled cause as requested in the praecipe filed herein, as the same appears from the originals of record and on file in my office as such Clerk.

And I further certify that there are also annexed to said transcript the original writ of error and the original citation issued in said cause.

And I further certify that the cost of the foregoing transcript, amounting to Sixty-six and 40/100 Dollars ($66.40), has been paid to me by the defendant (plaintiff in error).

WITNESS my hand and the seal of said Court, this 11th day of January, 1927.

[Seal] C. R. McFALL,
Clerk of the District Court of the United States for the District of Arizona.

By Paul Dickason,
Chief Deputy Clerk. [366]

CITATION.

United States of America,—ss.

To Roy Spencer, an Infant, by Sarah E. Spencer, His Guardian, *ad Litem,* Plaintiff, GREETING∴

You are hereby cited and admonished to be and appear at a United States Circuit Court of Appeals for the Ninth Circuit, to be held at the city of San Francisco, in the State of California, on the 8th day of December, A. D. 1926, pursuant to a writ of error on file in the Clerk's Office of the District Court of the United States, in and for the District of Arizona, in that certain Cause No. L.–170—Prescott, wherein The Atchison, Topeka and Santa Fe Railway Company is plaintiff in error and you are defendant in error, to show cause, if any there be, why the judgment rendered against the said plaintiff in error as in the said writ of error mentioned, should not be corrected, and speedy justice should not be done to the parties in that behalf.

WITNESS, the Honorable F. C. JACOBS, United States Judge for the District of Arizona, this 10th day of November, A. D. 1926, and of the Independence of the United States, the one hundred and fiftieth.

F. C. JACOBS,

U. S. District Judge for the District of Arizona.

[367]

WRIT OF ERROR.

United States of America,—ss.

The President of the United States of America, to the Judges of the District Court of the United States, for the District of Arizona, GREETING:

Because in the record and proceedings, and also in the rendition of the judgment of a plea which is in the said District Court, before you between The Atchison, Topeka and Santa Fe Railway Company, the plaintiff in error, and Roy Spencer, an infant, by Sarah E. Spencer, his guardian *ad litem,* defendant in error, No. L.–170—Prescott, a manifest error hath happened, to the great damage of the said plaintiff in error as by its complaint appears, and it being fit, that the error, if any there hath been, should be duly corrected, and full and speedy justice done to the parties aforesaid in this behalf, you are hereby commanded, if judgment be therein given, that then, under your seal, distinctly and openly, you send the record and proceedings aforesaid, with all things concerning the same, to the United States Circuit Court of Appeals for the Ninth Circuit, together with this writ, so that you have the same at the city of San Francisco, in the State of California, on the 8th day of December next, in the said United States Circuit Court of Appeals, to be there and then held, that the record and proceedings aforesaid be inspected, the said United States Circuit Court of Appeals may cause further to be done therein to correct that error, what of right and according to the law and custom of the United States should be done.

WITNESS, the Hon. WILLIAM HOWARD TAFT, Chief Justice of the United States, this 10th day of November, in the year of our Lord one thousand nine hundred and twenty-six and of the Independence of the United States the one hundred and fifty-first.

[Seal]   C. R. McFALL,

Clerk of the District Court of the United States of America in and for the District of Arizona.

By Paul Dickason,
Chief Deputy Clerk.

The above writ of error is hereby allowed.

F. C. JACOBS,
Judge.

RETURN ON WRIT OF ERROR.

United States District Court,
District of Arizona,—ss.

The answer of the Judge of the District Court of the United States for the District of Arizona, to the within writ of error;

As within commanded, I herewith transmit to the United States Circuit Court of Appeals for the Ninth Circuit within mentioned, a true and complete transcript of the record and proceedings in the foregoing cause, this 11th day of January, A. D. 1927.

C. R. McFALL,

Clerk of U. S. District Court, District of Ariz.

By Paul Dickason,
Chief Deputy Clerk.

[Endorsed]: No. 5038. United States Circuit Court of Appeals for the Ninth Circuit. The Atchison, Topeka & Santa Fe Railway Company, a Corporation, Plaintiff in Error, vs. Roy Spencer, an Infant, by Sarah E. Spencer, His Guardian *Ad Litem,* and Sarah E. Spencer, Individually, Defendants in Error. Transcript of Record. Upon Writ of Error to the United States District Court of the District of Arizona.

Filed January 13, 1927.

F. D. MONCKTON,

Clerk of the United States Circuit Court of Appeals for the Ninth Circuit.

By Paul P. O'Brien,

Deputy Clerk.

TOPICAL INDEX.

TABLE OF CASES.

IN THE

United States
Circuit Court of Appeals,
FOR THE NINTH CIRCUIT.

The Atchison, Topeka and Santa Fe
Railway Company, a corporation,
Plaintiff in Error,

vs.

Roy Spencer, an infant, by Sarah E.
Spencer, his guardian ad litem,
Defendant in Error.

BRIEF FOR PLAINTIFF IN ERROR.

(Note: Figures in brackets refer to pages of transcript)

Statement of the Case.

This action was brought in the Superior Court of the
State of Arizona, in and for the County of Navajo, by
Roy Spencer, an infant, by Sarah E. Spencer, his guardian
ad litem, defendant in error, and Sarah E. Spencer, individually, against plaintiff in error, to recover damages
arising out of injuries sustained by defendant in error in a
collision between an automobile driven by his father and
one of plaintiff in error's trains at a highway crossing in
the State of New Mexico.

On the petition of plaintiff in error, the case was removed to the District Court of the United States for the District of Arizona on the ground of diversity of citizenship, the defendant in error and said Sarah E. Spencer being residents and citizens of the State of New Mexico [52, 118] and plaintiff in error a resident and citizen of the State of Kansas [5]. At the trial the case was dismissed as to Sarah E. Spencer, individually.

For brevity, the parties will hereinafter be referred to as follows:

Roy Spencer, defendant in error, "plaintiff";

The Atchison, Topeka and Santa Fe Railway Company, plaintiff in error, "defendant."

The complaint alleged that on the 11th day of June, 1923, while plaintiff was riding in an automobile driven by Benjamin B. Spencer, his father, along a public highway which crossed the tracks of the defendant about one mile west of the village of Mountainair, in the County of Torrance, State of New Mexico, the said automobile was struck by the locomotive of a train of cars then being operated by defendant and the plaintiff received the injuries complained of.

The gravamen of the complaint is that the said collision and resulting injury to Roy Spencer were caused by the negligence of the defendant, its officers, agents, and servants in the following particulars:

(a) That the aforesaid crossing was one of peculiar and extraordinary danger in that the view or warning of approaching trains were cut off and obstructed from both directions by cuts, curves and the configuration of the ground, and the crossing so constructed as to make it very difficult to operate and drive a car over and across the same, or to learn of the danger from approaching trains in time to enable a reasonably careful person to protect himself and guard against said danger.

(b) That the defendant built its line of railroad across the said highway, and it became and was the duty of the defendant, under the laws of the State of New Mexico then existing, to construct the said crossing and to restore the highway to a reasonably safe condition for public traffic.

(c) That the construction of the said highway was improperly and carelessly and negligently done in that the railroad tracks were built upon a grade some eight or nine feet above the level of the said highway at the point of such crossing, and instead of conducting the said highway under the said tracks as could have been done, and in the exercise of reasonable care and prudence should have been done, by means of a bridge or trestle to conduct the railroad over the said highway, the highway crossing was improperly and negligently constructed by building a steep embankment on each side of the said tracks with a turn at approximately right angles on each side of such grade, and of such a character as to make it unreasonably difficult to operate a car over the said tracks.

(d) That view of approaching trains at the point where the said highway was carried across the railroad tracks was very largely concealed from both directions by deep cuts through which the railroad tracks were constructed and laid, concealing the views of approaching trains from persons passing across the said tracks, of such a nature and character that due to the rise in the grade of the highway and the position of the aforesaid cuts and the configuration of the adjacent land the approach of trains cannot be discovered by the exercise of reasonable care and prudence by a person lawfully attempting to use the said highway crossing until in a position of peril, which it is unreasonably difficult to avoid or escape from.

(e) That the condition of the said crossing, as the defendant constructed the same, and under the surrounding circumstances, was and is an improper, careless, unsafe and unreasonable method of carrying the highway across the railroad tracks due to the peculiar dangers and natural topographical features above recited, and it was readily and reasonably possible and practical, and the duty of the said defendant in restoring the said highway to a safe condition to carry the highway under the railway tracks and thus to avoid the dangers and difficulties so produced by the grade crossing.

(f) That due to the difficulty of discovering approaching trains because of the nature and condition of the highway and cuts above mentioned, it was the duty of the defendant to establish and maintain automatic signals or warnings of approaching trains so maintained and operated as to give signals and warnings to persons using the highway of the dangerous proximity of such trains, such as an automatic bell or other reasonable signalling device, the physical condition of such crossing requiring and making necessary extraordinary methods and means of warning persons using the highway of such danger.

(g) That the train which collided with the said plaintiff and injured him was then being carelessly and negligently operated in approaching said crossing at a very high rate of speed so as to be, and was, beyond the reasonable control of the servants of the defendant operating the same while approaching said crossing.

(h) That the agents and servants of the defendant operating the said train approached the said crossing without giving any signal by bell or whistle of its approach, within the distance reasonably calculated to be heard by or warn the plaintiff or other travelers of its approach, or within the proper, usual and customary distance from the said crossing, and the approach of said train being then concealed by dust stirred up by the breeze.

The answer fully traverses the complaint and denies all negligence therein charged both with respect to the manner in which said crossing was constructed, and the manner in which said train was operated. The answer affirmatively alleges that the crossing was in all respects constructed in compliance with the laws of the State of New Mexico, and that said highway was restored to a reasonably safe condition for public travel. The following special defenses were by the answer pleaded:

(a) That the injuries to plaintiff and the damages resulting therefrom were caused solely by and resulted wholly from the negligent acts and omissions of the plaintiff and his father who was driving the said automobile, in that the said plaintiff and his father failed to look or listen for the approach of defendant's engine and train of cars to said crossing, and that the plaintiff failed to take any precaution whatsoever to avoid said collision, and failed to observe and heed the warning signals given by the sounding of the whistle and the ringing of the bell of said locomotive engine as it approached the crossing; and

(b) That the accident and the injuries to plaintiff resulting therefrom were caused by the negligence of plaintiff, which negligence was the direct and proximate cause thereof, and were not caused by any negligence or carelessness of the defendant, or its agents, servants or employes.

The case was twice tried, but the question involved in the first trial which resulted in the second trial did not arise in the latter.

Trial.

The case was tried before a jury.

At the close of plaintiff's case, the defendant moved for a directed verdict [162 et seq], which motion will hereinafter be set forth at length. This motion was, by the court, denied, to which action of the court, defendant duly excepted [167]. At the close of all of the testimony in the case, the defendant, with the consent of the court, renewed its said motion upon all the grounds theretofore stated in support of the said motion made at the close of plaintiff's case [290], which motion was, by the court, again denied, and to which action of the court the defendant duly excepted [290].

The defendant requested the court to give certain instructions to the jury, which the court refused to give, and to such refusal the defendant excepted. And the court gave certain instructions to the jury to which the defendant excepted, as will hereinafter more fully appear.

A verdict was rendered in favor of plaintiff and against the defendant for $16,750.00; and judgment thereon was entered for said sum on the 15th day of August, 1926.

Within the time and in the manner provided by law, the defendant filed a motion for a new trial [322 et seq] which was denied [373].

The defendant now brings error from said judgment to this court.

Facts.

Plaintiff was injured at a railroad crossing near Mountainair, New Mexico, on the 11th day of June, 1923 [52]. It was shown by the testimony of the plaintiff and his witnesses that on the day mentioned, plaintiff, who was then 17 years of age [85], and his father were travelling along a highway in a northeasterly direction toward the town of Mountainair, New Mexico, and that the highway ran more or less parallel to defendant's line of railroad for a distance of a half mile before reaching the crossing [55], and from two to three hundred yards distant from the railroad [87]. The vehicle in which they were riding was a Ford truck which the father was driving [53]. The truck had no top on it [53]. Plaintiff's sight and hearing were good, as were also his father's [89]. The day was perfectly clear and there was not sufficient wind blowing to pick up dust or obscure plaintiff's view along defendant's track [83, 93, 94].

When plaintiff and his father had reached a point where the highway along which they were travelling turns northerly and passes over and across defendant's railroad, they stopped and plaintiff got out of truck [58, 87]. The distance from this point to he center of the track at the crossing is 100 feet [59, 87]. The highway along which they travelled runs in a northeasterly direction to the point where plaintiff and his father stopped and then it turns north [58], and crosses the track at about a 45 degree angle [93].

(Photo.)

Plaintiff's exhibit No. 2 shows the situation, and we have inset on the opposite page a copy of that exhibit on a reduced scale. This picture as shown on the exhibit, was taken from a point 50 feet south of the railroad right of way, looking north over the crossing.

After getting out of the car, plaintiff looked both ways for trains and could not see any [65]. After stopping for half a minute [87], plaintiff and his father started to go over the crossing, the father driving the car, and plaintiff pushing from behind. [65]. In so doing, plaintiff placed his hands on the back of the truck bed, with his head down [65]. When the car had proceeded to a point 50 feet from the track, plaintiff released his hold with the right hand from the back of the car, and stepped around to the right hand side. As he proceeded along the right hand side, his left hand was not released. He did, however, remove the position of his left hand as he made the change from the rear to the side of the car. When shifting his position to the side of the car, he placed his right hand against the back of the seat [65, 86]. He continued in that position until the front wheels of the car reached the first rail of the track, plaintiff, himself, then being six feet, or more, from the rail. [68]. He ceased to push, when the front wheels of the car had just gone over the first rail. [69]. At that time, he stepped up onto the running-board and into the car and turned around and looked eastwardly down the railroad track toward Mountainair, [69, 70] and saw an approaching west-bound passenger train about 50 yards away. [71]. The collision followed. [71].

When plaintiff and his father left their home on the day in question, the car was running well. [53]. But on

he above picture is taken from the point about 50 feet
e railroad right of way, looking north over the crossing
cident occurred. The automobile shown in the picture
e foot of the railroad embankment just at the corner
the south fence of the railroad right of way joing
long the road running over the tracks.

-37-

their return trip from Scholle, a town about 14 miles from Mountainair, to which they had gone earlier in the day, some trouble had developed in the mechanism of the car and it had been necessary when coming to grades in the road for plaintiff to assist the car up the grade by getting out and pushing from behind. [54.] This had occurred twice between Scholle and the crossing. [55].

As to the condition of the car, and the manner in which plaintiff's father operated it when going up grades, plaintiff testified as follows:

"The car started to missing after we had gone about two miles from [53] Scholle * * * That day the spark had been out, and he (plaintiff's father) had been working on the coils to get it to run. The lid was off the top of the coils and he worked, done something, worked on the platinums. I do not know just what he done. The coils were down in front of the car, under the windshield— right next to the bottom of the car. [54.]

Q. Show the jury what position you saw your father take in working with those coils when the car was running badly? [54]. A. Well, he would stoop over them little plates that work up and down that takes the spark, he would work them that way and look at them; I don't know just how he worked them. * * * Just exactly what my father was doing [54], and what was wrong, I am unable to state. I saw him bend down when his car would go bad and do something down under the windshield several times. The driving wheel of the Ford, or the steering wheel is on the left side of the car. My father would stoop down—he would stoop to the right. He would stoop his head at those times when he was fixing the car—probably to where the windshield connects." [55.]

Plaintiff did not observe the position of his father, or what he was doing while plaintiff was pushing on the car as they proceeded toward the crossing from the point 100 feet south of the track, at which they had stopped. Whether his father was sitting up or leaning over plaintiff did not know. [71.]

It took the car a full minute to travel from the place where plaintiff and his father stopped until it got onto the track. [68, 87, 88.] At no time did plaintiff look in the direction from which the train came, while he was travelling this space of 100 feet. The first time he looked after leaving the point 100 feet south of the track was when he had gotten into the car and was about to sit down [86, 91], and at no time did the car stop. [86.]

Plaintiff was thoroughly familiar with the railroad between the station of Mountainair and the crossing. [56.] The distance is one and three-quarter miles. [57.]

Plaintiff had sufficient knowledge to know that to get into the car as it was about to go onto the track and to do so without looking was highly dangerous, unless he had already looked. [91.]

When plaintiff was at the point 100 feet south of the track, he could have seen the whistling post which was located a quarter of a mile east of the crossing; as he approached the track, he could have seen farther and farther easterly; when 50 feet from the track, he could have seen beyond the whistling post quite a little ways; when 30 feet from the track, he could have seen somewhat farther; and when at a point 20 feet from the center line of the track, he could have seen beyond the whistling post. The whistling post is about 6 or 7 feet tall, and is located about 6 or 8 feet north of the track. A locomotive is taller than the whistling post. [91, 92.]

Plaintiff further testified that from the point 100 feet south of the track one could see a little of the smokestack of a locomotive when it got to the whistling post.

From the aforesaid point, a person standing thereat could see the greater part of a freight train of 52 cars extending from the crossing to a cut east thereof, as shown by plaintiff's "Exhibit No. 8." [64].

Plaintiff further testified that when he was about 500 yards from the crossing he saw a freight train passing over the crossing and going easterly toward Mountainair, [55, 57]. When plaintiff saw this train, it was moving at a speed of 20 miles per hour. There were 20 cars in the train, with a locomotive on the rear end pushing. [88.] Plaintiff and his father, after plaintiff first observed the freight train, travelled 75 or 100 yards along the highway while the train was going over the crossing [88, 89]. It took them 10 minutes to go from the point where plaintiff saw the train going over the crossing, to the point where they stopped [89]. In response to a question by the court, plaintiff further testified he judged that about 10 minutes elapsed from the time the freight train passed the crossing going toward Mountainair until he was struck by the passenger train going west [96].

Plaintiff listened as much as he could for the sound of any train while pushing on the car, but on account of the noise made by the car he could not hear very well [83].

Plaintiff further testified on direct examination:

"Q. Was there anything in the position which you had alongside of the car going up there that would obstruct your view of the train from the east, had you looked in that direction?" "A. Yes, there was. The right of way fence where these poles and boards set up there for a distance of four feet; a curve in the track down the road throws your head right against the fence on down the track. You also have to pass a distance of about eight feet before you could see the track. Exhibit No. 5, the photograph which shows a fence alongside of the roadway as you pass up, that is the fence which I am speaking of which would obstruct my vision." [82.]

(Photo.)

We have inset on the opposite page a copy of plaintiff's exhibit No. 5 showing the wing-fence and the view to the east along the track.

On cross-examination, plaintiff testified about the fence as follows:

"Referring to plaintiff's Exhibit No. 5, which you handed me, I have examined that. I stated on direct examination that there is a fence that shows in that picture, and that that fence obscured my vision somewhat. That fence is right here. That is just a white board fence there. It is right on the south side of the track [95].

Q. Is there not a wing fence at the cattle-guard? A. I think that is what it is. I think it extends out from the track, the south rail of the track, about six feet— something like that. That is the only fence there is along there at all. What I was referring to is the curve that goes around there. But as I approached the crossing, that is, while I travelled the distance of one hundred feet, I did not see that fence at all." [95, 96.]

Plaintiff further testified that the railroad enters a cut west of the crossing on a curve as shown on plaintiff's Exhibit 4, and that a train disappears in the cut between two and three telegraph poles west of the crossing [71, 72]. The telegraph poles are 150 feet apart [61].

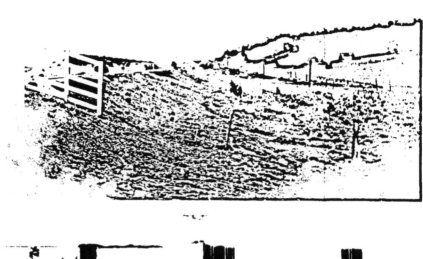

Admitted and Filed Aug 11 1926

C. R. McFadden, Clerk

By

Deputy Clerk

No. 5038

UNITED STATES CIRCUIT COURT OF APPEALS
FOR THE NINTH CIRCUIT

FILED

JAN 13 1927

F. D. MONCKTON,
CLERK.

Plaintiff's sole excuses for not looking in the direction from which the train approached while he was pushing on the car are found entirely in the following testimony:

"From what I know of that car, and the way it was running, I know what would have happened if I had ceased for a minute or an instant to push; it would stop. From my knowledge of the grade, I know whether or not stopping on that grade it could, have been started again without going back to the bottom of the grade; it would have to run back to the bottom of the grade, if it stopped, undoubtedly. I did not cease for an instant to push from the time that car started until it got up on the track. From the position that I was in while I was pushing that car at the side of it, I could not have turned my head around and looked behind me in the direction in which that train was coming. In order to have looked in the direction in which the train came at the time—from the time I came to the side of the car, until I came up on the track—I would had to have taken this arm down and look and quit pushing, and the car would have stopped and would have run down and it would have had to." [68, 69.]

"I have stated to the jury that while I was pushing the car up the hill I did not look to the left behind me at the track down toward Mountainair. The reason I did not was because I could not.

"Q. You have given that reason. Now, is there any other reason? A. One reason was, when I started up the hill I had already looked to the east; you can see a distance there, something like 200 yards, a little over. And this cut that comes out of the track on the west, it comes right out; when you are standing at the corner post you cannot see the train coming from the west more than a telegraph pole before it gets to the crossing. The reason why I did not look and see was I was watching toward the west, because it is more dangerous than the east of the track, the crossing there." [81, 82.]

Jesus Berreras, a witness for plaintiff, testified:

"Q. How far was the truck which Mr. Spencer and his son were operating when you first observed it; how far was it from the railway track? A. For the first time that I seen it it was standing there at the corner of the right of way. I don't know how long the truck stood at the corner. I don't know how long it had been there or how long it had been stopped when I observed it from my house. It was standing there, and I looked down towards the crossing. I don't know how long it stood there, but when I first observed it, until I seen it start, it might have been ten minutes; something like that. As to the length of time it took the car or truck after I observed it start in motion, to reach the railroad crossing— why, to my estimation, I don't know, but it seemed to me just about a minute from where it started until it hit the track." [103, 104.]

The foregoing is a complete statement of the relevant evidence in the case as to contributory negligence, and we shall argue therefrom that the plaintiff was guilty of contributory negligence as a matter of law and that the trial court erred in refusing to direct a verdict in defendant's favor.

Facts as to Location and Physical Character of Crossing.

Frank Sedillo, a witness for plaintiff, after testifying on direct examination that he was familiar with the highway before the railroad was built [41], testified as follows:

"Q. Whereabouts, with reference to where the present road crosses the railroad did that old road cross the line of the road before it was built? A. The old road crossed right in the same place, only it crossed kind of west, the old road; and this new road crosses right straight south. That is, the old road crossed at a much sharper angle than this road. [42.]"

On cross examination, the witness testified:

"I think that the present crossing is about at the place where the old road crossed the line. The old road made more of a diagonal crossing there than at present." [43.]

Juan Chavez, another witness for plaintiff, testified:

"I know the crossing about one mile west of Mountainair where the train struck him * * * the road before the railroad—I have known that road to cross at the point where it does now before the railroad was built * * *

"Q. Where did the road—the old road—cross the line that is now occupied by the Railroad? Was it at the place it now crosses, or some other place? A. Don't remember very well whether it was right at the same point.

Q. Was it approximately, more or less, right where it is now that it crossed that place? A. I don't remember. I could not testify whether it was right there, or farther [44]. One thing I know, when they put in the railroad there had been a crossing.

Q. A crossing right where? A. Right where it is. [45.]

It was stipulated between the parties that certain witnesses, viz: Juaquin Sisneras, Merijildo Sisneres, Miguel Garcia, Comita Contreres and Natividad Salas would, if produced on behalf of the plaintiff, testify in substance and to the same effect as the aforesaid witnesses Sedillo and Chavez [46, 47].

John Block, a witness for defendant, testified:

"Q. Do you know whether the highway, before the line of railroad was constructed across it, was in the same location, in the same position, as it is now? A. It was practically in the same place, yes, sir. It may be a little different variation in the railroad crossing right at present, as the road did before the railroad was established, it may be just a little bit of difference of a curve, but it is practically in the same place. I would say the exact point of crossing of the highway and the railroad is in exactly the same place as it was before the railroad was constructed." [181, 182.]

Grade of Highway at Crossing.

Plaintiff testified that there is a rise of 10 feet from the railroad fence, 100 feet south of the track, to the nearest rail [58, 59], and that the grade of the railroad extends on a level outside of the rail for a distance of 4 feet. [59.]

R. M. Noble, a civil engineer, and a witness for the defendant, testified that he made the map and profile [167] introduced in evidence as defendants exhibit "A" [169]; that the map shows the profile grade of the highway and of the tracks [167]; and that the profile shows the highway to be level for a distance of 20 feet out from the center of the track [168].

A copy of defendant's exhibit "A," on a reduced scale, is appended to this brief.

This witness further testified:

"I can state whether or not the grade of the highway, as it approaches the crossing from the south, that is, the portion of the highway extending from [176] the south right of way fence to the crossing, is on the same, was built on the surface of the earth. There was no grading of the highway at all any more than just to make a smooth running surface for the road. Of course, that is on the side of the hill, possibly it was cut on one side—filled a little on the other. Generally speaking the roadway was not built so there was a fill, not any more than I said. It was not a fill proposition. You see, it is the side of a hill—you borrow some and fill some, having the cut on one side. I think the highway people call it a balanced section, maybe. A balanced section is where they borrow enough dirt to make all the fill necessary, partly in cut, partly in fill. So the contour of the roadway is coincident with the contour of the land.

Q. Is there any ballast at all along under the track at the crossing? A. Yes.

Q. How much? A. Well, ten inches is standard ballast section." [177.]

John Block testified:

"The highway follows the natural grade of the earth, contour of land, as it approaches the crossing—the railroad crossing, from the south—no grade—only the natural lay of the ground as it was before there was any railroad constructed there." [182.]

From the foregoing facts, we shall further argue that the trial court erred in giving and refusing to give certain instructions hereinafter set out.

Specifications of Error.

The defendant now specifies the following errors, upon which it relies, and which it intends to urge, to-wit:

I.

The Court erred in denying the defendant's motion for a directed verdict, made by it at the close of all the evidence in the case, [290] said motion being made upon the grounds following:

First. Upon the ground that the plaintiff had not offered sufficient evidence to entitle the cause to be submitted to the jury.

Second. That there was no evidence of any negligence on the part of this defendant.

Third. That the undisputed evidence showed that the plaintiff was guilty of negligence himself, which was the sole and proximate cause of the injury complained of.

Fourth. That the undisputed evidence showed that the plaintiff was guilty of negligence and want of ordinary care for his own safety, which contributed to the injury complained of.

Fifth. That the evidence, considered in the light of physical conditions, showed that if the precautions required by law had been observed by the plaintiff, he would have seen the train in time to have avoided the accident.

Sixth. That the undisputed evidence showed that if plaintiff had looked and listened before attempting to cross the railroad track, he would have seen and heard the train in time to have avoided the accident.

Seventh. That the undisputed evidence showed, that if plaintiff had looked in the direction of the on-coming train before attempting to cross the railroad track, he would have seen the train in time to have avoided the accident. [163.]

Eighth. That the undisputed evidence showed that the plaintiff and the driver of the automobile in question, to-wit: Benjamin Spencer, were engaged in a joint enterprise at the time of said accident, and that if the driver of said automobile and the plaintiff had exercised ordinary care and caution before attempting to cross said railroad track, they could have seen or heard the train in time to have avoided the accident.

Ninth. That the undisputed evidence showed that the plaintiff and driver of the automobile in question, to-wit: Bejamin Spencer, were at the time of the accident engaged in a joint enterprise, and that if plaintiff and the driver of said automobile, or either of them, had exercised ordinary care and caution for their own safety before attempting to cross the said railroad track, they could have seen or heard the train in time to avoid the accident.

Tenth. That the undisputed evidence showed that if the plaintiff and/or the driver of said automobile had looked up the railroad track in the direction from which the train was approaching when they were thirty feet from the railroad track, they could have seen the train for a distance of not less than 1500 feet, and in time to have avoided the accident.

Eleventh. For the reason that the undisputed evidence showed that if plaintiff suffered any damage or injuries whatsoever at the time or in the manner or because of the facts alleged in plaintiff's complaint therein, such damages were directly, immediately and proximately caused and occasioned by the fault, neglect and carelessness of the plaintiff himself, and not by any fault, neglect or carelessness on the part of the defendant. [164.]

Twelfth. By the allegation of the complaint and the evidence adduced, it was conclusively shown that the plaintiff Roy Spencer, and his mother, are citizens and residents of the State of New Mexico, and that the alleged cause of action sued on herein accrued in that state; by reason whereof, the plaintiff's right to recover judgment against this defendant is controlled by the laws of New Mexico; that under the laws of New Mexico the plaintiff had not only not offered legally sufficient evidence of defendant's negligence, proximately causing his injury or damage, but, on the contrary, the evidence established such negligence on the part of Roy Spencer as to bar his right to verdict and judgment in his favor.

Thirteenth. That the Arizona constitutional provision, being Section 5, Article 18, was not applicable in this case, because of the residence and citizenship of Roy Spencer and his mother, and because the alleged cause of action sued on herein accrued in the State of New Mexico, and because the evidence of the plaintiff himself, viewing it in the most favorable light, showed that he himself was guilty of contributory negligence. [165.]

Fourteenth. That the evidence adduced herein showing that the negligence of Roy Spencer proximately contributed to his injury, there was no controverted question of fact to be determined by the jury, and, therefore the record presented solely a question of law for the Court, which, notwithstanding the provisions of Section 5, Article 18, of the Arizona Constitution, it was the right and the duty of the Court, under the law as administered by the Federal and United States Supreme Courts, and under Rule 60 of the rules of practice of the Court to decide. That said Rule 60 reads as follows:

"The defendant in an action at law, tried either with or without a jury, may either at the close of the plaintiff's case or at the close of the case on both sides, move for a nonsuit. The procedure on such motion shall be as follows:

"The defendant or his counsel, shall state orally in open court that he moves for a nonsuit on certain grounds, which shall be stated specifically. Such a motion shall be deemed and treated as assuming for the purposes of the motion (but for such purposes only) the truth of whatever the evidence tends to prove, to-wit: Whatever a jury might properly infer from it. If, upon the facts so assumed to be true as aforesaid, the Court shall be of opinion that the plaintiff has no case, the motion shall be granted and the action dismissed. The party against whom the decision on the motion is rendered may then and there take a general exception, and may have the same, together with such of the proceedings in the case as are material, embodied in a bill of exceptions. If evidence shall be introduced by either party after the decision on the motion has been made, the same shall operate as a superseding of the motion; but such motion may be renewed at the close of all the evidence." [166.]

II.

The Court erred in not granting said motion made at the close of all the evidence in the case on the third ground therein set out.

III.

The Court erred in not granting said motion made at the close of all the evidence in the case on the fourth ground therein set out.

IV.

The Court erred in not granting said motion made at the close of all the evidence in the case on the fifth ground therein set out.

V.

The Court erred in not granting said motion made at the close of all the evidence in the case on the sixth ground therein set out.

VI.

The Court erred in not granting said motion made at the close of all the evidence in the case on the seventh ground therein set out.

VII.

The Court erred in not granting said motion made at the close of all the evidence in the case on the tenth ground therein set out.

VIII.

The Court erred in not granting said motion made at the close of all the evidence in the case on the eleventh ground therein set out.

IX.

The Court erred in not granting said motion made at the close of all the evidence in the case on the twelfth ground therein set out.

X.

The Court erred in not granting said motion made at the close of all the evidence in the case on the fourteenth ground therein set out:

XI.

The Court erred in refusing to instruct the jury as follows:

"You are instructed to return a verdict in favor of the defendant, The Atchison, Topeka and Santa Fe Railway Company."

XII.

The Court erred in refusing to instruct the jury as follows:

"As the collision between plaintiff and defendant's railroad train occurred in the State of New Mexico, and as Roy Spencer is a resident of that state, the laws and construction thereof by the courts of New Mexico are controlling upon you in deciding this case."

"Under the system in New Mexico it is contributory negligence for a person to approach a railroad track without paying careful attention for the approach of trains on such track in both directions."

XIII.

The Court erred in refusing to instruct the jury as follows:

"You are instructed that under the laws of the State of New Mexico a railroad company has the right to construct its railroad at grade across any highway which its said railway shall intersect, but that such company shall restore such highway to its former state, as near as may be, so as not to unnecessarily impair its use. You are further instructed in this connection that the laws of the said State of New Mexico govern and fix the rights and duties of the defendant insofar as the construction of said crossing at which plaintiff was injured are concerned. You are, therefore, instructed that if you find from the evidence that in constructing its railroad across the highway on which the plaintiff was traveling when injured, the defendant restored said highway at the place intersected by its said railroad to its former state, as near as may be, so as not unnecessarily to impair its use, you cannot find the defendant negligent in respect to that matter."

XIV.

The Court erred in refusing to instruct the jury as follows:

"You are instructed that a railway track is of itself a warning. It is a place of danger. It can never be assumed that cars or trains are not approaching on a track, or that there is no danger therefrom; it is, therefore, the duty of every person who approaches a railroad track to exercise proper vigilance to ascertain and to know whether any trains or cars are approaching thereon before attempting to cross thereover. Proper vigilance requires every such person to look and listen before going upon said track or so close thereto as to be in danger. The exercise of ordinary care required of a traveler upon the highway under such circumstances, also requires such person or persons to stop, if necessary, in order to look and listen; and if for any reason, the view to any extent is obstructed, the person or persons endeavoring to cross such track or tracks, or going so close thereto as to be in danger either of their person or property, should exercise a greater precaution than ordinary prudence would dictate in such an exigency; that is, in stopping and looking and listening. The looking and listening and stopping thus required should be exercised at the last moment which ordinary prudence would dictate before passing from a place of safety to one of danger. If the ability either to look or listen be obstructed by natural or artificial objects, then the duty to stop before entering into a zone of danger in order that proper investigation may be made, becomes absolute.

"It is also equally the duty of one accompanying the driver of an automobile to exercise care for his own safety, and if you believe and find from a preponderance of the evidence that said Roy Spencer, without objection or protest, permitted the said Benjamin B. Spencer to drive upon the railroad crossing immediately in front of an approaching engine and train of cars without stopping to look or listen or without having the automobile under such control that it might be stopped in time to avoid going on the crossing in front of the approaching train, or without exercising any care on his own part to ascertain whether the crossing was safe, then you are instructed that the plaintiff, Roy Spencer, would be guilty of negligence contributing to his own injuries, and your verdict must be for the defendant."

XV.

The Court erred in refusing to instruct the jury as follows:

"You are instructed that even if you find from the evidence that Roy Spencer looked and listened for the approach of the train in a place of safety, yet your verdict must be for the defendant if you further find from the evidence that said plaintiff was not constantly and actively vigilant in approaching and listening for the approach of the train during the time occupied by him in going from the place where he may have stopped, looked or listened to and upon the track."

XVI.

The Court erred in refusing to instruct the jury as follows.

"The passenger train involved in the accident described in the complaint and referred to in the evidence was running in interstate commerce between states and it was the right of the defendant to operate said train at a high rate of speed and no presumption of negligence is to be indulged from the mere fact of the speed at which the train was operating. The reasonableness of that rule is apparent. The great purpose to be subserved by railroads is promptness, speed and dispatch in carrying passengers and freight. And under ordinary circumstances, in the open country, there is no duty resting upon the railway company to slacken the pace of its trains at crossings. To hold otherwise would, to a great extent, destroy the advantages derived from modern facilities for transportation. To impose upon the defendant the duty of operating said train at a rate of speed so that it could be stopped at every grade crossing in Arizona in time to prevent accident should any person or vehicle be crossing the track on the highway, would be invalid under U. S. Constitution, Article I, Section 8, as a direct and unreasonable burden upon interstate commerce when applied to said train and therefore you are instructed that the defendant had the right at the time and place and in the circumstances disclosed by the evidence to operate its said train in approaching the place of the alleged collision at a rate of speed of fifty miles an hour and you should not consider the speed of the train in this case as disclosed by the evidence as establishing the negligence of the defendant."

XVII.

The Court erred in refusing to instruct the jury as follows:

"If you find that the automobile in which plaintiff, Roy Spencer, was travelling with his father on the day in question was in bad repair or deficient, that fact is not to be considered by you as in any way excusing plaintiff from exercising due care and circumspection for his own safety in crossing the railroad track."

XVIII.

The Court erred in instructing the jury as follows:

"If the railroad grade at any given place where a railroad is being constructed across a highway is so far above or so far below the level of the road or crossing, or so near to extraordinary cuts or curves as to seriously interfere with its use or seriously increase the danger of such use by bringing the road crossing to the level of the tracks and the situation is such that the road can be carried over or under the tracks without unreasonable cost and expenditure and the danger or obstruction cannot otherwise be reasonably removed, then it is the duty of the railroad company to construct a crossing under its rails or carry the road above its tracks by a bridge, as the occasion requires. If it fails to perform such duty, and in an attempt to carry the road over its tracks at grade, that is, on a level with the rails, it seriously and unreasonably increases the danger and difficulty at the crossing, then the railroad company will be liable for any injury or damage proximately resulting because of such dangerous condition to one rightfully using the road."

XIX.

The Court erred in instructing the jury as follows:

"If a railroad crossing is so constructed at grade as to create a condition of extraordinary danger due to existence or adjacent cuts or other physical conditions concealing approach of trains so as to render it probable that persons using the road would not be warned by the usual and ordinary measures or discover by the exercise of reasonable care the approach of trains in time to avoid a collision, then it is the duty of the railroad company to use extraordinary and additional precautions such as automatic signalling devices or other like methods to give warning of the approach of its trains or to approach such crossing carefully and with the train under such reasonable degree of control as would enable them to discover and avoid injury to persons so placed in danger."

XX.

The Court erred in accepting the verdict of the jury and entering judgment thereon for the reason that the evidence shows affirmatively and conclusively that the plaintiff was guilty of negligence proximately contributing to his injury.

XXI.

The Court erred in accepting the verdict of the jury and entering judgment thereon for the reason that the jury failed to follow the instruction of the Court reading as follows:

"If, after considering all of the evidence in the case, you further believe that the plaintiff, Roy Spencer, was also negligent to any extent and that his negligence concurred and cooperated with the negligent act or acts of the defendant to the extent that both the negligence of the defendant and the negligence of the plaintiff was the proximate cause of the injury, then the plaintiff cannot recover."

XXII.

The Court erred in accepting the verdict of the jury and entering judgment thereon for the reason that the jury failed to follow the instruction of the Court reading as follows:

"You are instructed that a traveler on a highway approaching a railroad crossing must use ordinary care in selecting the time, and place to look and listen for oncoming trains and he should stop to make observations, when necessary. It is his duty to use all of his faculties and it is not enough if he carefully listen, believing those in charge of the approaching train will ring a bell or sound a whistle or operate the train at a reasonable rate of speed. He must take advantage of every reasonable opportunity to look and listen and he has no right to omit looking and listening because he has heard no appropriate signal of the approach of the train, even though the law requires that signals be given."

XXIII.

The Court erred in accepting the verdict of the jury and entering judgment thereon for the reason that the jury failed to follow the instruction of the Court reading as follows:

"You are instructed that even though you should find that the employes of the defendant in charge of the train involved in the accident described in the complaint failed to sound the whistle or ring a bell as a warning of the approach of said train toward the point where the collision is alleged to have happened or operated said train at an excessive rate of speed, still such failure to signal the approach of said train or the operation of such train at such an excessive rate of speed did not relieve the plaintiff from the duty of exercising ordinary care to look and listen."

XXIV

The Court erred in accepting the verdict of the jury and entering judgment thereon for the reason that the jury failed to follow the instruction of the Court reading as follows:

"You are instructed that a person traveling upon a street or highway crossed by the tracks of a railroad company have no right to presume that the railway employes engaged in running trains or cars upon such track will signal the approach of all trains, and plaintiff had no right to omit the performance of any part of his duty to look and listen for approaching trains, upon the presumption that the train men would do their duty."

XXV.

The Court erred in accepting the verdict of the jury and entering judgment thereon for the reason that the jury failed to follow the instruction of the Court reading as follows:

"You are instructed that while the defendant railway company and Roy Spencer each had the right to use the road across which the track of said defendant is and was laid, yet, because such trains and cars are run on a fixed track and with great momentum, it had the right of way or precedence upon or over those parts of the road occupied by the railway track, and it was the duty of Roy Spencer to give unobstructed passage to the approaching train."

XXVI.

The Court erred in accepting the verdict of the jury and entering judgment thereon for the reason that the jury failed to follow the instruction of the Court reading as follows:

"If you find that the defendant railroad company was free from negligence or that the plaintiff was negligent, however slight, which in anywise contributed to the collision, you need not consider at all the nature and extent of plaintiff's damages but you should, without hesitation, return a verdict for the defendant railway company."

XXVII

The Court erred in accepting the verdict of the jury and entering judgment thereon for the reason that the jury failed to follow the instruction of the Court reading as follows:

"It was the duty of the plaintiff, Roy Spencer to use that degree of care that a reasonably prudent person would use under like circumstances surrounding the case. It was his duty to look and listen for approaching trains before venturing on the tracks.'

XXVIII.

The Court erred in accepting the verdict of the jury and entering judgment thereon for the reason that the jury failed to follow the instruction of the Court reading as follows:

"You are instructed that a traveler on a highway approaching a railway crossing must use ordinary care in selecting a time and place to look and listen for oncoming trains and he should stop to make observations, when necessary. It is his duty to use all of his faculties and it is not enough that he carefully listened, believing that those in charge of the approaching train will ring a bell or sound a whistle or operate the train at a reasonable rate of speed. He must take advantage of every reasonable opportunity to look and listen and he has no right to omit to look or listen because he has heard no appropriate signal of the approach of the train, even though the law requires such signals to be given."

The foregoing errors were assigned in defendant's Assignment of Errors and are found at pages 376 to 387, both inclusive, of the transcript.

Argument.

I.

The first, second, third, fourth, fifth, sixth, seventh, eighth, ninth, tenth and eleventh specifications of error will be considered under the general heading: "DEFENDANT WAS ENTITLED TO A DIRECTED VERDICT;" and under the sub-headings: "PLAINTIFF'S EXCUSES FOR NOT LOOKING," "RULE OF FEDERAL COURTS," "RULE IN NEW MEXICO," "DIVERSION OF PLAINTIFF'S ATTENTION" and "CONSTITUTION OF ARIZONA," because they are more or less correlated and present in different phases our contention that under the evidence in the case the court should have directed a verdict for the defendant.

Defendant Was Entitled to a Directed Verdict.

It cannot successfully be denied that, as shown by plaintiff's own evidence, he at no time, after leaving the point one hundred feet south of the crossing, looked in the direction from which the train approached, until the front wheels of the car had gone over the first rail of defendant's track and he had stepped up into the car and was about to sit down. He then, for the first time, looked in that direction. Nor can it be denied that, according to his own testimony, the view was unobstructed for at least a quarter of a mile east of the crossing and that had plaintiff looked he could have seen the approaching train in time to have avoided the accident. We maintain that this evidence shows that plaintiff's own negligence proximately contributed to his injuries and entitled defendant to a directed verdict.

Plaintiff's Excuses for Not Looking.

To overcome plaintiff's unusual admission that he failed to look to the east, until he had stepped into the car and was about to sit down, he attempted to excuse such failure on three grounds:

First: That due to his position on the right hand side of the car while traveling the last 50 feet, it was physically impossible for him to look in that direction, without releasing his hold on the car;

Second: That on account of having seen the freight train going east as he approached the crossing, and on account of the character of the view to the west, he considered the danger from that direction the greater and therefore looked in that direction, only. He made no attempt to excuse his failure to look toward the east while traveling the first fifty feet; and

Third: That after releasing his hold on the car, when he, himself, was still 6 feet or more from the first rail, he still did not look toward the east until he had stepped into the car and was in the act of sitting down, because, as he testified, the car was about to go over the track and on down the other side of the crossing. [68-69.]

Neither excuse, offered by plaintiff, we submit, is valid in law. Moreover, the first and third, we say, are entirely frivolous.

There was no showing that plaintiff, prior to his injuries, was suffering from any physical disability which would have prevented him from turning his head, normally, nor any showing, except his mere statement, that the position which he assumed along side of the car prevented him from turning his head sufficiently to look down the track to the east. As shown by plaintiff's own evidence, he was approaching the crossing at a 45 degree angle and it was therefore necessary that he turn his head but slightly to look down the track. And had he approached the track even at a right angle thereto, there was, we contend, nothing in the situation to have prevented him from looking. Even assuming, as, of course, the jury had a right to assume, under the testimony, that plaintiff, in taking his position along side of the car, placed his left hand on the bed of the car, with his right hand against the back of the seat, still it is a physiological fact, of common knowledge, that there was, in assuming such position, nothing to prevent him from turning his head sufficiently to look in the direction from which the train came.

In the trial court, counsel for plaintiff contended, and no doubt will here insist, that because plaintiff had seen

an eastbound train going over the crossing toward Mount-
tainair as he approached the crossing, and, believing at
the time that the distance from the crossing to the switch
at Mountainair was about two miles, he therefore had a
right to assume that he and his father could safely pass
over the crossing before a train from the opposite direction
could reach that point, and that, therefore, it was a question for the jury to say whether the plaintiff was negligent
in acting on such assumption. The undisputed evidence
shows the utter absurdity and fallacy of this contention.

As already shown the plaintiff testified that he was 500
yards from the crossing when he first saw the eastbound
freight train which was then passing over the crossing
[55, 57]; that this train consisted of 20 cars and a
locomotive and was moving at a speed of 20 miles per hour
[88]; that it took plaintiff 10 minutes to travel from the
point where he saw the freight train to the point where he
and his father stopped [89]; that they stopped for half a
minute [87]; and that it took the car a full minute to
travel from the place where they stopped to the track
[68, 87, 88]. The mere statement of these facts is, we
submit, a complete refutation of such contention on
counsels part. Taking plaintiff's own testimony, it is
readily demonstrated that it would take the freight train
only six minutes to reach the switch, assuming that the
distance from the crossing to the switch was two full
miles, as he at the time believed, which would leave but
six minutes for a passenger train coming from the opposite direction and running even at no greater speed than
20 miles per hour to reach the crossing. As a matter of
fact, plaintiff testified that he was just guessing that the
distance was 2 miles [57]. It is common knowledge, of

which courts take judicial notice, that passenger trains ordinarily run at a greater speed than 20 miles per hour, but even assuming that speed here, and assuming the distance between the crossing and the switch to be two miles, it is apparent that plaintiff would, on his own statement, have had but one-half of a minute leeway of safety, for, as shown, 11½ minutes elapsed from the time he observed the freight train going over the crossing until the collision occurred.

We say that the law did not permit plaintiff to take the risk on such a narrow margin. A traveller is not allowed to indulge in mere speculation as to distances or the speed of trains.

Further, as heretofore shown, plaintiff did not observe the position of his father, or what he was doing while plaintiff was pushing on the car as they proceeded toward the crossing from the point where they had stopped, nor did he know whether his father was sitting up or bending down to manipulate the coils or other mechanism of the car to make it run. Accordingly, plaintiff did not know whether his father was looking or listening for trains, and paid not the slightest attention to what if any care his father was exercising in approaching the crossing. Surely, this alone, was a negligent act which contributed proximately to plaintiff's injuries.

Now, even conceiving for the sake of argument, the possibility of the law excusing plaintiff's actions, down to the instant he released his hold on the car, he then still being six feet, or more, from the tracks; as seen, upon releasing his hold, plaintiff stepped up onto the running board, and into the car; he then turned around, and *then* for the first time, *looked eastwardly* down the track. Thus, it appears that plaintiff left a place of perfect safety and voluntarily projected himself into a place of extreme danger without exercising the slightest degree of care. If such conduct on his part did not constitute the grossest negligence as a matter of law, then, indeed, we submit, it would be impossible to imagine a case of gross negligence on the part of a traveller at a railroad crossing; and the numerous decisions of this and other courts dealing with the care to be exercised by such a person would be wholly meaningless. ·

Let us see what excuse plaintiff offered for his conduct in this regard. He testified:

"I ceased to push, when the front wheels of the car had just went over the first rail. At that time, I was beside the car and I stepped on the rail—on the running-board, rather, and stepped in the car and turned around. *My purpose in stepping in the car when ceasing to push, stepping on the fender, was because it was going over the rails down the other hill. The car was in motion."* [68, 69.] (Italics ours.)

On cross-examination, he testified:

"Q. Have you ever measured or are you competent to measure the height or depth of the embankment over that crossing? A. At what point?

Q. From the middle of the crossing to the point you have described as the hill or cut east of the crossing? A. Why, I think so. It is dug off just like a wall; it is straight up and down like a wall. I never did measure it more than just looked at it.

Q. Roy, as a matter of fact, is not the land on the north side of the railroad track, from the crossing easterly, a considerable distance higher than the land on the south side of the track? A. Yes, it is more hilly.

Mr. Wood: Does your question mean immediately at the embankment, or farther away?

Mr. Reed: At the embankment, along the embankment, the north side of the embankment.

The Witness: You mean the railway?

Q. (By Mr. Reed): Yes, on the north side of the railway. [89.] A. Well, there is not much difference along there; it is just the same where they have made the fill, just scooped out along on both sides and made the fill just the same along there. The highway on the north side of the track is higher than the highway on the south side of the track. So, in approaching the railroad, when you were traveling westerly away from Mountainair, along the highway—you would not have much of an ascent or grade to climb to get up over the track. It is more than a couple of feet. I don't know just what it is, it is just a small hill.

Q. As you got up to the track and hopped up into the truck, why didn't you look east? A. I did.

Q. Why didn't you before getting up into the truck? A. Oh, before? Well, because the car was going over the track.

Q. Well, what was there in that to prevent you looking eastward before you stepped up on to the running-board? A. Well, the car was going over the track and down the other hill; naturally so, I stepped up on the running-board before, and then I looked down the track." [90.]

Thus it will be seen that at the crossing and for some distance eastwardly therefrom the railroad is constructed on ground that slopes somewhat to the south. A short distance west of the crossing the ground gradually becomes more of a hill along the south slope of and through which, eventually, by means of a cut, the railroad is built [71, 72]. The track is laid on a roadbed approximately

20 feet in width. Upon the roadbed and in the center thereof is laid 10 inches of ballast, in which the ties are embedded. The rails are fastened on the ties, of course, so that the top of the rail is approximately 16 inches above the level of the 20 foot roadbed. The wagon road going as the plaintiff went toward Mountainair, winds around the foot of and gradually ascends the slope of this hill on the side of and through which the railroad is constructed. It ascends to the level of the railroad roadbed upon the natural gradient of the hill. That gradient was not changed when the railroad was constructed. Up to the level of the railroad roadbed the wagon road follows the original natural contour of the ground. After reaching the level of the roadbed, the wagon road is given a gradient so as to reach the level of the top of the rails, crosses the track, and then descends slightly to the other edge of the roadbed, at about which point the road swings so as to be approximately parallel with the railroad and follows the railroad to Mountainair.

Thus in going from the point where the automobile stopped, namely at the foot of this slope, to where the automobile was struck by the train, the automobile had first in moving about 80 feet to ascend approximately 10 feet to the level of the railroad roadbed and from that point ascend approximately 16 inches to the level of the top of the rails.

The appended copy of Defendant's Exhibit "A" shows the situation.

So, in order that plaintiff's father might be spared the possible necessity of stopping the car after getting across the tracks, or in order, at the most, that plaintiff himself might be saved the possible necessity of accelerating his speed on foot sufficiently to overtake and get into the car, he chose the hazardous alternative of leaving his position of perfect safety and deliberately placing himself in a hazardous one.

Ordinary care would require more than this even from a mere child of average intelligence; plaintiff was an intelligent and exceedingly active and energetic young man. [91, 119.]

In the light of the foregoing facts, we are curious to know whether counsel for plaintiff will, in this court, have the temerity seriously, or otherwise, to even attempt to justify plaintiff's conduct after plaintiff's hold on the car was released.

So, we submit, that the evidence so clearly and indisputably shows that plaintiff was guilty of negligence, which proximately contributed to his injuries, that it was not a question on which reasonable minds might differ, and that, therefore, the court erred in refusing to direct a verdict, and, at this point, we venture the assertion that but for the constitutional provision of the State of Arizona on the question of contributory negligence, hereinafter considered, the motion for such a verdict would have been granted [290.]

RULE IN FEDERAL COURTS

While, under the evidence, it would seem to be a work of superfluity to call the attention of this learned court to any of the decisions supporting our contention that plaintiff was, as a matter of law, guilty of negligence proximately contributing to his injuries, entitling defendant to a directed verdict, we take the liberty of inviting the court's attention to a few of such cases.

Bradley v. Mo. Pac. R. Co., 288 Fed. 484 (C.C.A. 8th Circuit.) Plaintiff's intestate, Bradley, was killed while riding in an automobile driven by one Brown which was struck by one of defendant's trains at a grade crossing. Brown and Bradley proceeded toward and onto the crossing without stopping. The evidence showed that from a point on the highway forty feet from the track a train would be visible for a distance of 800 feet from the crossing. Before reaching the forty foot point in the highway an embankment of earth along the right of way made the sight of an approaching train difficult, if not impossible. The court, in the course of its opinion, said:

"If they (the occupants of the automobile) could not see the train it was their duty to listen. If they could not hear because of the noise of the automobile, it was their duty to stop. The undisputed evidence introduced on plaintiff's side of the case further shows that, from the time the automobile turned the corner of Garland Street to go west until it was struck at the crossing, it did not slow up, but kept on at the same speed, which one witness placed at 8 to 10 miles per hour.

The duty of a party approaching a railroad crossing, a known place of danger, is to look at an effective point for approaching trains. If one sees the approaching train, then, of course, there is no need for listening; but, if the view is obstructed, it is the duty to listen, and, if the noise of the automobile or conveyance is such as to prevent effective listening, then to stop and listen. A party might be excused from looking or listening at any particular point in approaching a railroad track by the intent to look or listen at a more advantageous one, but he cannot be excused from failure to look or listen at some effective place, and if by looking or listening he can ascertain the approach of a train, and fails to do so, he is guilty of contributory negligence. The necessity of stopping depends upon the question of being able to see or hear. This court in the case of Davis v. Chicago, R. I. and P. Ry. Co., 159 Fed. 10, 16, 88 C.C.A. 448, 494 (16 L. R. A. (N. S.) 424), said: 'The duty to stop is a relative one. It depends upon the situation of the particular case, the knowledge the traveler has of the situation, and the reliance he may reasonably place under the circumstances on his opportunities for seeing and hearing without taking the last precaution of stopping. The authorities are quite in accord on the proposition that if the view is unobstructed, so that an approaching train, before it reaches the crossing can be seen, there is no occasion for the special exercise, of the sense of hearing—listening, and therefore there is no reason why he should stop for that purpose. On the other hand, if the view is obstructed, interfering with the sense of sight, then he must bring into requisition the sense of listening carefully and attentively. And if there is any noise or confusion over which he has control, such as that of the noise of the horse's feet, or the grinding sound of the wheels, or the ordinary noise of the vehicle, interfering with the acuteness of the sense of hearing, it is his duty to stop such noise or interfering obstruction and listen for the train before going upon the track.'

Citing this rule with approval in the case of Chicago, M. & St. P. Ry Co., v Bennett, 181 Fed. 799, 103 C. C. A. 309, this court said:

'We are unwilling to depart from or relax it.'

The duty of a person approaching a railroad crossing is not an open question in this jurisdiction. As said by this court in Chicago Great Western Railway

Co. v. Smith, 141 Fed. 930, 931, 73 C. C. A. 164, 165;

"The law requires of one going into so dangerous a place the vigilant exercise of his faculties of sight and hearing at such short distance therefrom as will be effectual for his protection, and if this duty is neglected, and injury results, there can be no recovery, although the injury would not have occurred, but for the negligence of others.' Pyle v. Clark et al. 79 Fed. 744, 25 C. C. A. 190; Davis v. Chicago R. I. & P. Ry. Co. 159 Fed. 10, 88 C. C. A. 488, 16 L. R. A. (N. S.) 424; Chicago, M. & St. P. Ry. Co., v. Bennett, 181 Fed. 799, 104 C. C. A. 309; Chicago Great Western R. Co. v. Biwer, (C. C. A.) 266 Fed. 965.

The rule is no different in other jurisdictions. Grand Trunk Ry. Co. of Canada v. Cobleigh, 78 Fed. 784, 24 C. C. A. 342; McCrory, Adm'x etc. v. Chicago M. & St. P. Ry. Co. (C. C. A.) 31 Fed. 531; Railroad Co. v. Houston, 95 U. S. 697, 24 L. Ed. 542; St. Louis I. M. & S. Ry. Co. v. Chamberlain, 105 Ark. 180, 184, 150 S. W. 157. Clearly and indisputably Brown, the driver of the car, was guilty of negligence as a matter of law."

And the court held that Bradley was equally quilty of contributory negligence.

In this Bradley case the court, on page 487, made the timely and appropriate observation that

"If parties driving automobiles persist in gambling with death at railroad crossings, their estates should not be augmented by damages if death win. Care, not chance, is the requisite at railroad crossings."

In the case of Noble v. C. M. & St. P. Ry Co. 298 Fed. 381, (C. C. A. 8th Circuit, 1924), the plaintiff was injured at a railroad crossing while riding in an automobile with her husband, who was driving. At the end of her case, the court directed a verdict in defendant's favor.

The road upon which the automobile was traveling ran straight north a considerable distance to the track, which it crossed at right angles. The track running easterly from the intersection soon entered a cut, and at a point about 500 feet or a little more curved to the southeast. The road sloped down to the track from a point half a mile south thereof. Immediately adjacent to the track it was six or seven feet below the railroad grade, and finally raised again to reach the track. Both plaintiff and her husband were more or less familiar with the road and the crossing. Plaintiff testified that they drove up to the track at about 15 miles per hour, looking both ways for a train but that she did not see any or hear any noise or whistle. She further testified the first she saw of the train was when they started up the incline onto the track.

Plaintff's husband testified that they approached the crossing at the rate of 15 miles per hour. He claimed to have maintained a strict lookout, but did not see or hear anything that would indicate the approach of a train, and that he continued to drive on until he was 15 or 16 feet south of the track, at which point he started to put on a little more gas to make the elevation.

Other witnesses for the plaintiff testified that from a point on the road at least fifty feet south of the crossing a train approaching from the east could be seen 250 to 400 feet away.

The plaintiff, on being recalled, stated that they slowed down to 5 miles an hour and the automobile was just moving until they reached a point 16 feet from the track where it was necessary to accelerate the speed to make the grade. Her husband, on being recalled, testified that when they were about 50 or 60 feet from the track, he looked both ways, saw no train, and he "stepped on the gas"; that at that point one could see an engine coming down the track from the east 250 or 260 feet away from the crossing. Plaintiff testified that the train was going at the rate of 40 miles per hour, and that no whistle was blown or bell rung.

The court said:

"It is clear that, if the plaintiff was guilty of contributory negligence, no recovery can be had, irrespective of the manner in which the defendant operated this particular train or protected the crossing. The undisputed facts show that the driver of the car had full opportunity to observe all surrounding conditions at the time of the accident, with which he was already more or less familiar, and could by the proper and intelligent exercise of his senses, have seen or heard a train approaching from the east at 40 miles per hour, at a sufficient distance to enable him to stop his car and avoid the accident. This is especially true in this case, if, as he states, he approached the track at the rate of 5 miles per hour and had an up-grade to aid him in bringing the car to a standstill. The facts force the conclusion that there was at least one effective point in the course of this approach to the track at which he could have seen and heard the train and stopped in time to avoid an accident. If he did do this, he must have seen the train, and was guilty of gross carelessness in continuing on. If he did not, he was guilty of gross carelessness in failing to look or listen for the train. Under either alternative, he was guilty of contributory negligence, which absolves the defendant from any liability as to him.

This is the settled law in this circuit. Davis v. C. R. I. & P. Ry. Co., 159 Fed. 10, 88 C. C. A. 488, 15 L. R. A. (N. S.) 427; Wabash R. Co. v. Heulsmann (C. C. A.) 290 Fed. 165; Bradley v. Mo. Pac. R. Co. (C. C. A.) 299 Fed. 484, and cases cited, are a few of the many cases so holding."

The court held that the plaintiff, although not the driver of the car, was as a matter of law equally guilty of contributory negligence as her husband and affirmed the judgment of the lower court.

In Southern Railway v. Priester, 289 Fed. 945 (C. C. A. 4th Circuit), the plaintiff and her husband were injured by a train at a railroad crossing. The husband was driving the automobile. At the time of the accident, the automobile was approaching the railroad from the east on a public road, which crossed the track at a right angle. On the south side of the road, the view of the railroad was shut off by a field of tall corn to a point 33 feet from the track. On the other side of the road, near the edge of the cornfield, the railroad company had made holes in the road by digging up earth to fill a hole dug for a sign post. The railroad was higher than the level of the public road, the rise being 3½ feet in 15 feet. On both sides of the east rail thick boards had been placed. Against these boards earth had been thrown so as to make the ascent and crossing easier; but the earth had been washed or worn away so that the boards and rails were about 4 inches above the earth. From the end of the cornfield, 33 feet from the track, the view of the railroad was clear both ways, and one looking down the track could see a train approaching from the south at least a half mile. The road being narrow, driving across the hole required unusual attention and care. According to the testimony of plaintiffs the approaching train must have been in full view when they reached the end of the cornfield.

Priester, the driver, after driving to the end of the cornfield slowed down to pass the holes. He then put on speed to make the rise of 3½ feet and crossed the track. Still neither of the plaintiffs saw the train, or looked, or heard its noise. When plaintiffs did finally see the train, almost at the instant that the automobile reached the east rail, it was within 50 or 60 feet of them.

In holding that both plaintiffs were guilty of contributory negligence, the court, in the course of its opinion, said:

> "It is well settled that when a railroad company has done nothing to allay his (the traveller's) sense of danger, and there are no extraordinary conditions sufficient to distract the attention of a man of ordinary prudence and self-possession from the duty to take precaution, a traveller will be held guilty of contributory negligence, as a matter of law, when the evidence shows affirmatively beyond dispute that he walked or drove his vehicle on a crossing without taking any precaution whatever to ascertain if a train was approaching." Citing cases.

The court further said:

> "The condition of the crossing contributed in no way to the accident. For a distance of 33 feet after passing the edge of the cornfield, plaintiffs had an unobstructed view of the track for at least a half mile. From the hole where the automobile was slowed down, for a distance of at least 15 feet, plaintiffs still had the unobstructed half mile view of the track. Neither of them took the obvious precaution of look-

ing. It seems to us impossible to escape the conclusion that the failure to look was gross negligence. Surely it cannot be held that a hole in the road which requires slow driving puts a traveler in such an extraordinary situation as to excuse him from looking down a straight track."

In C. & S. I. Ry. Company v. Sellars, 5 Fed. Rep. (Second) 31, (C. C. A. 8th Circuit), the court said:

"A passenger who sits quietly in an automobile and allows the driver thereof to take him into a place of danger, without effectively exercising his senses for his own protection, or without warning the driver or making a protest, is guilty of contributory negligence and cannot recover."

Applying the rule enunciated in the Sellars case to the facts of the case at bar, we can see no distinction in principle between the passenger who sits supinely in an automobile without exercising any care for his own protection and the plaintiff in this case who knowing he was approaching a railroad track blindly proceeded onto the track and up into an automobile without exercising the slightest degree of care or caution for his safety.

For an admirable statement of the care required of the traveler in approaching a railroad grade crossing, and of the power and the duty of the federal courts to direct a verdict where the evidence discloses contributory negligence on the part of such traveler injured at such a crossing, we respectfully invite the court's particular attention to the cases of

Dernberger v. B. & O. Rd. Co., 234 Fed. 405;
And the same case, 243 Fed. 21.

In dealing with the care to be exercised by the traveler, the court, in the first case, among other things, pertinently said:

"As regards the duties and obligations of the traveler about to undertake to make the crossing, various and diverse conditions may arise in different cases. He may be afoot, on horseback, in a large, cumbersome vehicle, as, for example, a traction engine or a threshing machine, a road wagon making much noise, a light buggy, making little, or in an automobile, noisy or not. He may be hard of hearing or not; his vision may or may not be defective. He may have one or he may have several, tracks to cross; the approach to the crossing may be clear, the track straight, and his vision for many rods unobstructed; on the other hand it may be obstructed, either by natural objects, such as trees, curves, embankments, and buildings, or by cars temporarily placed on side tracks by the company itself. Again, his hearing may be interfered with by noises of machinery, waterfalls, other than those created by the company or by those of engines, machinery, etc. operated in the vicinity by the company itself.

In all this diversity of conditions there are, however, some legal principles generally applicable. He must never be unmindful that he is primarily responsible for his own safety; that the railroad, at a crossing, has the right of way; that at a railroad crossing "the track itself, as it seems necessary to iterate and reiterate, is itself a warning. It is a place of danger. It can never be assumed that cars are not approaching on a track, or that there is no danger therefrom." Elliott v. Chicago, M. & St. P. Ry. Co., 150 U. S. 245, 248, 14 Sup. Ct. 85, 86, (37 L. Ed. 1068). The presence of noises or obstructions to view require a greater degree of caution to be exercised by him, and, if he fails to meet these requirements he is, under the law, as it now exists, held to be guilty of contributory negligence, which will bar any recovery for damages incurred, as against the railroad, whether the latter was negligent in the premises or not. Beyel v. Newport News & M. V. R. Co. 34 W. Va. 538, 12 S. E. 532; Berkeley v. C. & O. Ry. Co., 43 W. Va. 11, 16, 26 S. E. 349.

The only way the ordinary normal man can take these precautions is by control of his steps or the means he is using for locomotion, and the exercise of his senses of sight and hearing. Hence the rule to stop, look and listen."

The foregoing federal cases, and many others, which we deem it unnecessary to cite, firmly establish the rule, with which this court is entirely familiar, that the guest or passenger in an automobile, on approaching a railroad crossing, is required ordinarily vigilantly to exercise his faculties of sight and hearing at such short distances therefrom as will be effectual for his protection and if this duty is neglected, and injury results, he cannot recover.

We maintain that the principles laid down in the foregoing cases apply with fullest force and effect to the facts in the case now before this court. Plaintiff's conduct, we assert, fell woefully short of what the law required under the circumstances.

Rule in New Mexico

Under the rule established in the State of New Mexico, the plaintiff herein was, we contend, guilty of contributory negligence entitling defendant to a directed verdict.

In the early case of Candelaria v. Atchison etc. R. Co. 6 N. M. 266, 27 Pac. 497, the Supreme Court of that state, in holding that the trial court did not err in directing a verdict in favor of the defendant, enunciated the rule in the following pertinent language:

"Juries in this territory are the judges of the facts, and in all cases where there is a real conflict of evidence upon material issues in the case, where the material issues are disputed, and the testimony of witnesses must be weighed and their credibility determined, the court certainly would not be authorized to withhold from the jury the determination of the facts in such cases; but where the facts as to material issues are practically undisputed and are therefore established, the law is that in case the court is satisfied from the facts established that there is no right of recovery in the plaintiff, to the extent that the court would be compelled to set aside the verdict, the court may, without error, instruct the jury to find for the defendant. The former rule that all questions of fact must be submitted to the jury, if there is a 'scintilla' of evidence, has been materially relaxed, and from the more recent cases a more liberal interpretation of the rule has been obtained."

Following this language, the court, in adverting to the rule of the United States Supreme Court on the subject, continued:

"In the case of Railroad Co. v. Houston, 95, U. S. 697, from which we have already above quoted, the court, upon the state of facts there presented said:

'Upon the facts disclosed by the undisputed evidence in the case, we can see no ground for a recovery by the plaintiff. Not even a plausible pretext for a verdict can be suggested, unless we wander from the evidence into the region of conjecture and speculation. Under these circumstances the court would not have erred had it instructed the jury, as requested, to render a verdict for the defendant.'

In the case of Schofield v. Railroad Co. 114 U. S. 619, 5 Sup. Ct. Rep. 1125, the Court said: 'It is the settled law of this court that when the evidence given at the trial, with all the inferences which the jury could justifiably draw from it, is insufficient to support a verdict for the plaintiff, so that such a verdict, if returned, must be set aside, the court is not bound to submit the case to the jury, but may direct a verdict for the defendant.' (Citing cases.)

In Railroad Co. v. Jones, 95 U. S. 439, it was held that 'the plaintiff was not entitled to recover. It follows that the court erred in refusing the instruction asked upon this subject. If the company had prayed the court to instruct the jury to return a verdict for the defendant, it would have been the duty of the court to give such direction, and error to refuse.'

In the case of Railroad Co. v. Converse, 139 U. S. 469, 11 Sup. Ct. Rep. 569, which was a case in which it is stated by the court that there was some conflict in the evidence relating to certain matters, but certain facts were clearly established, the court says: 'It is contended that the court erred in not submitting to the jury the issue as to the defendant's negligence. Undoubtedly, questions of negligence in actions like the present are ordinarily for the jury, under proper instructions as to the parts of the law by which they should be controlled. But it is well settled that the court may withdraw a case from that body, and direct a verdict for the plaintiff or the defendant, as the one or the other may be proper, where the evidence is undisputed, and is of such a conclusive character that the court, in the exercise of its judicial discretion, would be compelled to set aside the verdict returned in opposition to it."

In the case of Morehead v. Atchison etc. Company, 27 N. M. 349, 201 Pac. 1048, the plaintiff, as he was approaching a railroad crossing in an automobile, according to his own testimony, stopped, looked and listened at a distance of 57 feet from the track, and did not thereafter again stop, look or listen but drove upon the track and was struck by the defendant's train. On these facts, the court held that the plaintiff was guilty of contributory negligence and that the trial court should have directed a verdict.

See also the case of Sandoval v. Atchison etc. Co. 30 N. M. 343, 233 Pac. 840.

That under the law, then, of the State of New Mexico, the state in which the injury to plaintiff in the case at bar occurred, and the right of action, if any, arose, contributory negligence is a defense of which a defendant is given a *right* to avail itself, and that the court has the power and is under the *duty,* when properly requested, of directing a verdict where the evidence shows the plaintiff to have been guilty of such negligence, will, of course, not be controverted by counsel for plaintiff. We therefore content ourselves on this proposition by calling the court's attention to the three cases last hereinabove cited.

DIVERSION OF PLAINTIFF'S ATTENTION

Should counsel for plaintiff argue to this court, as he did to the trial court, that plaintiff, on account of his belief that were he to release his hold on the car, the car would run back down to the bottom of the grade; that, thereby, his attention was diverted to and centered on his efforts toward assisting the car over the crossing; and that therefore, he was excused from looking; our answer to such argument will be, and is, that the only rule that, under the physical facts and the circumstances surrounding this case, could possibly excuse the plaintiff from looking has no application here. That is the rule which excuses a person's conduct when acting in a sudden emergency, or when his attention is suddenly distracted or diverted. The circumstances of this case do not, we submit, bring plaintiff within that rule. Plaintiff was not acting in a sudden emergency, nor was his attention suddenly distracted or diverted.

The two following cases aptly illustrate the foregoing rule:

In the case of Kanass v. Chicago etc. R. Co., 180 Wis. 49, 192 N. W. 383, plaintiff's tractor was struck by a train at a private crossing. At the time of the collision plaintiff was driving the tractor, to which was attached a trailer loaded with gravel. The tracks extended in a northerly and southerly direction, and the road used by plaintiff crossed the tracks approximately at right angles. Beginning at a point about 40 feet west of the west rail the road inclined toward the tracks, making a rise of 4 feet and 4 inches at the location of the crossing.

Plaintiff testified *that because of the steep grade it was necessary for him to cross the track in second speed; and that the poor condition of the road and the difficulty of taking a heavy load up the grade made it necessary for him to devote all his attention to managing the tractor.* (Italics ours.)

Plaintiff further testified that on the day of the accident he stopped the tractor at a point 60 feet west of the crossing and looked in both directions for a train and listened for a warning; that he saw no trains and heard no sound of any; that he then shifted into second gear and approached the crossing at a speed of from 8 to 10 miles an hour; that as he neared the rails the speed gradually slackened, and that when he reached the tracks, the tractor was barely moving; and when he was crossing the track he saw the engine approaching from the north only a few feet away and had only time to jump before the collision. In his testimony, the plaintiff said:

> "At the time I stopped, I looked and then I went ahead, and kept on going. I looked at my tractor all the time. It occupied all my attention. I never looked to the right or to the left."

It was contended on behalf of the plaintiff that the facts proved brought the case within the exception to the "look and listen" rule. It was claimed that

> "the existence of the grade; the fact that plaintiff exercised judgment in selecting the most favorable point to make the last stop; that he could not get the tractor over the grade if he started from this point on low speed; that if he had stopped to shift gears and look and listen for the train within a shorter distance than 60 feet from the track he could not have made the crossing; that the roadway was narrow and somewhat rough: that if he had proceeded at a higher rate of speed there was danger of accident on the other side of the crossing; that he was driving a loaded tractor and not a team of horses; that if he had looked for the train he would have had to lessen his power and slacken his speed and would not have made the crossing; and that he could not have made the crossing on low—were facts which brought the case within the doctrine of irresistible diversion of attention."

It was argued that plaintiff was required to give his undivided attention to the business before him and that from the time he started, 60 feet from the track, he was not bound to look or listen, and in support of this argument certain cases were relied upon. After analysing the cases so relied upon, the court disposed of the argument and the cases in the following language.

"We think that all these cases are plainly distinguishable from the one before us. They are cases in which circumstances beyond the control of the traveler so diverted his attention that he was prevented from using the care which ordinarily prudent persons would use under similar circumstances, and which the law requires. The mere fact that one is managing a loaded vehicle, and is in a hurry, and ascending a grade, and desires to make the crossing in a particular manner, and that he gives his whole attention to managing his vehicle, does not absolve him from the duty of looking and listening for an approaching train, when he is knowingly attempting to cross a railroad track. *To hold that any of those circumstances, or all of them combined, excuse inattention to an approaching train would introduce a new exception to the well-settled rule.* (Italics ours.)

Although the look and listen rule is not absolute and unbending, we are convinced that this case comes within none of the exceptions recognized in the case relied on by plaintiff's counsel, and that there was no irresistible diversion of attention. On the contrary, the case is ruled by numerous decisions of this court, the facts of which it is not necessary to discuss."

In Lee v. Davis, etc. 247 Pac. 1094 (Mont. 1926), the plaintiff and the driver of a wagon were engaged in hauling a heavy load of machinery. The driver sat upon a box on the front part of the load, and plaintiff occupied another box on the load but towards the rear of the wagon.

On account of the heavy load they traveled slowly—not to exceed 1½ miles per hour. When they approached the railroad crossing at which plaintiff was injured and when the team was 100 to 125 feet north of the track, plaintiff looked and listened for any approaching train and observed that the driver did the same thing. As they did not see or hear a train, they proceeded and when the front wheels of the wagon were on the track one of defendant's locomotives collided with the wagon resulting in plaintiff's injury.

The locomotive came into plain, unobstructed view of persons in the situation of plaintiff and the driver when it was a third of a mile east of the crossing. The evidence further showed that during the time that the locomotive traveled over that distance of one-third of a mile, the team advanced about 66 feet, or, in other words, when the locomotive came into view the team was about 66 feet north of the crossing and in a place of safety.

At the conclusion of plaintiff's case, the court granted a nonsuit, on the ground that, as a matter of law, plaintiff was guilty of contributory negligence.

After holding that it was unnecessary to determine whether the negligence of the driver should or should not be imputed to the plaintiff, but that the question for it to decide was whether plaintiff himself was guilty of negligence which contributed proximately to his injury, the court said:

> "The only deduction from his testimony is that he did not look for a train after the team left the point 100 or 125 feet north of the crossing, and if he had looked when the team had advanced one-half that distance, he would have seen the approaching locomotive

and would have had ample time and opportunity to alight from the wagon in safety if the driver then failed or refused to stop the team.

It is urged that he was engrossed in the duties of his position watching the machinery that it did not slip off the wagon, and because of this fact should be excused for not giving further attention to the danger signal—the crossing of a public road over a railroad track. He testified that when the team was 100 or 125 feet north of the crossing, he discovered that the lunch buckets had overturned and that the contents of the buckets were on the boards between the two 6x6 timbers, and that he engaged in picking up and replacing in the buckets the articles constituting the lunches. He testified:

'I was down doing that; I guess the machinery did not slip very much right then, but I was busy picking the stuff up.'

In answer to the direct question of his own counsel 'Were you watching the machinery at the same time?' he replied:

'I was looking at all of it as near as I could. I was picking up our dinners off the bottom of the wagon and putting it in the buckets again, and when the front wheels of the wagon, it raised when it struck the plank that took it up over the rail, why something said to me, "Look out." I looked up and seen the engine'. "

Continuing the court said:

"It is perfectly apparent from this testimony that there is not any merit in the contention that plaintiff was engrossed in looking after the machinery. During the time the team was advancing 100 or 125 feet to and upon the crossing, he was engaged in looking after the lunches. In other words, he was more concerned for the safety of the lunches than he was for the security of his life or limbs, or stated differently, he simply did not look when he came into a place of actual peril, but shifted the responsibility for his safety to the driver. This he could not do and relieve himself of the consequences of his own neglect. Brown v. McAdoo, 195 Iowa, 286, 188 N. W. 7; Knight v. Atchison etc. R. Co., 111 Kan. 308, 206 P. 893; Virginia & S. W. R. Co. v. Skinner 119 Va. 843, 89, S. E. 887. The law of contributory negligence would be meaningless if one approaching a known place of danger, and charged with the duty of making vigilant use of his natural faculties for his own safety, were permitted to excuse his negligence by saying 'My attention was irresistibly diverted by the fact that my lunch was in danger of being soiled or lost.' No decided cases can be found where such a flimsy excuse was held sufficient to exonerate one from the imputation of negligence. The authorities are all to the contrary. Tannehill v. Kansas City, C. & S. Ry. Co., 279 Mo. 158, 213 S. W. 818; Kansas v. Chicago, M. & St. P. Ry. Co., 180 Wis. 49, 192 N. W. 383; Southern Ry. Co. v. Priester (C. C. A.), 289 F. 945."

If the rule did not apply to the *driver* of the heavy tractor, in the first case, or to the plaintiff in the second case, it certainly has no application here. (Italics ours.)

Constitution of Arizona

As already stated, the cause of action, if any, arose in the state of New Mexico and is prosecuted by citizens of that state. We apprehend that the plaintiff will argue that the court had no power to grant a nonsuit or to direct a verdict on the ground that the plaintiff's negligence proximately contributed to his injuries. This contention, if made, will be based on Section 5 of Article 18 of the Constitution of Arizona, which reads as follows:

"The defense of contributory negligence or of assumption of risk shall, in all cases whatsoever, be a question of fact and shall, at all times, be left to the jury."

Indeed why, but in the hope of getting the advantage of this section, did plaintiffs' counsel bring this action in Arizona and not at the homes of the plaintiffs and their witnesses, in New Mexico?

As against this contention, we submit three propositions:

First: That the section applies only when contributory negligence is attempted to be proved by the defendant and does not apply where contributory negligence appears on the face of the plaintiff's own evidence.

Second: That if the section establishes a rule of substantive law it does not apply to an accident arising in another state.

Third: That if it be held that the section does not establish a rule of substantive law, but a law limiting the powers of the judge in jury trials, then the state rule, as directed by this section, does not apply to a trial in the federal court either under the constitution and laws of the United States or under the rules of the United States District Court for the District of Arizona.

As to the first of these propositions it is settled in Arizona by a recent decision that if the plaintiff in his own case shows that his negligence contributed to the injury he cannot recover although the defendant may not have pleaded contributory negligence as a defense. Bruno v. Grande, 251 Pac. 550, decided last December, but referring to and approving language used in the case of DeAmada v. Friedman, 11 Arizona 56, 89 Pac. 588· And it is our contention in this case that contributory negligence appeared from the plaintiff's own showing.

Therefore, at the close of the plaintiff's case [Tr. p. 163] defendant moved for a directed verdict on this ground and again when the defendant rested and the testimony closed [Tr. p. 290] defendant moved for a directed verdict upon the same ground. [See Tr. p. 31.]

Such is, of course, the practice in the federal courts. See Bowker v. Donnell, 226 Fed. 359; Long Island, etc. Co. v. Darnell, 221 Fed. 194.

We submit first that when the constitution of Arizona speaks of the *defense* of contributory negligence it refers only to cases in which contributory negligence is sought to be proved by the defendant upon a plea of contributory negligence by producing evidence to meet a *prima facie* case made by the plaintiff and does not apply to cases where the plaintiff's own evidence establishes contributory negligence so that the defendant may avail himself of that defect in the plaintiff's case on a motion for a nonsuit or a directed verdict on the ground that the plaintiff has failed to make out a *prima facie* case.

Thompson on Negligence, Section 369, says that contributory negligence is an affirmative defense and that the burden of showing it is on the defendant except where it is shown by the plaintiff's own evidence. The author says that some of the cases so state the rule, but that what is intended to be said is that where the plaintiff's own evidence shows that he was guilty of negligence contributing to the injury there can be no recovery whether the defense of contributory negligence be pleaded or not, and he says that this rule is strongest in its application where the inference of contributory negligence arises out of the testimony delivered by the plaintiff himself on the stand, for this testimony, so far as it works against him, stands on the footing of an admission and unless explained becomes conclusive.

Section 432 of Thompson's work on Negligence says, for example; An administrator suing under the statute for the death of his intestate caused by negligence or wrongful act, whose testimony shows that the negligence of the deceased contributed directly to the injuries resulting in his death, has failed to make out a *prima facie* right to recover and the demurrer to the evidence should be sustained.

In White's supplement to Thompson's Negligence, the section supplementing Section 369 says: The plaintiff's recovery would be defeated if his own evidence showed him to have been negligent. "So where the plaintiff's evidence shows circumstances of contributory negligence which would defeat his right of recovery the defendant may take advantage thereof, though his plea of contributory negligence has been stricken."

In Silcock v. Rio Grande, 22 Utah 179, 61 Pac. 565, the court said:

> "Where the plaintiff in a suit to recover damages for injuries shows by his own evidence that he was guilty of contributory negligence which was the proximate case of such injuries, the defense is relieved from the burden of proving such negligence and the plaintiff cannot recover."

And the court quotes from earlier cases as follows:

"Where the testimony on the part of the plaintiff who seeks to recover damages for injuries resulting from negligence shows conclusively that his own negligence or want of ordinary care was the proximate cause of the injury, he will not be permitted to recover even though the answer contains no averment of contributory negligence."

In Missouri etc. Co. v. Merrill, 61 Kans., 671, 60 Pac. 819, the court said:

"If the testimony introduced on behalf of Merrill showed that the injury was the result of his own negligence, then there could be no recovery, even if the opposite party introduced no evidence upon that subject."

In Dewald v. Kansas City etc. Co., 44 Kans. 586, 24 Pac. 1101, the court said:

"Where an action is brought to recover for personal injuries and the plaintiff's testimony shows that his own negligence contributed directly to the injury, he has failed to make out a *prima facie* right of recovery and a demurrer interposed to this evidence should be sustained."

See also Brennen v. Front Street Cable Co., 8 Wash. 363, 36 Pac. 272.

In Engelking v. Kansas City etc. Co., 187 Mo. 158, 86 S. W. 89, a plea of contributory negligence had been stricken out, but the court said: "It has always been ruled by this court that such advantage may be taken of the plaintiff's evidence regardless of whether the special defense is pleaded or not." And they sustained the lower court's nonsuit.

The word "defense" has, of course, been before the courts for definition many times. In Whitfield v. Aetna Life Insurance Co., 125 Fed. 269, the opinion says:

> "In law, 'defense' is that which is offered and alleged by the party proceeded against * * * What is put forward to defeat an action; and it has also been defined as the denial of the truth or validity of the complaint. A general assertion that the plaintiff has no ground of action."

In Wehle v. Rutter, 43 How. Prac. 5, it is held that a defense in legal language is a full answer to the whole or to some part of the plaintiff's demand.

In Donovan v. Maine, 77 N. Y. S. 229, it was said that defenses are of two classes, first, those which deny some material allegation on the part of the plaintiff, and, second, those which confess and avoid those allegations.

In Jewett etc. Co. v. Kirkpatrick, 107 Fed. 622, it is said that the means employed to defend against actions were called defenses. Originally, the word "defense," as used in the common law, meant simply the denial of the truth of the declaration. Later on it came to mean whatever was offered by the defendant as sufficient to defeat the cause of action stated in the declaration—offered by way of denial, justification, or confession and avoidance.

Certainly, to say that the *defense* of contributory negligence shall be for the jury is not equivalent to saying that the jury shall in all cases pass upon the question of contributory negligence. That question may be presented as a matter of defense or it may come into the plaintiff's own case so that he is prevented from making out a *prima facie* case. Certainly where the plaintiff himself fails to make out a *prima facie* case there is no question of defense involved at all, but purely a question under the old practice of demurrer to the evidence, or under the modern practice of nonsuit.

It is true that the Supreme Court of Arizona in discussing Section 5 of Article 18 of the constitution has used very broad language, but it was in every case *obiter*. The first time that the section was discussed was in the case of Inspiration v. Conwell, 21 Ariz. 480, 190 Pac. 88. It did not appear that the plaintiff's contributory negligence was demonstrated by his own testimony or testimony introduced on his behalf. The court said:

> "The language of the provision (Section 5) is plain and unambiguous and to our minds clearly indicates that the power or duty to finally and conclusively settle the question of contributory negligence or assumption of risk is by its terms transferred from the court to the jury * * * We think that the evident purpose and intent of the provision is to make the jury the sole arbiter of the existence or non-existence of contributory negligence or assumption of risk in all actions for personal injuries."

And the court refers to the Dickinson case, 74 Okla. 79, 177 Pac. 570, where the court used very broad language, but it does not appear that the court was ruling or considering what would be the effect of the constitutional provision in case of a motion for a nonsuit at the close of the plaintiff's own case.

The next case is Davis v. Boggs, 22 Ariz. 497, 199 Pac. 116. It cannot be gathered from that opinion that the evidence of contributory negligence was furnished by the plaintiff in making out his own case. On the contrary it appears that evidence was introduced by the defendant to establish contributory negligence as a defense.

In Calumet & Arizona etc. v. Gardner, 21 Arizona 206, 187 Pac. 563, the court held that in a case brought under Chapter 6 of Title XIV of the Civil Code known as the "Employers Liability Law," the court should direct verdict where the undisputed evidence showed that the accident was due solely to the employee's negligence. The effect of this ruling is that in a case involving Master and Servant where contributory negligence is not a defense but goes only to mitigate damages, the court may not say to the jury that contributory negligence has been established and that it is for them to diminish the damages as they see fit; but where the plaintiff's negligence is not contributory but is the sole cause, the court may direct a verdict for the defendant.

It is almost absurd to have the law in such a condition; the court having no power to direct a verdict as to contributory negligence but having power to direct a verdict where the plaintiff's negligence is the only negligence involved.

The latest case before the Supreme Court of Arizona is Pacific Construction Co. v. Cochran, 243 Pac. 405, but we do not gather from the opinion there that the whole case of contributory negligence was made out from the plaintiff's testimony or from testimony introduced on his behalf.

So we say that the precise question on which we are asking for a ruling is open in the state of Arizona.

Suppose what has happened more than once, namely, that the plaintiff in attempting to state his cause of action sets out in his complaint facts that showed the plaintiff guilty of contributory negligence, surely the question whether the pleading made out a case of contributory negligence cannot be submitted to the jury as a question of

fact in Arizona any more than in Kentucky. See Durham
v. Louisville etc. Co., 29 S. W. 737; Indianapolis etc. Co.
v. Wilson, 134 Ind. 95, 33 N. E. 793. The question there
arises not as a question of defense, but as a failure of the
complainant to state a case. The same must be true where
the plaintiff's own evidence fails to make out a case and
it can make no difference whether the failure consists in
the failure to prove the defendant's negligence or in the
affirmative proof of the plaintiff's own negligence.

See further Van Winkle v. New York etc. Co., 34 Ind.
App. 476, 73 N. E. 157; Slattery v. Colgate, 25 R. I. 220,
55 Atl. 639; Stillwell, Admr., v. South Louisville etc. Co.,
22 Ky. Law Rep. 785, 58 S. W. 696; Abrams v. Way-
cross, 114, Ga. 712, 40 S. E. 699; Lafayette v. Fitch, 32
Ind. App. 134, 69 N. E. 414.

We submit secondly that if the constitutional provision
is a part of the substantive law of Arizona then, of course,
it has no application to a cause of action which arose in
the state of New Mexico.

> Delaware etc. v. Nahas, 14 Fed. (2) 56;
> St. Louis etc. v. Rogers, 290 S. W. 74.

And it is to be remembered that

> "the law of any state of the Union, whether de-
> pending upon statute or upon judicial opinions, is a
> matter of which the courts of the United States are
> bound to take judicial notice without plea or proof."
> Lamar v. Micou, 114 U. S. 225.

So a federal court sitting in Alabama takes notice of
the laws of Ohio.

> Bohlander v. Heikes, 168 Fed. 886,

and the federal court sitting in Colorado takes notice of the laws of New Mexico.

> Denver v. Wagner, 167 Fed. 75.

Therefore the trial court knew judicially that New Mexico has no law similar to that of the Arizona Constitution but that on the contrary in New Mexico a court may direct a verdict for defendant when the evidence of contributory negligence is clear and convincing.

> Thompson v. Traction Co., 15 N. M. 407, 110 Pac. 552;
>
> Morehead v. Atchison, 27 N. M. 349, 201 Pac. 1048, *supra*;;
>
> Candelaria v. Atchison, 6 N. M. 266, 27 Pac. 497, *supra*.

A number of decisions hold that a rule such as that of the Constitution of Arizona is a part of the substantive law.

This seems to be the basis of the ruling in Illinois etc. v. Johnston, 205 Alabama 1, 87 So. 867.

That was a case under the Federal Employers Liability Act.

The Court said:

> "As respects matters of practice and procedure only —matters pertaining to the remedy merely—the local law, the procedural law of the forum is applicable and is due to be observed in the conduct of causes subject to the Federal Employers' Liability Act."

The Court then said:

> "The federal rule as to the amount of the evidence and not the rule prevailing in Alabama, commonly called the 'scintilla rule,' should be observed

in passing upon requested instructions such as would withdraw the case from the jury's consideration."

The Court said that if appellant's stated contention involved "practice and procedure," merely, it could not be well founded. And again:

"It is manifest that the gist of the inquiry thus made is whether the plaintiff has discharged the burden of proof assumed by him in his pleading and imposed by the federal act in respect of the material averment that the contractural relation of master and servant existed at the time alleged."

The Court then quotes from Central vs. White, 238 U. S. 507, as follows:

"The question of burden of proof is a matter of substance and not subject to control by laws of the several states"—

that is, with reference to cases under the Federal Employers' Liability Act; and the Alabama Court then proceeds:

"Under these decisions the matter of burden of proof is regarded as of the substance of the right created by the federal act and is, hence, without the category of 'practice and procedure.'"

In the very recent case of Delaware vs. Nahas, 14 Federal (2nd) 56, the report shows that the case had been tried in the Eastern District of Pennsylvania and that it was an action for damages for personal injuries at a grade crossing—precisely the same as the case at bar. The accident had occurred in the state of New York. The Circuit Court of Appeals of the Third Circuit said:

> "If a *tort* was involved it was committed in New York. Obviously, therefore, the plaintiff's right of action and the defendant's liability depend on the law of that state."

So the court discusses the New York law, and says that in actions for injury, as distinguished from actions for death, the burden of proof under the law of New York is on the plaintiff to show freedom from contributory negligence, and the court held that the law of New York as to this burden of proof applied, although the case was brought and tried in Pennsylvania. In other words, the court treated the burden of proof as a part of the substantive law and not as merely procedure law.

In the case of Southern Pacific vs. Martinez, 270 Federal 770, this court had occasion to consider the constitutional provision now in question. There is nothing in the opinion indicating clearly whether the court took the view that the constitutional provision was one of substantive law or of adjective law, but perhaps an inference may be drawn from the fact that the only decision cited in the court's opinion is Chicago etc. v. Cole, 251 U. S. 54. In that case the same provision as found in the Constitution of Oklahoma was involved, and Mr. Justice Holmes, writing the opinion, said:

> "The State Constitution was in force when the death occurred and therefore the defendant had only such right to the defense of contributory negligence as that Constitution allowed."

Now, Mr. Justice Holmes is not in the habit of wasting words in his opinions. If the constitutional provision affected procedure only it would have applied just the same whether the death had occurred before or after the adoption of the constitution, but if the constitutional provision is regarded as substantive law, then, of course, it would apply only to causes of action arising after the adoption of the instrument. The language which we have just quoted from the opinion has therefore no proper place in the opinion, unless we assume that Justice Holmes and the Supreme Court regarded the constitutional provision as establishing a rule of the substantive law.

The Supreme Court of the United States in that case was reviewing the decision of the Supreme Court of Oklahoma, in Dickinson vs. Cole 74, Oklahoma 79, 177 Pacific 570, and the Supreme Court of Oklahoma said:

> "In no event would there be any merit in the contention that the constitutional provision takes away a vested right from the receiver, since the facts on which the action is based arose after adoption of the Constitution."

Thus the state court views the provision as one of substantive law.

But there are certain other cases in the Oklahoma reports which show even more clearly that the Supreme Court of Oklahoma regards the constitutional provision in question as one of substantive law.

St. Louis etc. v. Snowden, 48 Okla. 115, 149 Pac. 1083, involved a case arising under the Federal Employer's Liability Act, relating to railroads. The court said:

> "Under the Federal Employer's Liability Act, the law of assumption of risk is that of the common law as it existed prior to the passage of said act, except where the common carrier violates the provisions of any statute enacted for the safety of its employes; and the assumption of risk under the facts in this case

(the injury not being caused by any violation of such act providing for the protection of employes), was a question of law for the court." Citing case:

The court proceeds:

"Where the uncontroverted evidence discloses the fact, as it does in this case, that the danger was apparent to an ordinarily prudent person, and that the services were rendered without complaint, the defence of assumption of risk is conclusively established, and there is no question for the jury, and the court should instruct the jury to return a verdict for the master."

This case shows clearly that the Supreme Court of Oklahoma treats the constitutional provision of that state as a part of the substantive law of the case and not involving a mere question of procedure, for there is no doubt that with respect to merely procedural questions cases arising under the Employer's Liability Act are tried in the state courts in accordance with the law of the forum.

C. R. I. & P. v. Jackson, 61 Oklahoma 146, 160 Pac. 736, was another case arising under the same Federal law. The court said:

"Whatever may be the apparent force of the contention that the provision of the Oklahoma Constitution (Const. Art. 23, Sec. 6) requiring the submission of the defence of assumption of the risk as a question of fact to the jury, applies to cases such as the present one, upon the theory that the right to the defence is preserved under such provision and only the method of determining it varied thereby and of the reasoning by analogy from the decision in Minn. etc. v. Bomboli, 241 U. S. 211, and of St. Louis etc. v.

Brown, 241 U. S. 223, the decision of this court in St. Louis v. Snowden (149 Pac. 1083) is conclusive that these provisions cannot affect a suit brought under the federal statute, upon the ground that if the defence permitted is 'that of the common law,' then in cases where the evidence is undisputed and the circumstances permit of but one conclusion, the question must be decided by the court as a matter of law, and not by the jury as a matter of fact, since such is the common law, and such must be the result in our courts in these cases where the Federal act creating the liability likewise allows the common law defence."

And the court further said:

"In this case the court submitted the question of assumption of the risk to the jury. The plaintiff's evidence clearly established the defence. Under the doctrine of the Snowden case the cause ought not to have been submitted to the jury. Having been submitted, it was error to sustain the verdict for the plaintiff."

In Chicago etc. v. Hessenflow, 69 Okla. 185, 170 Pac. 1161, it was stipulated that the case came under the Federal Act. The court said:

"The provision of Sec. 6, Art. 23, making the defence of assumption of risks in all cases whatsoever a question of fact for the jury, has no application in this case, and under the operation of said act the law applicable to such defence is that of the common law as it existed prior to the passage of said act, except where the carrier violates the provisions of some statute enacted for the safety of its employes."

So the court held that under the evidence it was the duty of the court to declare as a question of law that the plaintiff had assumed the risk and could not recover.

Along with the state decisions on the Oklahoma Constitution the opinion of the Circuit Court of Appeals in Atchison v. Wyer, 8 Fed. (2d) 30, should be considered. This case requires careful study and some knowledge of the situation outside the report of it in the volume referred to.

That action was begun in the state court of Oklahoma and plaintiff alleged that at the time of his injury he was engaged in interstate commerce. The suit was removed to the Federal Court upon a petition which alleged that the allegations of the complaint itself showed that plaintiff was not engaged in interstate commerce. He was engaged in dismantling a locomotive engine, so there is no doubt that he was not engaged in interstate commerce and the case was properly removed and the District Court refused to remand it. Of course, had the case actually involved the Federal Employer's Liability Act, it would not have been removable.

It was further insisted that if the plaintiff was not engaged in interstate commerce, he was nevertheless engaged in commerce and that therefore a provision of the Constitution of New Mexico applied which took over into the law of New Mexico the provisions of the Federal Employer's Liability Act. The court held that it was unnecessary to determine whether that provision of the New Mexico Constitution applied or not. Now under the law of New Mexico, the defence of assumption of risk is a complete defence, whether the provision extending the

Federal Employer's Liability Act applied or not. So whichever law of New Mexico applied, assumption of risk was a complete defence, as it was also under the law of Oklahoma, in which state the case was tried. The only difference was that under the law of Oklahoma the issue of assumption of risk must always be submitted to the jury, whereas under the law of New Mexico a court may direct a verdict on that issue, if established by convincing evidence.

Now in this case of Atchison v. Wyer, the Court of Appeals held that on the issue of assumption of risk it was the duty of the court upon the showing made to direct a verdict for the defendant. Was this because the court took the same view as the Supreme Court of Oklahoma, namely that the Constitution of Oklahoma established a rule of substantive law which was not applicable to a cause of action arising in another state? Or was it because the Circuit Court of Appeals of the 8th Circuit regarded the Oklahoma rule as a rule regulating the conduct of judges in the trial of causes and therefore not applicable in the Federal Court? Under either view of the case, it sustains our position here.

There are cases from other state reports taking substantially the same view in dealing with analogous situations. Thus in Hiatt v. St. Louis etc., 308 Mo. 77, 271 S. W. 806, the cause of action arose in Arkansas. In Arkansas there is a statute establishing what is known as the comparative negligence rule. The court said that the doctrine of comparative negligence is by statute made a part of the law of Arkansas and it was for the jury to determine the relative negligence of defendant and plaintiff. The court said:

"This is a case which is governed wholly by Arkansas law. We should give to such foreign statute the meaning given to it by the highest tribunal of Arkansas. * * * The universal construction has been that, where it is shown that an injury to either property or person has been occasioned by a moving train, proof of the fact that the injury was so inflicted, and of the injury, a statutory presumption arises that such injury was the result of the negligence of the railroad. * * * The statute itself says nothing about presumption of negligence, but the Supreme Court of Arkansas has ruled from first to last that such presumption inheres in such statute. It is a statutory right granted by substantive law to every person injured by a moving train."

And again referring to the rule of the presumption, the court said later on:

"This is a substantive statutory law, and not a mere rule of procedure. The presumption inheres in the cause of action itself. The court (that is the court of Arkansas) calls it a statutory presumption created by the statute which gives the right of action and fixes the liability."

And again:

"Much is said in the briefs that the statutory presumption of negligence, ruled in all Arkansas cases, is not applicable to a trial of the issues in Missouri, because it is a mere matter of procedure, and is adjective law rather than substantive law, and is therefore governed by the law of the forum. The trouble with this contention lies in the fact that the statute has been overlooked. As said, this statute (which is the foundation of this case) was enacted in 1875. * * * It does not say a word about negligent running of trains. Were it given a literal construction it would make the railroads of Arkansas insurers of persons and property as against all mishaps occurring through the running of their trains. It would be similar to our fire statute, which allows a recovery from a railroad for damages by fire 'communicated directly or indirectly by locomotive engines in use upon the railroad' owned by the corporation. We ruled that recovery could be had without proof of negligence. In other words, the only thing required to be shown is that the fire was cummunicated from the engine of the railroad corporation. * * * The Arkansas statute is as unqualified in terms as is our fire statute. Our fire statute gave to the citizen substantive right, not a mere procedural remedy. So likewise does this Arkansas statute. It is upon this substantive right that plaintiff in this case founds her action. But this statute has been universally construed by the Arkansas Supreme Court to mean that proof of injury by a moving train only makes *prima facie* case for the plaintiff, without proof of negligence, and that the defendant railroad can only overcome this *prima facie* case by showing that it was not at fault. This construction reads into the very statute words which are not found there, but by such construction we are bound. The construction given makes this statute read that a *prima facie* case is made for plaintiff by proof of injury without proof of negligence. This is substantive statutory law, and not procedure merely. It is a substantive right conferred by the statute, when the construction of the statute is read into the law."

And again:

"We rule that the statute, as construed by the Arkansas Supreme Court (which construction becomes a part and parcel of the statute so far as Missouri courts are concerned in trying a case under it), gives to plaintiff a substantial and substantive right, and is not a procedural statute at all."

Recurring to the Arkansas rule of comparative negligence, the court said:

"A peremptory instruction on the question of contributory negligence could not have been given, so that the sole duty of the jury, in this regard, was to compare the negligence of the parties."

In Morris v. C. R. I. & P., 251 S. W. 763 (Mo. App.), the cause of action for a personal injury arose in Iowa. The trial court directed a verdict for defendant. The court said:

"The cause of action arose in Iowa and is based upon Iowa law, and the defendant invokes that law and claims the protection of Iowa decisions, upon the pleadings of which no attack has been made. The liability of defendant must therefore be determined according to Iowa law."

The court says:

"Was a submissible case made under the charge of negligence in failing to give the signals? There was, unless, under the evidence and the rules of decision in Iowa, the plaintiff must be deemed to have been conclusively guilty of contributory negligence."

The court then considers various Iowa decisions and says:

"The effect of the foregoing decisions leads us to the view that, under the rules laid down by them, we could not say that plaintiff should be held conclusively negligent, but that the question of whether he was or not should be left to the jury."

And the court concludes as follows:

"As we view the question under the Iowa decisions, the peremptory instruction should not have been given. Accordingly, judgment is reversed."

In Reilly v. K. C., 256 Mo. 596, 165 S. W. 1043, the cause of action, which was for a personal injury, arose in Kansas. The court said:

"Counsel agree that the law of Kansas must govern the disposition of this case. In that state the rule is that: 'If only one conclusion can be drawn from the undisputed facts, the question of negligence is one of law. If reasonable minds might differ upon that question, the jury must decide.'"

The court then said:

"With respect to the application of this rule, the Kansas decisions must control in this case. In that state the general rule prevails as formulated by the encyclopedists: 'Any one who goes upon or near a railroad track is bound, at his peril, to make diligent use of his senses of sight and hearing in order to detect the approach of trains, and if, in disregard of this duty to his own safety, he steps upon the track without looking or listening, he is guilty of such negligence as to bar an action for the injury.'"

The court said:

"Applying these principles, approved by the Supreme Court of Kansas, plaintiff was clearly guilty of contributory negligence in taking a position so near the track, etc."

And again:

"Under the decision quoted, his contributory negligence bars recovery on the only theory upon which the case was submitted to the jury."

And again:

"The trial court was right in refusing to submit the case upon the doctrine of the last clear chance, since that doctrine as it is applied by the Supreme Court of Kansas cannot aid plaintiff in the circumstances of this case."

In Caine v. St. Louis etc., 209 Ala. 181, 95 So. 876, the action was for a personal injury resulting in death, the cause of action having arisen in Oklahoma. The court said:

"Among other defenses interposed, the defendant pleaded contributory negligence. * * * At the conclusion of the testimony the court gave the affirmative charge in favor of the defendant, evidently upon the theory, as we gather from the record, that the pleas of contributory negligence had been sustained by the proof."

The court said:

"This cause of action arose under the laws of Oklahoma, where the accident occurred, which resulted in the death of plaintiff's intestate. While it is well recognized that the statutes of another state have no extra territorial force, yet rights acquired thereunder will always, in comity, be enforced, if not against the public policy of the laws of the state where redress is sought."

The court said that under the law of Oklahoma contributory negligence presents merely a question of fact for the jury, and quoted from an Oklahoma case in which the court had declared that under the Constitution of Oklahoma contributory negligence was always a question of fact for the jury. The court then proceeds:

"The question, therefore, presented by the action of the court in overruling the demurrer to the pleas of contributory negligence, is whether or not upon the question of contributory negligence the law of Oklahoma should control."

The court quotes as follows from a text:

"All matters of defense to an action such as the fellow servant rule, contributory negligence, assumption of risk, etc., are to be determined in accordance with the *lex loci delicti.*"

The court then proceeds:

"There is some difficulty at times in drawing the distinction between those matters which inhere in and pertain to the right of action itself and those which pertain to the remedy and procedure merely. The direct question here presented was before the Supreme Court of Kentucky in the case of L. & N. v. Whitlow, 105 Ky. 1, 43 S. W. 711. In that case the suit was by an administrator and was brought in a Kentucky court upon a cause of action, *ex delicto,* arising in the state of Tennessee. Under the law of Tennessee contributory negligence did not bar a recovery, but merely served to reduce or mitigate the damages, while under the law of Kentucky contributory negligence was a complete defense. In a well reasoned opinion the court held that the law of Tennessee as to contributory negligence should be applied. * * * The issues presented as to the question of negligence are matters for submission to the jury under the rule of law prevailing in the state of Oklahoma."

It is true that in the Caine case just quoted from the court makes a curious slip in saying that under the law of Oklahoma the rule of comparative negligence prevails. There is no such rule of law in Oklahoma, as will be seen by referring to Mo. etc. v. Parker, 50 Okla. 491, 151 Pac. 325, but that slip in language does not affect the ruling which the Supreme Court of Alabama made.

We have now given the authorities with their arguments, holding or indicating that the constitutional provision establishes substantive law. Perhaps we ought to say no more on this head, yet frankness compels us to add that but for these opinions and expressions of the various courts, and especially the implications of the Martinez case, we should have thought that the constitutional provision is not a part of the substantive law but is rather the latest and the worst step in that course of legislation which has prevailed in the western states especially for more than half a century, tending and intended to limit and humble the power of the judges in the conduct of their courts in jury trials; and to exalt the functions of the jury and release them so far as possible from all control or guidance by the judge. The first step in this process was to forbid judges to charge the jury on questions of fact. In some states the judges were forbidden to direct the verdict if there was a scintilla of evidence on the other side.

That the results of this process have been disastrous is proved by the strong and general movement now under way to reform our civil and especially our criminal practice in state courts so as to restore to the judges in the state courts some of the powers which have been exercised, and wisely exercised, by the federal judges ever since 1789.

We, therefore, submit as our third proposition under this heading that if the constitutional provision of Arizona establishes a rule limiting and weakening the powers of the judge in jury trials, then that rule will not be followed and cannot be followed in trials in the federal court sitting in Arizona.

At the outset we admit that if the decision of this court in Southern Pacific v. Martinez, 270 Fed. 770, is to be read as adopting in the federal courts a state rule limiting the powers of judges, regulating the conduct of the judges, then we are foreclosed unless this court will give further consideration to the question. And we submit that the question should deserve further consideration in view of the long and unanimous line of authority to the contrary which we shall now cite.

The 7th amendment to the Constitution of the United States reads:

> "In suits at common law where the value in controversy shall exceed $20.00, the right of trial by jury shall be preserved, etc."

The cases clearly indicate that that amendment forbids the application in the federal courts of such a rule as that which we are now assuming to have been established by the Arizona Constitution.

In Hughey v. Sullivan, 80 Fed. 72, it appeared that a statute of Ohio forbade the granting of new trials on account of smallness of damages in any action for injuries to person or reputation. Judge Hammond wrote the opinion and said:

> "It may be doubtful if this Ohio statute * * * could be held constitutional if it were binding upon the federal courts."

Then he quotes the 7th amendment and says:

> "This does not mean only and barely that there shall be a verdict of twelve men under any conditions

that may be prescribed, but that there shall be a trial by jury as understood at common law. The control of the court over the verdict after it is given is as much a part of the trial by jury as the giving of the verdict itself, and the right to have the issues tried by a second jury, or even a third jury, when the verdict of the first jury is affected by some infirmity for which the common law required the trial court to set that verdict aside, is as much a right of 'trial by jury' preserved by the Constitution as the first trial."

Now, manifestly, if the Constitution of Arizona establishes a rule binding on the federal court, the federal court could not grant a new trial on the ground that the jury had determined the issue of contributory negligence contrary to the evidence.

Further on in Judge Hammond's opinion he says:

"This and other considerations attending the subject show the importance of the part that is taken by the trial judge in the process of a trial by jury. He alone can, on the application for a new trial, correct the errors that are made by the jury; and, if legislation may control his judgment, or prohibit him in the exercise of it, the right of trial by jury is to that extent impaired and restricted, and not preserved as it was known at common law. Congress clearly has no plenary power to thus impair, restrict, or destroy the right of trial by jury in any of its parts; and what Congress cannot do, surely the Legislature of the state cannot do in the application of their legislation to the federal courts."

Further on in the opinion he said:

"If a statute should be passed requiring the minority of a jury to conform their judgment to the majority, and return a verdict accordingly, it would be conceded everywhere that this would be an impairment of the right of trial by jury, because it would be imposing by law upon the jury a rule of judgment not known to the common law. So, when the trial judge comes to receive the verdict and, on proper motion, to inspect it, and determine whether or not it is affected by any infirmity which would authorize the court to set it aside, would it not be just as much impairment of the right of trial by jury if the Legislature should say that he should either set it aside, or let it stand, upon some rule that it should prescribe to control his judgment? While the Legislature may prescribe any rule of property or any rule of pleading or any rule of practice or any form of procedure, it cannot invade the domain of judgment either of the jury or its presiding judge, and direct what that judgment shall be, in the discharge of the respective or joint functions of either. These must remain under the federal Constitution, at least, to the government of the common law. * * * If, therefore, this statute of Ohio should be held to be a rule of practice or form of procedure, I should say that it was inconsistent with the Constitution of the United States, and, therefore, not binding on us."

In Maxwell v. Dow, 176 U. S. 581, it was held that as a right of trial by jury is preserved by the 7th amendment, it is implied that there should be a unanimous verdict of twelve jurors in all federal courts where a jury trial is had.

In California, in the state courts, nine jurors may render a verdict in civil cases. That is not followed in the federal courts because to do so would be a violation of the 7th amendment.

See also American Publishing Co. v. Fisher, 166 U. S. 464.

In Springville v. Thomas, 166 U. S. 707, the court said:

"In our opinion the 7th amendment secured unanimity in finding a verdict as an essential feature of trial by jury in common law cases, and the act of Congress could not impart the power to change the constitutional rule, and could not be treated as attempting to do so."

In Evans v. Lehigh, 205 Fed. 637, it appeared that an act of Pennsylvania authorized the court in cases where a jury disagreed, to certify the evidence so that the same might become a part of the record and then enter judgment on the whole record, if either party was entitled thereto, whenever a request for binding instructions had been reserved or declined by the trial court. The federal court held that this practice could not apply in the federal court sitting in Pennsylvania, citing Slocum v. Insurance Company, 228 U. S. 364, where the Supreme Court said:

"The court cannot dispense with a verdict, or disregard one when given, and itself pass on the issues of fact."

In other words, the constitutional guaranty operates to require that the issues be settled by the verdict of a jury unless the right thereto be waived.

In McKeon v. Central, 264 Fed. 385, the trial had been held in the state of New Jersey and had resulted in a verdict for the plaintiff. The judge thought that the trial had been without error but that the damages were not sufficiently large. He did not order a new trial generally, but ordered one on the sole issue of the amount of damages. He followed in this a practice of the state courts under an act of the state that when a new trial should be ordered because damages are excessive or inadequate and for no other reason, the verdict shall be set aside only in respect to damages and should stand good in all other respects. The court referred to the 7th amendment and said:

> "Preservation is not creation; it is the saving that which already exists. Seeing, then, we have here a word which implies the existence of something already in being, we find the call of the words 'preserve,' 'shall be preserved,' answered by a definite thing, to-wit, 'the right of trial by jury.' And it will be noted that we are not dealing with an indefinite, but a very definite, concrete, well understood thing, that at that time (1789) had such a recognized meaning that it was described, not as 'a right of trial by jury,' not as some or any right of trial by jury, but as 'the right of trial by jury.' The Declaration of Independence charged that the Crown had deprived us in many cases of trial by jury."

The court referred to Parsons v. Bedford, 3 Pet. 45, and said:

> "The jury at common law consisted of twelve men, heard all the evidence pertinent to the issue raised by the pleading, returned a single verdict and upon such verdict judgment was entered and if a new trial is granted it is a trial of the whole case."

Further along the court said:

> "The 7th amendment taken as a whole makes it clear that the jury trial thus preserved is the jury trial of English Common Law."

In the case of Indianapolis v. Horst, 93 U. S. 291, the court had been requested to direct the jury to find specially upon particular questions of fact if they found a general verdict. It was the law of the state of Indiana, where the case was tried, that the judge might be required to submit special questions to the jury. The court refers to the case of Nudd v. Burrows and says as to the trenching upon the common law powers of the court:

"Whether Congress could do the latter was left open to doubt. It was not then and is not now necessary to decide that question."

The courts are likely to avoid a construction of section 721 and section 914 of the Revised Statutes which would compel the court thereupon to determine whether or not those sections as so construed were in conflict with the 7th amendment. For a construction will, if practicable, be avoided which would raise a serious constitutional question.

U. S. v. B. & H., 213 U. S. 366 (480);
Federal Trade Com'n v. American etc., 264 U. S. 298 (370);
M. P. v. Boone, 270 U. S. 466.

We consider next the effect of sections 721 and 914 of the Revised Statutes. Section 721 reads as follows:

"The laws of the several states, except where the Constitution, treaties, or statutes of the United States otherwise require or provide, shall be regarded as rules of decision in trials at common law, in the courts of the United States, in cases where they apply."

In Wayman v. Southard, 10 Wheat. 1, it was held that section 34 of the Judiciary Act, being the same as section 721, "does not apply to the process and practice of the federal courts; it is a mere legislative recognition of the principles of universal jurisprudence as to the operation of the *lex loci"*—in this case the law of New Mexico. We have quoted section 721 because it has been so often referred to in connection with section 914, which is the one pertinent to this particular part of our argument, and reads as follows:

> "The practice, pleadings, and forms and modes of proceeding in civil causes, other than equity and admiralty causes, in the circuit and district courts, shall conform, as near as may be, to the practice, pleadings, and forms and modes of proceeding existing at the time in like causes in the courts of record of the state within which such circuit or district courts are held, any rule of court to the contrary notwithstanding."

Nudd v. Burrows, 91 U. S. 426, is a leading case. The trial judge had been requested to instruct the jury only as to the law of the case, without commenting on the facts, and had been requested that no instructions be given except in writing, nor modified except in writing, and that the instructions should be taken by the jury to the jury room and returned with the verdict, and that any papers read in evidence, except depositions, might be taken by the jury. These were statutory provisions in Illinois, where the court was sitting. The trial court had declined to follow this practice in any respect and the Supreme Court said:

> "The purpose of the provision (section 914) is apparent upon its face. No analysis is necessary to reach it: It was to bring about uniformity in the law of procedure in the federal and state courts of the same locality. It had its origin in the code enactments of many of the states. While in the fed-

eral tribunals the common law pleadings, forms and practice were adhered to, in the state courts of the same district the simpler forms of the local code prevailed. This involved the necessity on the part of the bar of studying two distinct systems of remedial law, and of practicing according to the wholly dissimilar requirements of both. The inconvenience of such a state of things is obvious. The evil was a serious one. It was the aim of the provision in question to remove it. This was done by bringing about the conformity in the courts of the United States which it prescribes. The remedy was complete. *The personal administration by a judge of his duties while sitting upon the bench was not complained of.* No one objected or sought a remedy in that direction. We see nothing in the act to warrant the conclusion that it was intended to have such an application. *If the proposition of the counsel for the plaintiff in error be correct, the powers of the judge, as defined by the common law, were largely trenched upon.*

"A statute claimed to work this effect must be strictly construed. But no severity of construction is necessary to harmonize the language employed with the view we have expressed. The identity required is to be in 'the practice, pleadings and forms and modes of proceeding.' *The personal conduct and administration of the judge in the discharge of his separate functions is, in our judgment, neither practice, pleading nor a form or mode of proceeding within the meaning of those terms as found in the context.* The subject of these exceptions is, therefore, not within the act as we understand it.

"There are certain powers inherent in the judicial office. How far the legislative department of the government can impair them, or dictate the manner of their exercise, are interesting questions, but it is unnecessary in this case to consider them."

In Indianapolis v. Horst, 93 U. S. 291, the trial court had been asked to direct the jury to find specially upon particular questions of fact if they found a general verdict, and this was in accordance with the law of Indiana. The Supreme Court held that the federal courts were not required by section 914 or section 721 to adopt this state practice.

In Vicksburg v. Putnam, 118 U. S. 545, the court held that in the federal courts the judges may comment on the evidence, call attention to the parts of it which they think important, and express their opinion on the facts, and the court said:

> "The powers of the courts of the United States in this respect are not controlled by the statutes of the state forbidding judges to express any opinion upon the facts."

St. Louis v. Vickers, 122 U. S. 360. The case had been tried in the federal court sitting in Arkansas, the Constitution of which state provides that judges shall not charge juries with respect to matters of fact but should declare the law and should reduce their charge or instruction to writing on request of either party. Chief Justice Waite, in the opinion, said.:

> "This judgment is affirmed on the authority of Vicksburg etc. v. Putnam, 118 U. S. 545; Nudd v. Burrows, 91 U. S. 426; Indianapolis etc. v. Horst, 93 U. S. 291. A state constitution cannot, any more than a state statute, prohibit the judges of the courts of the United States from charging juries with regard to matters of fact."

In Graham v. U. S., 231 U. S. 474, Mr. Justice Holmes said in his opinion at page 480:

"It is objected that the judge called the jury's attention to Graham's testimony concerning his expectation when he contracted. The judge had a right to do more than that if he left the decision to them. Universal distrust creates universal incompetence. In the courts of the United States the judge and jury are assumed to be competent to play the parts that always have belonged to them in the country in which the modern jury trial had its birth." Citing Reucker v. Wheeler, 127 U. S. 85, 32 L. Ed. 102.

Referring to this case of Graham v. United States, Mr. Justice Holmes in his dissenting opinion in the very recent case of Tyson v. Banton, February 28, 1927, Advance Opinions March 15, page 493, said at page 502:

"We fear to grant power and are unwilling to recognize it when it exists. The states very generally have stripped jury trials of one of their most important characteristics by forbidding the judges to advise the jury upon the facts."

In Chateaugay Ore & Iron Co., 128 U. S. 544, the court held that section 914 does not make the laws of the states part of federal procedure with respect to bills of exception and new trial. The court said the object of section 914 was to assimilate the form and manner in which the parties should present their claims and defense. It does not apply to a motion for new trial nor affect the power of the federal court to grant or refuse a new trial at its discretion. It does not cover any other means of enforcing or revising a decision once made by the federal court.

In U. S. v. Barry, 131 U. S. 100, where it appeared that the law of Wisconsin, where the case was tried, required the court to submit a special verdict, the court said:

"It is, however, conceded, in the brief of its counsel, that the refusal to submit a special question in connection with the general verdict, was not error, in view of the ruling of this court in Indianapolis etc. v. Horst, 93 U. S. 291. * * * *Section 914 of the revised statutes * * * was not intended to fetter the judge in the personal discharge of his accustomed duties or to trench upon the common law powers with which in that respect he is clothed.*"

City of Lincoln v. Power, 151 U. S. 436. The case had been tried by a federal court sitting in Nebraska, where the statutes require that all instructions be in writing unless waived. The Supreme Court held that the federal courts were not controlled in their manner of charging juries by state regulations.

Beutler v. Grand Trunk etc., 224 U. S. 85, 56 L. Ed. 679. The cause of action arose in the state of Illinois. Justice Holmes wrote the opinion. The question was whether the fellow servant rule applied. The court held that it did, and said:

"It may be that in the state court the question would be left to the jury (citing two cases), but whether certain facts do or do not constitute a ground of liability is in its nature a question of law. To leave it uncertain is to leave the law uncertain. If the law is bad, the Legislature, not juries, must make a change."

Barrett v. Virginia, 250 U. S. 473, arose under the federal employer's liability act and the question was whether in a case tried in the state of Virginia plaintiff could take a voluntary nonsuit after a motion had been made by the defendant for a directed verdict. The court said:

"It is now a settled rule in the courts of the United States that whenever, in the trial of a civil case, it is clear that the state of the evidence is such as not to warrant a verdict for a party, and that if such a verdict were rendered the other party would be entitled to a new trial, it is the right and duty of the judge to direct the jury to find according to the views of the court. Such is the constant practice, and it is a convenient one. It saves time and expense. It gives scientific certainty to the law in its application to the facts, and promotes the ends of justice."

And the court said:

"This rule is not subject to modification by state statutes or constitutions."

It would seem needless, in addition to these referencs, to cite opinions given in the District Courts and Circuit Courts of Appeal. But at the risk of making this argument too long we shall refer to some cases from these courts.

Mutual Building Fund v. Bossieux, 1 Hughes 386, 17 Fed. Cases 1076, No. 9977. There had been delay in filing declarations and they had' been filed at a time too late under the state statute and at a time when the defendant was entitled to a dismissal under the state statute and when under that statute it was beyond the power of the court to permit the filing. The opinion says:

"The question is, whether these two sections of the laws of Virginia, operating together, are not more than rules of practice and such as would take away from the United States courts that discretionary power over the proceedings of its officers which they have been decided to possess by the Supreme Court in the cases cited by plaintiff's counsel. Section 2 of this chapter of the code, it will be observed, is more than a rule of practice. It is a rule depriving a citizen of a right enjoyed before this law was enacted. It is, therefore, a rule of right. It may be said of every rule merely of practice, that it is for the government of the officers of court and does not deprive a court of its discretion to modify the application of the rule for sufficient cause. Any law which takes away that discretion from the court ceases thereby to be a mere rule of practice. * * * It would deprive a United States court of a discretionary power over orders of dismissal. * * * As such it does not fall within the meaning of the 914th section of the United States Revised Statutes. That section was not intended as more than a regulation of the practice in the courts of the United States. By rules of practice are meant the rules prescribed for the government of the officers of court respecting the times, forms, and methods of orderly proceeding in a court; and I repeat that rules of practice are not superior to the discretion of a court so as to deprive it of the power to secure the trial of causes on their merits."

In Erstein v. Rochelle, 22 Fed. 61, Mr. Justice Matthews said:

"It can hardly be supposed that it was the intent of this legislation (R. S. Sec. 914) to place the courts of the United States in each state, in reference to their own practice and procedure, upon the footing merely of subordinate state courts, required to look from time to time to the supreme court of the state for authoritative rules for their guidance in those details. To do so would be, in many cases, to trench in important particulars, not easy to foresee, upon substantial rights, protected by the peculiar constitution of the federal judiciary, and which might seriously affect, in cases easily supposed, the proper correlation and independence of the two systems of federal and state judicial tribunals."

Peoples Bank v. Aetna, 74 Fed. 507. The court said at page 512:

"It is the duty of the trial judge in the courts of the United States, in cases like this, where he would be compelled to set aside a verdict for the plaintiff if one should be returned by the jury, to direct a verdict for the defendant."

And the court said further:

"Even if the contention of the plaintiff in error be correct, which we do not find, that, under the 'practice and mode of proceeding' in the courts of South Carolina, a nonsuit was not authorized in this case, nevertheless, it is submitted that it was not the intention of the legislation now known as section 914 of the Revised Statutes, which requires the courts of the United States to conform 'as near as may be' to the practice existing in the courts of the state within which the trial is held, to change the now universal rule of procedure in the federal courts, to which we have alluded, and which has been commended by the Supreme Court, since the passage of the enactment mentioned."

Hughey v. Sullivan, 80 Fed. 72, cited and quoted *supra*.

In Sloss v. South Carolina, 85 Fed. 113, the court said, at page 138:

"If the court is satisfied that, conceding all the inferences which the jury could justifiably draw from the testimony, the evidence is insufficient to warrant a verdict for the plaintiff, the court should say so to the jury. (Citing cases.) This is a well established rule in the federal court. It is essentially different from the rule in the courts of many of the states, but is uniform in the courts of the United States; and it has been decided by this court that section 914 of the revised statutes does not require the federal courts to change this rule of procedure so as to conform to the practice existing in the courts of the states."

See, also:

Abbott *et al.* v. Curtis Mfg. Co., 25 Fed. 406.

Chicago v. Price, 97 Fed. 423. The cause of action arose in Illinois. The case was tried in Iowa. The state rule in Illinois is that the complainant must allege and prove that he was without fault. The court said:

"The rule of the national courts is settled and uniform that contributory negligence is an affirmative defense, which must be established by a preponderance of evidence. It may appear from the testimony introduced by the plaintiff, or from that presented by the defendant, but, in the absence of all evidence on the subject, it is no fault or defect of the plaintiff's case that he fails to plead or prove that the defense of contributory negligence does not exist. In the jurisprudence of the national courts he is not called upon to establish the negative."

The law of Illinois is very well settled that contributory negligence must be negatived and its absence established by the plaintiff. See Village etc. v. Licht, 77 N. E. 581.

Evans v. Lehigh, 205 Fed. 637, where a state law permitted a court in certain cases where the jury had disagreed, to enter judgment itself without a new trial.

Spokane v. Campbell, 217 Fed. 518 (9th Circuit). At page 523 the opinion says:

> "It is strongly urged that the Federal courts will adopt the local practice with respect to the taking of special and general verdicts and the local rules and procedure for construing the same in determining their potency and effect. It has been distinctly held that the local law with respect to submitting special findings along with a general verdict does not control the Federal courts in respect to the mode in which causes shall be submitted to the jury. (Citing cases.)

> "This being so, it is a logical consequence that the Federal courts will not be bound by the rules obtaining in local courts for interpreting such verdicts."

In Yates v. Whyel Coke Co., 221 Fed. 603, the case had been tried in the federal court sitting in Ohio. From the opinion it appears that the jury returned and asked for further instructions which were given in the absence of counsel. Under the Ohio code it is error for a judge, in the absence of party and counsel and without notice to them to give instructions. This rule of practice is not applicable in the federal courts under Section 914, Revised Statutes. The power of federal judges as defined by the common law in the submission of cases and the control of the deliberation of juries still remains.

McKeon v. Central, 264 Fed. 385, where a state law permitted a new trial to be granted on the sole issue of the amount of damages.

In Ellicott etc. v. Vogt, 267 Fed. 945, C. C. A. 6th Cir., is considered the state statute which provided that the proceedings of each day should be drawn up by the clerk from his minutes in a plain, legible manner, which, after being corrected as ordered by the court and read in an audible voice, shall be signed by the presiding judge. The state courts held that there could be no valid judgment until the date the judgment is signed by the judge on the order book. It was contended that Sec. 914 carried this provision over into the federal practice. The court said:

"It has now come to be the settled rule of the federal courts that the conformity statute does not apply to the manner in which the judge shall perform his personal duties on the bench, * * * but only to the manner in which the parties shall bring the case to issue and to trial, or to substantial rights which the state statute gives to a party."

And again:

"Sec. 914 directs that the state practice shall be followed by the federal courts as near as may be, any rule of the federal court notwithstanding. Revised Statutes, Sec. 918, expressly grants to the federal trial courts power to make rules to 'regulate their own practice as may be necessary or convenient for the advancement of justice and the prevention of delays in proceedings.' "

The court says:

"There seems no way of reconciling these seemingly inconsistent provisions, unless through the effect of the 'as near as may be' clause. There is ample room to say that where, in the judgment of the federal *nisi prius* court, the adoption of a particular practice is necessary or convenient to meet the particular needs of that court, and when there is reasonable basis for that judgment, the 'as near as may be' clause will take effect, precise conformity will not be required, and the particular federal practice or rule will prevail."

' In Fugitt v. Lake Erie, 287 Fed. 556, the state law provided that an action might be dismissed without prejudice by the court when the plaintiff failed to appear at the trial and that in all other cases the decision should be upon the merits. The federal court had, upon the calling of the case for trial and the failure of the plaintiff to appear, impaneled a jury, heard defendant's witnesses, received a verdict for the defendant, and entered judgment accordingly. The court refers to the case of Indiana v Horst and says:

> "The line of discrimination appears to be drawn where the state statute assumes to control the discretion of the court in the final disposition of the case. By settled adjudication under the conformity statute, the court had the power to disregard such procedure where it seemed to be in the interest of justice to do so. We do not believe that the conformity act requires the federal court to submit to the dictation of the state practice a matter which involves its discretion respecting the final disposition of the case over which, both as to facts and parties, it has complete control.

In West Tenn. etc. v. Shaffer, 299 Fed. 197, the Circuit Court of Appeals of the 6th Circuit held that a state law permitting introduction of evidence to prove the good character and reputation of witnesses whose veracity had not been attacked, would not be followed in the federal court.

The court said:

> "It is not easy to see how the receiving or excluding of such evidence as that involved here is not one of those matters pertaining to the conduct of the trial, which are left to the discretion of the trial judge."

In the last preceding section of this part of the brief we have referred at length to the recent decision of the Circuit Court of Appeals for the 8th Circuit in the case of Atchison etc. v. Wyer, 8 Fed. (2d) 30. Under one view of it that opinion is in point here as under another view it is in point where cited *supra*.

A very recent case bearing closely on the question now under discussion is Seestad v. Post etc. Co., 15 Fed (2) 595. That was an action to recover damages for libel. The plaintiff asked the court to give the following instruction:

> "The court instructs the jury that under the Constitution of the state of Colorado, in this case the jury are the sole judges of both the law and the facts."

The court refused that instruction and gave this:

> "The court * * * has decided and instructs you that the article is libelous *per se;* and the question left for you to decide is what damages, if any, the plaintiff, Frank Seestad, is entitled to."

On a motion for new trial Judge Symes reviewed the cases in an opinion which deserves reading, in the course of which he says:

> "It is proper and necessary to consider the consequences that would logically follow the adoption of the view urged by counsel for the defendant. They say there is no danger but that the court's views of the law would prevail in any given case, in view of the court's power to set the verdict aside, and grant a new trial, etc. However, if the jury was entitled to decide the law, the judge, to be consistent, should give up these prerogatives for the same reasons

that now render him powerless to control the views of a jury on the facts. Otherwise it would be an invasion of their province."

He proceeds:

"Every litigant in the federal court has an inherent right to have the law of the case determined by the judge. It is not hard to imagine why this right is most valuable and necessary for the protection of his property as well as his liberty. If the jury were to declare the law, who could say in advance what the law on any given question is? The judge would or should, at least, abstain from the responsibility of stating the law to them, and would not be justified in directing a verdict, irrespective of the weakness of the prosecution, or of the plaintiff's testimony, as the case might be. The inevitable result would be that the rights and liberties of our citizens would be determined by an irresponsible tribunal, ignorant of the law, instead of one trained in jurisprudence and discipline by study and experience, or to put it in another way, by a tribunal that makes its own law as distinguished from one that would follow the law impartially as laid down by statute, eminent jurists, or decided cases, irrespective of persons, times or popular agitation."

The Constitution of Arizona by making the defence of contributory negligence always a question of fact for the jury does in effect make the jury the judges of the law and the fact, like the libel section in Colorado; so that the case last cited is in substance the same as the case at bar.

The clause which we have quoted from the Constitution of Arizona is not the only provision of that constitution which regulates the functions of the judge in the trial of jury causes. Article VI, Section 12, reads:

"'Judges shall not charge juries with respect to matters of fact nor comment hereon, but shall declare the law."

Now why, in a trial in the federal court sitting in Arizona, should the constitutional provision respecting contributory negligence be followed and this provision in regard to charging on the facts or commenting on the facts be disregarded, as declared in several cases hereinbefore cited?

So far as concerns the adoption of state practice, the statutes of the state are just as much controlling as the state constitution. Now, turning to the laws of Arizona, we find in the Civil Code of 1913, Section 514, the following:

"After the evidence is closed and before the commencement of the argument, the court shall charge the jury."

Must this order of procedure be followed in the federal courts?

The same section provides further:

"The charge shall be taken down by the court reporter, and at the request of either party shall be written out, signed by the judge, and filed with the clerk. If the court reporter be not present the charge shall be in writing, unless waived, and signed by the judge."

Surely the federal courts are not required to follow this section.

Section 515 says:

"The court shall not charge the jury with respect to matters of fact, nor comment thereon, or upon the weight of the evidence, but shall declare the law

arising on the points, and the charge shall submit all
controverted questions of fact solely to the decision
of the jury."

That seems to go a long ways toward depriving the
court of the power to direct a verdict in any case. Or is
the word "controverted" to be construed to mean controverted by the evidence instead of controverted by
allegation?

Section 516 says:

"Either party may, before the charge is given,
present to the court, in writing, such instructions as
'he desires to have given to the jury, and for this
purpose a reasonable time therefor shall be given;
The court shall pass on the same, and either give or
refuse the same as asked, or modify the same to
conform to the court's view of the law, plainly
indicating the modifications made, and give the same
as modified. If given, the judge shall, on the margin
of each instruction, write the word 'given' in ink,
and sign his name thereto in ink; and if refused,
he shall, in like manner, write the word 'refused', and
sign his name thereto. If given as modified, he shall
in like manner write the words 'given as modified,'
and sign his name thereto in ink."

Is this very specific statutory direction to be followed in
the federal courts, even down to the use of ink?

And if not, why should the provision here in question
be followed?

Section 532 provides:

"In all trials of civil cases in the Superior Courts
where a jury of twelve persons shall be impaneled to
try such cause, the concurrence of nine or more
jurors shall be sufficient to render a verdict therein."

This provision follows a provision of the constitution, Article II, Section 23, which says:

> "The right of trial by jury shall remain inviolate, but provision may be made by law for a jury of a number of less than twelve in courts not of record, and for a verdict by nine or more jurors in civil cases in any court of record."

And yet by repeated rulings of the Supreme Court of the United States, hereinbefore cited, this rule of procedure, statutory though it is, is not followed in the federal courts.

A law, whether constitutional or statutory, that takes away from the court its power to direct a verdict in jury trials, seems to go further than those laws which the courts of the United States have declined to follow in the cases cited above, and the reasoning of those cases applies quite as well to the case at bar as to those in which the language was used.

Section 918, Revised Statutes of the United States, reads as follows:

> "The several Circuit and District Courts may, from time to time, and in any manner not inconsistent with any law of the United States, or with any rule prescribed by the Supreme Court under the preceding section, make rules and orders directing the returning of writs and processes, the filing of pleadings, the taking of rules, the entering and making up of judgments by default, and other matters in vacation, and otherwise regulate their own practice as may be necessary or convenient for the advancement of justice and the prevention of delays in proceedings."

We admit that the powers granted by Section 918 are subject to the provisions of Section 914, already quoted.

In pursuance of the powers granted by Section 918 and possibly under the inherent powers of courts, the District Court of the United States for the District of Arizona established Rule 60 of its Rules of Practice, reading as follows:

"The defendant in an action at law, tried either with or without a jury, may either at the close of the plaintiff's case or at the close of the case on both sides, move for a nonsuit. The procedure on such motion shall be as follows:

"The defendant or his counsel, shall state orally in open court that he moves for a nonsuit on certain grounds, which shall be stated specifically. Such a motion shall be deemed and treated as assuming for the purposes of the motion (but for such purposes only) the truth of whatever the evidence tends to prove, to-wit: Whatever a jury might properly infer from it. If, upon the facts so assumed to be true as aforesaid, the Court shall be of opinion that the plaintiff has no case, the motion shall be granted and the action dismissed. The party against whom the decision on the motion is rendered may then and there take a general exception, and may have the same, together with such of the proceedings in the case as are material, embodied in a bill of exceptions. If evidence shall be introduced by either party after the decision on the motion has been made, the same shall operate as a superseding of the motion; but such motion may be renewed at the close of all the evidence."

The motion for a directed verdict which was made in this action at the close of the plaintiff's case and repeated at the close of all the evidence, was made pursuant to this Rule 60 and we submit that under that rule, taking the evidence as it stood, it became the duty of the court to

The cases which we have cited in considering Section 914 of the Revised Statutes are, we think, in point here and nothing would be gained by reviewing them further or calling attention to other cases (and they are very numerous) where the federal courts have ordered verdicts for defendants on the ground of clear proof of contributory negligence of the plaintiff.

II.

Defendant's Requested Instructions.

Under this heading, we will consider the 12th, 13th, 14th, 15th, 16th and 17th Specifications of Error.

The first paragraph of the instruction set out under the *12th* specification is so generally recognized to state the law that in support thereof we do no more than refer the court to the following cases:

> Penn. R. Co. v. Cutting, 5 Fed. (2nd) 936;
>
> Delaware & Hudson Co. v. Nahas, 14 Fed. (2nd) 56, *supra;*
>
> Hall v. Hamel, 244 Mass. 464, 138 N. E. 925;
>
> Norfolk etc. R. Co. v. Norton etc., 279 Fed. 32;
>
> Northern Pac. R. Co. v. Babcock, 154 U. S. 190; 38 L. Ed. 958;
>
> Cleveland etc. R. Co. v. Wolf, 189 Ind. 585, 128 N. E. 38;
>
> Caine v. St. Louis etc. R. Co., 209 Ala. 181; 95 So. 876;
>
> Marshall v. B. & M. R. Co., 124 A. 550.

In support of the second paragraph of said instruction, we invite the court's attention to the following cases:

In Morehead v. Atchison etc. Ry. Co., 27 N. M. 349, 201 Pac. 1048, the court, in the course of its opinion,

among other cases, referred to that of Chicago Great
Western Railway Company vs. Smith, 141 Fed. 930, at
page 931, and quoted with approval therefrom as follows:

> "The duty to look and listen requires the traveler to
> exercise care to select a position from which an effec-
> tive observation can be made. The mere fact of look-
> ing and listening is not always a performance of the
> duty incumbent upon the traveler, for he must also
> exercise care to make the act of looking and listening
> reasonably effective. Elliott on Railroads, vol. 3, par.
> 1166."

> "The doctrine has been repeatedly stated by this
> court that a traveler approaching a railroad crossing
> must take notice of the fact that it is a place of danger,
> and must not only look and listen for the approach
> of trains before he goes upon the track, but must
> continue to look and listen until he has passed the
> point of danger. He must continue his vigilance until
> the danger is passed, *and must look both ways up and
> down the track."* (Italics ours.)

To the same effect is the New Mexico case of Sandoval
v. Atchison etc. Ry. Co., 30 N. M. 343, 233 Pac. 840.

This rule has also been recognized and approved in the
following cases:

> Bates v. San Pedro etc. R. Co., 38 Utah 568, 114
> Pac. 527;
> Holton v. Kinston etc. Ry. Co., 188 N. C. 277,
> 124 S. E. 307.

Elliott on Railroads, vol. 3 (2nd Ed.) section 1166,
declares the rule to be as stated in the foregoing cases.

The instruction set forth under the *thirteenth* specification of error, contained, we submit, a correct statement of
the law upon the phase of the case to which it was

addressed, and, we maintain, that the court's refusal to give the instruction was highly prejudicial to defendant, particularly in view of the instruction given by the court on the same subject hereinafter considered under the *eighteenth* specification of error.

The instruction now under consideration pertained to the location and physical character of the crossing. As shown by the undisputed evidence, the defendant, when constructing its railroad across the highway, installed the crossing at the exact place where the rails intersected the highway as the latter then existed. The angle of the approaches to the crossing was perhaps slightly changed, but the location of that part of the highway occupied by the rails was in nowise altered [42, 43, 45, 46, 47, 181, 182]. In other words, the defendant built its line of railroad across the highway exactly where the highway was found at the time the railroad was constructed.

Further, as already stated, the evidence shows clearly and undisputably that the highway approaches to the crossing followed the natural surface and contour of the land; that there was no grading of the highway any more than just enough to make its surface smooth [176, 177, 182]. Underneath the tracks themselves at the crossing there was only the standard ballast 10 inches in depth [177].

Subdivision 5 of Section 4697, New Mexico Statutes (Annotated) 1915, empowers a railroad corporation

> "to construct its railroads and telegraphs across, along, or upon any * * * highway * * * which its railroad * * * shall intersect, cross or run along but such corporation shall restore such * * * highways * * * so intersected, to their

former state, as near as may be, so as not to unnecessarily impair their use or injure their franchises * * *."

We submit that the evidence in the case shows that defendant fully complied with this law, and that it was entitled to the instruction requested, and that the court erred to defendant's prejudice in its refusal to give it. That this is so, is amply shown by the following authorities.

In the cases of Morris v. Chicago etc. R. Co., 26 Fed. 23, and Lapsley v. Union Pacific R. Co., 50 Fed. 172, it was held that under the laws of the state of Iowa which conferred upon railroad companies the right to construct lines across highways, railways and public highways may lawfully be built so as to intersect or cross each other upon the same grade or level.

The case of De Lucca v. City, etc., 142 Fed. 595, involved the statute of Arkansas reading in part as follows:

"Whenever any railroad company has constructed, or shall hereafter construct, a railroad across any public road or highway or street in any incorporated city or town in this state now established or hereafter to be established, * * * such railway company shall be required to so construct said railroad crossing, or so alter or construct the roadbed of such public road or highway or street in any incorporated town or city, that the approaches to the roadbed on either side shall be made and kept at no greater elevation or depression than one perpendicular foot for every 10 feet of horizontal distance, such elevation or depression being caused by reason of the construction of said railroad; provided, at any crossing of any public highway or street in any incorporated city or town, such railroad may be crossed by good and safe bridge to be built and maintained in good repair by the railroad company owning or operating such road."

The suit was by a citizen to enjoin the city of North Littlerock from constructing a viaduct over one of its streets crossing the tracks of a railroad, it being contended by plaintiff, among other things, that the Arkansas statute above quoted made it the duty of the railroad company to construct such a viaduct, and that the city did not possess that power.

In answering this contention, the court said:

"The claim is that the word 'may' in the proviso should be construed as 'shall' or 'must', or, in other words, that this proviso is mandatory, and not merely permissive or even directory. During the argument, learned counsel for complainant insisted that, under this proviso, the state, through its Attorney General, can, by mandamus proceedings, compel every railroad in the state to build and maintain, at its own expense, bridges over every highway in the state or street in any incorporated city or town across its tracks, and for this reason the city has no power to build such a bridge at its own expense. In the opinion of this court there is nothing in that act to warrant such a construction. The intention of the legislature is so clearly expressed and is so free from ambiguity that there is nothing for the courts to construe. The act itself, leaving out the proviso, provides that every railroad company 'shall be required to so construct such railroad crossing, or so alter or construct the roadbed of such public road or highway or street in any incorporated town or city, that the approaches to the roadbed on either side shall be made and kept at no greater elevation or depression than one perpendicular foot for every 10 feet of horizontal distance.'

This clearly shows that at all grade crossings, whether across the county road or a street of a city, the corporation must construct the approaches so that there should not be a greater grade than ten per cent. By the proviso the road is granted the option, without any further act on the part of the county court, if it is a county road, or the municipal authorities, if it is a street in a town or city, to build a bridge across its tracks, if it deems it best to do so * * * In many instances the railroad may cross a road or street in a deep cut or on a high embankment which would make it more expensive to build the approaches to its roadbed at a 10 per cent. grade than the construction of a bridge, or in populous cities where there is a great deal of travel across the railway tracks, as is charged in the bill is the case in the city of North Littlerock on this street, the railroad may deem it best to build the bridge, owing to the great danger to persons and teams crossing the street. If so, the proviso grants it that privilege. The fact that in the main body of the act the word 'shall' is used, while in the proviso the word 'may' is used, shows conclusively that the Legislature intended to make a distinction. To give it the construction claimed by the complainant the proviso would nullify the act itself and at the same time place a burden on the railroads, which, if the contention of complainant is sustained, would be required to build bridges over more than 10,000 crossings in this state. It would be unreasonable to presume the Legislature intended to impose such burdens on them unless the language used is so plain and unambiguous that no other conclusion could be reached."

The case of Leitch v. C. & N. W. R. Co., 93 Wis. 79, 67 N. W. 21, was a suit to recover for personal injuries at a highway crossing of the defendant's railway track. The plaintiff was approaching the track from the westward. The conformation of the ground was such that plaintiff's view to the left was obstructed until she got over close to the railroad track, by reason of a high bank within the right of way, most of which consisted of a natural hill, through the foot of which the railroad ran, making a small cut. One of the charges of negligence was that the highway crossing was not properly restored and maintained in proper condition, the claim being that the defendant should have removed the embankment within its right of way so as to permit a clear and unobstructed view of the railroad track for a long distance upon each side of the crossing.

The defendant requested an instruction to the effect that no recovery could be had on the ground of failure to restore the highway to its proper condition under the Wisconsin statute at the time of building the railroad, which instruction was refused.

The statute of Wisconsin relied upon by the defendant (Sec. 1836 Rev. Stat.), provides:

> "Every corporation constructing, owning or using a railroad shall restore every highway, etc., across, along or upon which such railroad may be constructed to its former state or to such condition as that its usefulness shall not be materially impaired and thereafter maintain the same in such condition against any effects in any manner produced by such railroad."

Respecting said alleged ground of negligence, the court said:

> "There was absolutely no evidence of any failure to restore the highway to its former condition, as

required by Section 1836, Rev. Statutes, so far as it is physically possible to do so when a railroad is constructed across a highway. The only defect or change in the condition of the highway seriously claimed is that a bank existed along the right of way, and entirely outside of the highway, which obstructed the view of an approaching train to travellers approaching the track from the direction in which the plaintiff approached. This defect consisted almost entirely, if not, entirely, of a natural hill through the base of which the track was laid in a cut. · The failure to remove this hill cannot, in any fair or reasonable sense, be said to be a failure to restore the highway to its former condition of usefulness * * * We are aware of no rule of law, either statutory or otherwise, which requires a railway company to remove natural hills within its right of way."

In Ill. Cent. Rd. v. Bentley *et al.*, 64 Ill. 438, it appeared that the Railroad Company's charter gave it the right to construct its road over or across highways, etc. which the route of its road might intersect, but required it to restore the highway etc. thus intersected to its former state, or in a sufficient manner not to have impaired its usefulness. The charter further provided that "Whenever the track of said railroad shall cross a road or highway, such road or highway may be carried under or over said track, as may be found most expedient; and in case where an embankment or cutting shall make a change in the line of such road or highway desirable, with a view to a more easy ascent or descent, the said company may take such additional lands for the construction of such roads or highway as may be deemed requisite by said corporation. * * *"

In constructing its railroad, the company changed the location of the highway, and instead of making a crossing at the point where the highway intersected the company's track, it made a sharp curve in the highway, carrying it westward to a point about ten rods west of the original line of the road, and there made a crossing, then taking the highway back eastward upon the other side, to connect with the original line of the highway.

Suit was brought by several individuals who owned lands adjoining the highway and near the crossing, plaintiffs claiming that travel upon the highway was rendered dangerous, and that the usefulness of the highway was thereby greatly impaired, and praying that the highway and crossing be restored to the original line of the highway, and that defendant erect a bridge, with proper approaches, over and across its track on the original line of the highway. The trial court entered a decree accordingly.

The Supreme Court of Illinois in reversing the decree held that the company under its *charter* had the right to change the line of the highway. Then the court, calling attention to the provision of the charter reading "Whenever the track of said railroad shall cross a road or highway, such road or highway may be carried *under* or *over* said track, as may be found most expedient," said:

"An option is here clearly given to the corporation, which may depend wholly upon a question of engineering, *and it is to be determined* by the company. The courts have power to put the corporation in motion, but not to determine the option." (Italics ours.)

In Cowles v. N. Y. etc. R. Co., 80 Conn. 48, 66 Atl. 1020, it was shown that the tracks at the crossing were laid upon an embankment some 18 feet high. In answer to plaintiff's claim that defendant was negligent in establishing the grade crossing, the court said:

"A difference is to be noted between the duties imposed by law upon a railroad corporation as a body politic acting as the agent of the state in pursuance of the state's directions in the establishment, construction, and maintenance of a railroad highway for a public use as well as its private profit, and as a private corporation conducting the business of a carrier of goods and passengers through the operation of cars upon the highway thus established. The former are mainly created by statute, which also determines the consequences of their violation. The latter are mainly created by force of the common law, and are similar in character and consequences of violation to those imposed by the common law upon all persons engaged in a similar business. The location and establishment of a railroad highway, as well as of other highways, is an act of state sovereignty exercised by state agents appointed for that purpose. When a railroad highway and a carriage highway intersect at a common grade, the "grade-crossing" thus formed is established by the state through its agents. A grade crossing or the condition existing by reason of such an intersection of two highways is in the nature of a nuisance, and cannot lawfully exist, unless in pursuance of state authority. State v. Branford, 59 Conn. 402, 405, 22 Atl. 336; New York & N. E. R. Co.'s Appear, 62 Conn. 527, 530, 540, 26 Atl. 122. Such a subjection of public safety in the use of both highways to public convenience and necessity can only be directed by the state. When so directed, neither the railroad corporation as charged with the maintenance of the railroad highway, nor the town or other corporation as charged with the maintenance of the carriage highway, are responsible for the dangers resulting solely from such a construction of the two highways. Hoyt v. Danbury, 69 Conn. 341, 352, 37 Atl. 1051; Newton v. N. Y. & H. R. Co., 72 Conn. 420, 429, 44 Atl. 813. At the time of such a construction, or at any time thereafter, the state may order these corporations as its agents to alter the construction by a separation of grade or otherwise, and may impose upon them new duties in respect to the performance of which they may be made liable as far as the statute points out; *but without such state action, they do not, so far as concerns the condition of the roadbed, the crossing, and of its approaches, owe any legal duty to travellers using the highway beyond the statutory duty of maintaining in safe condition the highway as established by the state."* (Italics ours.)

In Elliott on Railroads, Sec. 1108, 2nd Edition, the rule is stated as follows:

"When a railway company has the right to construct its track across public highways and there is no statute prescribing how the crossing shall be made it seems that the company may exercise its discretion as to the manner in which the crossing shall be constructed. As a rule railway lines are constructed so as to conform in a great degree to the surface of the ground over which they pass and where the country through which a railroad passes is comparatively level nearly all crossings will be found to be grade crossings. *Unless there is a statute forbidding grade crossings the railway company may so construct its line as to cross the highway on the same level."* (Italics ours.)

FOURTEENTH AND FIFTEENTH SPECIFICATIONS OF ERROR.

We respectfully submit that the instructions to which both the fourteenth and fifteenth specifications of error relate were, under the evidence and the law applicable thereto, correct in their every element and, while certain phases of such requested instructions may have been covered in part by the court's general charge, we maintain that defendant was, on the facts and the law, entitled to these two requested instructions *in toto*. That this is so is borne out by the decisions cited and the argument presented under the general heading, "Defendant Was Entitled to a Directed Verdict," and sub-headings, "Plaintiff's Excuses for not Looking," "Rule in Federal Courts," "Rule in New Mexico" and "Diversion of Plaintiff's Attention," *supra,* which argument and decisions are applicable here, and by the following additional cases to which we invite the court's attention.

In the case of Northern Pacific Ry. Co. v. Alderson, 199 Fed. 735 (Ninth Circuit), plaintiff and her husband were riding in a wagon being driven by the latter. Owing to certain obstructions to the view at the crossing at which plaintiff was injured the wagon was stopped within 20 to 25 or 30 feet of the track and plaintiff and her husband listened for a train; hearing none, they then drove upon the track. Plaintiff looked continuously from the time they stopped until they got onto the track.

While this court in that case held that under the facts there shown, the question of plaintiff's negligence was for the jury, it is apparent from the opinion that such holding was predicated solely upon the evidence that the wagon was stopped at such a short distance from the track, for the court, after referring to the obstruction along the track said:

"That imposed upon them (plaintiff and her husband) the precaution of *stopping* within a short distance of the track and listening for an approaching train." (Italics ours.)

In the case at bar it is undisputed that plaintiff and his father did not stop after leaving a point 100 feet from the crossing, and we say therefore, that under the physical facts and all the circumstances shown in this case plaintiff and his father did not "stop within a short distance of the track" and that therefore they did not comply with the rule of this court announced in the foregoing case.

This court, in the Alderson case, further employed the following pertinent language:

> "It is undoubtedly true that travelers upon the public highway, approaching a railroad crossing where passing trains are to be expected, are required to use their senses, of both seeing and hearing, to detect the approach of such trains, and that, when the track is obscured to the sight, greater care is devolved upon them in the use of their sense of hearing, because the capacity for detecting the danger has been diminished."

According to plaintiff's own testimony in the case now before this court, plaintiff's hearing was interfered with by the noise of the car and his view, to the west, was obscured by a cut in the railroad; and yet he did not once look toward the east while travelling the entire distance of 100 feet immediately before reaching the track.

In Griffin v. San Pedro etc. R. Co., 170 Cal. 772, the court said:

> "A person approaching a railway track, which is itself a warning of danger, must take advantage of every reasonable opportunity to look and listen.

(Citing cases.) A traveller who is about to cross a railway track at a place where ordinarily the engineer gives appropriate signals of the approach of the train, may not depend upon such signals. 'He has no right not to look or listen because he has heard no such signals.' " (Citing cases.)

To the same effect is the case of Chicago etc. R. Co. v. Pounds, 82 Fed. 217. It was said by the court in that case, at page 218:

"The doctrine is too well settled to admit of controversy that a person is guilty of culpable negligence if he walks or drives upon a railroad crossing in close proximity to an approaching train, which is in plain view, and might have been seen for a considerable distance before he reached the track. The precautions which a person travelling upon the highway must take when he approaches a railroad crossing are so well defined that it is no longer the province of a jury to decide whether such person was guilty of negligence, in those cases where it is obvious that in approaching the crossing he failed to look *up* and *down* the track as he might have done, and thereby avoided all risk of injury. It is universally conceded that a person omits not only a reasonable but a necessary precaution when he drives upon a railroad crossing, at a place where his view is unobstructed, without looking along the track with sufficient care to ascertain with certainty whether a train is coming *from either direction. A railroad track is in itself a warning of danger, because trains may be expected to pass at any moment.* Therefore the courts have repeatedly declared that a person is, as a matter of law, guilty of contributory negligence if he drives upon a crossing without making a vigilant use of his senses of sight and hearing. If either of these senses is impaired, or for any reason cannot be exercised to advantage, he ought to be more vigilant in the use of the other." (Italics ours.)

SIXTEENTH SPECIFICATION OF ERROR.

The instruction set out under the sixteenth specification of error related to the speed of the train involved in the collision, and, we maintain, that it was applicable to the facts in this case and that the court's refusal to include the instruction in the charge to the jury was prejudicial to the defendant.

The evidence shows, without conflict, that the train was one of defendant's transcontinental passenger trains being run in interstate commerce between states; [201, 202]; that the crossing was an ordinary country crossing; [41, 42, 44, 45] and unpaved, as shown by plaintiff's Exhibit No. 2. There is no evidence in the case to show that the speed of the train was in excess of 50 miles per hour.

That, under the evidence, the instruction was a proper one and one which defendant was legally entitled to is shown by the following, among other, decisions:

> Seaboard Air Line R. Co. v. Blackwell 244 U. S. 310; 61 L. Ed. 1161;
>
> Larrabee v. Western Pac. Ry. Co., 173 Cal. 743.

SEVENTEENTH SPECIFICATION OF ERROR.

The instruction to which the seventeenth specification of error relates ought not, it seems to us, require argument or citation in its support. Its purpose was to tell the jury that even though they found that the automobile in which plaintiff was travelling was in bad repair or deficient, that fact was not to be considered in any way *excusing plaintiff from exercising due care and circumspection* for his own safety in crossing the track. (Italics ours.) On plaintiff's theory of the case, the condition of the car was an extremely important factor. And we therefore contend that as the requested instruction is good in law, and applicable to the facts of this case, and was not covered by any of the instructions given, the court's refusal to give the instruction in question was highly prejudicial to the defendant.

III.

Instructions Given by the Court.

Under this heading, we will deal with the eighteenth and nineteenth specifications of error.

The instruction set out under the eighteenth specification of error told the jury that under the conditions and circumstances therein stated, it is the duty of a railroad company to construct a crossing under its rails or carry the road above its tracks by a bridge, failure to do which renders it liable for injury or damage resulting therefrom to one rightfully using the road.

We submit that, even conceding, in the first place, for the sake of argument, that the instruction contains a correct statement of the law as an abstract proposition, we say that it was not warranted under the evidence in this case. Moreover, we submit, as shown by our argument and the cases in support thereof under the thirteenth specification of error, which argument and cases are clearly applicable here, the instruction did not contain a correct statement of the law.

We say that in view of the absence of any law in the State of New Mexico requiring grade separation at highway railroad crossings, and in view of the foregoing decisions of the courts of other states construing laws similar to that of New Mexico regulating the manner in which such crossings shall be constructed, the court committed prejudicial error in leaving to the determination of the jury the question of whether defendant was or was not negligent in failing to separate the grade at the crossing involved in this case. And this error is all the more apparent when we consider that the evidence not only does not show that defendant failed to comply with the law of the state of New Mexico, when building its line of railroad across the highway, but, on the contrary, the evidence clearly and undisputably shows that it fully complied therewith. And that, we submit, was all that could legally be required, and we further submit that a jury should not be permitted the opportunity of requiring more.

This court will take judicial notice of the fact that the state of California has a railroad commission, and that many other states have similar bodies created by law, vested with authority to say when, where and in what manner grade separation shall be accomplished. There are thousands of highway grade crossings throughout the country, similar in character, and no better or no worse than the crossing involved in this case, and we submit that the law does not and should not leave the decision of the public necessity and/or feasibility of separation of grade at a given point to the uncertain and varied whims and prejudices of juries.

Since laws, both federal and state, regulate and control the railroads in the matter of rates, income, construction of new and abandonment of old lines, issuance of stocks and bonds; in short, regulate and control them in their every essential activity, it is, therefore, we contend, but just that the determination of the question of the necessity and advisability of making large expenditures for grade separation should be left to the body through which the state exercises such regulation and control, and not to juries.

That the usual heavy additional cost of grade separation must, both in theory and in fact, in the last analysis, be paid by the shipping and travelling public in freight rates and passenger fares is recognized by the enactment of such laws as the states of California, Arizona and others have on the subject.

In the court's instruction to the jury set out in full under the nineteenth specification of error, the jury was told that when certain physical conditions surround a railroad crossing, it is the duty of a railroad company to use extraordinary care and additional precautions such as automatic signalling devices or other like methods to give warning of the approach of trains or to approach such crossing carefully and with the train under such reasonable degree of control as would enable them to discover and avoid injury to persons.

The vice in this instruction, even assuming that it is a correct statement of an abstract legal principle, is that, like the instruction last hereinabove discussed, it was not applicable to this case, and, like the other instruction mentioned, was essentially harmful and prejudicial to defendant, for the reason that no evidence was offered to substantiate the allegations of the complaint that the crossing created a condition of extraordinary danger or that persons using the highway would not be warned by the usual and ordinary measures or discover by the exercise of reasonable care the approach of trains in time to avoid collision.

On the contrary, as shown by the evidence, the crossing here involved, (so far, at least, as the character of the view to the east, and plaintiff's opportunity, in the exercise of reasonable care and prudence, to see a train in ample time to avoid injury, are concerned) did not fall within the category of crossings to which the instruction here complained of purported to apply.

Furthermore, we do not concede that the instruction contains a correct statement of the law in any case involving a country crossing such as we have here. As authority for this view, we refer the court to the very recent case of Henry v. Boston & M. R. Co., 134 A. 193 (Maine, 1926.)

That case involved a collision between a motor truck and an express train at a crossing. The defendant was charged with negligence in two particulars:

First, that it ran its train in a reckless, careless and negligent manner; and

Second, that it maintained no guards or warnings at the crossing.

In support of the former charge, it was shown that the train was moving at a high rate of speed, that the day was foggy, and that the crossing by reason of obstructions to vision, was a peculiarly dangerous one.

In disposing of the first charge, the court, among other things, said:

> "In running its train at forty miles an hour at the time and place of the accident the defendant

company was doing no more than its duty to its passengers and no less than its duty to the plaintiff's intestate, a traveller upon the highway."

As to the second charge, the court, after saying that the ordinary crossing signs were in place, but that there was no flagman or automatic signal, said:

"In the absence of official mandate that the crossing be so guarded, the absence of such flagman or signal does not, under the conditions shown to exist in this case, prove negligence."

The court then refers to a Kansas case, Cross v. Railroad Company, 120 Kan. 58, 242 P. 469, and quotes approvingly therefrom as follows:

"Cases may arise where the speed of a train may be considered by a jury in connection with the location and other surrounding circumstances upon a question of negligence. In densely populated districts such as towns and cities public safety requires the speed to be moderated. This crossing as we have seen however is in the country where there was no statutory or municipal regulation with respect to the speed of trains. In such cases there is no limit upon the speed at which trains may be run except that of a careful regard for the safety of trains and passengers."

The Kansas case disclosed the running of the defendant's train at the rate of 45 to 60 miles an hour over a country crossing not guarded by flagman or automatic signal.

In the case now before this court, as in the Cook case, defendant maintained the standard railroad crossing sign at the crossing. This is plainly shown on plaintiff's Exhibit No. 2.

So, we submit, that in the absence of any law in the state of New Mexico requiring railroads to maintain automatic signalling devices or other like methods to give warning of approaching trains, defendant cannot be charged with negligence in failing to maintain such instrumentalities as those in question.

IV.

Verdict.

Under this heading, we will consider the 20th, 21st, 22nd, 23rd, 24th, 25th, 26th, 27th and 28th specifications of error.

These specifications are, for the reasons therein stated, each and all predicated on and complain of the court's acceptance of the verdict and the entry of judgment thereon.

What we have hereinbefore said with relation to the First, Second, Third, Fourth, Fifth, Sixth, Seventh, Eighth, Ninth, Tenth, Eleventh, Twelfth, Fourteenth and Fifteenth specifications of error, and the cases we have cited in support thereof, apply with equal force to the specifications immediately under consideration. Hence, to avoid repitition, we here but add that the instructions set out in the 21st, 22nd, 23rd, 24th, 25th, 26th, 27th and 28th specifications are based upon and apply fully to the evidence, and it is obvious that had the jury followed such instructions, as, considering the evidence, they were bound to do, the verdict must necessarily have been for the defendant. Accordingly, we say that, in view of the evidence, the court erred in accepting the verdict and entering judgment thereon.

Summary.

Here, then, as is conclusively shown by the evidence, we have the case of an intelligent young man who, we submit, despite the flimsy and frivolous excuses tendered, knowingly and, not only heedlessly but recklessly, brought about his injuries by approaching and going upon a railroad track in the face of an on-coming train, with the view unobstructed for at least a quarter of a mile and without stopping and without looking at any point at which looking and listening would have been effective, for his protection. If, under such circumstances, the traveller is not, as a matter of law, guilty of contributory negligence, it would, we submit, be utterly impossible to conceive of such a case.

We, therefore, respectfully submit that the trial court erred in refusing to direct a verdict in favor of the defendant; in giving and refusing to give certain instructions as hereinbefore set out; in accepting the verdict of the jury and entering judgment thereon; and that said judgment should be reversed with directions to set it aside and render judgment in favor of the defendant.

Respectfully submitted,

E. W. CAMP,
M. W. REED,
ROBERT BRENNAN,

436 Kerckhoff Building,
Los Angeles, California.

CHALMERS, STAHL, FENNEMORE & LONGAN,
Phoenix, Arizona.
Attorneys for Plaintiff in Error.

E. E. McINNIS,
Of Counsel.

(Defendants' Exhibit A—Photo.)

In the

United States
Circuit Court of Appeals,
FOR THE NINTH CIRCUIT.

The Atchison, Topeka and Santa Fe
Railway Company, a corporation,
Plaintiff in Error,

vs.

Roy Spencer, an infant, by Sarah E.
Spencer, his guardian ad litem,
Defendant in Error.

BRIEF FOR DEFENDANT IN ERROR.

O. N. Marron,
Francis E. Wood,
Albuquerque, N. M.
Attorneys for Defendant in Error.

VALLIANT PRINTING CO., ALBUQUERQUE, N. M.

INDEX OF TOPICS

TABLE CASES CITED

TABLE CASES CITED (Continued)

No. 5038

THE ATCHISON, TOPEKA AND SANTA FE
RAILWAY COMPANY, A CORPORATION,
PLAINTIFF IN ERROR,

VERSUS

ROY SPENCER, AN INFANT,
BY SARAH E. SPENCER, HIS GUARDIAN AD LITEM,
DEFENDANT IN ERROR.

BRIEF FOR DEFENDANT IN ERROR
STATEMENT

The pleadings, issues and result of the trial are sufficiently stated in the brief of the plaintiff in error, but we
are not content with their statement of facts. For convenience we will also use the terms "plaintiff" and "defendant" to refer to the parties as they were in the trial
Court. Figures in brackets refer to the printed transcript.

FACTS

On June 11th, 1923, the plaintiff, a boy of 17, was struck
by defendant's train at a highway crossing about two miles
west of the village of Mountainair in central New Mexico,
and suffered injuries which will totally disable him for
life. (160).

The highway was a very old and apparently much travelled road (41, 44, 181) leading from Mountainair westward some forty miles to the Rio Grande Valley. The
railroad was built about twenty years ago. It crossed the
old highway line in the same general location as now. In
the plains country of New Mexico, unfenced and uncon-

fined, a road may have been "in the same place" though rods removed from the exact spot it now occupies; and we believe the evidence does not sustain the defendant's claim (Brief pg. 110) that the old road crossed their track line in the identical spot as now.

We submit the evidence warrants a finding that defendant changed the road line and constructed a "cheap" crossing at a point and manner to create the maximum of danger and hazard to the public, when a perfectly safe crossing could have been constructed at trifling cost where the road ran. The old road followed a low valley or "draw" between two hills, at this point something less than 500 feet apart. (180). The witness *Chavez* had known the road since 1873. Testifying by deposition (through an interpreter) he said (45):

"Q. Do you remember, more or less, whether that road ran up near the railroad crossing before they built the railroad?

"A. *The valley* that goes on the *other side of the hill* I know where that road runs, now, from here—*the valley* here up to the crossing where Spencer was killed and from here up to there it follows, *more or less* the line of this old road I was speaking of. * * * and *on down the valley* crossing the railroad, and on down." (Italics ours).

The witness *Sedillo* testified (42):

"The old road crossed right in the same place, only it crossed kind of west the old road; and this new road crosses right straight south. * * * *there is a mark at or near the crossing* outside the present line to show where the old road ran. * * * *It shows the wagon track inside the fence of the railroad* right of way on the south side of the crossing and to the west of the road." (Italics ours).

The marks referred to by the witness appear on the photograph "Exhibit 3" identified by the witness Spencer

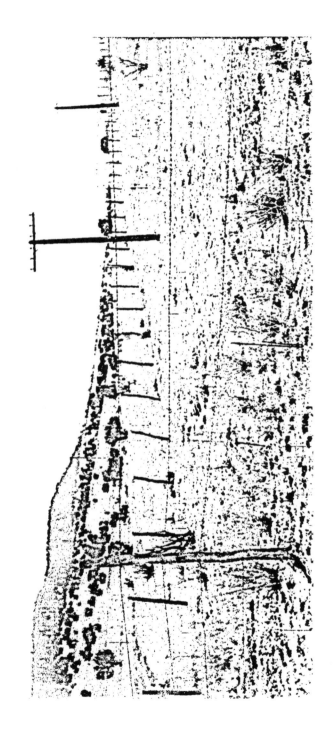

(60.) On the opposite page we have inserted an enlarged print from the film or negative from which Exhibit 3 was printed. It shows Exhibit 3 as it would appear under a strong reading glass. This picture shows the view looking west along the railroad right of way at the telegraph pole on the road leading up to the crossing. It shows, though somewhat vaguely the line of the old road running near the first fence post and second telegraph pole "just inside the corner" (59-60) made by the right of way fence and the fence leading up to the crossing. The track clearly appears *following the valley close to the foot of the hill,* more or less paralleling the bottom of the hill and the railroad, and the line of the new road down the valley outside the railroad fence.

The witness *Luna* said (46):

"The road followed the lower part of that section generally as it passed thru."

The witness *Noble,* defendant's engineer, testified (174):

"That draw which takes the general direction that the road approaches, and which the car approached, continues right on thru under the tracks and right up beyond in the direction of Mountainair."

Speaking of the road as now existing the witness said (178):

"The railroad is located down Abo Canyon (we have called it 'draw') and this particular point is on the north side of the canyon. Now, on this hill there will be little projections or noses sticking out from the side of it, and that is the actual contour at this point—indicated right here on the track profile—see? See this little knob there? The road crosses on that knob. * * * The road comes up the draw until it gets to the point on this knob which extends down from the north, and then it climbs up the knob *to bring itself up to the* *level of the railroad track.* (Italics ours) * * * and then goes on over the knob on the other side of the track, and then more or less paralleling the track, comes down to the level of the draw that is shown in that profile.''

The profile on the map Defendant's Exhibit A (appended to defendant's brief) shows that the ''draw'' extends right up to the crossing on the east as it now exists, (the picture exhibit 2 in defendant's brief shows the same), and is deepest immediately adjoining the crossing and within the limits of the original road way if extended at the same angle as it approached the point.

Notwithstanding this *demonstration,* the defendant produced the witness *Block,* who testified in substance (186-7) that the road, before the railroad, came up the bottom of this draw to the point where it now turns and crosses, *then left the draw as it does now, climbed the knob to the same level it does now, and then turning back as it does now,* after crossing at the identical point as now returned to the draw and continued on its way, as now. Taking Block's testimony as gospel, even against their own engineer, the pictures, and common horse sense to the contrary, defendant premises as follows in their brief at page 110:

"As shown by the undisputed evidence, the defendant, when constructing its railroad across the highway installed the crossing *at the exact place* where the rails intersected the highways as the latter then existed.''

and then founds a major charge of error on the basis of facts so assumed to be ''undisputed.''

When defendant built its road it constructed a bridge in this draw about 350 feet from the crossing with an opening 6 feet wide by 7½ high (174). It was intended to allow the water falling in the drainage area north of the road to pass thru (19). Rainfall in that country is light, there is no visible stream bed there, and enough water has not passed thru since the road was built to make a showing or disturb the grass or verbage in its direct path below the culvert or bridge (63-4-179 Exhibits 6 and 7). The railroad grade in the draw is at least 12 feet above the surface, the profile in the map shows more. Move this bridge a few yards west, or the road a few east, double its opening and you have a perfectly safe crossing. It is common knowledge that in that country even streams carry water only a few hours in any year, and crossings on shallow water courses are commonly constructed by laying a concrete pavement in the bottom to prevent washing. The one opening here could have been made and used to perfectly accommodate water and road at a trifling additional cost without altering the grade of either road or railroad. That would have perfectly "restored the road to its former condition of safety and usefulness." But it was cheaper to carry the road up the abrupt natural knob across in the death trap right at the mouth of the deep curving cut into the hillside. And there the crossing was built.

Having thus created a death trap, the company took no precaution whatever to lessen the hazards they themselves had created. Automatic signals at less dangerous crossings have been in common use by that company for years (109-10-14) only a few miles west in the Rio Grande Valley and are designed for use at just such crossings; but none were located there.

Then, to cap the climax, the defendant's train crew in charge, ran their train out of the near-by cut and over this crossing, without any signal or warning, (48,100) at a speed so great as to frighten passengers accustomed to railroading, riding in the train (47,100) and well calculated to produce the tragic result—swoop unperceived on the highway travellers.

The plaintiff, a farmer boy of 17, living at home with his

parents (52) at the request of his father accompanied him to assist on an errand to a village some miles west. The boy knew and trusted his father to be an unusually careful man in driving and watching out for danger (82). The Ford truck which the father was driving became out of condition during the trip (54) and the boy was directed by his father on at least two occasions to get out and push the car in climbing small elevations; and it was necessary to do so to keep it in operation (53). When they came to the railroad crossing they stopped at the foot of the grade at the right of way, looked and listened. There was no train in sight, or hearing. The boy then went behind the car to help push it up the grade. He testified it was necessary to exert, and he did exert, all his strength (65-6). He pushed with both hands against the back of the car and his body stretched out behind in the position to exert the greatest power. Half way up the grade (68, 98) he slipped quickly around to the right side of the car, his left hand at the back corner and his right on the body toward the seat (86). This placed him with his back almost directly toward the approaching train. The testimony shows that a full minute was occupied in climbing the grade from the edge of the right of way to the tracks (68), a distance of one hundred feet. The last ten or twelve feet of this distance, to a person in plaintiff's position, the view to the east was obstructed by a cattle guard fence, shown in Exhibit 5, slanting to the edge of the ties (82).

A person occupied as was the plaintiff in order to look in the direction the train was coming would be obliged to temporarily cease pushing and turn about (69, 817). With the help of the last ounce of pushing energy the lad could exert the car moved at a snail's pace on the steep incline. To relax the force meant the car would stop, could not be started again on the grade and would have to slide back to the bottom of the hill for a new start. He testified: (86 9)

"A. I pushed just like that (indicating) of course, I went as slow as I could, pushing as hard as I could. From the time I went around the end of the car, as I was pushing up the hill on the side of the car, I was facing to the west. I expect we were half way up the hill, or from the place where we started, when I changed my position from the rear of the car to the side, I should judge it took a full minute to get the car to move from the point where it was at rest from the time it started at the bottom of the hill, until I got up on the track. From what I know of that car, and the way it was running, I know what would have happened if I had ceased for a minute or an instant to push; it would stop. From my knowledge of the grade, I know whether or not stopping on that grade it would have been started again without going back to the bottom of the grade; it would have run back to the bottom of the grade, if it stopped, undoubtedly. I did not cease for an instant to push from the time that car started until it got up on the track. From the position that I was in while I was pushing that car at the side of it, I could not have turned my head around and looked behind me in the direction in which that train was coming. In order to have looked in the direction in which the train came at the time—from the time I came to the side of the car, until I came up on the track—I would had to have taken this arm down and look and quit pushing and the car would have stopped and would have run down and it would have had to. * * * I ceased to push when the front wheels of the car had just went over the first rail. At that time, I was beside the car and I stepped on the rail—on the running board, rather, and stepped in the car and turned around. My purpose in stepping in the car just when ceasing to push, stepping on the fender, was because it was going over the rails down the other hill. The car was in motion. From the time I raised up and ceased to push it was just a second or so before I got a view of the train; just as quick as I could step in and look, step in and face to the west, when I turned around and throwed my head down the track.''

As the boy and his father approached the crossing and when they were about 500 yards away a freight train passed over the crossing going toward Mountainair at about twenty miles per hour (55). They knew there was no place nearer than the station about 2 miles distant (57) at which a west bound train could pass (56). He said:

"It seemed like six or eight minutes, maybe ten, had elapsed from the time I saw the train pass the cut going toward Mountainair, and the time when I started up the grade. (72)."

The engineer testified that his train waited at Mountainair for this freight, or work train (217) which was blocking its track and, therefore, no doubt hurrying. When Mrs. White spoke to the conductor about the high speed they were making he replied they were fifteen minutes late (47-8). This condition well supports the plaintiff's claim that he feared and watched for danger from the west, and expected none from the east (81-2). It is apparent that whatever the actual minutes which elapsed, the passenger was waiting at the switching place for the freight, both were hurrying and the space was covered by both trains in much less time than any person could have reasonably anticipated, or expected. Yet defendant's counsel asks this Court to conclude that (Brief pg 36):

"The undisputed evidence shows the utter absurdity and fallacy of this contention."

The physical facts as to the movement and speed of both trains, and the reason for the speed in each case, are more eloquent and convincing of the time occupied than any witnesses' estimate of the minutes that elapsed.

So, watching the west, and unable to turn around and look east, the boy pushed with all his might until the front wheels crossed the rail, when straightening up, he stepped on the running board, looked to the east and saw the engine almost upon him. It hit him before he could collect his thoughts (71).

The mouth of the cut from which the train emerged was about 850 feet (170, 57) from the crossing. A stiff breeze was blowing in the direction of the cut, carrying, as is usual in that dry country, some dust and sand; not enough to conceal an object in the open, but ample to create a haze on distant objects (83) and hide altogether any object while in the cut (95). At the rate of speed testified by the witness White, above 50 miles an hour, (49) the train would cover the distance from the entrance to the cut to where the plaintiff saw it in less than 10 seconds; just about the time the boy's view would have been obstructed by the cattle guard fence (82). It is apparent that had he looked east all the way up the hill until opposite that fence it would have done him no good, the train would not have been in sight. Trusting to his father's known carefulness (82) and obeying his directions to push, lulled into some security from the east by the freight train that had just passed (55-6) on a single track, the nearest switch nearly two miles away, (57) and a condition of real danger presented demanding constant vigilance to the west where a train was hidden until you were right at the track and the train right upon you (81-2) a perfect case is presented of the boy's attention and watchfulness being necessarily and excusably fixed on something else and obstructed and prevented, or excusably distracted from the direction of the train.

POINTS AND AUTHORITIES.

POINT I.

The bill of exceptions shows upon its face that it does not contain all the evidence; and this Court, therefore, cannot consider any question dependent upon the sufficiency of the evidence.

Numerous exhibits, maps and photographs bearing directly upon the question of the defendants' negligence and the alleged contributory negligence of the plaintiff, were received in evidence. Copies of these exhibits are not included in, or made a part of, the bill of exceptions, but instead, the original exhibits are sent here pursuant to an order of the trial judge (393) directing:

"That defendant need not insert in its bill of exceptions, nor in the transcript of record, on writ of error, any of the above mentioned exhibits."

Rule 14, Subs. 3 and 4 of the rules of this Court provide:

"3. No case will be heard until a complete record containing, in itself and not by reference, all papers, exhibits, depositions, and other proceedings, which are necessary to the hearing in this Court, shall be filed."

"4. Whenever it shall be necessary, or proper, in the opinion of the presiding judge in any Circuit, or District Court, that original papers of any kind should be inspected in this Court, upon writ of error, or appeal, such presiding judge may make such rule, or order for the safe keeping, transporting, and return of such original papers as to him may seem proper; and this Court will receive and consider such original papers *in connection with* the transcript of the proceedings." (Italics ours).

Counsel has apparently proceeded upon the theory that original exhibits and records sent to this Court under the provisions of the preceding rule, take the place of the copies required to be in the bill of exceptions, and excuses the necessity of including them therein. It will be noted that this is not a case of including such exhibits in the bill of exceptions by definite and proper reference and annexation to the bill. Exhibits so annexed and referred to became part of the bill, and are thus within the certificate of the trial judge attesting its authenticity. In this case there is no certificate of the trial judge to the authenticity of any original exhibit except such as may be inferred from the Clerk's certificate, and their having been filed with him. Defendant sought and procured an order that these exhibits should not be included either in the transcript or the bill of exceptions, and they were omitted from both.

We submit that this condition of the record prohibits this Court from considering such exhibits. The purpose of transmitting original documents, or exhibits, to this Court is that the Court may inspect evidence like mechanical exhibits incapable of being included in a bill of exceptions; or documents like alleged forgeries, where a physical inspection of the exhibit itself is aidfull. It is not intended thereby to substitute original documents for what can and should be included in the printed transcript. The bill of exceptions is intended to, and should, contain all matters capable of being copied or transcribed, which are to be considered upon the appeal.

In the instant case the exhibits omitted here are as easily and readily copied and included in the bill of exceptions as are all other matters therein. In this day and age it is no more difficult to copy a map or photograph than a written or printed document. Indeed, in his printed brief counsel for plaintiff in error has easily and appropriately copied all the photographs and exhibits which he deems necessary to illustrate his part of the argument. But in order for his opponent, or the individual members of the Court, to have the use of the other exhibits which he does not copy in his brief, we must proceed to the Clerk's office in San Francisco and examine them one by one. The Clerk has quite properly omitted to put those exhibits in the printed record and would quite properly refuse permission to remove them from his office to be copied for use in making our brief. It must be manifest that if these photographic exhibits may be omitted from the bill of exceptions, and the originals sent in place to this Court, then any other exhibit or document printed or written, may likewise be omitted and the record sent here, as was actually attempted in Courier-Journal vs. Columbia Fire Ins. Co., 54 S. W. 966 as a trunk full of miscellaneous documents.

In Craig vs. Smith, 100 U. S. 226, substantially the same thing was attempted to be done. The trial Court there ordered:

"That the Clerk of this Court transmit to the Supreme Court of the United States the original exhibits, patent certificates, schedules, drawings and models on file along with and as a part of the record and transcript in this cause."

At that time it was provided by Statute as appears in the opinion:

"That either the Court below, or the Supreme Court, may order any original document, or other evidence, to be sent up, in addition to the copy of the record, *or in lieu of a copy* of a part thereof." (Italics ours).

The rules of the Supreme Court contained, as Part 4, Rule 8, a provision in the identical language of Sub. 3 of Rule 14 above quoted. The Court says:

"Construing this Statute in the light of those from which it was taken and the practice that had prevailed in the Courts, which it was, undoubtedly, intended to confirm, we think the power of the Courts below and of this Court, over the transmission of original papers to this Court on appeal, is and should be confined to such as require actual inspection as originals in order to give them their full effect in the determination of the suit. * * * The papers which have come up are what were used below as ordinary depositions and there certainly appears to be no good reason why they should not be copied into the transcript no complaint was made of their authenticity, and so far as any representatives have been made to us, there can be no possible necessity for their inspections."

It will be noted in the above case that the Statute expressly permitted the Court to:

"Order any original documents, or other evidence to be sent up * * * in lieu of a copy of a part thereof."

Yet, the Supreme Court, moved by the reasons we have attempted to outline, refused to consult them and held that only when forced by clear, definite terms of the certificate would they assume that the Court intended to make exhibits capable of copying and not required for physical inspection to be included in the record.

In Chicago Great Western Ry. Co. vs. Le Valley, 233 Fed. 384 (8th C. C. A.) the Court had before it a question closely allied to the one here considered. It was there sought to review the action of the trial Court refusing to direct a verdict in a personal injury case. The motion for a directed verdict and the order of the Court denying the same, were filed and entered by the Clerk upon his records and returned by him as a part of the record on appeal, but were not included in the bill of exceptions. The Court refused to consider the motion and order so appearing. It says:

"A federal appellate court considers only such matters as appear in the record. 'From time immemorial,' says the Supreme Court, 'That has been held to include the pleadings, the process, the verdict, and the judgment, and such other matters as by some statutory or recognized method have been made a part of it.' Metropolitan R. Co. vs. District of Columbia, 195 U. S. 322, 332, 25 Sup. Ct. 32 (49 L. Ed. 219). Exceptions to the rulings of the court in the course of the trial of a case, exceptions to the rulings of the court upon motions conditioned upon the evidence, may be made a part of the record by a bill of exceptions certified and signed by the judge, but the clerk of the court may not make them part of the record by writing them into his minutes or journal of the proceedings. In Young vs. Martin, 75 U. S. (8 Wall) 354, 357, (19 L. Ed. 418), the Supreme Court said:

'It is no part of the duty of the clerk to note in his entries the exceptions taken, or to note any other proceedings of counsel, except as they are preliminary to, or the basis of, the orders or judgment of the Court. To be of any avail, exceptions must not only be drawn up so as to present distinctly the ruling of the Court upon the points raised, but they must be signed and sealed by the presiding judge. Unless so signed and sealed, they do not constitute any part of the record which can be considered by an appellate Court.'

"The sealing of such bills of exception is no longer necessary, but the authentication of them by the signature of the judge is still required. * * *

"And neither the filing of a written motion founded on evidence, nor the recital by the clerk in his record of the proceedings of the trial of a case of such a motion, or of an exception to a ruling upon it, makes them a part of the record in the case, so that an appellate court may review the ruling. Thus in Dietz vs. Lymer, 61 Fed. 749. 10 C. C. A. 71, and Ghost vs. United States, 168 Fed. 841, 842, 94 C. C. A. 253, 254, this Court held that a bill of exceptions was indispensable to make motions to strike out parts of pleadings parts of the records. In Hildreth vs. Grandin, 97 Fed. 870, 872, 38 C. C. A. 516, 518, where an attempt was made to review an order on a motion founded on an affidavit and a judgment, this Court declared that:

'When a motion is presented to a trial Court which presents issues of fact for determination by that Court adduced by the respective parties, the action of the trial Court cannot be reviewed on a writ of error, unless a proper bill of exceptions, embodying the motion and the proofs, is duly settled, signed, and filed, so as to show to this Court, in an authentic form, on what state of facts the action of the trial Court was predicated.'

"In England vs. Gebhart, 112 U. S. 502, 503, 505, 5 Sup. Ct. 287, 288 (28 L. Ed. 811), the Supreme Court refused to review an order based on affidavits and the opinion of the Court, all of which were filed in the Court below and set forth in the transcript, and held that a bill of exceptions was essential to bring each of them into the record, although by one of the rules of that Court a copy of the opinion filed was required to be and was annexed to and transmitted with the record. The Court declared that the attachment and transmission of the opinion did not of itself make it a part of the record below and that 'the mere fact that a paper is found among the files in a cause does not of itself make it a part of the record. If not a part of the pleadings or process in a cause, it must be put into the record by some action of the Court.' To the same effect is Evans vs. Stettnisch, 149 U. S. 605, 607, 13 Sup. Ct. 931, 37 L. Ed. 866.''

In *Pauchet vs. Bujac*, 281 *Fed.* 962, the question was again before that Court. Defendant, in a law case being tried without a jury, sought to raise a question as to the sufficiency of the evidence. At the close of the trial he renewed a motion to dismiss made at close of plaintiff's evidence, stating to the Court that he would file it in form. The bill of exceptions contained that offer and statement which the Court treated as a sufficient renewal of the motion, but did not contain a statement of the Court's ruling or an exception thereto.

The record certified by the Clerk, however, did contain such a motion and an order signed by the Judge overruling it, and allowing an exception. It was held that this was not a sufficient showing to permit the Court to review it. They say:

"There is printed in the record a copy of a document filed with the clerk purporting to be a written motion of the defendant, made at the close of all the testimony, for a dismissal of the action on the ground of the insufficiency of the evidence to support a judgment in plaintiff's favor. Following this motion in the printed record is what purports to be a copy of an order, signed by the judge, overruling and denying the motion, with an exception by the defendant noted thereon. It is claimed by the defendant that these two documents constitute a bill of exceptions, which should be treated as supplementing the duly authenticated bill appearing in the record, and that they authorize this court to review the evidence under the assignment above quoted. We cannot concede this contention.

'It is a familiar and established rule of practice of
the federal courts that in actions at law a bill of exceptions, stating the ruling and the exception, settled
and signed by the trial judge, is indispensable to the
review of rulings upon motions, oral or written.''

''The papers above mentioned were never settled
and allowed as a bill of exceptions, or authenticated
by the trial judge as such, and, in addition to these
defects, they contradict the certificate of the trial
judge which is attached to the authenticated bill in the
record. That certificate recites that the bill of exceptions to which it is attached contains all the evidence and all other proceedings had at the trial of the
cause. The assignment of error based upon these un-
authenticated documents printed in the record cannot
be considered.''

In *Arizona & N. M. Ry. Co. vs. Clark,* 207 *Fed.* 817, this
Court refused to consider a deposition certified up by the
Clerk and printed in the record, because not certified in
the bill of exceptions.

In *Metropolitan Ry Co. vs. MacFarland,* 195 *U. S.* 322,
the transcript purported to contain certain instructions
asked and refused and marked ''Filed'' by the Clerk, and
certain other papers pertaining to the evidence stipulated
as correct by counsel; and while the facts are distinguishable from those at bar, the Court emphasized the necessity
of having such matters authenticated by the signature of
the trial judge and not by the Clerk of the Court.

In Dowagiac Mfg. Co. vs. Brennen Co. 156 Fed. 213, the
question was examined with some care by District Judge
Evans on an application to certify exhibits to the Circuit
Court of Appeals. Denying the motion he said:

''The inspection of the copies and the inspection of
the originals would be exactly as useful, the one with

the other. So that we conclude also that within the proper interpretation of the rule the inspection of the originals as distinguished from the copies would not be 'proper' because neither 'useful' nor 'aidful.' Besides it must have been the policy of congress, as well as of the appellate court, not to require original papers and records to be taken out of the custody of the Courts of original jurisdiction and of their clerks, and sent to distant points with all the attendant risks of loss and destruction unless there were some useful or necessary purpose to be subserved, in which event, of course, other considerations ought to give way. Familiar instances suggest themselves in this connection, namely, where there might be a conflict over hand writing or authenticity of documents or their age. Also, there may have been exhibited machines or models or other things, but these are provided for by the thirty-fourth rule of the Circuit Court of Appeals, and refer to 'material' exhibits. The least attention to that rule would show that it has no reference whatever to written evidence. In appeals it has always been intended that copies should be transmitted, and, as the record is usually printed for the Appellate Court, the judges of that court need not otherwise, and most probably would not otherwise, see the originals, and the danger of losing or mutilating important papers if they are sent to the printer in the originals would be very considerable.''

The question here presented was before the Supreme Court of New Mexico in Baca v. Unknown Heirs, et al, 20 N. M. 1. Construing a rule in the identical language of subdivision 4 of Rule 14 above quoted, the Court said:

''The second section of the rule was not designed to obviate the necessity of incorporating copies of ex-

· hibits in the transcript of the record, but its only purpose was to authorize the sending of original exhibits
to this Court, whenever, in the opinion of the district
judge, an inspection of the original paper would disclose some fact which could not be made to appear by
a copy thereof. Such, for example, as a comparison
of handwriting, an attempted erasure, alteration, authenticity of documents or their age. Many other familiar illustrations might be suggested.

"The use of such exhibits in this Court is only temporary, and, when they have been received and inspected, of course they will be returned to the files of
the district court, where they properly belong. They
are only withdrawn from the files of the district court,
by order of the judge of that Court, and sent to this
Court when in the opinion of the judge of that Court,
it is necessary and proper that this Court should have
before it the original exhibit or exhibits when it considers the case. Such exhibits do not become a part
of the files of this court, but remain, as is proper, a
part of the files of the district Court." * * *

"Without express statutory authority, no original
paper, document, or entry in a cause can be incorporated in the transcript filed on appeal in the Supreme
Court, but all papers, documents, and entries must be
copied into the transcript, and if any such original
paper, document, or entry is incorporated in the tran
script it will be disregarded. Mankin vs. Pennsylvania Co., 160 Ind. 447, 67 N. E. 299; Bottigliero et al
vs. Cozzi, 176 Ill. App. 311; Cornell vs. Matthews, 28
Mont. 457, 72 Pac. 975; Wallace vs. Coons, 48 Ind.
App. 511, 95 N. E. 132; Courier Journal Job Printing
Co. vs. Columbia Fire Ins. Co. (Ky.) 54 S. W. 966.
In this case the exhibits in question were not incor-

porated in the transcript at all, but were sent to the clerk of the Court in a box.

"In the case of Leach vs. Mattix, 149 Ind., 48 N. E. 791, the Court said:

'It has been held by this Court that, in the absence of statutory authority, an original paper or document cannot be certified to this Court so as to become a part of the record. Goodwine vs. Crane, 41 Ind. 355; Reid vs. Houston, 49 Ind. 181.

"For the reasons stated, this court could not receive and consider the original exhibits certified to this Court in this case, even though we were not required to dismiss the appeal.''

If the rule be as stated in the foregoing cases, then it follows that it affirmatively appears in the bill of exceptions that very material evidence received upon the trial and pertinent to the issues that are now up for consideration in this Court, is not included in the bill of exceptions, but intentionally omitted therefrom; and as this Court can not consider evidence not certified in the bill of exceptions it cannot, in the absence of all the evidence, pass upon any of the questions raised in appellant's brief, which must necessarily result in the affirmance of this judgment:

United States vs. Copper Queen, etc. Co. 185 U. S. 495.

A. T. & S. F. Ry. Co. vs. Meyers, 63 Fed. 793.

POINT II

Under the evidence, the Court properly submitted the question of plaintiff's contributory negligence to the jury.

Defendant intimates that except for the Arizona Constitutional provision the trial Judge would have directed a verdict in its favor and, therefore, puts the gravamen of his argument on the proposition that it does not apply to the case at bar. The record does not bear him out in that contention. The trial Judge expressly refused to limit his ruling to that ground which necessarily implies that he thought that independent of the Constitution, the plaintiff's contributory negligence was for the jury. The record shows the following:

> "MR. REED. Am I right in my understanding, your Honor, that the reason or grounds for denying the motion is that it is based on the constitutional provision of the State of Arizona?
>
> THE COURT: I don't care to limit the reason for the ruling. That is one of the reasons." [290].

After the verdict, the defendant moved for a new trial on the ground, among others, that the evidence did not justify the verdict (323) which the Court denied (373).

We will first consider the evidence independent of the constitutional provision. By the weight of authority the measure of care and diligence required of a guest, passenger, or child, accompanying the driver of a vehicle, to watch out for trains at crossings is quite different and not so exacting as that required of the person driving or controlling the vehicle. Defendant in his brief insists they are the same, at least in New Mexico, whose decisions he contends must govern the plaintiff's rights in this respect. The question has never been before the New Mexico Supreme Court and no rule exists, or has been established in that State on the subject. This Court and the trial Court are, therefore, not bound, or hampered, in any event by a New Mexico rule on that subject.

We think counsel in quoting authorities from other jurisdictions overlooked the fact that this Court is already committed to the rule as above stated by us.

In Southern Pac. Co. vs. Wright, 248 Fed. 261, the deceased was riding with one Tucker, a competent chauffeur assigned to him to demonstrate a truck. Tucker, with

Wright sitting beside him, drove on the track under circumstances quite clearly constituting negligence in not looking for a train in plain view. Holding that the deceased
was not chargeable with Tucker's negligence the Court
continues:

"Was he himself guilty of negligence? Can we say
as a matter of law that he failed to use the degree of
care for his own safety which an ordinarily prudent
person under like circumstances would have exercised?
If so, in what respect was he careless? What did he
leave undone, that he should have done? We must bear
in mind that there is no evidence that he was an experienced driver; fair inferences are to the contrary.
Insofar as we are advised, Tucker was competent, and
by Wright was believed to be competent, to operate the
truck skillfully and safely. As was aptly said by Judge
Marshall, in Pyle v. Clark (C. C.) 75 Fed. 647:

'It is a matter of common experience that passengers
in a vehicle trust to the driver to avoid the ordinary
dangers of the road, and I do not know of any principle
of law which requires them to tender advice, unless
conscious of the driver's ignorance or want of care.'

It is not a case where the passenger, knowing the
danger, voluntarily takes the risk, as where he drives
in the night time over a perilous road or without lights.
There was no circumstance to warn Wright of danger,
or to suggest the need of assistance, or the advisability
of cautioning Tucker. True, the railroad tracks were
there, and they always warn of danger; and he may
have known that the passenger train was about due. But
until he had some reason to suspect that Tucker was
incompetent, or careless, or was unable safely to operate the truck, he not only had the right, but it was his
duty, to assume that he would not rashly or carelessly
go into peril. Generally it is the duty of the passenger

to sit still and say nothing. It is his duty, because any other course is fraught with danger. Interference, by laying hold of an operating lever, or by exclamation, or even by direction of enquiry, is generally deprecated; in the long run, the greater safety lies in letting the driver alone. And there is nothing in the circumstances here suggesting an exception to the rule. Tucker was sober, awake, and had the truck completely under con-control; and, so far as we know, he had shown no disposition to be reckless or venturesome. It was a bright morning, and the view of the track was unobstructed, so that in an instant he could glance either way and see an approaching train. There were no other sources of possible danger requiring vigilance, and plainly Wright's assistance as a lookout was not required. In short, it was a case where a reasonably competent, vigilant driver needed no assistance, and would in the long run be better off if left alone.''

Counsel cites liberally from the 8th C. C. A. and it must be admitted, that so far as concerns the driver, that Court has pretty well committed itself to the proposition that a person injured in a crossing collision can appeal only to God for redress. But even that Court relaxes the rule in favor of a passenger not in control of the vehicle.

In Pyle vs. Clark, 79 Fed. 744, after ruling that Pyle the driver was guilty of contributory negligence barring a recovery because he drove leisurely on the track without looking when the view was unobstructed for over 2000 feet Judge Sanborn speaking for the Court said:

''On the other hand, Wright was sitting on the south side of the wagon, and he exercised no control over the movements of the team. The wagon and horses were Pyle's, and he was driving them. It was his act of starting them forward upon the track without looking out for the train that came from his side of the vehicle that was the active, moving cause of the disaster. Wright was not responsible for this act. The negligence of the owner and driver of a vehicle cannot be imputed to one who is riding with him gratuitously, so as to defeat a recovery for an injury caused by the concurring negligence of the driver and the third person. Railway Co. vs. Lapsley, 4 U. S. App. 542, 2 C. C. A. 149, and 51 Fed. 174, 178, and cases there cited; Little vs. Hackett, 116 U. S. 366, 6 Sup. Ct. 391. It may be that a person of ordinary prudence riding with another under such circumstances as existed in this case would put a certain trust in the driver—would naturally expect that he would watch for the approach of danger from his side of the vehicle, and that he would not drive forward unless he was assured that there was none in that direction; and that in this way one might be lulled into some degree of security, and led to watch for danger from his own side, and be less cautious about its approach from the opposite direction than he would be if he were the driver. The question was whether Wright exercised such care as a person of ordinary prudence would have used under the circumstances of his case. We hesitate to say that the facts in Wright's case were such that all reasonable men, in the exercise of their deliberate judgment, must come to the conclusion that he did not exercise ordinary care. In our opinion, there was sufficient doubt about this question to warrant its submission to the jury. "

I cannot find that the Court has ever departed from or criticized this ruling.

The rule as laid down by the trial Court in his instructions to the jury is the rule substantially adopted and accepted in all jurisdictions and no exception was noted to it by the defendant in this case. He said: (306-7)

"It was the duty of the plaintiff Roy Spencer to use that degree of care that a reasonably prudent person would use under like circumstances surrounding the case. * * * What would be due care in that respect under one set of surrounding circumstances might be negligence under another. The fundamental rule and test is what should a person of ordinary care and purdence, situated as was the plaintiff, surrounded with all the circumstances and duties that he was, have done; and would the plaintiff's conduct measure up to that requirement. In determining whether or not plaintiff was negligent, it is for the jury to determine, under all the circumstances surrounding the plaintiff at the time, and you have a right to take them into consideration, and it is for you to say whether, under all of those circumstances a reasonably prudent person would have acted as did the plaintiff. If you find that such a reasonably prudent person, under those circumstances, would not have so acted, but would and should have discovered the approaching train in time to have avoided the danger, then your verdict should be for the defendant. If, on the other hand, you find that, considering all the circumstances surrounding this plaintiff, he used as high a degree of care as a reasonably prudent person would have used and conducted himself as such a reasonably prudent person would under all the circumstances in which he was placed at the time, he was not chargeable with any negligence which should deprive him of his right to recover."

The rule as stated is in general terms, and is applied to passengers, guests, or members of a family riding in a vehicle under control of another competent person. A strong current of authority supports the rule laid down by this Court in S. P. Co. vs. Wright, supra.

In the State of New York the rule is perhaps as severe as
in any State in the Union in negligence cases. Until re-
cently modified by Statute the burden was upon the plaintiff in all negligence cases to allege and prove, not only the
negligence of the defendant, but also that the plaintiff
himself was free from contributory negligence. Yet in
that State the rule is well settled that the duties of a guest
or passenger are much less exacting than those of the
driver.

In Turck vs. N. Y., 108 App. Div. 142, it appeared that
two boys were driving together in an automobile, and while
the railroad track was somewhat concealed, there was still
an opportunity for them to have looked and discovered the
train if they had done so at the proper place. The accident was witnessed by the occupants of a carriage just
ahead of the automobile. Both boys were killed and the
testimony as to their movements was supplied by the occupants of the carriage ahead of them. The Turck case
was brought by the administrator of the boy who was driving the car and the Court ruled that the evidence failed to
show that he was free from contributory negligence, but
indicated instead that he failed to look and listen at the
point where he must have seen the train had he done so,
and reversed a verdict in favor of his administrator.

In Sherwood vs. N. Y. C., 120 App. Div. 169; 105 N. Y.
Sup. 547, the Court considered the identical evidence in an
action brought by the administrator of the other party or
boy about 17 years old who was riding along side of the
driver, and sustained a verdict in his favor. Distinguishing
the Turck case, they said:

> "It is claimed, however, that, within the decision of
> this court in the Turck Case, the plaintiff's intestate
> was not shown free from contributory negligence. His
> tender years and entire want of knowledge as to an
> automobile, the fact that the machine was being driven

by an older person whom he knew was familiar with its operation and was fully able to manage and control it, and that Dr. Sahler, with his family, was driving in his carriage immediately ahead of him and was making the crossing, together with the entire situation at the crossing, present a case where it cannot be said as a matter of law that the plaintiff's intestate was guilty of contributory negligence, or that he showed less care than a lad of his years would ordinarily exercise under like conditions.''

In Terwilliger vs. L. I. R. Co., 152 App. Div. 168, the deceased was riding in the front seat of a roadster automobile when the driver ran up onto the tracks in front of a train and he was killed in the collision. He knew that they were approaching the railroad, it was in the day time, he was familiar with the locality; but, he had turned around as they approached the crossing and was engaged in talking with another guest who occupied the ''rumble'' seat in the rear. The Court said:

"There is nothing in the record to show that Dr. Terwilliger had any reason to suppose that Mr. Welsh was not a careful man, or that he had any reason to exercise any other degree of care than that which is incumbent upon all men, and that is that reasonable care which a reasonably prudent man should or would have exercised under the same circumstances. He was a passenger—a guest—in a car operated by a friend, and, while he should not close his eyes to an obvious or wellknown danger, he was not called upon to exercise any active vigilence to guard against a danger which was not known to him or which was not likely to befall one situated as he was in this car. He had a right to assume that Mr. Welsh, his friend, would exercise reasonable care in the operation of his car;

and unless he was aware of the railroad crossing, and had reason to apprehend that Mr. Welsh would run his car into a position of danger, the jury might properly find that he was in the exercise of that reasonable degree of care which an ordinarily prudent man would exercise under like circumstances, by merely sitting still in his seat and talking with a fellow passenger. That is probably what 75 per cent of the persons who go out for a drive with their friends do under the circumstances. They have no power over the car, no authority over the driver; and while they would not be free to ride with a reckless driver, knowing the fact, and charge their misfortune upon others, we do not think it can be said that there is a failure to produce evidence of a lack of contributory negligence where it appears, as it does here, that the plaintiff's intestate was sitting in his seat engaged in a conversation with a fellow passenger who was occupying the seat behind him, and who was leaning forward for the purpose of carrying on the conversation, leaving the driver free to manage the car. Of course, if the passenger was familiar with a known danger, if he was better informed of the circumstances than the driver, it might be his duty to watch and point out the danger, but here the car was being driven upon a flat land, in broad daylight, and at an angle with the railroad track, which had been crossed some distance back and which was to be crossed again at grade. It was an open country, and, assuming that the plaintiff's intestate was familiar with the country, which is the most favorable view for the defendant, it cannot be said as a matter of law that he was bound to anticipate that Mr. Welsh would drive upon the crossing without observing the situation and taking the necessary precautions. If there had been nothing to divert his attention, he might have been called upon to observe the situation in common with Mr. Welsh and to protest against taking any risks, but the evidence shows that he was engaged in talking with Mr. Rhoades, who was behind him, and he certainly was not called upon to be alert and active to see to it that Mr. Welsh did not drive upon a railroad track which was in his immediate view in the face of an oncoming train. True, the jury might have held that this was not that reasonable degree of care which the circumstances demanded, but we cannot say as a matter of law that taking the most favorable view of the evidence, the plaintiff's intestate was guilty of contributory negligence, nor can we say that there was not evidence in the facts and circumstances to justify a jury in saying that he was in the exercise of reasonable care.''

In Noakes vs. N. Y. C. 106 N. Y. Sup. 522, the deceased, a bright, intelligent girl of 16, was injured when the automobile in which she was riding was struck by the defendant's train at a crossing. Her father and an experienced chauffeur were riding in the front seat of the automobile. The girl was seated with two other passengers in the rear seat. There were a number of people at the crossing; it was near the railroad station, and a ball game was just over. The approaching train could have been seen a distance of 2000 feet away. The girl testified that as they approached the tracks she was seated in the back of the automobile doing nothing; not looking at anything at all, but was watching the people. She was also asked whether she did not look either to the right or the left along the railroad tracks to which she answered that she did not. After stating the general rule as to the duty of persons approaching the railroad crossing the Court said:

"But it seems to me evident that in determining in every particular case whether or not a failure to look or listen was negligence that contributed to the accident; the age, condition, and situation of the plaintiff, the existing circumstances, and the condition in which the plaintiff was as she approached the track, are to be considered in determining whether under the particular circumstances of the case a failure to look and listen was as a matter of law contributory negligence. It is clear that it is not in every case that a failure to look or listen would be negligence, as in the case of a passenger in a street car approaching a railroad track, where the car is entirely under the control and management of those charged with its management, or in the case of a very young child in a conveyance approaching the track; for, as I view it, it must not only appear that there was a failure to look and listen to constitute contributory negligence as a matter of law, but it must also appear that there was nothing in the age or condition of the person injured, or in the attending circumstances, which excused or would have rendered unavailing any knowledge that was acquired by the person injured.

"The plaintiff, a girl 16 years of age, was riding in an automobile owned by her father and controlled by her father's servant. No relation of master or servant or principal or agent existed between either the chauffeur who had control of the machine, or her father, under whose direction it was being operated. She was not in a position that she could give orders to either, and she was not responsible for the management or control of the conveyance. An examination of a few of the many cases which have discussed this subject will, I think, make it plain, considering the age and sex of the plaintiff and the circumstances under which she was riding in this automobile, that it was a question for the jury as to whether she was guilty of negligence which contributed to the accident.''

Both the cases last referred to were affirmed by the New York Court of Appeals, the latter in 195 N. Y. 543 and the former in 209 N. Y. 522.

In Howe vs. Minneapolis etc. RR. Co. 62 Minn. 71; 30 L. R. A. 684, the plaintiff was taking a gratuitous ride upon the invitation of one Pomeroy, the owner and driver of the team. The wagon was struck by defendant's train at a grade crossing. The Court says:

"We think that it would hardly occur to a man of ordinary prudence, when riding as a passenger with a competent driver, whom he had no reason to suppose was neglecting his duty, that he was required, when approaching a railroad crossing, to exercise the same degree of vigilance in looking and listening for approaching trains that he would if he himself had the control and management of the team. And our conclusion is that a court cannot hold, as a matter of law, that a passenger having no control over the team or its management is guilty of negligence merely because he does not exercise the same degree of vigilance in 'looking and listening' on approaching a railroad crossing which is required of the one having the control and management of the team. It is a matter of common knowledge that under ordinary circumstances passengers do largely rely on the driver, who has exclusive control and management of the team, exercising the required care when approaching a railway crossing, and we do not think that the courts are justified in adopting a hard and fast rule that they are guilty of negligence in doing so. Every case must depend upon its own particular facts.''

In Carnegie vs. Gt. N. Ry. Co., 150 N. W. 164, the same

Court, in a case where two boys were killed in a crossing collision, passing on the duty of a passenger, said:

"A person of ordinary prudence riding with another, upon his invitation, will naturally put a certain trust in his judgment, and will rely in some measure on the assumption that he will use care to avoid the ordinary dangers of the road. In order to conclusively charge a mere passenger with contributory negligence in failing to see an approaching train, something more than ability to see and a failure to look must be shown. His failure to look is evidence to be considered on the question of his negligence, but it is not conclusive against him. In general, the primary duty of caring for the safety of the vehicle and its passengers rests upon the driver, and a mere gratuitous passenger should not be found guilty of contributory negligence as a matter of law, unless he in some way actively participates in the negligence of the driver, or is aware either that the driver is incompetent or careless, or unmindful of some danger known to or apparent to the passenger, or that the driver is not taking proper precautions in approaching a place of danger, and, being so aware, fails to warn or admonish the driver, or to take proper steps to preserve his own safety."

In Glanville vs. C. R. I. & P. Ry. Co., 180 N. W. 152, the Supreme Court of Iowa, speaking of a collision in which the plaintiff rode in the back seat of the automobile with her husband at her right; her daughters, with a young man operating the car in the front seat, says:

"In the nature of things one in the back seat of an automobile will not be as vigilant as he would be were he in the front seat. He may see better from the latter and is likely to be more at attention than when in the back seat, and there is a degree of relaxation and reliance on the lookout of those ahead, and this is somewhat enhanced by the fact of being a guest. These are matters to be taken into account in connection with plaintiff's situation in the automobile, wrapped in a heavy coat with the sacks and vegetables about her, the curtains down together with her story of what she did and the lawful speed at which trains should move, and, as we think, left the issue open whether plaintiff exercised such a degree of vigilance as a person of ordinary prudence would under like circumstances.''

In Weidlich vs. N. Y. N. H. & H. Ry. Co. 106 Atl. 323, the Supreme Court of Errors of Connecticut, in the suit for damages for killing a person riding as a guest in the rear seat of an automobile controlled and driven by his brother, who was on the front seat, said:

"The intestate must exercise due care, but this is the care that may be reasonably inferred from the circumstances.

''The guest on the rear seat of the automobile owes a very limited degree of care. He is not expected to direct the driver, nor to keep a lookout. Dangers or threatened dangers known to him he must warn the driver of, and for his failure to do so be chargeable with having proximately contributed to the accident, unless a reasonable person under all circumstances would not have given the warning.''

In Chicago R. I. & G. Ry. Co. vs. Johnson, 224 S. W. 277 (Texas), the Court said:

-    ''We think the evidence tends to show that the plaintiff and her husband relied upon the driver to operate the car in a careful manner, and to take the necessary precautions to avoid accidents; it being a matter of common knowledge that guests in such cars usually so act. There is no testimony showing that Simmons was a reckless driver, and we cannot say as a matter of law that either plaintiff or her husband was guilty of negligence in relying upon him to keep a proper lookout for a train at the crossing.

"Accordingly, even tho it could be said, which it is unnecessary in this case to decide, that aside from the question whether or not Simmons violated the requirements of the statute quoted above, he was, as a conclusion of law, guilty of contributory negligence in failing to discover the approach of the train, and after such discovery to stop or so slacken the speed of his car as to avoid the collision, yet we cannot reach that conclusion as to either plaintiff or her husband."

Turning to the cases relied on by the defendant as establishing the negligence of the plaintiff as a matter of law, and especially the cases in the Federal Court, we respectfully submit that they do not establish, or countenance, the rule for which defendant contends as applicable to passengers, or guests; except such as are found to be in a position equally advantageous with the driver and unaccompanied by circumstances distracting their attention from the tracks, or approaching trains.

In the Bradley case, 288 Fed. 484, the deceased, though a passenger in the automobile, was seated in an open car with nothing to distract his attention and with the same opportunities of observing the train that were possessed by the driver. Even more, as the driver's attention would necessarily be somewhat distracted by the management of his machine. There were no facts or circumstances whatever to distinguish the guest's situation from that of the driver, except such as would be a greater burden on the guest. But, in the Bradley case the Court did not direct a verdict for the defendant, but, on the contrary, submitted the case to a jury who found the facts in favor of the defendant, and the Court is really discussing the question as to whether or not the evidence *warranted the jury* in finding the plaintiff negligent.

In the Noble case, 298 Fed. 381, the facts as stated by the Court

> "Mrs. Noble was riding in the front seat of the car, on the right side of her husband, who was driving. They were both more or less familiar with the road and crossing. It is undisputed that she was on the side from which the train approached and that there were no curtains to obstruct her view, or that of her husband."

She further testified that she was looking both ways for a train but did not see any. The Court very correctly rules that there was certainly nothing to distract her attention and nothing to obstruct her view within a reasonable space to have discovered the train and warned her husband, who was apparently relying somewhat upon her watchfulness. The Court followed the well established rule that the testimony of a witness that they looked and saw nothing when there was nothing to obstruct their view will not be credited if the train must have been seen had the person looked, as they testified to.

> In the Priester case, 289 Fed. 945 (4th C. C. A.)
> "The automobile was the property of Mrs. Priester. Priester drove it when he and his wife were together, but she often drove in his absence. Both frequently drove over the crossing and were familiar with it."

As they approached the track their view was entirely hidden by a field of tall corn. They neither stopped, listened, nor did they look after passing the corn field, although

"From the end of the corn field 33 feet from the track the view of the railroad was clear both ways and one looking down the track would see a train approaching from the south at least a half mile. * * * That neither of them looked south down the track at this point is evident from their own testimony and from the fact that they did not see the train."

The husband's attention was somewhat distracted by holes in the road at this point that he was endeavoring to avoid in driving the car. *But there was nothing at all to distract the attention of the wife* and the Court concluded:

"She, equally with her husband, was guilty of gross contributory negligence because she had the same opportunity to look out for the train and warn him of its approach."

The Dernberger case, 243 Fed. 21, was not the case of a guest at all, but of the driver and we need give it no further attention; though it falls far short of ruling the driver negligent if the facts bore any resemblance to those in the case at bar.

In the Sellers case, (5 Fed. 2nd 31), far from finding that the evidence established contributory negligence of Mrs. Sellers, who was seated in the rear seat of the automobile with two small children while her husband and Mrs. Burris, who was driving, occupied the front seat, the Court said:

"Considering all the facts, *the jury might be justified in finding* that Mrs. Sellers did not exercise the care for her own safety that the decisions of this Court impose upon a passenger in an automobile."

The case was reversed, not on the legal proposition of negligence, but for misdirection of the Court as to the law, on other subjects.

It results then that no decision in the Federal Courts is

cited on behalf of the defendant which detracts from, or modifies, the rule established by the case cited by us, applicable to the condition of the driver.

The Kanass case, 180 Wis. 49, cited by defendant, is also the case of a driver, with no question of the guest, or passenger, involved.

Lee vs. Davis, 247 Pac. 1094 (Mont. 1926) does give some countenance to the contention of the defendant here, and seems to be the only case ever decided in the United States that does. Even at that, it is quite distinguishable from the case at bar. There the injured man was seated on a load of machinery for the purpose of watching and guarding the machines. The truck had moved at a snail's pace over a distance of some 125 feet since he last looked at the tracks, and during at least one-half that distance he could have seen the engine. He was not distracted by watching the machinery, but, as he expressed it, about that time the lunch bucket of himself and the driver upset and spilled the contents in the bottom of the wagon and he was down gathering up and replacing the lunch. The Court seems to concede that if his attention had been distracted in watching the machinery it would have furnished a possible excuse, but they balk at permitting him to be distracted by his lunch. We think there is a flaw in their reasoning as to what a "reasonably prudent man" might have done under the same circumstances. It may be that in the exhilarating air of the Mountain State, with its health producing climate and robust well-fed people, no "reasonably prudent man" would for a moment allow his attention to be distracted at the prospect of merely losing his lunch, or some similar trifling cause; such for example as would occur if a sliver or nail, had penetrated his pants and was tearing them and clamored for attention and they were new pants. To make the illustration clearer let us assume that just at the crucial moment, when his eyes could and should have been on the approaching train his pant leg had caught on the hub of the wheel and he was about to be deprived of that necessary article if he did not give it immediate attention. We take it that in Montana a "reasonably prudent man" would have kept his eyes on the railroad and let the pants go hang. Certainly otherwise his Supreme Court would have told him with equal logic, that

"He was more concerned for the safety of his breeches then he was for the security of his life, or limb."

Perhaps after all the basis of that decision is to be found in the sentence last quoted from the opinion:

"No decided cases can be found where such a flimsy excuse was held sufficient to exonerate him from the imputation of negligence."

Counsel for the injured man must have neglected to call the Court's attention to such cases as the New York, Minnesota, Iowa and other cases cited by us in this brief, and the multitude of cases on similar facts to be found in the books. We are led to surmise that lawyers in Montana are as negligent in their search for cases as their Supreme Court would have a "reasonably prudent" hungry workman be in trying to save his lunch. Unlike the case of the person in charge of the vehicle, it is the distracting tendency or quality of a given instance rather than its seriousness that the Court takes into consideration in passing on the contributory negligence of a passenger or guest.

Turning to the New Mexico cases cited by defendant the same situation is found. None of them consider, or establish, any rule with reference to the amount of care requisite in a guest, or passenger, or what conduct, or failure to look, will establish negligence of such a party as a matter of law.

The Candelaria case, 6 N. M. 266, has no bearing whatever upon the question. The person injured and who sued was a trespasser on the railroad tracks not at a highway crossing at all. We need no citation of authorities that in the absence of some positive prohibition it is the duty of the trial Court to direct the verdict when the undisputed facts, or the only inference to be reasonably drawn from the undisputed facts requires a verdict in favor of one or the other parties.

In the Morehead case, 27 N. M. 349, the driver of the car struck sued the railroad company for the damages. Of course, it does not enunciate any rule which would apply to a passenger, or a guest. As against the driver the language of the Court appearing in the opinion probably carries the rule to an extreme as to what constitutes contributory negligence of the driver as a matter of law. But like many another opinion, until one knows the facts contained in the record, and incompletely or inadequately, stated in the opinion, one cannot know the exact limits of the rule that the Court is establishing and applying. The record in the Morehead case shows (in harmony with the facts contained in the opinion) that the plaintiff was driving along the right side of the street in the populous little city of Roswell, New Mexico. The railroad tracks cross this street at practically right angles in the midst of the built up portion of the town. The train which struck the plaintiff's car came from his right. The plaintiff testified that he stopped at a restaurant 57 feet from the tracks, for the purpose of getting a late lunch; found the restaurant closed, listened at that point, and looked for a train, but did not thereafter stop, look, or listen while driving the additional 57 feet on to the tracks when he was struck. The record, however, shows the additional fact that the restaurant building before which the plaintiff stopped his car, extended on in front of him and completely hid his view of the track, except for a very few feet below the crossing, at the point where he stopped; and he could not have seen the train from that point. These facts being added to the opinion would merely show that the Court was applying the well-nigh universal rule that the traveller does not perform his duty by looking at a place where he could not see and, ceasing to look between the obstruction and the track, when to have so looked when he had passed the obstruction would have disclosed the train and prevented the injury.

In the Sandoval case, 30 N. M. 343, also, the driver and not the guest complained. The real holding of the case was that no negligence of the defendant was proven warranting submission to the jury. The discussion on the question of contributory negligence demontsrates that as said by the Court:

"It is a worse case than the Morehead"
case. A high bank of earth obscured the crossing as the plaintiff approached parallel to the track, the road coming through this bank to cross the railroad at grade. This bank concealed all view of the track until they had passed it, and came to about 33 feet of the track, when the view became unobstructed. The evidence showed that the plaintiff did not look at all after passing this obstruction, but continued right on to the track until warned by one of his companions that the train was *then passing in front of him,* when only 10 feet from the track, and his automobile ran into the side of the train before he could stop it after being warned by his companion. It seems rather far fetched to cite the Sandoval case as throwing any light upon the question we are considering.

We submit that neither the federal cases, nor the New Mexico cases sustain the rule now contended for by the dedendant. To say that the condition of this boy established by the evidence does not distinguish his case from that of a passenger "seated supinely," as some of the Courts say, beside the driver; in a position as much, if not more, advantageous to discover the approaching train, is to close our eyes and our mentality to common sense. This plaintiff never has and could not operate an automobile (91). He was accompanying his father at his father's request on an errand. He knows his father to be a good driver, a careful, watchful man for danger and for trains. At his father's direction he gets out of the car at the foot of the railroad grade for the purpose of using, and he then continues to use, the very maximum of his strength to help push that car up the grade. His attention is necessarily distracted and concentrated upon the energy that he is using, and upon his task to keep the car in motion and get it over the tracks. His attention is again distracted and held by the dangerous cut to the west.

It is no answer to say that the train came from the other way. The condition so created commanded and compelled constant and continuous watch to the west and thus interfered with and prevented ordinary watchfulness to the east. An east bound train at that point would have been on him in *one second* after coming in sight, and counsel would then be here insisting that under the circumstances ordinary prudence required him to give his undivided attention to the west.

A train has passed some few minutes before going east through that cut and the way is clear from that direction for another train to follow it. A train from that direction as you approach the track from the position the boy was in, is hidden by the nearest edge of the cut and will be on you in an instant; a condition of danger created by the defendant itself. The train approaching from the east cannot be seen until it emerges from that cut 850 feet away. At the speed it is travelling it covers the distance to where the plaintiff saw it in less than 10 seconds, during which time he is pushing the car in front of the cattle guard fence which is between him and the train and would quite effectively blur, if not conceal, its approach even had he looked. If we are to give any credit at all to the evidence of the engineer (213-216) in which he stated that although he was occupying his seat on the engine with his eyes fixed on this crossing all the way from the whistling post, yet he did not see the automobile until after he had struck it, it must aid in demonstrating that during the greater part of that time the automobile was hidden behind this cattle guard fence. In no other way can we credit the engineer's testimony with any truthfulness. As the front wheels of the Ford went over the first rail of the track the boy ceased to push and with the same motion stepped up on the car; when turning his head he sees the train upon him! and as he states, the next thing he remembers is when they picked him off the cow-catcher to carry him into the car. That is probably the next thing that most any ordinary person, more especially a boy, unaccustomed to driving would remember.

It may be that some person placed as was the boy, performing the duties he was under the circumstances, might through some fortuitous circumstance have ceased to push and turned around and looked toward the east during this 10 seconds the train was in sight and thus discovered it. It might be that some person exercising the maximum of possible care when he ceased to push would have stepped back, let his father remain in the position of danger on the tracks and looked both ways for his own security as he stood on the end of the ties; and then as suggested by defendant's counsel, have chased the automobile down on the other side. But, if so, we respectfully submit that that would be the unenlargeable maximum of care that could be used by the most careful individual that can be conceived of; whose attention is fixed on railroad trains undistracted and unhampered by any circumstance. It is to this measure of care that the defendant asks this Court to commit itself and to enjoin on this boy, as a matter of law; and not to the measure laid down by the trial Court in language to which no exception was taken. The question is, can the jury say that no reasonably prudent, careful person situate in the boy's shoes; his age, his surroundings, his duties, his difficulties, his necessity to watch, to put his mind on the exertion of the last ounce of his power to propel the car; would have done as the boy did. On the contrary, must the jury find that no such person in his shoes would have failed to look for and to discover this train before stepping up on the automobile. We respectfully submit that no rule intended or intimated, by any Court; no facts appearing in any reported case will carry, or justify the application of the rule of contributory negligence to the point contended for by defendant's counsel here.

POINT III

The Arizona, and not the New Mexico practice governed the trial of this cause.

Defendant contends that the provision of the Arizona Constitution requiring that the question of contributory negligence be submitted to a jury does not apply to an action originating in New Mexico; but instead, the Court must try it by the same rules and methods that would prevail in New Mexico if it had been tried there.

Of course, if we are correct in our contention under Point II this question is entirely moot. However we submit that the rule is otherwise.

In 5 R. C. L. 1043-4 it is stated generally as follows:
"Among the matters which have been held to pertain to the remedy and consequently to be governed by the *lex fori,* are the mode of procedure * * * * . The rules of pleading and evidence; whether the determination of a particular matter is for the Court or the jury.

In Mass. B. L. Assn. vs. Robinson, 42 L. R. A. 261, (Ga.) the question was as to the materiality of the representations in the insurance contract. Under the laws of Massachusetts where the contract was made, those were treated as questions to be decided by the Court. Under the rule in

Georgia, where the case was being tried, the materiality
of the representations was a question for the jury. The
Court held that as this question was being tried in the
Georgia Courts it was governed by Georgia law and must
go to the jury. They said:

"The Courts of the State of Georgia will recognize
this contract as a valid contract, because it appears to
be such under the laws of Massachusetts, and is clearly
such under the laws of this state, but will give the
plaintiff and the defendant, respectively, for the purpose of enforcing it, on the one hand, or defeating it,
on the other, such remedies as are given to other persons who sue or are sued in the Courts of this state.
It is immaterial, therefore, for us to consider what is
the law of Massachusetts in reference to the tribunal,
or that part of the tribunal, that determines the materiality of the misrepresentations relied upon to defeat the contract of insurance which is the subject of
a suit in this state. These are questions which each
state is entitled to decide for itself, and to that end
erect tribunals, and lay down rules of procedure therein. The law of Georgia can declare what questions
shall be passed upon by the Court, and what questions
shall be passed upon by the jury. Persons seeking
either to enforce or defeat contracts made in another
state with citizens of this state, when they sue or are
sued in the Courts of this state, have no right to say
that the tribunal fixed by its laws is not satisfactory
to them, and to demand a tribunal erected in accordance with the law of the state in which the contract
is made. * * *

"Mr. Justice Story states the rule correctly when
he says: 'Whenever a remedy is sought, it is to be
administered according to the *lex fori*, and such a judgment is to be given as the laws of the state where the

suit is brought authorize and allow, and not such a judgment as the laws of other states authorize or require. Story Confl. L. p. 954, 573: De. a Vega v. Vianna, 1 Bran. 'Ad. 284; Whittemore vs. Adams, 2 Cow. 626. When a party comes into the Courts of this state to enforce his remedy upon his contract, that remedy will be enforced in accordance with the laws of this state regulating that remedy, and not according to the remedy provided for the enforcement of similar contracts in the state of South Carolina, although the contract may have been made in the latter state.' See also South Carolina R. Co. vs. Nix, 68 Ga. 572. As has been held by this Court in the case of Phoenix Ins. Co. vs. Fulton, 80 Ga. 224, that it was proper to submit to the jury the question as to whether or not a misstatement made in the application for a policy of fire insurance was material, and would have the effect of avoiding the policy, and this being so, as long as that decision stands, as the established procedure to be followed in such cases, there was no error in the present case in submitting to the jury the question of the materiality of the misrepresentation alleged to have been made by the insured in his applications for reinstatement.''

In S. F. & W. Ry Co. vs. Evans, 41 S. E. 631, a negligence suit, the injury occurred in Florida and the trial had in Georgia. Applying the rule laid down in the Robinson case the Court said:

"Under the law of this state, in the trial of cases of the character now under consideration, the question as to what acts do or do not constitute negligence is exclusively for determination by the jury, except in those cases where a particular act is declared to be negligence, either by statute or by a valid ordinance of a municipal corporation. See Railway Co. vs. Bryant, 110 Ga. 247, 34 S. E. 350, and cases cited; Railroad Co. vs. Vaughn, 113 Ga. 354, 38 S. E. 851. While the present case, so far as the right of the plaintiff to recover and the measure of damages in the event of a recovery were concerned, was to be tried according to the law of the state of Florida, and on these subjects the Courts of this state would apply the law of Florida in exactly the same way it would be applied if the case were pending in one of the courts of the state, our laws would, of course, control in reference to the procedure to be followed. It is immaterial, therefore, for us to consider what would be the practice under the law of Florida in such cases.''

In Johnson vs. C. & N. Y. Ry. Co. 59 N. W. 66, the injury complained of occurred in the State of Illinois and suit was brought in the State of Iowa. A statute of the State of Illinois was pleaded that where a train was operated at more than a given speed negligence should be presumed, and also the rule of comparative negligence prevailed in that State. On motion these allegations were stricken from the complaint. Sustaining this ruling the Court said:

"There is no question as to the right to maintain an action in this state. But the claim that the courts of this State must, in any case, adopt rules of practice of another state which pertain merely to the weight of evidence,—as that a fact may be presumed without evidence, or that a recovery may be had because the defendant was chargeable with a greater degree of negligence than the plaintiff,—can have no application to the trial of cases in our Courts, because they pertain to the remedy, and are in no sense causes of action. In Knight vs. Ry. Co. 108 Pa. St. 250, it is said: 'The statute of another state has no extraterritorial force, but rights under it will always, in comity be enforced, if not against the policy of the laws of the forum. In such cases the law of the place where the right was acquired or the liability was incurred will govern as to the right of action, while all that pertains merely to the remedy will be controlled by the law of the state where the action is brought.' ''

In Jones vs. Louisiana Western Ry. Co. 243 S. W. 976, the injury occurred in the State of Louisiana; the suit was in Texas. The Court said:

"The laws of this state must be applied in determining what Courts have jurisdiction of this case; the rules of evidence with reference to the burden of proof, the weight of the evidence, and whether a particular issue is one of law or fact; the respective functions of Court and jury in the trial thereof, and the nature and extent of review on appeal. * * * The issues of whether deceased, Walter Jones, in approaching said crossing and in attempting to cross the track used ordinary care to discover the approach of the train and protect himself from injury, and whether, before attempting to enter on the railroad track, he stopped, looked, and listened and exercised ordinary care in doing so, are to be determined according to the rules of evidence and under the established procedure in our Courts."

The authorities cited by the defendant to show that the quantum of evidence to establish negligence, or contributory negligence depends upon the *lex loci;* either do not go into that question with care, or do not so decide; and in many cases the matter the court confuses adjective and substantive law in its decision. Such, for instance, is the case of Delaware vs. Nahas, 14 Fed. 2nd. 256, cited by defendant at page 70. That opinion presents some very particular features. It assumes, or rather states:

> "The plaintiff's rights of action and the defendant's
> liability depend on the law of that state (New York)
> and to invoke the law of that state—distant from the
> trial forum—it must be proved as a fact."

In this case the defendant asserts the law and correctly,
in his brief on the preceeding page 68, that the Federal
Courts take judicial notice of the laws of all the States.
Next that case assumes, without arguing, or reasoning, and
states as a fact as one of the laws which follow the case
that

> "Quite opposed to general rule the plaintiff must
> affirmatively show his freedom from contributory neg-
> ligence."

They entirely overlook the rule stated by the Supreme
Court of the United States in the case of Central Vermont
Ry. Co. vs. White, 238 U. S. 507 at 512. After referring to
this rule that in many States, the plaintiff, to recover for
negligence, must, as a part of his case, show himself free
from contributory negligence under the State rule they say:

> "But the United States Courts have uniformly held
> that, as a matter of general law the burden of proving
> contributory negligence is on the defendant. The
> Federal Courts having forced that principle even in
> trials in States which held that the burden is on the
> plaintiff." Citing numerous cases.

It seems apparent that that case was not well argued, or
well considered, and is entitled to little, or no, weight as an
authority.

Cases brought under the Federal Employers' Liability
Act, of course, have no bearing. All the rights claimed, or
asserted, under that act are dependent upon the provisions
of the act itself and not to be varied by State laws, or
practices.

The case of Atchison vs. Wyer, 8 Fed. 2nd 30, rests upon

the same ground. The first Federal Employers' Liability Act was enacted while New Mexico was a Territory and, of course, all rights claimed under that law were subject to the same rules of construction as would apply to the Federal Act itself. This Act as originally passed and as applied to the territories was not limited to interstate commerce, but by its language included all railroad employees, whether engaged in interstate commerce, or not. The Supreme Court later held that law unconstitutional except as affected interstate commerce as to claims arising in States; but the law remained in full force so far as it affected the Territories and the District of Columbia and still remained a Federal Statute. That Statute was adopted and continued in force by express provision of the New Mexico Constitution, and continued after New Mexico became a State. Being adopted in its entirety the same rules were applied to its construction and enforcement after as before statehood. While it is true that the Court in the Wyer case said it was unnecessary to determine the question and it probably was, but judging from the opinion, the question now presented was not urged upon them, or considered in any manner.

In Hiatt vs. St. Louis, 308 Mo. 77, the suit was in Missouri for an injury happening in Arkansas. By the Arkansas Statute existing at the time of the injury it was expressly provided

> "Contributory negligence shall not prevent a recovery where the negligence of the person so injured, or killed, is of less degree than the negligence of the * * * railroad."

This, of course, means that contributory was no defence to the suit in Arkansas, or rather, it adopted the comparative negligence rule and abolished the defence of contributory negligence, except as so preserved. This unquestioned substantive law followed the controversy into whatsoever Court it might go.

In Morris vs. C. R. I. & P. 251 S. W. 763, the question as to whether the rules being considered were adjective, or substantive, was not in controversy; both parties apparently conceded that the entire law of Iowa, where the accident happened, followed the action and made no distinction between adjective and substantive law, and the case is not an authority upon the point.

In Caine vs. St. Louis, etc., 209 Ala. 151, the "curious slip" referred to in the brief at page 82 really explains the whole decision. The Court went upon the assumption that the Oklahoma Constitution abolished the defence of contributory negligence and substituted in place of it the comparative negligence rule that prevails in Arkansas. That such was the ruling of the Court appears plainly from the Kentucky case which they cite as covering "the direct question here presented."

The Kentucky Court was considering a cause that arose in the State of Tennessee in which State "contributory negligence did not bar a recovery."

If the "curious slip" referred to by counsel was a mistake of the Court, nevertheless, the rule announced by them was based upon the same mistake, which deprives it of all value as an authority here.

We submit, therefore, that none of the cases cited by counsel in his brief, rule that a Court trying a case for personal injuries suffered in another State is governed by *lex loci* upon the question of the weight of evidence, burden of proof, or whether or not any given case shall be submitted to the jury. All such questions, as we have shown, are matters of adjective law and governed by the law of the forum, in this case Arizona.

POINT IV

The Court's instruction, and refusal to instruct, as to the duty of the railroad company in restoring the highway to its former condition of usefulness, was correct.

The New Mexico Statute governing the duty of defendant, Sub-division 5 of Section 4697, compiled laws 1915, authorized the company

"To construct its railroads and telegraphs across, along, or upon any * * * highway * * * which its railroad * * * shall intersect, cross or run along; but such corporation shall restore such * * * highways * * * so intersected, to their former state, as near as may be, so as not to unnecessarily impair their use or injure their franchises. * * * "

This Statute is merely declaratory of the common law duty of defendant.

The plaintiff contended, and the evidence tended to show, that before the railroad was constructed the highway ran along the bottom of the "draw," near the base of the western ridge. The railroad, as built, emerged from the east ridge in a deep cut, crossed the "draw" somewhat diagonally on an embankment about 12 feet high (84-5) and disappeared in the other ridge on a curve in another deep cut. The highway, as now constructed, follows the bottom of the "draw" to the point of a knob, or projection, of the westerly hill just where the tracks penetrate that hill in the cut. Then the road is made to turn abruptly and climb the hill on its natural slope, rising 10 feet in 80 to the level of the track at the mouth of the cut, when it crosses the track, turns again at right angles and descends more gradually along the hill slope to the bottom of the "draw," where it continues on toward Mountainair. It needs no argument to demonstrate that a crossing right at the mouth of a deep curving cut presents the maximum of danger.

The reason for the construction now existing speaks for

itself. It carried the highway up to and across the railroad grade and track, with little or no expense, using the natural slope of the hill for the purpose. The defendant claimed and produced one witness who testified that the old road *left the "draw" and climbed that knob,* in the same place and manner before the railroad was built. The defendant was obliged to leave a culvert or opening in the "draw" for the water to escape. By constructing this culvert where the road formerly ran, or moving the road opening a few yards and slightly enlarging the opening in the 12 foot embankment, a perfectly safe crossing could have been made at trifling additional cost. Due to the proximity of the two cuts this could hardly have been done in any other way. These facts are shown and the evidence summarized in our statement of facts on pages 2-5 supra. With this question and condition of the evidence confronting it, the Court charged the jury as follows: [303]

"If the railroad grade at any given place where the railroad is being constructed across a highway is so far above or so far below the level of the road or crossing, or so near to extraordinary cuts or curves, as to seriously interfere with its use and seriously increase the danger of such use by bringing the road crossing to the level of the tracks and the situation is such that the road can be carried over or under the tracks without unreasonable cost and expenriture, and the danger or obstruction cannot otherwise be reasonably removed, then it is the duty of the railroad company to construct a crossing under its rails or carry the road above its tracks by a bridge, as the occasion requires. If it fails to perform such duty, "and in an attempt to carry the road over its tracks at grade, that is, on a level with the rails, it seriously and unreasonably increases the danger and difficulty at the crossing, then the railroad company will be liable for any injury or damage proximately resulting because of such dangerous condition to one rightfully using the road."

Defendant excepted to this instruction and to the court's refusal to give the following in place of it.

"You are instructed that under the laws of the State of New Mexico a railroad company has the right to construct its railroad at grade across any highway which its said railway shall intersect, but that such company shall restore such highway to its former state, as near as may be, so as not to unnecessarily impair its use. You are further instructed in this connection that the laws of the said State of New Mexico govern and fix the rights and duties of the defendant insofar as the construction of said crossing at which plaintiff was injured are concerned. You are, therefore, instructed that if you find from the evidence that in constructing its railroad across the highway on which the plaintiff was traveling when injured, *the defendant restored said highway at the place intersected* by its said railroad *to its former state, as near as may be*, so as not unnecessarily to impair its use, you cannot find the defendant negligent in respect to that matter." (Italics ours).

Instead the Court gave the following *to which no exception was taken:*

"It is for you to say from the evidence whether or not the defendant railroad company, in constructing its road at the point where the accident occurred, did restore the crossing as near as was reasonably possible and practical to its former condition of safety and usefulness and, if you find that they did not and that the plaintiff suffered his injuries as a proximate result of such failure, then the company is liable to respond in damage for the injuries they so caused." * * * * [305]

The requested instruction, so far as it contains proper matter was covered by the instruction given. It was unnecessary to tell the jury that "New Mexico" law governed, as long as the legal principles by which they should be guided, were correctly and sufficiently stated.

The vice of the instruction requested is that it did not define, or explain, what was meant, or intended, by the law in the phrase "restored to its former state"—the entire point at issue. It was calculated, and defendant's argument here demonstrates it was intended, to convey to the jury the impression, or idea, that all the defendant was required to do in constructing the crossing was to restore the road to the *exact spot, condition, level* and *grade* it formerly occupied as near as may be. The thought was to be conveyed to them that if the defendant did that, it had performed its full duty, *even though the danger to travellers on the highway was multiplied a hundred fold by so doing,* and could be altogether eliminated and the road made entirely safe and passable by reasonably altering the conditions.

That such was the construction and purpose of the request is shown by the evidence of the witness Block and the conclusion drawn therefrom by defendant in his brief at page 110. He says:

"As shown by the undisputed evidence, the defendant, when constructing its railroad across the highway installed a crossing at the exact place where the rails intersected the highway as the latter then existed."

"The angle of the approaches to the crossing was *perhaps* slightly changed but the location of that part of the highway occupied by the rails was in no wise altered."

We submit that the instruction given guardedly and correctly measured the defendant's duty as applied to the facts under consideration, and the instruction requested in the language of the Statute unexplained and unclarified, would have mislead the jury as to the correct application of the law. The defendant insisted at the trial and now requests this Court to rule that the duty of the defendant, enjoined by the Statute, is to restore and leave the crossing *in the exact spot* and grade they found it when they built the road, regardless of the fact that it do so would increase the danger to traffic a hundred fold; and altho the danger could be entirely eliminated by changing the position of the crossing a few feet, or yards, or by carrying it above, or below, the tracks in the same spot. We submit that such a construction is supported neither by authority, logic, nor common sense. As was well said by this Court in the Martinez case, 270 Fed. 770, quoting the general rule from 22 R. C. L. 991:

> "It is the duty of a railroad so to construct and maintain its crossings that they may be safely used by persons travelling the highway, and for the negligent breach of this duty it must answer in damages to one injured thereby while exercising ordinary care provided such breach was the proximate cause of the injuries."

In the same work the duty of the company to restore the highway to its former condition gathered from the authorities, is stated as follows:

> Restoration or Repair of Highway.—At common law the duty rests upon a railway corporation, when it occupies or crosses a public thoroughfare with its tracks, to restore the same, by some reasonably safe and convenient means, to its former condition of usefulness, and such duty is also often expressly imposed by statute or charter. This includes the doing of whatever is necessary to be done to restore the highway to such condition, as for instance in case of a bridge the approaches or lateral embankments without which the bridge itself would be useless. The duty is founded upon the equitable principle that it was the act of the railroad, done in pursuit of its own advantage, which rendered this work necessary, and therefore the road, and not the public, should be burdened with its expense. The construction of a railroad over a highway makes

it practically impossible to restore the highway to its former condition, but it should be restored to its former *state of safety for public travel*. It may not be important or necessary that it should be brought as nearly as possible to its original condition, but it is important that it should be restored so that the danger of its use would be reduced as much as possible. Where the situation is such that a railway company is under a legal obligation to restore a highway across its track to a reasonably safe condition for travel, and this result cannot be otherwise accomplished, a court may require the construction of a subway, but only where the necessity is clearly shown and supported by competent expert evidence. * * * * In a suit in damages for personal injuries, based upon the theory that the defendant had failed to restore a highway to its former state, or to such a condition that its usefulness would not be materially impaired, where there is evidence reasonably tending to prove the condition of the highway at the place of the accident, prior to the construction of the crossing at which the injury occurred, and that the same had not been restored, the court does not err in submitting the question of whether it had been restored to the jury. 22 R. C. L., 889.

It may well be that it should not be left to a jury in any ordinary case to determine whether or not a crossing should have been at grade; but that was not the question presented by the defendant in this case. He asked the Court to charge in the language of the Statute that it was the company's duty to "restore such highway to its former state as near as may be." His complaint is because the Court explained what that meant. He now tells us what he thought it meant and what he expected the jury would understand from his proposed charge. That the only duty resting upon the railroad in building its line across a highway is to put it at the exact spot it formerly occupied, vertically and horizontally, and if this is done, the railroad has

fully performed its duty, though the crossing could be made
absolutely safe by carrying the road above, or below the
tracks, or by moving the crossing a few feet, or yards,
whereas to construct on the spot would make it a veritable
death trap. We do not think any decision could be found
that carefully considered justifies such a construction and
if it does, we submit it should not be followed by the Federal
Courts.

POINT V

The Court's instructions as to the duty of the company to either use extraordinary means, like automatic signals, etc., at exceptionally dangerous, or cause their trains to approach the crossing under control, was likewise correst.

It is true that ordinarily the speed of railroad trains
in rural districts and crossings is neither negligence, nor
evidence of negligence; but where the road itself has constructed a crossing so as to produce a condition extraordinarily dangerous, they then take upon themselves, create
for themselves, the duty to take extraordinary methods of
protecting against the danger. The rule was so well stated
in the early New Jersey case Penn. Ry. Co. vs. Matthews,
36 N. J. L. 533, that we copy its language as follows:

> If one of them chooses to build its track in such a
> mode as to unnecessarily make the use of a public road
> which it crosses, greatly dangerous, I think such com-
> pany, by its own action, must be held to have assumed
> the obligation of compensating the public for the in-
> creased danger, by the use of additional safeguards.
> The reasonable and indispensable implication is, that
> the railway is to be constructed so as not unnecessarily
> to interfere with the safe use of the public roads;
> and *if a railroad, for its own convenience curves its
> track as it leaves a deep cut, within a few feet of a
> highway*, (Italics ours) and also sees fit to put up

buildings close along such track and by these means, or either of them, heightening the danger in the use of such highway, it seems to me very clear, that such company must be held to have taken upon itself the duty of averting such danger, by the employment of every reasonable precaution within its power. On such occasions as this, or whenever the situation is embraced within the principle stated, the presence of a flagman, or some equivalent safeguard can be demanded of the company. The rule is, as I understand it, that when the company has created extra danger, it is bound to use extra precautions. If, the track is put in a position where the trains, when close to their transit over a public street or road, cannot be seen, this is an extra danger which calls for more than the ordinary cautionary signals. I can see no difficulty in applying this rule; it will, obviously, be very much under the control of the court.

In Cincinnati, N. O. & T. P. Ry. Co. vs. Champ, 104 S. W. 988, the Kentucky Court of Appeals says:

In the numerous cases involving crossing accidents that have come before this Court, the central idea in all of them is that the company must use such care and precautions for the safety of travelers as the character of the crossing makes reasonably necessary for their safety and protection. What this degree of care is must depend upon the facts of each case, and is a question for the jury. At one crossing ringing the bell and sounding the whistle might be amply sufficient; at another, it would be wholly inadequate, and a flagman or other safety device be necessary. This does not necessarily mean that the speed of trains must be slackened, as no rate of speed at ordinary crossings is usually negligence; but at exceptionally dangerous crossings, if the company does not choose to have a flagman or other safety device, and the statutory signals are not sufficient, the speed of the train must be so regulated as not to unnecessarily imperil the safety of persons using the highway. The duty of observing such degree of care as the situation and surroundings of the crossing may reasonably demand to prevent accidents is not alone imposed on the company, but applies as well to the traveler, who must use such care as might usually be expected of an ordinarily prudent person to learn of the approach of the train and keep out of its way. In short, the obligations to avoid injury and accident are reciprocal, and must be commensurate with the danger. Thompson on Negligence Secs. 1535, 1553.

The subject is covered by note in Annotated cases 1914 B at page 602. The following appropriate headings well sustained by the authorities, show the holdings gleaned from the cases there collated :

"The recent cases are in accord with the general rule that while no rate of speed is of itself negligence, it may be negligent to run a train at a high rate of speed through a populous community and over a much frequented crossing."

Whether a given rate of speed over a crossing is negligent depends on all the surrounding circumstances.

The fact that the view of one approaching a railroad crossing is obstructed may make it negligence to propel a train across such a crossing at a high rate of speed.

The recent cases are in accord to the effect that where no warning is given of the approach of a train to a crossing the speed at which the train is moving is a material consideration on the question of negligence.

Likewise where warning signals are given but the

speed of the train is such as to render them useless, such speed is negligent.

We submit the instruction given was correct.

POINT VI

Defendant's other requested instructions.

Following the manner and outline of defendant's brief we will notice his proposed instructions and criticisms in the order in which they are made.

The instructions given by the Court were very complete and covered generally every feature of the law of negligence having any bearing on the case. What constituted negligence and contributory negligence were clearly and definitely defined (298-9). The jury were told that the degree of care and caution to be exercised by a person approaching a railroad crossing is dependent upon the circumstances and conditions existing at and near such crossing at the time, and that the greater the danger from such circumstances and conditions the greater must be the care and caution required to be exercised. (301). They were instructed that it was the duty of a traveller to use care in selecting the time and place to look and listen for coming trains and should stop to make observations when necessary; that he must use all his faculties and take advantage of every reasonable opportunity to look and listen and may not rely on the failure to give signals (301); and has no right to presume that the railroad employees would give such signals, or omit to perform any part of his duty upon the presumption that they would (302). They were told that the railroad had the right of way and precedence at the crossing and it was the duty of the plaintiff to respect this right of way. (302). They were told that ordinarily the rate of speed at which a train is travelling is not evidence of, nor is it to be considered as negligence, but if the railroad company in constructing its line created a condition of extraordinary danger to travellers at any point, then it was their duty to use more than the ordinary methods of warning against the danger and that the duty of the company in that respect was increased proportionately to the danger they had so created. They were told it was the duty of the plaintiff (306-7)

> "To use that degree of care that a reasonably prudent person would use under like circumstances surrounding the case. It was his duty to look and listen for approaching trains before venturing on the tracks:"

that this duty did not require him to neglect all other duties and occupations, or to keep his eyes fixed on the track during the whole time to the exclusion of such other necessary duties.

> "That what would be due care in that respect under one set of circumstances might be negligence under another. The fundamental rule and test is what should a person of ordinary care and prudence, situate as was the plaintiff, surrounded with all the circumstances and duties that he was, have done; and did the plaintiff's conduct measure up to that requirement."
> (307).

The jury was cautioned a second time (308) that the plaintiff should use care in selecting the time and place to look and listen and that he must take advantage of every reasonable opportunity to do so. (308)

These instructions fully and fairly covered the ground so far as it was correct presented by the defendant's requested instructions. The instruction refused in the 12th Assignment of Error was fully covered so far as it was proper. It would have given the jury no additional help to have told them that the substantive rules in question were New Mexico instead of Arizona laws.

The 13th instruction we have already covered.

The 14th Assignment of Error, so far as it contains the correct rule, was fully covered; but it contained propositions that were not correct and would have been error. The Court was requested to instruct the jury that regardless of surrounding circumstances

"The looking and listening and stopping should be exercised at the last moment which ordinary prudence would dictate"

and that

"If the ability either to look or listen be obstructed by natural, or artificial objects, then the duty to stop before entering in the zone of danger in order that proper investigation may be made became absolute."

If that instruction means anything more than what was contained in the charge as given, as applied to the case at bar, it would have instructed the jury that it was the absolute duty of the plaintiff to have caused his father to stop the car, or himself to cease pushing and stop on the hillside before passing behind the cattle guard fence, failing in which he would be guilty of negligence as a matter of law. There were no other obstructions except the cuts and the proposed instruction did not refer to them. The defendant's same request further asked the Court to charge that if the plaintiff

"Without objection, or protest permitted"

his father to drive up on the railroad crossing in front of the approaching engine that their verdict must be for the defendant.

It is an elementary rule that where a proposed instruction contains any improper matter it may be rejected as a whole, tho proper matters are contained thereon.

So, in the proposed instruction contained in the 15th Assignment of Error. The matter, so far as requested and was proper, was covered by the instruction given. So far as it exceeded that it was an attempt to tell the jury that it was the duty of the plaintiff to give attention to nothing else but the track and the train as he approached the crossing; a proposition which, as we have seen, is not a law.

The proposition contained in the 16th Assignment of Error, so far as it stated the law was fully and correctly covered by the Court's charge. The argument contained therein was, of course, improper and the final observation by which the jury were told as a matter of law that this crossing was so safe, correct and ordinary that they were not permitted to take into consideration the speed of the train, was plainly erroneous.

The proposition contained in the 17th specification of error is, of course, erroneous. It would have instructed the jury that the fact that the automobile was working badly and that as a result the plaintiff was obliged to get out and use his efforts and energy in pushing the car was "Not to be considered by you in any way."

It needs no argument to demonstrate the error of that proposition.

POINT VII
The Arizona constitutional provision.

The defendant spends a considerable portion of the brief in attacking the decision of this Court in Southern Pacific vs. Martinez, 270 Fed. 770, heading his argument with the proposition (Brief 84)

"At the outset we admit that if the decision of this Court in Southern Pacific vs. Martinez, 270 Fed. 770, is to be read as adopting in the Federal Courts a state rule limiting the powers of judges, limiting the conduct of judges; then we are foreclosed unless this Court will give further consideration to the subject."

And they proceed to argue that the rule in the Martinez case commits this Court to the proposition that all the other regulations of matters on trials contained in the

Arizona Constitution, would have to be adopted likewise. We do not understand the Martinez case as establishing any such rule. Instead, it adopts and applies the principles laid down by the United States Supreme Court in C. R. P. Ry. Co. vs. Cole, 251 U. S. 54, and if that decision be erroneous, or upon erroneous grounds, it is for the court to correct it.

We think in this case, however, it would supply no useful purpose to go into an extensive consideration of the arguments and authorities contained in defendant's brief upon that question.

If, as we think we have demonstrated, the plaintiff's contributory negligence in this case was for the jury in any event, the Arizona constitutional provision is of no importance. Besides the trial Court did not put his ruling on that ground and refused to ground it on the Arizona constitution; which means necessarily, that he submitted the case to the jury because, independent of that constitutional provision, he thought a question of fact was presented for the jury.

Counsel argues that the ruling of this Court in the Martinez case would necessarily mean that a Federal Judge was prohibited from granting new trial upon the ground that in his judgment the jury erred in not finding the plaintiff guilty of contributory negligence. We find nothing in the Martinez case, or in the constitutional provision requiring such a decision. In this case the defendant did present to the trial Court in his motion for a new trial, just such a request and included all the evidence upon that question in his motion and the trial judge overruling the motion in no manner, so far as the record discloses, grounded his refusal on the Arizona constitutional provision. We have then the situation where the Judge upon the trial denied the request of the defendant to rule that the evidence showed the plaintiff guilty of contributory negligence and *expressly declined to rest that ruling on the constitutional provision,* and then refuses a motion for new trial, on the same matter.

We submit that the ruling was right regardless of the constitutional provision and that this Court is not required, under the facts in this case, to re-examine it.

We respectfully submit that there is no error in the record and the judgment of the lower Court should be affirmed.

Albuquerque, New Mexico.
Members of the Bar of the
United States Supreme Court.
Attorneys for Defendant in Error.